Smart Innovation, Systems and Technologies

Volume 198

The Smart Innovation, Systems and Technologies book series encompasses the topics of knowledge, intelligence, innovation and sustainability. The aim of the series is to make available a platform for the publication of books on all aspects of single and multi-disciplinary research on these themes in order to make the latest results available in a readily-accessible form. Volumes on interdisciplinary research combining two or more of these areas is particularly sought.

The series covers systems and paradigms that employ knowledge and intelligence in a broad sense. Its scope is systems having embedded knowledge and intelligence, which may be applied to the solution of world problems in industry, the environment and the community. It also focusses on the knowledge-transfer methodologies and innovation strategies employed to make this happen effectively. The combination of intelligent systems tools and a broad range of applications introduces a need for a synergy of disciplines from science, technology, business and the humanities. The series will include conference proceedings, edited collections, monographs, handbooks, reference books, and other relevant types of book in areas of science and technology where smart systems and technologies can offer innovative solutions.

High quality content is an essential feature for all book proposals accepted for the series. It is expected that editors of all accepted volumes will ensure that contributions are subjected to an appropriate level of reviewing process and adhere to KES quality principles.

**** Indexing: The books of this series are submitted to ISI Proceedings, EI-Compendex, SCOPUS, Google Scholar and Springerlink ****

More information about this series at http://www.springer.com/series/8767

Luciana Pereira · José Reginaldo Hughes Carvalho ·
Petter Krus · Magnus Klofsten ·
Victor Juliano De Negri
Editors

Proceedings of IDEAS 2019

The Interdisciplinary Conference on Innovation, Design, Entrepreneurship, And Sustainable Systems

Springer

Editors
Luciana Pereira
Federal University of ABC
Santo Andre, Brazil

José Reginaldo Hughes Carvalho
Federal University of Amazonas
Manaus, Brazil

Petter Krus
Linköping University
Linköping, Sweden

Magnus Klofsten
Linköping University
Linköping, Sweden

Victor Juliano De Negri
Fedderal University of Santa Catarina
Florianópolis, Brazil

ISSN 2190-3018 ISSN 2190-3026 (electronic)
Smart Innovation, Systems and Technologies
ISBN 978-3-030-55373-9 ISBN 978-3-030-55374-6 (eBook)
https://doi.org/10.1007/978-3-030-55374-6

This Springer imprint is published by the registered company Springer Nature Switzerland AG
The registered company address is: Gewerbestrasse 11, 6330 Cham, Switzerland

Preface

Over the past decades, innovations have emerged from problem-oriented research based on knowledge integration at the intersection of different disciplines. Therefore, interdisciplinary collaboration has been critical to find solutions for complex societal challenges. However, there are little spaces within and beyond the academy where researchers are encouraged to defy their field's boundaries—leveraging their disciplinary mindset into contributions to broad domains, particularly within the Science, Technology, Engineering, Entrepreneurship, and Management (STEEM) and across Social Sciences and the Humanities.

IDEAS conference aims to stimulate novel questions and challenge existing policy and practices on more innovative and sustainable systems. Throughout history, technological innovation has led to the most significant dimensions of development—economic, social, and environmental—underpinning core changes in the way we live.

How to turn interdisciplinary dialogues into action? The foundations of innovation dynamics lay on integrating knowledge to propose new systems and constantly refining our approach to meet the needs of a rapidly changing world. There is an increasing realization that this level is hard to address properly without communication between the respective scientific communities.

Organization

Organizing Committee

José Reginaldo Hughes Carvalhoc	Federal University of Amazonas, Brazil
Magnus Klofsten	Linköping University, Sweden
Petter Krus	Linköping University, Sweden
Luciana Pereira	Federal University of ABC, Brazil

Local Committee

Janaina Maciel Braga	Nilton Lins University, Brazil
Nelson Kuwahra	Federal University of Amazonas, Brazil
Marcelo Albuquerque de Oliveira	Federal University of Amazonas, Brazil

Scientific Committee

Alessandra Akkari	Mackenzie Presbiterian University, Brazil
Daniel Capaldo Amaral	University of São Paulo, Brazil
Glauco Arbix	University of São Paulo, Brazil
Karina Assis	Federal University of São Carlos, Brazil
Thorsten Bartel	Fraunhofer-Institut LBF, Germany
Janaina Maciel Braga	Nilton Lins University, Brazil
Iara Chao	São Paulo Water and Waste Co., Brazil
Diogo Rosenthal Coutinho	University of São Paulo, Brazil
Mattias Elg	Linlöping University, Sweden
Renato Garcia	State University of Campinas, Brazil
Luiz Carlos Goes	Aeronautics Institute of Technology, Brazil
Mario Aguirre Gonzales	Federal University of Rio Grande do Norte, Brazil
William Grosky	University of Michigan, USA
Kerstin Johansen	Jonköping University, Sweden

Deny Willy Junaidy	Bandung Institute of Technology, Indonesia
Wisdom Kanda	Linköping University, Sweden
Solmaz Filiz Karabag	Linköping University, Sweden
Henik Kock	Linköping University, Sweden
Paulo Kurka	State University of Campinas, Brazil
Mattias Lindahl	Linköping University, Sweden
José Roberto Dale Luche	State University of São Paulo, Brazil
Nelson Tavares Matias	Teresa Dávlia University Center, Brazil
Akane Matsumae	Kyushu University, Japan
Adriana Marotti de Mello	University of São Paulo, Brazil
Josivaldo Cesar Modesto	Mamirauá Institute for Sustainable Development, Brazil
Lina Moreira	GENYZ, Brazil
Edleno Silva de Moura	Fedeal University of Amazonas, Brasil
André Ogliar	Federal University of Santa Catarina, Brazil
Johan Ölvander	Linköping University, Sweden
Luis Carlos Paschoarelli	State University of São Paulo, Brazil
Aderson Campos Passos	Militar Institute of Engineering, Brazil
Marcelo Caldeira Pedroso	University of São Paulo, Brazil
Marcela Savia Pessoa	Amazonas State University, Brazil
Mikael Román	Linköping University, Sweden
Anna Öhrwall Rönnbäck	Luleå Technical University, Sweden
Suzana Russo	Federal University of Sergipe, Brazil
Mario Sergio Salerno	University of São Paulo, Brazil
Kristine Samson	Roskilde University, Denmark
Gabriela Scur	FEI University Center, Brazil
Andrés Serbin	CRIES, Argentina
Ricardo Serudo	Amazonas State University, Brazil
Thiago Caliari Silva	Aeronautics Institute of Technology, Brazil
Louise Trygg	Linköping University, Sweden
Emilia Villani	Aeronautics Institute of Technology, Brazil
Dalton Chaves Vilela Jr.	Federal University of Amazonas, Brazil
Irlan von Linsingen	Federal University of Santa Catarina, Brazil
Juarez Xavier	São Paulo State University, Brazil
Eduardo Zancul	University of Sao Paulo, Brazil

Support

Ministry of Education-Coordination for the Improvement of Higher Education Personnel (CAPES), Brazil
Ministry of Science, Technology, Innovation, and Communication-National Council for Scientific and Technological Development (CNPq), Brazil

Contents

Innovation Systems Under Technological Paradigms

Barriers and Facilitators of Reverse Innovation: An Integrative Review 3
Tatiana Tombini Wittmann, Daniela de Oliveira Massad, Gertrudes Aparecida Dandolini, and João Arthur de Souza

System Architecture of a Robotics Airship 13
José Reginaldo H. Carvalho, Miguel Á. C. Rueda, José R. Azinheira, Alexandra Moutinho, Luiz G. B. Mirisola, Ely C. de Paiva, Lucas A. C. O. Nogueira, Gustavo A. Fonseca, Josué Jr. G. Ramos, Mauro F. Koyama, Samuel S. Bueno, and Christian Amaral

The New Brazilian Legal Framework of Science & Technology: Barriers, Borders and Opportunities for Innovation 23
Dércio Luiz Reis, Marcelo Albuquerque de Oliveira, Sicy Rusalka Goes de Melo Barreto, Joaquim Maciel da Costa Craveiro, and Ana Nubia dos Santos de Oliveira

A Cyber Physical System Approach to Customer Services of Home Appliances 34
Ana Carolina Fragal, Anderson Orzari Ribeiro, and Crhistian Raffaelo Baldo

Improvement in the Method of Assembly of a New Production Process in the PIM 44
Matheus Oliveira, Karina Cavalcante, Gabriela Veroneze, Marcelo Albuquerque de Oliveira, and Joaquim Maciel da Costa Craveiro

The Strategic Role of MES Systems in the Context of Industry 4.0 52
José Eduardo da Costa Dias, Fernando Gonçalves de Castro Filho, Alexandre Acácio de Andrade, and Julio Francisco Blumetti Facó

New Venture Creation, Business Growth and Cases

Locational Dynamics of Academic Spin-Offs: Evidence from Brazil ... 65
Filipe Scorsatto, Bruno Fischer, and Paola Rücker Schaeffer

An Application for Automatic Classification of Unconventional Food Plants ... 76
Karla Okada, Eulanda M. Santos, José Reginaldo H. Carvalho, Carlos Gustavo Nunes-Silva, and Max Vasconcelos

Tactical Design: Understanding, Experimenting, and Learning from Design to Organizational Growth ... 86
Bruno Raphael de Carvalho Santos, Luana Bittencourt Saraiva, and Claudete Barbosa Ruschival

State-of-the-Art on Furniture Design: A Visual Review ... 95
Ana Carolina Correa de Medeiros, Roger Pamponet da Fonseca, and Augusto César Barreto Rocha

Study on the Relationship Between Design Management and Business Project Management ... 105
Úlima Santos, Alef Santos, and Claudete Barbosa Ruschival

Leadership in Social Business: A Needed Context ... 115
Ricardo Pereira, Daniela de Oliveira Massad, Gertrudes Aparecida Dandolini, Gregory Falavigna, João Arthur de Souza, and Édis Mafra Lapolli

The Positive Impacts of Using the Service Design Approach for Expanding the Innovative Potential on Business ... 126
Giovanna Silva, Ricardo Gaspar, Júlio Francisco Blumetti Facó, and Alexandre Acácio de Andrade

Low Cost Solution for Home Brewing and Small Brewing Business Using Raspberry Pi ... 136
Alexandre Acácio de Andrade, Álvaro Batista Dietrich, Júlio Francisco Blumetti Facó, and Ricardo Reolon Jorge

Design, and Systems Development with Data Analytics Management in the Digital Age

Relating Design Management and Project Management: Application of the PM Mind Map Tool in the Creative Design Process ... 149
Bruno Perdigão Pacheco, Mateus da Silva Bento, Almir de Souza Pacheco, and Magnólia Grangeiro Quirino

Development of the Innovative Design of an Automatic Equipment to Aid in Physical Rehabilitation ... 160
Roberta Goergen, Marianna Gioppo de Souza, Maurício Oberdörfer, Matias Alles Hubert, Jocarly Patrocínio de Souza, Luiz Antônio Rasia, and Antonio Carlos Valdiero

Stuttering and the Use of Facebook as a Tool for Interaction Between People Who Stutter: A Content Analysis 168
Gabriela Costa Arouck de Souza and Franciane da Silva Falcão

Exploratory Study on the Behavior of the Brazilian Financial Market Using Google Trends 179
Fernando Gonçalves de Castro Filho, José Eduardo da Costa Dias, Alexandre Acácio de Andrade, and Julio Francisco Blumetti Facó

Systematic Review of Data that Uses Project E in the Development of Digital Products Focusing on User Experience 189
Paula Silva and Franciane Falcão

Affordance Based Design: The Affordance Digital Stimuli Tool to Stimulate Creativity During Product Design 197
Karuliny Marques, André Ogliari, Rodrigo Bastos Fernandes, and Matheus Gomes

Efficient Methods for System Design Analytics 208
Petter Krus

Analysis of the Application of a Product Development Method Called Meta Project 219
Aline Fernandes and Franciane Falcão

Systematic Approach to Develop Physical Models as Creative Stimuli in Conceptual Design Phase 228
Matheus Gomes, André Ogliari, Rodrigo Bastos Fernandes, and Karuliny Marques

Development of an Automatic Machine for Sensor Manufacturing by the GOP Technique 239
Edmilton Stein, Carlos Valdiero, Matias Alles Hubert, Roberta Goergen, Marianna Souza, Luiz Antônio Rasia, and Antonio Carlos Valdiero

Innovation in Education with Inclusion, New Technologies, and Civil Society Engagement

Educating City: A Media for Social Innovation 251
Yohani Dominik dos Santos Figueiredo, Márcia Aparecida Prim, and Gertrudes Aparecida Dandolini

Virtual Reality as an Educational Tool for Elementary School 261
Marcos Figueiredo, Rovilson Mafalda, and Andrea Kamensky

The "Mount Your School" Case and the Contribution of Design Thinking in Public Education in Brazil 268
Leandro Vasconcelos and Andres Amoedo

Faculdade Zumbi dos Palmares Case Study of Racial Integration in the Advertising Area 280
Lina Maria Moreira Garai da Silva, Alexandre Acácio de Andrade, and Patrícia Gonzaga Cesar

***Engrena ITA*: Alliance for Innovation Prospection on Gear Technologies** 290
Ronnie Rodrigo Rego

How to Map Employees' Competencies for More Innovative Higher Education Institutions? The Case of a Brazilian Interdisciplinary University 298
Mauricio Wojslaw, Luciana Pereira, and Júlio Francisco Blumetti Facó

A Case Study on Hackathon # Develops and a Hackathons Contribution to Innovation 314
Deise Carolina de Souza Silva, Clarissa Stefani Teixeira, and Jatyr Ranzolin Junior

Editorial Production of Scientific Journals: The Influence of Technological Development on Scientific Certification, Editorial Management, and Post-publication 322
Tito Lívio do Nascimento Fernandes and Franciane da Silva Falcão

Design of Sustainable Business, Products, and Services for Rural, Forestry, and Agriculture Innovation

Knowledge Sharing Structure of Agricultural Products: Case of *KokuzoYuzu* (Citrus) 335
Shintarou Mori, Akane Matsumae, and Yukari Nagai

Blockchain and Conservation: Why Does It Matter 346
Luiz Villares

Development of a Human–Machine Interface Implemented in Smartphone for a Variable Rate Fertilizer Applicator 356
Ben-Hur Maciel, Ivan Mantovani, Matias Alles Hubert, Roberta Goergen, Carla Rannov, Luiz Antônio Rasia, and Antonio Carlos Valdiero

Comparing Aerial Platforms for the Amazon Application Scenarios . . . 363
José Reginaldo H. Carvalho, Samuel S. Bueno, and Josivaldo Modesto

Designing Efficient Energy Policies, Systems, and Tools

The Diffusion of Solar Photovoltaics in Brazil: A Technological Innovation System Approach 377
Mauricio Uriona-Maldonado, Thiago Caliari, Luiz H. de Souza Costa, and Caroline Rodrigues Vaz

Design Management for Sustainable Development: Concepts and Examples 386
Ana Carolina Correa de Medeiros, Claudete Barbosa Ruschival, Eminy Laís Silva da Costa, and Luciana Kurack da Silva Misucochi

Investigation of the Classical Control Design Methodology to Improve the Performance of the Photovoltaic System on Grid 396
Marenice Melo de Carvalho and Renan Landau Paiva de Medeiros

Investigation of the Performance of the Control Methodologies to Mitigate the Undesirable Oscillation in a Smart Grid System 406
Isaías V. de Bessa and Renan Landau Paiva de Medeiros

Low-Cost Photovoltaic Maximum Power Point Tracking Project for Autonomous Electric Vehicle Prototype 416
Matias Alles Hubert, Antonio Carlos Valdiero, Roberta Goergen, Edmilton Stein, Rosângela Rommel Regner, and Ben-Hur Maciel

Smart Cities with Sustainable Management of Natural Resources, Green Mobility, and Entrepreneurship

Collaboration Networks for Social Innovation in the Context of Social Incubators: Constitutive Elements 427
Márcia Aparecida Prim and Gertrudes Aparecida Dandolini

Comparison of Alternative Wastewater Treatment Plants Using Life Cycle Assessment (LCA) 437
Anna Carolina Dall'Aneze Ferreira, Sueli Oliveira, and Roseli Frederigi Benassi

The Diagnosis of Solid Residues Produced in Free Trade Shows: A Case Study 447
Glaucia Oliveira, Deivison Molinari, Marcelo Albuquerque de Oliveira, Gabriela Veroneze, and Dércio Luiz Reis

Bicycle as a Mode of Transport in Brazil: Joint Action of Society, Sustainability and Innovation 457
Thiago Caliari, Ernesto Cordeiro Marujo, and Thais Marzola Zara

Management and Entrepreneurship Orientation in the Fisheries of the Micro-region of Rio Preto da Eva – AM 467
Patrícia dos Santos, Marcelo Albuquerque de Oliveira, Joaquim Maciel da Costa Craveiro, Dercio Luiz Reis, and Ely Senna de Almeida

The Interrelationship Between Sustainable Development and Social Innovation: A Bibliometric Study 480
Ricardo Pereira, Daniela de Oliveira Massad, Fábio Lorensi do Canto, and Gertrudes Aparecida Dandolini

Author Index 491

Innovation Systems Under Technological Paradigms

Barriers and Facilitators of Reverse Innovation: An Integrative Review

Tatiana Tombini Wittmann(✉), Daniela de Oliveira Massad, Gertrudes Aparecida Dandolini, and João Arthur de Souza

Federal University of Santa Catarina, Florianópolis, SC, Brazil
tatianaw@gmail.com

Abstract. The reverse innovation mainly refers to innovations developed in developing or emerging markets, and that has the potential to be disseminated to developed countries. This new locus of innovation demands that organizations implement new business models and new management strategies. In this context, this article aims to identify the barriers and the facilitators to the generation of successful reverse innovations. For this, an integrative review of the literature was carried out from articles selected in the Scopus database and the Google Scholar tool, together with others considered relevant by the authors. Among the main barriers to reverse innovation found in the literature are the dominant logic of the strategic actions of multinational companies based in developed markets; the prejudgment in relation to reverse innovations, considered as of low quality; the institutional distance between the headquarters of multinationals and their subsidiaries; the fear of cannibalization of products; and regulatory barriers. Among the facilitators are the growing demand of emerging markets for new products; increasing role of technology as a facilitator of the development of economical and easy-to-use innovations; the need for multinational companies to expand their markets; the increasing internationalization of companies and the consequent increase in the decentralization of research and development (R&D) areas.

Keywords: Reverse innovation · Emerging markets · Trickle-up innovation · Multinational companies · Constraint-based innovations

1 Introduction

For a long time, one of the main strategies of multinational companies has been the flow of innovation from developed to developing countries. The common was, and still is, to develop products for domestic markets, and then companies to change them to distribute them around the world at lower prices [1, 2].

However, there has been a considerable growth in innovation processes conceived in countries with scarce conditions in the last two decades, with a focus mainly on emerging markets and bottom of the wealth pyramid clients [3]. There are different types of constraint-based innovations, from innovations that generate frugal products (frugal innovation) to innovations developed in developing countries that subsequently

L. Pereira et al. (Eds.): IDEAS 2018, SIST 198, pp. 3–12, 2021.
https://doi.org/10.1007/978-3-030-55374-6_1

gain market share in developed countries (reverse innovation). Literature in this field, however, has resulted in the inaccuracy of terminology [3].

The academic literature has shown an increase in the popularity of the term reverse innovation and a growing interest in it. The relation of reverse innovation to organizational strategy and its managerial implications justifies the academic interest [4].

Companies that want to invest in reverse innovations need to be open to a paradigm shift, maintaining organizational strategies aimed at managing problems and uncertainties that may emerge in that context. The country of origin of a product, for example, may negatively influence its classification, especially if the product is from a low-income country [5, 6]. On the other hand, reverse innovation is being identified as one of the leading emerging trends in global health systems, in part because resource-poor environments are excellent laboratories for low-cost and high-impact innovations, that are developed by necessity [7].

The recent growth of interest in reverse innovation and the scarcity of studies on this practice stimulated the production of this article. Aiming to carry out an exploratory study that provided insights for the actors of innovation, the research question that this article intends to answer is "What factors are barriers and the facilitators for companies to generate successful reverse innovations?". For that, the methodological approach used was the integrative review of literature, through a critical search of publications adhering to the research question. In this way, it is hoped to support a better understanding of reverse innovation, to list its barriers and facilitators, and to identify knowledge gaps to support future research in the field.

The next section of this article presents a theoretical background. The third section deals with methodological procedures. The fourth section brings the results of the analyzes of the publications raised and, in the last section, are the final considerations. The final section presents bibliographic references.

2 Background

There is an imprecision in the literature of the use of terminologies associated with constraint-based innovations, and sometimes the concepts of frugal innovation, reverse innovation and trickle-up innovation, for example, are confounded [8–10].

This is an era of rapid change, with companies operating in a highly competitive, complex, and dynamic environment, and emerging economies are increasing as sources of innovation, not just receivers. Several authors - such as Govindarajan and Trimble, Hart and Christensen, Immelt, Kenney, Massini and Murtha, and Vernon - are investigating the phenomenon of innovation in emerging economies [11]. The rise of these economies, such as China and India, and the "flattening" of the world, is provoking a shift in the locus of innovation [12].

In emerging markets, organizations can not adopt the same strategies and business models developed for mature traditional markets, as the path of development is not always the same [13]. Specific market characteristics, such as the institutional, cultural and economic context, need to be considered and local subsidiaries should not be seen only as executive branches, but as learning centers, where locally accumulated knowledge can be transferred to developed markets [14].

Innovation from developing markets can change the level of competitiveness of foreign multinationals, mainly because they allow the entry and exploration of a market not previously exploited [15].

There is no consensus in the literature of the definition of reverse innovation. It usually refers to innovation originating in a low-income country, or emerging market, that has the potential to solve global problems and boost the profits of multinational corporations [16].

Specific constraints of the emerging or developing markets mainly drive reverse innovation, such as cost constraints, lack of infrastructure, and cultural differences between developed and emerging market consumers [4].

Supporters of reverse innovations celebrate the potential of reverse innovation partly because they fear that the dramatic growth rates of their "glocalization" model are slowing [17]. The glocalization is presented as opposed to the reverse innovation, since it refers to the development of technological products in the rich countries, followed by their global distribution with some adaptations to local conditions [16]. The epicenter of global growth is shifting from developed economies to emerging economies, which may cause glocalization not to deliver the expected results [18]. On the other hand, it is expected that glocalization continues to be the largest provider of profits [19]. The two models - glocalization and reverse innovation - need not only to coexist but also to collaborate [18].

Prahalad was one of the first to challenge the dominant logic of thinking about innovation. He introduced the term trickle-up innovation, which refers to any innovation developed for the basis of the pyramid and which subsequently reaches high-income countries [20]. Innovation in trickle-up differs from reverse innovation, as reverse innovation must meet the new or different needs of emerging or developing markets, regardless of income levels [4].

Reverse innovation not necessarily should be developed in a developing country. Nowadays, the concept is related to the flow of ideas and products from low- and middle-income countries to high-income [21].

Well-known examples of emerging-market innovations include Tata Nano, Grameen Bank (microfinance), GE ultrasound, Embraer regional jets, BYD electric cars, Bharti Airtel's ultra-cheap wireless phone, and Nokia's cheap mobile phones. Some have already been made available in wealthy countries [12].

In addition to studies that focus mainly on reverse product innovation, it is important to highlight the existence of studies that focus on the reverse innovation of strategies [22], processes and business models [23].

3 Methodology

This research is characterized by an integrative review of the literature, which "is a form of research that reviews, critiques, and synthesizes representative literature on a topic in an integrated way such that new frameworks and perspectives on the topic are generated" [24] (p. 356). The studies included in an integrative review can be both qualitative and quantitative methodologies, to present the state of science, contribute to the development of the theory and provide practical application [25]. This is also exploratory research because it has the purpose of clarifying concepts and providing an overview of a theme, and descriptive, for detailing the characteristics of a phenomenon [26].

The integrative review of the literature was carried out by adapting the research steps proposed by different authors who researched it [24, 25, 27]. They are: 1) Formulation of the problem; 2) Definition of objectives; 3) Construction of protocol and research design; 4) Search for studies in databases; 5) Evaluation and selection of studies pertinent to the subject under study; 6) Reading the selected documents; 7) Data analysis; 8) Writing the article.

The research planning was elaborated through a protocol, defining the research question, objectives, inclusion and exclusion criteria, as well as search strategies and critical evaluation criteria. Thus, the research question of this review is: "What factors are the barriers, and the facilitators for companies to generate successful reverse innovations?".

The integrative review of the literature used the Scopus database as a priority and directed research using the Google Scholar tool. The survey of publications adhering to the research considered terms in English and Portuguese, in article title, abstract, and keywords. In the Scopus database, in September 2018, research was conducted using the terms (i) *inovação reversa*/*reverse innovation*, applying as a filter the term (ii) *barreira/barrier*, which resulted in the identification of 21 publications. A second research used the terms (i) *inovação reversa*/*reverse innovation*, applying as a filter the term (iii) *enabler/facilitador*, which resulted in the identification of 05 publications. Finally, a research was performed using the terms (i) *inovação reversa*/*reverse innovation*, applying as a filter the term (iv) *facilitador/facilitator*, resulting in the identification of 1 publication. Excluding duplicate and unavailable publications, the Scopus database identified 21 publications.

To guarantee a better comprehensiveness of the research, a targeted search was conducted in the Google Scholar search tool, as well as publications already known by the authors. In this second stage, 24 documents were identified, including articles and books, totaling 45 publications. It is important to emphasize that the research had no focus on a specific subject area, neither on industry size nor sector.

After the detailed reading of the titles, abstracts, and keywords, it was identified that 34 publications were adherent to the research question and objectives. However, after further reading, only 29 were part of the descriptive analysis of this article.

4 Results

In this section, the 29 publications from the data collection carried out are analyzed, aiming to promote and advance the discussions that involve the factors that act as barriers or facilitators of reverse innovation.

4.1 Barriers

One of the main barriers for multinationals to invest in reverse innovation is the need to overcome the dominant logic and go against institutionalized thinking that guides their strategic actions based on developed markets [2]. Companies must overcome, for example, resistance to the transfer of power and control away from the multinational headquarters and must be willing to reshape the organizational models and expectations of local teams [28]. They need to implement new business models and new management strategies [4].

In the establishment of international health partnerships, for example, although there is a focus on two-way learning, this emphasis is not institutionally visible, and the subtle bias with partner brokerage in low- and middle-income countries seems to stifle the search for reverse innovation [29].

Allied to this is the fact that reverse innovation is confounded as a developer of low-tech products and limited resources [3] and viewed with skepticism by professionals in high-income environments [30]. The term may pass the connotation of frugality, a result of poor leadership and, or associated with poor quality, hindering the potential translation to other contexts [8].

It is argued that reverse innovation is not popular within corporations because of the institutional distance barrier between developed countries (multinational headquarters) and emerging markets (location of subsidiaries) [2, 31]. Organizational structures that combine strategic integration between matrix and subsidiaries, support subsidiaries and enable them to have the autonomy, lessen the effect of institutional distance, favor the generation of reverse innovation, and make it a competitive advantage for the corporation [2, 31].

Another critical aspect of the internationalization of research and development (R&D) that deserves attention is the diversity of people and cultures involved in product development. Diversity is considered an element of complexity and leads to new challenges for companies [1].

In health, a barrier to the success of reverse innovation is the dispute between a low-cost care policy originated in developing countries and a profit policy promoted by multinational pharmaceutical corporations to continually expand their markets [16]. The collapse of the first Kangaroo Maternal Care Program (KMC), a neonatal care practice created in 1978 in Colombia, is one of the examples. Although the United Nations Children's Fund (UNICEF) praised the revolutionary and zero-cost technique and supported the worldwide dissemination of the program, new for-profit insurance schemes developed in Colombia by the neoliberalist reforms carried out in the country undermined the KMC program [16].

Two mechanisms can assist managers in deciding whether or not to do reverse innovation: the degree of product adaptation required and the perceived risk of cannibalization [32]. The cannibalization of products is considered a substantial barrier to reverse innovation. It occurs when cheaper products reduce sales of other higher priced products and higher profit margins. This has made companies like Philips Electronics resist to market low-cost lighting, powered by solar energy, to protect their existing product line, and thus maintain their profit margin [33].

Legal and regulatory barriers have also been identified in the transfer of health interventions to the United States, contributing to an anti-innovation atmosphere, discouraging reverse innovation [7]. The regulatory burden observed was demanded by international medical licensing, reimbursement, and problem in the definitions of tasks and scope of work among non-professionals as community health agents [7].

4.2 Facilitators

Reverse innovation gained momentum as demand and supply of local innovation met in emerging markets, and the needs of middle-class consumers in these markets have, thus, become to have strategic importance to businesses everywhere [12]. A motivator for the growth of reverse innovation is the infrastructure gap faced by developing countries, such as unreliable electricity. This reality sometimes imposes particular characteristics to products or even solutions to different products [4].

Among the main facilitators of reverse innovation is "the increasing role of technology as an enabler of developing cost-effective and easy-to-use innovations, specifically in the service industry, given its close dependence on information technology" [3] (p. 11).

Economic liberalization in many emerging markets, together with a "flat" world, due to the technological change, has also enabled local firms in emerging economies to draw on local and global resources to innovate in their markets [12]. Like this, to survive and thrive in global markets, multinational corporations must also shift their glocalization paradigms, being reverse innovation an option to increase their markets and profit [16].

The increasing internationalization of companies and the consequent increase in the decentralization of research and development (R&D) areas have therefore motivated the advent of the reverse transfer of innovation, particularly of subsidiaries of multinational companies in developed countries [2]. Sources of knowledge are increasingly sought after in emerging countries (in particular China and India), and teams immersed in local reality, that understand market needs and constraints, are an essential aspect of the success of reverse innovation [4]. Thus, it is suggested that the physical design and product development are carried out in the emerging market subsidiary [34].

Regarding China, advantages are reported for both supply and demand innovation. Offering low-cost workforce for engineers and scientists - absorbing more skilled and committed people than multinational competitors, government support, the entrepreneurial spirit of Chinese entrepreneurs, and the speed and creativity of the local internet industry, allow excellent results for innovation [35]. Some facilitators of demand are the rapid growth and fault tolerance of the Chinese market, little inheritance of Chinese habits due to the restrictions of the early Communist years, the search for simpler and cheaper products, and the rapid progress of large-scale government projects [35].

An example is GE, which developed originally for the Chinese market a cheap ultrasound device, which became the basis of a global business, with avid customers in both developing and developed countries [17]. At Siemens, the healthcare industry has been a trendsetter for reverse innovation and has launched other products, such as the Multix Select DR, a digital X-ray machine. It was produced not only for emerging markets but was also designed to serve global markets [36].

Some management strategies need to be considered by companies that wish to implement reverse innovations in their processes [19]. The participation of local collaboration networks, regardless of the origin of the company and in which market (emerging or developed) it is trying to enter, is an example of organizational strategy relevant to reverse innovation [4]. Local partners assist foreign companies in accessing distribution channels and administrative management, intermingling relations with government agencies, retailers, and sub-distributors [22].

In the health area, international partnerships have increasingly targeted mutual responsibility. To establish a joint agenda between countries, however, it is crucial that partners in high-income countries remain open-minded about the origin of learning, which needs to be two-way. This practice enables them to benefit from the experience of partners in low- and middle-income countries, thus facilitating the reverse innovation [29]. Learn relevant lessons from the less developed health systems of emerging markets, for example, has been identified as an opportunity to strengthen Canada's health care system, which is struggling against rising costs and difficulties in meeting the current and future demands of the population [37].

Another outstanding facilitator for reverse innovation is the involvement and support of communities' infrastructure, health professionals, and local policymakers. One example is the implementation of a model of prevention of cardiovascular diseases developed in Kenyan favelas in a community of migrants in the Netherlands [38]. Studies show that open innovation may also be useful in stimulating frugal and reverse innovations, especially in the health sector [9, 10].

The co-localization of foreign and domestic firms can act as excellent drivers of reverse innovation [39], corroborating with the thought that "reverse innovation isn't optional; it's oxygen" [17] (p. 5).

Innovating in the reverse direction generates numerous benefits for companies. First, they can take advantage of economies of scale, since they must work with large volumes of production to generate profits and create cost efficiencies throughout the process. Besides, innovations developed in emerging markets can provide organizations with a competitive advantage, which can further extract value from innovation in developed markets [40].

Finally, an exploratory study on health innovations transferred from the south (developing countries) to the north (mainly in the United Kingdom) concluded that it is a particularly appropriate moment in order to analyze reverse innovation, since cost-effectiveness is a prerequisite for any industrially developed world, however, "any transfer must be done with cultural sensitivity, with care not to lose detail" [41] (p. 6).

Table 1 presents the barriers and facilitators found in the analysis of the articles.

5 Final Considerations

This article aimed to answer "what factors are barriers and the facilitators for companies to generate successful reverse innovations." Through an integrative review of the literature, which included the analysis of 29 publications, we were able to compile a series of barriers and facilitators, which are listed in Table 1, in the results section.

The examples found in the literature show that reverse innovation has been developed mainly in the health sector. The Institute for Global Health Innovation, for example, was commissioned by the United States-based Commonwealth Fund to identify low- or mid-income frugal innovations (low-cost innovations developed in developing countries) that could become reverse innovations being transferred to the US, aiming to expand access to care and drastically reduce costs [30]. However, reverse innovation does not focus only on the health sector, much less just on products. It can also be applied to companies' distribution strategies in emerging markets.

Table 1. Barriers and facilitators of reverse innovation.

Barriers	Facilitators
• The dominant logic and institutionalized thinking of multinationals; • The resistance to transfer power and control away from the matrix; • The preconception of the professionals of the matrix concerning the peers of the subsidiaries of developing countries; • The perception that reverse innovation develops low technology products and limited resources; • The skepticism of health professionals in high-income environments; • The institutional distance between developed countries and emerging markets; • To manage the diversity of people and cultures in the R&D process; • The need of the multinational corporations to expand their markets and not to lose space for low-cost solutions; • To change-resistant culture; • The fear of products cannibalization; • The legal and regulatory barriers	• Growing demand from emerging markets for new products; • The advancement of the use of technology to economic and easy-to-use innovations; • Economic liberalization in emerging markets; • The multinational's need to expand their markets and profits to survive in the global market; • Decentralization increase of R&D areas; • Low-cost workforce offer, government support, entrepreneurial spirit and speed and creativity of the Chinese Internet industry; • Search for simpler/cheaper products, and the fast pace of large-scale Chinese government projects; • Possibility for multinationals to strengthen their competitive advantage and economic growth; • Openness to participate in collaboration networks; • Open innovation; • Support of infrastructure, professionals and local policymakers in developing countries; • International partnerships movement in the health area, aiming the mutual responsibility and definition of a joint agenda; • The co-location of foreign and domestic firms; • The diversity in the early stages of product development to reduce uncertainties

Source: Prepared by the authors (2018).

It is evident from this review that the literature on barriers and facilitators to reverse innovation is limited. The barriers and facilitators presented in this article were identified in studies that did not have them as a focus of study in their genesis. Empirical studies that address the practice of reverse innovation, with an emphasis on barriers and facilitators and drivers of innovation, are needed to assist in the compilation of a set of good practices of reverse innovation. Analyzes to check the potential of each barrier and facilitator on reverse innovation is also essential so that efforts can be made on the most relevant factors. The behavioral aspect of prior rejection of reverse innovation by developed countries and the ways to overcome it are important research topics.

References

1. Ehrenmann, S., Warschat, J.: Diversity in the early phases of product development. Int. J. Ind. Eng. Manag. (IJIEM) **4**(1), 19–26 (2013)
2. Borini, F.M., Costa, S., de Miranda Oliveira Junior, M.: Reverse innovation antecedents. Int. J. Emerg. Markets **11**(2), 175–189 (2016)
3. Agarwal, N., et al.: A systematic literature review of constraint-based innovations: state of the art and future perspectives. IEEE Trans. Eng. Manag. **64**(1), 3–15 (2017)
4. Hadengue, M., De Marcellis-Warin, N., Warin, T.: Reverse innovation: a systematic literature review. Int. J. Emerg. Markets **12**(2), 142–182 (2017)
5. Harris, M., et al.: They hear 'Africa' and they think that there can't be any good services - perceived context in cross-national learning: a qualitative study of the barriers to reverse innovation. Glob. Health **11**(1), 10–17 (2015)
6. Harris, M., et al.: Measuring the bias against low-income country research: an implicit association test. Glob. Health **13**(1), 1–9 (2017)
7. Rowthorn, V., Plum, A.J., Zervos, J.: Legal and regulatory barriers to reverse innovation. Ann. Global Health **82**(6), 991–1000 (2016)
8. Harris, M., et al.: That's not how the learning works - the paradox of reverse innovation: a qualitative study. Glob. Health **12**(1), 1–8 (2016)
9. Hossain, M.: Frugal innovation: a systematic literature review. SSRN 2768254 (2016)
10. Hossain, M.: Mapping the frugal innovation phenomenon. Technol. Soc. **51**, 199–208 (2017)
11. Corsi, S., Di Minin, A.: Disruptive innovation in reverse: adding a geographical dimension to disruptive innovation theory. Creat. Innov. Manag. **23**(1), 76–90 (2014)
12. Govindarajan, V., Ramamurti, R.: Reverse innovation in emerging markets. In: Haar, J., Ernst, R. (eds.) Innovation in Emerging Markets, pp. 140–157. Palgrave Macmillan, London (2016)
13. Prahalad, C.K., Lieberthal, K.: The end of corporate imperialism. Harvard Bus. Rev. (2003). https://hbr.org/2003/08/the-end-of-corporate-imperialism
14. Dawar, N., Chattopadhyay, A.: Rethinking marketing programs for emerging markets. Long Range Plan. **35**(5), 457–474 (2002)
15. Costa, S., et al.: A Relação entre Inovação Reversa e Centros de Excelência em Subsidiárias Estrangeiras de Mercados Emergentes. Anais ANPAD - VI Encontro de Estudos em Estratégia, pp. 1–4 (2013)
16. Abadía-Barrero, C.E.: Kangaroo mother care in Colombia: a subaltern health innovation against for-profit biomedicine. Med. Anthropol. Q. **32**(3), 384–403 (2018)
17. Immelt, J.R., Govindarajan, V., Trimble, C.: How GE is disrupting itself. Harvard Bus. Rev. **87**(10), 56–65 (2009)
18. Gautam, N.: Reverse innovation—enablers and opportunities. Auto Tech Rev. **2**(1), 54–56 (2013)
19. Govindarajan, V.: A reverse-innovation playbook: insights from a company that developed products for emerging markets and then brought them back home. Harvard Bus. Rev. (2012). https://hbr.org/2012/04/a-reverse-innovation-playbook
20. Prahalad, C.K.: The Fortune at the Bottom of the Pyramid: Eradicating Poverty Through Profits, 1st edn. Wharton School Publishing, Upper Saddle River (2004)
21. Ibe, C.A., et al.: From Kisiizi to Baltimore: cultivating knowledge brokers to support global innovation for community engagement in healthcare. Glob. Health **14**(1), 19 (2018)
22. Hu, L.: The Chinese market as an opportunity to innovate distribution strategies? Evidences from Italian firms. Eur. Bus. Rev. **30**(5), 607–626 (2018)
23. Chittoor, R., Aulakh, P.S.: Organizational landscape in India: historical development, multiplicity of forms and implications for practice and research. Long Range Plan. **48**(5), 291–300 (2015)

24. Torraco, R.J.: Writing integrative literature reviews: guidelines and examples. Hum. Resour. Dev. Rev. **4**(3), 356–367 (2005)
25. Whittemore, R., Knafl, K.: The integrative review: updated methodology. J. Adv. Nurs. **52**(5), 546–553 (2005)
26. Gil, A.C.: Métodos e técnicas de pesquisa social, 6teen edn. Atlas, São Paulo (2008)
27. Botelho, L.L.R., de Almeida Cunha, C.C., Macedo, M.: O método da revisão integrativa nos estudos organizacionais. Gestão e sociedade **5**(11), 121–136 (2011)
28. Govindarajan, V., Trible, C.: Is reverse innovation like disruptive innovation. Harvard Bus. Rev. Blog Netw. **30** (2009). https://hbr.org/2009/09/is-reverse-innovation-like-dis.html
29. Kulasabanathan, K., et al.: Do international health partnerships contribute to reverse innovation? A mixed methods study of THET-supported partnerships in the UK. Glob. Health **13**(1), 1–11 (2017)
30. Bhatti, Y.A., et al.: The search for the Holy Grail: frugal innovation in healthcare from low-income or middle-income countries for reverse innovation to developed countries. BMJ Innov. **3**(4), 212–220 (2017)
31. Borini, F.M., de Miranda Oliveira Junior, M., Silveira, F.F., de Oliveira Concer, R.: The reverse transfer of innovation of foreign subsidiaries of Brazilian multinationals. Eur. Manag. J. **30**(3), 219–231 (2012)
32. Zhu, F., Zou, S., Xu, H.: Launching reverse-innovated product from emerging markets to MNC's home market: a theoretical framework for MNC's decisions. Int. Bus. Rev. **26**(1), 156–163 (2017)
33. Jana, R.: Innovation trickles in a new direction. Bloomberg Business Week 11 (2009). http://www.businessweek.com/magazine/content/09_12/b4124038287365.htm
34. Zeschky, M., Widenmayer, B., Gassmann, O.: Organising for reverse innovation in Western MNCs: the role of frugal product innovation capabilities. Int. J. Technol. Manag. **64**(2–4), 255–275 (2014)
35. Yio, G., McKern, B.: Innovation in emerging markets - the case of China. Int. J. Emerg. Markets **9**(1), 2–10 (2014)
36. Agarwal, N., Brem, A.: Frugal and reverse innovation-literature overview and case study insights from a German MNC in India and China. In: 2012 Proceedings of the 18th International Conference on Engineering, Technology and Innovation, pp. 1–11 (2012)
37. Snowdon, A.W., et al.: Reverse innovation: an opportunity for strengthening health systems. Glob. Health **11**(1), 2 (2015)
38. van de Vijver, S., et al.: Cardiovascular prevention model from Kenyan slums to migrants in the Netherlands. Glob. Health **11**(1), 1–6 (2015)
39. Li, H., Zhang, Y.A., Lyles, M.: Knowledge spillovers, search, and creation in China's emerging market. Manag. Organ. Rev. **9**(3), 395–412 (2013)
40. Shankar, V., Hanson, N.: How emerging markets are reshaping the innovation architecture of global firms. In: Review of Marketing Research, pp. 191–212. Emerald Group Publishing Limited (2013)
41. Fry, C.V., et al.: Health Innovation Transfer from South to North, 1–79 (2011)

System Architecture of a Robotics Airship

José Reginaldo H. Carvalho[1](✉), Miguel Á. C. Rueda[2], José R. Azinheira[3], Alexandra Moutinho[3], Luiz G. B. Mirisola[4], Ely C. de Paiva[5], Lucas A. C. O. Nogueira[2], Gustavo A. Fonseca[2], Josué Jr. G. Ramos[2], Mauro F. Koyama[2], Samuel S. Bueno[2], and Christian Amaral[6]

[1] Instituto de Computação, UFAM, Av. Rodrigo Otávio, 6200, Manaus, AM 69077-000, Brazil
reginaldo@icomp.ufam.edu.br
[2] CTI Renato Archer, Rod. D.Pedro I (SP-65) km.143.6, Campinas, SP 13069-901, Brazil
[3] IDMEC/LAETA/IST, Universidade de Lisboa, Av. Rovisco Pais, 1049-001 Lisbon, Portugal
[4] IEC/ITA, Pç Marechal Eduardo Gomes, 50, São José dos Campos, SP 12228-900, Brazil
[5] FEM/UNICAMP, Rua Mendeleyev 200, Campinas, SP 13083-970, Brazil
[6] Omega Aerosystems, R. Domingos Cordeiro, 389, Campo Largo, PR 83601-120, Brazil

Abstract. This paper aims to present the conception, design, and realization of a robotics airship. Differently, from the classic bi-propulsion, this vehicle has four vectored thrusters. Tail surfaces complete the directional actuation. The project has the code-name DRONI, the acronym for "Dirigível Robótico de Concepção Inovadora" (Robotic Airship with an Innovative Design in free translation). The initial motivation to start the project was the demand for an aerial platform for environmental monitoring of flooded areas in the Amazon region. This operation scenario imposes severe restrictions on the practical usage of quadrotors and fixed wings. High maneuverability, medium to long-endurance, flexible flight modes (hover, vertical takeoff and landing) are the most essential requirements for an effective aerial platform to overfly the vast Amazon canopy. Medium to long-endurance surveillance, cargo, and telecommunication relay are other common application of airships. The paper presents the current vehicle's architecture. The main subsystems shown are the robotic embedded infrastructure, the ground station and the communication system. The 6 kg payload enables the use of several types of cameras and sensors. Moreover, the paper presents the dynamic modeling with a control based on Incremental Nonlinear Dynamics Inverse. The inaugural flight is also shown, and the results are promising, towards a multipurpose aerial platform, that can be applied beyond environmental monitoring.

Keywords: Unmanned aerial vehicle · System architecture · Environmental monitoring

1 Introduction

Unmanned aerial vehicles (UAVs), which have been used in a wide range of applications, can be classified by the way the lift force for sustained flight is generated: (i) by fixed wings (various aircraft configurations), (ii) by rotating wings (helicopters, multi-rotors), (iii) by aerostatic principle (airships).

L. Pereira et al. (Eds.): IDEAS 2018, SIST 198, pp. 13–22, 2021.
https://doi.org/10.1007/978-3-030-55374-6_2

In the case of airships, the aerostatic lift comes from a lighter than air gas conditioned in an envelope, whereas its maneuverability is typically ensured by two controllable propellers mounted on a gondola under the envelope and aerodynamic actuation surfaces mounted on the tail. Interesting characteristics of these aircraft are greater intrinsic stability, greater flight autonomy, from low altitudes up to the stratosphere, at low speeds, their ability to hover, and to perform vertical takeoff and landing. On the other hand, some difficulties are less maneuverability due to the slower dynamics, under actuation, mainly in the lateral direction to the movement of the aircraft and more pronounced at low speeds, and the abrupt and strongly non-linear transition between low speed and cruising flights. The development of control systems for airships is, therefore, a complicated task. In this scenario, from the 1990s, interest in robotic airships increased, for example: in Brazil - the AURORA project [1], Germany [2], France [3], Korea [4], USA [5], Portugal [6], China [7]. With the evolution of electric propulsion, a small prototype equipped with four vectoring propellers, but dispensing the tail surfaces, was presented in [8]. Applications include long to medium term surveillance, telecommunication relay, cargo ships, and, as in the present case, environmental monitoring.

This article addresses the development of an unmanned robotic airship, with an innovative arrangement of four independent vectored electric propellers. This new configuration shows better stabilization and maneuverability at low speeds (hovering and vertical takeoff and landing), whereas maintains the efficiency of the aerodynamic control surfaces at the tail in the averages to high speeds (cruise flight). The airship is named NOAMAY - a word in the language spoken by the Yanomami Indians in the Amazon in Brazil, which means to take care of, to protect. Indeed, the purpose of the robotic airship is the realization of pilot-scale applications of environmental observation in the Amazon [9]. NOAMAY comprises the airship itself, its robotic infrastructure (onboard system, control station and communication system), as well as control and autonomous guidance system.

Figure 1 shows the first airship prototype. It has 11 m of length, 43 m^3 of volume and 6 kg as payload. The envelope has two layers: the inner one is composed of polyethylene polymer film, while the outer one is composed of high tensile rip-stop nylon synthetic fabric. At the bottom of the envelope, an aluminum frame receives the front and back propellers structure and, at its center, a rack that houses the embedded system. A distinct vectoring mechanism of $\pm 225°$ couples each of the four propellers. The aerodynamic surfaces, as actuators, can be arranged in "+" configuration (as in the figure) or in "x"

Fig. 1. NOAMAY airship during its inaugural flight.

pattern. The power system consists of 4 packs of LiPo batteries, which provide 22.2 V, 40 Ah, 888 Wh.

In this ongoing research context, this article focuses on the development of the robotics infrastructure associated with the aircraft and on the modeling and design of a nonlinear control system. Thus, after this introductory section, Sect. 2 summarizes the dynamic modeling of the aircraft and presents a control strategy based on inverse nonlinear incremental dynamics. Then, Sect. 3 describes the robotic infrastructure developed for the aircraft. Section 4 presents sensory data acquired during the airship's inaugural flight, in a remotely controlled way, demonstrating the capabilities of the airship and its robotic infrastructure. Finally, Sect. 5 concludes the article.

2 Robotics Infrastructure

Typically unmanned aerial vehicle holds at least three, well defined, subsystems i) the vehicle itself, avionics and embedded electronics within; ii) a ground station with mission planning capabilities iii) a reliable communication system to establish a secure link between the other sub-systems.

NOAMAY's robotics infrastructure, which is based on AURORA's project airship [10], is composed by:

- The onboard system, including autopilot, application processor, sensors and application-specific hardware for commuting between automatic operation mode (autopilot, autonomous flight mode), and manual (radio-controlled mode);
- A ground station, installed on a notebook.
- A communication system via radio modem.

NOAMAY solution consists of the same three subsystems, concerning to its functionalities, but with updated components (AURORA infrastructure is ten years old). Figure 2 shows the new infrastructure, also detailed in the sequence.

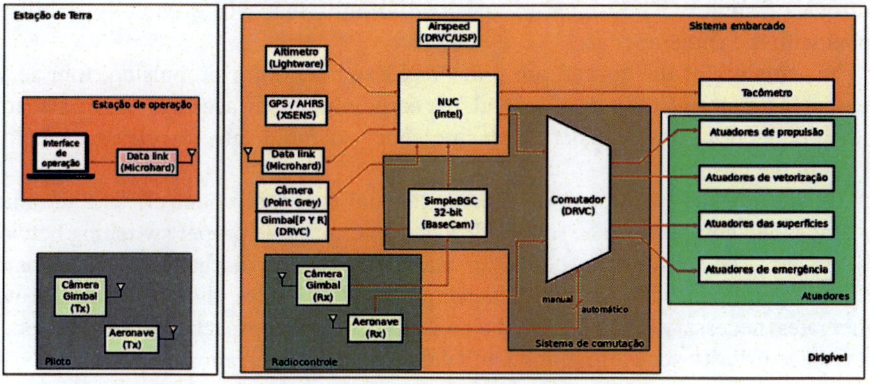

Fig. 2. Airship general architecture.

2.1 Embedded System

The embedded system integrates off-the-shelf components and overlaps a conventional radio control solution with PWM (pulse width modulation) commands. In addition to its primary functionality (ensuring the calculation on the embedded computer of control signals for the actuators when in automatic mode of operation), an additional feature is the logging of all commands sent to the actuators when in the manual operation mode. This peculiarity is useful, especially in the development phase, for example, when executing specific maneuvers to derive the dynamic behavior of the airship and collect data for the identification of model parameters. This solution is also attractive for the possibility of "robotization" of other radio-controlled vehicles.

The embedded system consists of an Intel NUC form factor computer, and a set of sensors integrated with USB, Ethernet and serial (RS-232) interfaces. This computer operates with Linux 14.04 Ubuntu and the Robot Operating System framework (ROS) [11], Indigo version, as *middleware*. Characteristics such as weight, power consumption, and interfacing capacity were the criteria for choosing the robotic components explained in Fig. 2, and listed as follows:

- GPS/AHRS MTi-G-700 (Xsens);
- Altímetro Laser SF11/C (Lightware);
- CMOS Flea 3 USB (Point Grey) 3.2MP camera;
- SimpleBGC 32-bit Gimbal Stabilizer (Basecam electronics);
- Airspeed/wind speed sensor
- Standard tachometer (rotating velocity of the propellers);
- Battery charge sensor.

The last three are self-made and are in development. As for the wind speed sensor, it uses a low-pressure sensor SDP-1000 (Sensirion), in the −50 to 500 Pa range and communicates with the on-board computer via Xbee at 900 MHz; a self-made pitot tube, designed for final calibration in a wind tunnel, completes the hardware.

A driver as ROS node was created for each of the sensors. There is also a Basecam electronics SimpleBGC 32-bit card and inertial units that will eventually stabilize a gimbal with the camera.

The actuators of the aircraft are four brushless motors for propulsion, four servo-actuators for vectoring the propellers and four aerodynamic surfaces with servo-actuators for their deflection. Besides, there is an emergency solenoid valve to release the helium gas out of the envelope.

The aircraft has two operation modes: manual (via radio control) and automatic (via embedded computer). The self-made electronics provide proper switching between these modes. The sensor data acquisition and the command dispatcher to the actuators are fully operational via ROS nodes. Also, in the ROS nodes, the control cycles with higher rates, necessary to implement the nonlinear control approach discussed in Sect. 3, will receive priority scheduling, reducing delays of communication.

2.2 Ground Station

The ground station uses a laptop computer with Linux Ubuntu and ROS Indigo. It is coupled to the onboard computer by a remote radio link (see Sect. 2.3), using the MAVLink protocol, which is computationally light enough to run on low-performance microcontrollers. Due to this, MAVLink is widely adopted by the UAV R&D community, especially the small ones [12]. Concerning the Ground Station core software, it was decided to use the program QGroundControl [13]. It was the first to implement the MAVLink protocol, built in synergy with the protocol development team. In addition, one can run it on Windows, Mac, and Linux. Then, it was only necessary to develop software elements to perform the communication between the ROS system in the vehicle and the MAVLink protocol. The user interface, shown in Fig. 3, enables:

- The mission schedule in terms of waypoints (latitude and longitude coordinates and altitude profile) and flight attributes (vertical takeoff or landing, hovering or cruising).
- The selection of control and navigation algorithms and the definition of their adjustment parameters.
- The reception and display, during the flight, of telemetry data (sensory information, aircraft status, power supply systems, etc.) as well as images from the camera.

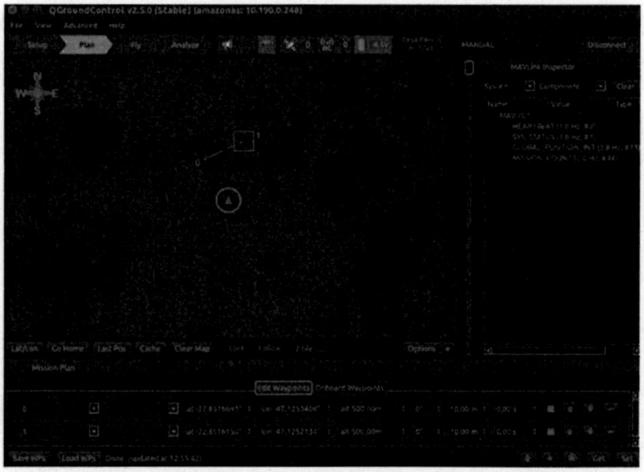

Fig. 3. Ground station main software window.

2.3 Communication System

The communication system consists of two setups. One is intended for manual piloting and uses two conventional 2.4 GHz radios-controls (Hitec Ltd). One of the radios is in charge of controlling the aircraft's actuators, while the other controls the camera's gimbal.

The second solution establishes the communication between the Ground Station and the onboard system in the aircraft. It is based on a radio link IPnDDL2450 (Microhard), with an arrangement of 8 dBi omnidirectional antennas mounted on the aircraft and, next to the Ground Station, a 17 dBi Yagi directional antenna guided by a pan-tilt tracker (Flir D48-E). The arrangement has a maximum nominal range of 15 km, with a bandwidth of 12 MBps in two channels, one for data and another one for command and control.

3 Airship Modeling and Control

The airship dynamic model is defined considering a local coordinate system as its reference, whose origin is at the center of volume of the envelope. The vector of linear and angular velocities is represented by $V = [u, v, w, p, q, r]'$, and the dynamics of the aircraft is expressed as:

$$M\acute{V} = F_d(V) + F_a(V) + F_p(V) + F_g(V) + F_w(V)$$

where M is the 6×6 mass matrix which includes the elements of real and virtual inertia, typical of floating air vehicle dynamics, and F_d, F_a, F_p, F_g and F_w are, respectively, inertial, aerodynamic, propulsive, gravitational/thrust and wind-influenced forces. The main actuators in the airship (Fig. 4) are four aerodynamic surfaces whose deflection function as rudder, elevator, aileron, and four vectorizable electric thrusters that compensate the excess weight of the aircraft at low speeds, besides improving maneuverability throughout the flight profile.

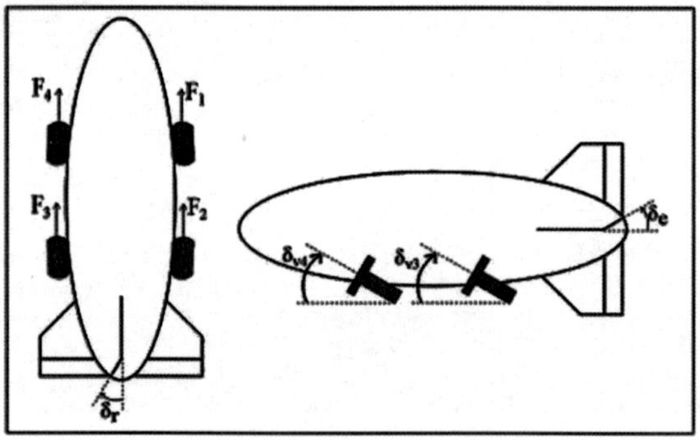

Fig. 4. Airship main actuators.

The vector of control inputs is $u = [\delta_e, \delta_a, \delta_r, \delta_1, \delta_2, \delta_3, \delta_4, \mu_1, \mu_2, \mu_3, \mu_4]'$, where δ_e, δ_a, and δ_r are the deflections of elevator, aileron and rudder, δ_i is the normalized input voltage of the *i-th* thruster and μ_i is the vectoring angle of the *i-th* propeller. To make the controller design more natural, the propeller powers were configured for joint use at forward power $\delta_f = (\delta_1 + \delta_2 + \delta_3 + \delta_4)/4$; left/right differential $\delta_{lr} = (\delta_1 + \delta_2 - \delta_3 - \delta_4)/4$, front/back differential $\delta_{fb} = (\delta_1 - \delta_2 - \delta_3 + \delta_4)/4$, and cross-differential $\delta_c = (\delta_1 - \delta_2 + \delta_3 - \delta_4)/4$. Similar configurations have been set for vectorizations.

This dynamic model of 6 degrees of freedom, evolved from [14], forms the basis of a simulator in MATLAB/Simulink [15] that allows the design and validation of airship control and navigation strategies. Among several possible approaches to control the NOAMAY airship flight, one that has been very promising is the *Incremental Non-linear Dynamic Inversion* (INDI). This automatic control technique considers that the vector of states x and its derivative $\acute{x}$ are observable and measurable at a sufficiently rapid sampling rate, when then the control law for a request for change of states is predominantly dependent on the respective non-linear input function F_u being given by:

$$u = u_0 - F_u^{\{-1\}}(\acute{x}_d - x_0)$$

in which $\acute{x}_d = K(x_d - x_0)$, where x_d is the desired state, F_u is a function of the state vector, and K is a gain matrix that stabilizes the closed-loop system. Figure 5 depicts the closed-loop block diagram of an INDI controller for any dynamic system.

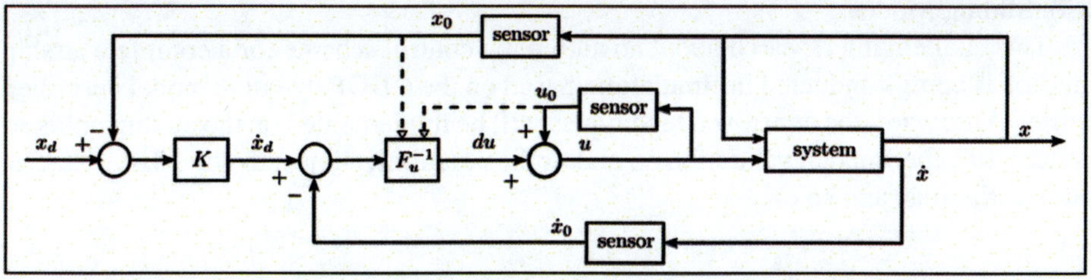

Fig. 5. Incremental Non-linear Dynamic Inversion (INDI) block diagram.

Eventually, for the particular case of a linearized model around an equilibrium condition, the input function F_u will correspond to the input matrix B of the linearized dynamics. This is the case when considering the decoupled models of the lateral and longitudinal dynamics of the NOAMAY airship, resulting from the linearization around a trimming condition, where we obtain:

$$\acute{x}_{lon} = A_{lon}x_{lon} + B_{lon}u_{lon}$$

$$\acute{x}_{lat} = A_{lat}x_{lat} + B_{lat}u_{lat}$$

and the matrices $A_{lon}, B_{lon}, A_{lat}, B_{lat}$ mainly depend on the trimming conditions (airspeed and altitude).

Longitudinal Dynamics: The state vector $x_{lon} = [u, w, q, \theta, h]$' comprises, respectively, the longitudinal and vertical velocities, rate of attack angle, attack angle, and altitude. The inputs $u_{lon} = [\delta_e, \delta_f, \delta_{fb}, \mu_f]$' are the elevator deflection, forward power, forward differential power, and synchronized vectorization (same angles).

Lateral Dynamics: The state vector $x_{lon} = [\beta, p, r, \varphi, \psi]$' comprises the lateral slip angle, roll, and yaw rates, and roll and yaw angles respectively. The inputs $u_{lat} =$

$[\delta_a, \delta_r, \delta_{lr}, \delta_c, \mu_{lr}]'$ include aileron, and rudder deflections, left/right differential power, cross-differential power and front/back differential vectoring, respectively.

A detailed description of INDI control for the lateral movement of the NOAMAY airship is given in [16], where are stressed the advantages of the approach, including its simplicity of design and tuning, and its efficiency. It should be emphasized that the INDI approach dispenses the knowledge aerodynamic parameters that are state-dependent, as these parameters usually are challenging to measure or to estimate.

Aiming the automatic control of both lateral and longitudinal airship dynamics, in order to cover all the phases of a complete mission (vertical takeoff, hovering, cruise flight, hovering and vertical landing) the INDI methodology is being extended, as well as a second control technique based on gain scheduling, similar to the one developed in [1]; this latter technique is also adopted in the incorporation of a planning level for airship flight [17]. Additional approaches being considered for the NOAMAY control use the LQR gain scheduling for the longitudinal mode, combined with the sliding modes for lateral movement [18], as well as a multivariate approach to Backstepping combined with Sliding Modes.

This challenging research about an automatic control scheme for a complete airship mission is being conducted in simulation, based on the 6 DOF dynamic model described earlier. Afterward, the conceived techniques will be implemented on the airship onboard system for experimental validation on real flight situations, subject to wind disturbances, mass variations, and so on.

4 Experimental Results

To validate the airship along with its robotic infrastructure, a set of flight tests were performed. The airship was flown under remote control, whereas sensory data was acquired by the onboard system and transmitted to the ground control station. A video of the ground station, showing the airship movement and other data received from the embedded system, is available at https://youtu.be/Ld8Hsmeak2U.

Figure 6 presents a screenshot of the user interface at the ground operation station. One can also see the trajectory performed by the aircraft in one of the flights, determined by the Xsens Mti-G-700 sensor. To analyze the behavior of the sampling period of each acquired sensory data, with respect to their nominal rates, for the total acquisition period, we computed the increments $\Delta t_k = t_k - t_{\{k-1\}}$ where t_k and $t_{\{k-1\}}$ are the time intervals between two consecutive acquisitions, and also the metrics $\Delta t_{\{MED\}} = mean(\Delta t_k)$ and $\Delta t_{\{MAX\}} = max(\Delta t_k)/\Delta t_{\{MED\}}$.

As an example, consider the Xsens Mti-G-700 sensor (GPS, inertial, altimetry and fused GPS + inertial information) whose data were acquired by the "ROS node" software at a nominal rate of 100 Hz - the highest sampling rate used in the robotic infrastructure. The Fig. 6 also shows the behavior of the $\Delta t_k/\Delta t_{\{MED\}}$ ratio, revealing a uniformity in acquisition at the specified 100 Hz, without loss of information, since the $\Delta t_k/\Delta t_{\{MED\}}$ ratio reaches at most a value of 1.7, staying below 1.2 most of the time. For the other sensors, whose sampling rates are much lower, the dispersion of the $\Delta t_k/\Delta t_{\{MED\}}$ around the unit was even lower.

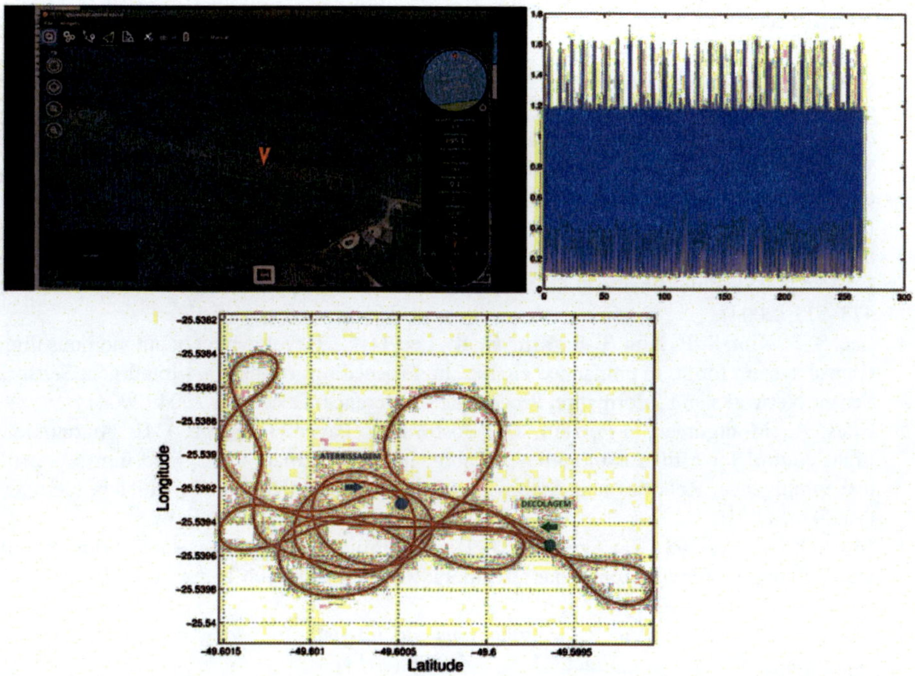

Fig. 6. Top-left: Screenshot of the Ground Station, with the airship current position marked over a map. Attitude, speed, compass and other variables are displayed on a panel on the right side of the map. Top-right: dispersion of the $\Delta t_k/\Delta t_{\{MED\}}$ ratio as measured by the Xsens sensor. Bottom: trajectory (red) of the airship, with the takeoff (green) and landing (blue) dots.

5 Conclusions

Unmanned airships are interesting platforms; both for its potential applications and for the research challenges they pose in control and other areas of robotics. This paper described the robotic infrastructure for a lighter than air aircraft, equipped with four vectoring electric propellers and aerodynamic surfaces, and presented a control architecture based on Nonlinear Incremental Inverse Dynamics.

The experimental flight was encouraging. In addition to proving the validity of the project as a whole, the vehicle presented maneuverability beyond expectations.

The work continuity includes aircraft optimization, with a general review of the project aiming at weight reduction, implementation, and experimental validation of control strategies, complementary to the existing ones. Moreover, there will have additional tests, to exercise the vehicle in flight conditions closer to the field operation, including multiple transitions between hovering and aerodynamic flight.

Acknowledgments. This work is sponsored by DRONI (CNPQ 402112/13-0), INCT (CNPQ 465755/14-3, FAPESP 2014/50851-0), Brasil; FCT (LAETA UID/EMS/50022/2013), Portugal.

References

1. Elfes, A., Bueno, S.S., Bergerman, M., Ramos, J.G.: A semi-autonomous robotic airship for environmental monitoring missions. In: Proceedings of 1998 IEEE International Conference on Robotics and Automation, Leuven, Belgium, May 1998 (1998)
2. Wimmer, D.-A., Bildstein, M., Well, K.: Research airship 'lotte' make: development and operation controllers for autonomous flight phases. In: IEEE International Conference on Intelligent Robots and Systems, pp. 55–68, Lausanne, Switzerland, October 2002 (2002)
3. Hygounenc, E., Soueres, P., Lacroix, S., Jung, I.-K.: The autonomous blimp project at LAAS/CNRS: achievements in flight control and terrain mapping. Int. J. Robot. Res. **23**(4), 473–511 (2004)
4. Lee, S.-J., Kim, S.-P., Kim, T.-S., Kim, H.-K., Lee, H.-C.: Development of autonomous flight control system for 50 m unmanned airship. In: Proceedings of the 2004 Intelligent Sensors, Sensor Networks and Information Processing Conference, December 2004 (2004)
5. Elfes, A., Montgomery, J.F., Hall, J.L., Joshi, S.S., Payne, J., Bergh, C.F.: Autonomous flight control for a titan exploration Aerobot. In: 8th International Symposium on Artificial Intelligence, Robotics and Automation in Space, i-SAIRAS 2005, Munich, Germany (2005)
6. Moutinho, A., Mirisola, L., Azinheira, J., Dias, J.: Project diva: guidance and vision surveillance techniques for an autonomous airship. In: Robotics Research Trends, pp. 77–120. Nova Science Publishers (2008)
7. Zheng, Z., Huo, W., Wu, Z.: Autonomous airship path following control: theory and experiments. Control Eng. Pract. **21**(6), 769–788 (2013)
8. Earon, E., Rabbath, C., Apkarian, J.: Design and control of a novel hybrid vehicle concept. In: AIAA Guidance, Navigation, and Control Conference and Exhibit, South Carolina, USA, August 2007 (2007)
9. Carvalho, J., Bueno, S., Modesto, J.: Sistemas aéreos não-tripulados para o monitoramento e gestão de risco do bioma amazônico. Revista Computacão Brasil, March 2014 (2014)
10. OSRF: ROS - Robot Operating System (2009). www.ros.org. Accessed 29 Jan 2018
11. Meier, L., Tanskanen, P., Heng, L., Lee, G.H., Fraundorfer, F., Pollefeys, M.: The Pixhawk open-source computer vision framework for MAVs (2011). www.pixhawk.org. Accessed 29 Jan 2018
12. Meier, L., Tanskanen, P., Heng, L.: Qgroundcontrol: ground control station for small air-land-water autonomous unmanned systems (2010). www.qgroundcontrol.org. Accessed 29 Jan 2018
13. Azinheira, J.R., de Paiva, E.C., Bueno, S.S.: Influence of wind speed on airship dynamics. J. Guid. Control Dyn. **25**(6), 1116–1124 (2002)
14. Arias, R.: Modelagem de um dirigıvel robótico com propulsão elétrica de quatro motores. Master's thesis, Universidade Estadual de Campinas, SP, Brasil (2014)
15. Azinheira, J., Moutinho, A., Carvalho, J.: Lateral control of airship with uncertain dynamics using incremental nonlinear dynamics inversion. In: 11th IFAC Symposium on Robot Control SYROCO, Salvador - BA, Brazil, August 2015 (2015)
16. Carvalho, J., Moutinho, A., Azinheira, J.R.: Integrating mission planner to the flight control system of a robotic airship, XXI Congresso Brasileiro de Automática - CBA (2016)
17. Vieira, H., Azinheira, J.R., Moutinho, A., de Paiva, E.C.: Controladores não lineares para um dirigível robótico de propulsão quádrupla. XXI Congresso Brasileiro de Automática – CBA (2016)

The New Brazilian Legal Framework of Science & Technology: Barriers, Borders and Opportunities for Innovation

Dércio Luiz Reis[1(✉)], Marcelo Albuquerque de Oliveira[1(✉)], Sicy Rusalka Goes de Melo Barreto[2(✉)], Joaquim Maciel da Costa Craveiro[1(✉)], and Ana Nubia dos Santos de Oliveira[3(✉)]

[1] Faculdade de Tecnologia (FT), Departamento de Engenharia de Produção, Universidade Federal do Amazonas, Avenida General Rodrigo Otávio, 6200 - Coroado I, Manaus, AM 69080-900, Brazil
derciolr@hotmail.com, marcelooliveira@ufam.edu.br, jgabael26@gmail.com

[2] Assessoria de Relações Internacionais e Interinstitucionais - ARII, Universidade Federal do Amazonas, Avenida General Rodrigo Otávio, 6200 - Coroado I, Manaus, AM 69080-900, Brazil
sicizinha@hotmail.com

[3] Faculdade La Salle, Avenida Dom Pedro, 151 - Dom Pedro I, Manaus, AM 69040-690, Brazil
ananubia_oliveira@lasalle.org.br

Abstract. The recognition of science and technology as a risk activity, focusing on results rather than procedures, means that researchers are more effectively engaged in activities involving innovation. The purpose of this article is to analyze the applicability of law known as the Legal Framework of Science and Technology, and it was constructed with bibliographical support seeking to contribute to a different view of the control organs regarding the research. The new Brazilian legislation brings with it the expectation that research and market have a process of approximation, reducing the distance between the knowledge produced in universities and their transformation into wealth. The possibilities arising from the new legislation tend to have effects in solving problems of quality, productivity, cost reduction, with the possibility of incorporating benefits to production and competitiveness, with the introduction of technology, methods and processes aligned with lean production. It concludes that the Legal Framework for Science and Technology, with its specific purpose of reducing bureaucracy in the country's research and innovation activities in general, is an important instrument in the integration of the academic and scientific community at all levels, and companies, representing a new path to boost the process of education.

Keywords: Innovation · Knowledge management · Legal framework of science and technology · Lean culture

1 Introduction

The new Brazilian landmark of science, technology, and innovation represents a turning point for universities and for the development of the Brazilian economy in the renewal

L. Pereira et al. (Eds.): IDEAS 2018, SIST 198, pp. 23–33, 2021.
https://doi.org/10.1007/978-3-030-55374-6_3

of its market capacity because it provides for stimuli to scientific development, research, scientific and technological training and innovation. The old legislation hindered the paid work of researchers from public institutions in projects related to the productive sector, and the new legislation allows private initiative, research institutes, academy members and public educational institutions to work together to market scientific production of which in many cases was lost in laboratories. This approach tends to make it possible for knowledge restricted to the academic environment and larger organizations to be effectively incorporated into small and medium-sized enterprises, which tend to have greater difficulties in incorporating knowledge into their operations.

The Decree No. 9.283 [1], which regulated the Legal Framework, since it brought the forms of concessions of benefits and incentives to R&D and Innovation activities, the legal formatting appropriate for the achievement of these instruments, and established the tools that allow Public Entities to grant legal privileges for research, development, and innovation in the technological field is analyzed. Brazil is now ranked 69th in the Global Innovation Index, based on a study by the University of Cornell at the World Intellectual Property Organization (WIPO) [2]. Among the 18 countries in Latin America, Brazil ranks 7th. The biggest problem identified is the downward trend, since the country ranks 47th in 2011.

In the implementation of the new law, the collaboration between universities and research institutes was prevented by the current regulations restricting the participation of researchers from public universities in remunerated activities of innovation or applied research. Professors from public universities, with exclusive work contracts, were prevented from promoting paid research in collaboration with companies. In addition, the control mechanisms did not offer legal certainty to the companies requesting the research, away from universities and research institutes. The Innovation Survey (PINTEC), which is a research carried out every three years, by the Brazilian Institute of Geography and Statistics (IBGE) [3], was used to give a brief overview of the scenario of innovation activities undertaken by Brazilian companies. PINTEC follows the Oslo Manual recommendation, in which product and process innovation are defined as the implementation of products (goods or services), new or substantially improved processes [4].

According to Damanpour [5], an innovation can be a new good or service, a new production process, a new structure or administrative system, or a new plan or program adopted by the organization. In this sense, innovation implies the generation, development and implementation of new ideas and behaviors, paying particular attention to their usefulness. The documentary research carried out in the previous legislation, the results presented in PINTEC and the comparison with the new possibilities introduced in the new Legal Framework, made it possible to outline a prospective scenario for the new moment presented for relations between the Academy, Governments, and Companies.

The need for organizations to promote systematic value creation for customers and thus position their products to the standards required by global market competition drives organizations on a constant search for innovation. This requirement positions Lean as the management philosophy that, for consistent results, is capable of promoting significant changes in the way companies handle their processes. Growing interest in Lean has provided the development and sharing of new techniques and experiences that enhance results and enable a growing learning curve. The Lean set of knowledge essentially aims to eliminate waste and solve problems, which implies that processes and products are constantly undergoing evaluations and changes, which are the basis of the innovation processes necessary to maintain the competitiveness of organizations and adapt to market needs.

This applies to all types of organizations, whether private or public. It is precisely this need to eliminate barriers to innovation where the new Brazilian Science, Technology and Innovation legislation will play a fundamental role when fully implemented. Since there is no way to dissociate the Lean from Innovation, it is in this view that the article is based.

In his discussion of the transformation from what he defined as static activities (consumption and production) to dynamic activities (innovation), Cooter [6] argues that laws aimed at innovation must create an open competition structure for innovators to develop their ideas. Indeed, laws that increase the pace of economic innovation must support laws that increase entrepreneurial profits as a driving force for innovation, as well as prioritize their protection and results.

Stiglitz [7] argues that the production of knowledge is not free, so there has to be some way to finance it. In the field of innovation, laws and procedures for regulating intellectual property must be re-evaluated, and one issue is the difficulty of determining the boundaries of property rights, as well as the degree of novelty that the idea envisages. The author also maintains the need to redesign the intellectual property regime to increase its benefits and reduce its costs in order to increase economic efficiency and probably drive the pace of innovation. In this context, Greenwald and Stiglitz [8] argue that industrial and economic policies must be aligned with government policies to avoid conflicts and collapse of more robust economic models based on innovation.

Granieri and Renda [9] emphasize that innovation is perceived as the main way to achieve economic growth and competitiveness, where technological innovation is an essential factor for long-term well-being, as it is attributed to the improvement of wellbeing and benefits to future generations. However, innovation is perceived as complex under the prism of public policies, since it involves both governmental and private investment and has a broad scope and interests, ranging from the existence of public policies to encourage innovation, from education to intellectual property, among other demands, as well as their degree of reach, which may be local, regional or international. The authors further explore the changing meaning of innovation and innovation policy and how the modes of innovation have changed, depending on the characteristics of the models, i.e. from traditional patterns to systemic and collaborative patterns, proprietary models, modular and granular models, innovation based on user offer and innovation, from closed to semi-open and open business models. The need to involve all actors in this process requires an adequate degree of governance, in order to meet the real needs of companies, entrepreneurs, universities, research institutions, among others, each one with its purposes. With the aim of boosting innovation levels, Europe has launched The Framework Program as a promoter of the European innovation system. Frietsch *et al.* [10] argue that such initiatives could provide stability and growth, both in terms of funding and in terms of a political message that placed a high priority on science and technology. The most recent initiative is the new Framework Programme Horizon

2020 that integrates various EU funding activities for research and innovation, which are properly regulated [11].

2 Literature Review

In the context of innovation, be it product, technology or processes, we present below a brief discussion about the Brazilian scenario in the context of innovation and the new legal framework, in order to understand the new possibilities related to this process.

2.1 Brazilian Scenario

The legislation in force before the new Legal Framework was a deterrent to Brazil's innovation actions. In order to be able to visualize the possibilities brought in by this new moment, an analysis of the results of the PINTEC 2014 survey is necessary. The study shows that there was a growth in 2012 of 1.90% of GDP in relation to the previous year and, respectively, of 3.00% and 0.50% in the years 2013 and 2014. The survey indicates that in Industry, 14.5% of companies promoted product and process innovation and 18.2% only process. In the Services sector, 15.9% made innovations in product and process 11.8%; only in process. The innovation in processes in the Brazilian industry was 32.7% and in the services sector 27.7%.

Table 1. Innovation in processes by company size

Busy person tracks	Innovation in processes (in%)	
	Industry	Services
10 to 49	30,00	25,80
50 to 99	40,10	35,70
100 to 249	44,80	35,40
250 to 499	49,40	44,30
Over 500	59,80	43,60
Total	32,70	27,70

Source: (IBGE 2014)

An analysis of Table 1 shows that the smallest companies have the lowest rate of innovation in processes and this is due to the budget constraints of smaller companies to invest in materials, equipment, hire skilled labor, and have access to Science and Technology Institutions (STIs) and the lines of development of government agencies, such as BNDES and FINEP. These would be the major beneficiaries of the new legislation because they can benefit from the approach to universities and institutes and research, and can access techniques and methodologies capable of introducing significant improvements in their products and processes, at a cost compatible with their capacity, or even without no cost of implementation, possible in partnerships where they are objects of academic research. For companies that implemented some innovation in the period from 2012 to 2014, research shows that the introduction of technological innovations in the market is important for 30.8% of companies in the industrial sector and 46.4% of service companies. The acquisition of external R&D and Innovation is considered important only for 5.4% of the industries and 5.3 for the service companies. Internal R&D and Innovation development are important for 15.2% of industries and for 39.2% of service companies. Training of human resources is considered important by 61.5% of the industries while the services sector has a percentage of 67.5%. In their manifestations of importance, Brazilian companies do not consider research institutes and technology centers as sources of information for innovation. The industry index stood at 18.5% and the service sector at 25.5%. Similarly, universities are considered sources of information for innovation by 16.4% of the national industry and by 25.7% of the service sector, which indicates the distance between academia and the market.

Considering the sectors of industry, electricity and gas and services, the investment in innovative activities and internal R&D and Innovation of Brazilian companies was 3.31% of the companies' net revenue. The service sector was the one that most invested in innovation and R&D and Innovation, totaling 9.94%, followed by the manufacturing industry with 2.84%, the extractive industry with 1.85% and the electricity and gas sector with 0, 74%. Sectors that depend fundamentally on intellectual capital, such as software development, have investments of more than 10%. Architectural and engineering services, which have not yet incorporated Building Information Modeling (BIM) into their processes, have investments of less than 2.5% in innovation. It is not the purpose of this study to evaluate these numbers, but the percentages presented show that R&D and Innovation activities are outside the immediate objectives of a large number of Brazilian companies.

It is of great concern to the industrial sector that, instead of aiming to meet world quality standards, especially those related to the implementation of systems aimed at Industry 4.0, demonstrates the preference for the acquisition of innovation incorporated in machinery and equipment, with 40, 2% of the expenses for this item and, despite considering important human resources training, in 2014 only 1.0% of the expenses had this destination. The internal expenses with research and development were 31.5% and the external expenses were 8.0%.

Considering that part of these expenditures involves the purchase of knowledge, this is a clear option for the innovation incorporated into equipment, which allows manufacturing with better quality and more quantity without, however, having the development of innovation from R&D and Innovation as a goal. This can be a dangerous strategy for the medium and long term, especially when one observes the low percentage of revenue invested in innovation.

Even more worrisome for Brazil is the low number of post-graduate human resources in research and development activities in industrial and service companies. The last survey (Innovation Research: 2014/IBGE, 2016) identified 146,000 professionals working in internal research and development activities in full or part-time companies and presents another worrying factor that is the trend of displacement of professionals from the integral period for a partial period in the activity. When analyzing the qualification of these professionals, it is verified that only 7.9% in industry and 15.3% in the service sector have a post-graduate level education. The largest contingent of researchers has an undergraduate degree, with 60.5% in industry and 63.2% in the service sector. Certainly, this affects the quality of the results of the investigation processes.

Keeping a professional with a Ph.D. in his or her workforce is not feasible for smaller companies and considered expensive by medium and even larger companies. The solution to this problem is the partnerships between companies, universities, and research centers who did not happen due to the obstacles of previous legislation. The results above show a low level of investment in innovation, coupled with the low qualification of the human resources involved, as a result of improvements in products and processes introduced. The approach of the companies with the Academy is fundamental for an organized and continuous updating of the business processes with the best practices and technologies being implemented. The promotion of this interaction tends to introduce state-of-the-art practices into production.

2.2 The New Brazilian Legal Framework

First of all, it is necessary to say that there has always been a segregation between the Public and Private Authorities in Brazil. The Federal Decree no. 9.283 [1] regulated the provisions of Federal Law no. 13.243 [12], known as the Legal Framework for Science, Technology, and Innovation, and promoted significant changes regarding this separation. A framework was inserted in the Brazilian Federal Constitution through Constitutional Amendment no 85, which changed and added provisions in the Constitution to update the treatment of Science, Technology and Innovation activities.

The new set of legal rules has as its main mission to encourage the creation, implantation, and consolidation of environments that promote innovation. The purpose is to promote technological development, remove bureaucratic barriers that hamper the activity of innovative researchers and entrepreneurs, increase competitiveness and interact among the characters involved in this R&D and Innovation ecosystem, including Public Agencies, Development Agencies, Technology Parks, Institutions Scientific and Technological Research, Private Companies, among others.

The aim is to stimulate and support strategic alliances, development of cooperation projects between companies and STIs for the generation of innovative products, processes and services, transfer and diffusion of technology, contemplate networks and international projects of technological research, technological entrepreneurship and creation of innovation, incubators, and technology parks, training and qualification of qualified human resources.

Universities and public STIs may share intellectual capital, laboratories, equipment, instruments, materials and other facilities with companies and individuals for research, development and innovation activities, if such permission does not directly interfere with their activity or conflict with it. In addition, it may directly assign to companies the use of real estate for the installation of environments that promote innovation. The legislation prior to the Legal Framework explicitly prohibited that such actions were implemented, characterizing this type of collaboration as misuse of public resources, subjecting managers to penalties such as reimbursement of values to public coffers and large fines.

This unfavorable environment of cooperation has brought significant damage to all concerned and to the quality of human-resource training at all levels. By preventing the academy from having a closer approximation to the productive environment, the ability of students and researchers to actually implement the objects resulting from academic research was restricted, restricting the learning process based on practical experience resulting from interventions in productive processes, which is a fundamental step for the training of human resources qualified to work in the productive sector.

The new law extended the mechanisms of economic subsidy to micro, small and medium enterprises, implementing, among other measures, the technological bonus, destined to the payment of contracting technological services. It authorizes the law even though the economic subsidy can be used by companies both for the financing of research activities and for capital expenditures. Before, the economic subsidy could not be used for the acquisition of capital goods.

It is important to emphasize that this possibility brought by the norm comes in many ways to contribute to the interaction between the public and the private since few companies have enough capital and expertise to set up their own laboratories. The universities have the infrastructure already installed, and the use of this by the companies is obviously a huge advantage for both parties.

There is no point in innovation if you do not leave the university benches for production environments. In order for this migration from academia to industry, it is necessary that the technology transfer process developed in STI take place.

The new law also allows the Scientific and Technological Institutions (STIs) to sign agreements with companies for the development of joint research, and STI may assign to the private partner all intellectual property rights through financial or non-financial compensation, provided that economically measurable, and technology transfer should be formalized through a technology transfer agreement.

According to the Innovation Law, all ICTs must have a Nucleus of Technological Innovation (NIT) that will be responsible for the management of the innovation policies that should be able to negotiate and manage these contracts, which need for a technical and legal body to supervise the execution of these contracts, in order to verify if they are in accordance with the contractual clauses. It is important to note that this change in the standard is very attractive for the formalization of a public/private partnership since the ownership of the results of research is no more property of the public institution, who before was forced to open a public competition to license the technology. This system was completely discouraging to private companies since there was no way to be sure that the company that financed the research would license the result.

Likewise, the public government allowed to directly fomenting technological innovation in companies and STIs through various mechanisms, including the direct contracting of research projects involving technological risk, for the solution of a specific technical problem or obtaining of product, service or an innovative process, without the obligation of the acquisition process.

The use of intellectual capital is seen with the possibility of public service researchers in an exclusive dedication to engage in paid research, development and innovation activity in STI or company, if assured to the continuity of their teaching and research activities.

Public institutions may be minority quotas in the social capital of technology based companies and investment funds to innovation, either direct or indirect, through investment funds constituted with own or third-party resources, among other actions.

The federal government may also become a partner of startups. Universities, besides the well-known vocations for teaching, research, and extension, can collaborate for the emergence of companies with the participation of their teachers and students. Usually, academic startups are more successful in transferring knowledge than other companies. This is because they have the technical advice of academics that integrate the corporate structure, knowing the technical aspects of the inventions.

However, it is important to emphasize that the Brazilian Securities and Exchange Commission still needs to regulate this type of investment, which is fundamental for the sector, since it is a measure adopted by the main universities and research centers of the world. The technological order is also one of the innovations brought with the new Legal Framework. Through it, organs and entities of the public administration may contract directly public or private STI, private nonprofit entities or companies, alone or in a consortium, directed to research activities and recognized technological qualification in the sector, with a view to carrying out R&D and Innovation activities involving technological risk, for solving a specific technical problem or obtaining an innovative product, service or process.

The supply of the product or the innovative process resulting from the research, development and innovation activities ordered may be contracted by no-bid acquisition, including with the developer of the order, subject to the provisions of specific regulations. This possibility brought by the norm allows the public sector to break the bureaucratic barriers, until then in force, that established complex rules for the contracting of technological services, which required the elaboration of previous detailed projects and the observation of deadlines for the processing of acquisition, proposals and resources to the results that sometimes took years and did not always reach a satisfactory result.

This extreme bureaucracy has made the Brazilian State an unfavorable environment to adopt technologies that allow the optimization of resources and their availability in the right place at the right time. Therefore, lean philosophies need to be incorporated into the provision of public health services, education, connectivity, among other important actions in a country with more than 220 million inhabitants and a continental size. In parallel with this new contracting format, it is important to highlight that the use of the incorporation of knowledge and technologies of social impact will only have the expected effects if the human resources of the Brazilian public sector is duly qualified. This need will promote a strong acceleration in inclusive Lean programs in the public sector in areas where they are usually not included in the academic curriculum of the courses, such as in the social sciences and health, generating new opportunities for new players and new career opportunities.

Another relevant point brought with the Legal Framework is that the accounting procedures of resources passed on under the Law should follow simplified and standardized forms and, in order to ensure governance and transparency of information, be carried out annually, preferably through electronic sending of information, according to the regulation.

3 Discussion

Every process of change is an excellent time to put the wheel of the transformation to spin. It is also an opportunity to initiate actions that promote discussions about the role of knowledge in the implementation of new paths for the economy of the state of Amazonas. When you look at the business environment that drives the world's major economies, you can identify what a slogan is innovation. In a simplified way, to generate innovation, you need to have an infrastructure of facilities capable of supporting the necessary research and qualified human resources. Laboratories can be purchased, but qualified human resources, with masters and doctorates, depend on time and investment, and are available in our institutions. The higher the qualification, the greater the likelihood of expressive results leaving the laboratories.

The simplification of procedures and the closer relationship between companies and government research institutions is one of the advances brought by Federal Law 13.243, of 2016, regulated by Decree No. 9, 283, of 2018, known as the Legal Framework for Science, Technology and innovation. Through the Legal Framework, the Brazilian Government's obligation to foster innovation in companies and non-profit entities was generated. The mentioned legal norm extended the mechanisms of economic subsidy to micro, small and medium enterprise, especially the participation in companies.

Public institutions may be minority share holders in the social capital of technology based companies and investment funds for innovation, either directly or indirectly, through investment funds constituted with own or third-party resources, among other actions, for the purpose of developing products or innovative processes that are in accordance with the guidelines and priorities defined in science, technology, innovation and industrial development policies.

The legislation establishes that minority participation will be a financial contribution or not, and it must be economically measurable and may be accepted as a form of remuneration for the transfer of technology and for the granting of the right to use or to exploit the creation of ownership of the Union and its entities. One of the attractions for the legal entity to build the relationship with public institutions is that the intellectual property on the results obtained will belong to the company, in the form of the current legislation and its constitutive acts.

The assets of the equity interest may be sold without the need for public bidding, and the funds received as a result of the sale of the equity interest should be invested in research and development or in new equity interests. It is imperative to emphasize that the permit discussed here allows a significant opening in international relations, since the legal norm did not limit the companies to the national territory. Several international companies have innovative processes that would boost the development of Brazilian science, technology and innovation. In order for this to happen, it is essential to promote joint PD&I activities between the public and private sectors of the Brazilian economy, through relationships that articulate knowledge and resources in a clear and safe way for both.

This ensures greater predictability and, as a consequence, attractiveness to investments in the area of CT&I.

4 Conclusion

We can conclude that the new set of legal rules will significantly influence the production of science through articulation with technology and innovation, which will imply the increase of additional resources coming from the private sector. In the same way, it will affect the quantity, quality, and flexibility of scientific production. In order for this to happen, it is essential to promote joint R&D and Innovation activities between the public and private sectors of the Brazilian economy, through relationships that articulate knowledge and resources in a clear and safe way for both. This ensures greater predictability and, consequently, attractiveness to investments in the area of Technology and Innovation.

At the regional level, the state of Amazonas can benefit significantly if actions are taken towards building an environment conducive to the approximation of companies with local institutions, which have a large amount of research on the potential to generate important products economically, but who need investment, management and marketing, so that they can contribute to the formation of a new matrix for the state economy. The restructuring of the Amazon Biotechnology Center (CBA) is a great opportunity, which can transform the region's biodiversity into products to integrate a new 10 economic matrix for the State, and other similar structures can be designed so that other areas of knowledge can contribute for this construction.

For this to happen, it is essential that joint activities be fostered between the public and private sectors, through relationships that articulate knowledge and resources in a clear and secure way for both, thus ensuring greater predictability and, consequently, attractiveness to investments. Under the new law, universities and public research institutions are now allowed to share their intellectual capital, laboratories, equipment, instruments, materials and other facilities with companies and individuals for research, development and innovation activities, provided that such permission does not directly interfere with its end-activity or conflicts with it. This approach is the quickest solution to overcome the challenge of making research results turn into products that can deliver real benefits to society, generating jobs throughout the production chain, and results that promote welfare or any other factor that serves society.

Basic research is fundamental to the advancement of science, but applied research, which transforms knowledge into product, is essential for the well-being of society.

References

1. Brasil: Lei no 9.283, de 07 de fevereiro de 2018. Regulamenta o Marco legal da Ciência e Tecnologia. Brasília (2018)
2. WIPO: The Global Innovation Index 2018: Energizing the World with Innovation. Ithaca, Fontainebleau, and Geneva. Cornell University (2018)
3. IBGE: Pesquisa de inovação Coordenação de Indústria. IBGE, Rio de Janeiro (2014)
4. OCDE – Manual de Oslo: Proposta de diretrizes para coleta e interpretação de dados sobre inovação tecnológica. FINEP, Brasília (2004)
5. Damanpour, F.: Organizational innovation: a meta-analysis of effect of determinants and moderators. Acad. Manag. J. **34**(3), 555–590 (1991)
6. Cooter, R.: The Falcon's Gyre: Legal Foundations of Economic Innovation and Growth. Berkeley Law Books. Book 1 (2014). http://scholarship.law.berkeley.edu/books/1

7. Stiglitz, J.E.: Economic foundations of intellectual property rights. Duke LJ **57**, 1693 (2007)
8. Greenwald, B., Stiglitz, J.E.: Industrial policies, the creation of a learning society, and economic development. In: The Industrial Policy Revolution I, pp. 43–71. Palgrave Macmillan, London (2013)
9. Granieri, M., Renda, A.: Innovation Law and Policy in the European Union: Towards Horizon 2020. Springer, Heidelberg (2012)
10. Frietsch, R., Rammer, C., Schubert, T.: Heterogeneity of innovation systems in Europe and horizon 2020. Leibniz information centre for economics. Forum the impact of horizon 2020 on innovation in Europe. Intereconomics **50**, 4–30 (2015). https://doi.org/10.1007/s10272-015-0521-7
11. Europe: Regulation (EU) No 1291/2013 of The European Parliament and of the council of 11 December 2013 - Establishing Horizon 2020 - the Framework Programme for Research and Innovation (2014–2020) and repealing Decision No 1982/2006/EC. Official Journal of the European Union (2013)
12. Brasil: Lei no 13.243, de 11 de janeiro de 2016. Marco legal da Ciência e Tecnologia. Brasília (2016)

Check for updates

A Cyber Physical System Approach to Customer Services of Home Appliances

Ana Carolina Fragal(✉), Anderson Orzari Ribeiro, and Crhistian Raffaelo Baldo

Federal University of ABC, Av. dos Estados, 5001 - Bangú, Santo André, SP, Brazil
ana.fragal@gmail.com

Abstract. To survive in the innovative and competitive home appliance market, the manufacturers have heavily invested in digital technologies that improve the quality of their products and after-sales services. In this context, Cyber Physical Systems (CPS) technologies, which allows integration between physical and digital worlds, can be a successful bet for this industry. CPS enables gathering field data that feedbacks different stages of product lifecycle and use, allowing assertive investments to improve products, processes and services, as well as opening up new possibilities for the business model. Considering this scenario, here we present an overview of home appliance industry in Brazil and a brief explanation of CPS. Then, we discuss the potential applications of CPS in home appliances and how CPS can improve products and after-sales services.

Keywords: Cyber Physical Systems · Home appliances · Customer services

1 Introduction

Some factors such as globalization, information access, technological advances and consumer habits are increasing competition in product and service sectors. Moreover, consumers have become increasingly selective when concerning products, which means that companies need to seek differentiation and sustainable competitive advantages in their segment [1].

Besides this, due to low cost labor, many companies are facing competitors with similar products at lower prices. Consequently, it has become difficult to maintain profits with the same products [2]. In this context, organizations have been developing the perception that customer service is one of the few ways to improve competitive performance and differentiate themselves from competitors [3].

To overcome this challenge, the use of digital technologies is inevitable. According to a study by the McKinsey Global Institute, the application of new technologies such as sensors and connectivity can lead physical products manufactories to an increase of 2.5% to 5% in productivity by 2025 [4].

By 2025, about 80% of manufacturing industries will be using Cyber Physical Systems (CPS) technology [4]. This technology allows integration between physical and digital worlds, through embedded hardware devices (sensors and actuators), connected to a computer system that controls and monitors physical processes. CPS enables field

L. Pereira et al. (Eds.): IDEAS 2018, SIST 198, pp. 34–43, 2021.
https://doi.org/10.1007/978-3-030-55374-6_4

data collection, that feedbacks the different stages of product life cycle and use, allowing assertive investments in order to improve products, processes and services, as well as open up new possibilities for companies' current business model.

One sector that presents a great potential for application of the CPS technology is the home appliances sector. The use of CPS in these products can provide monitoring of field data, allowing a greater understanding of operation, use patterns and consumers' preferences.

Connected home appliances can generate data that allows diagnosis improvements, anticipates possible failures and solves problems remotely [5]. In addition, data could be used to develop new products that address the weaknesses of their predecessors or that have value propositions more aligned with consumer's usage patterns and expectations.

2 Home Appliances

2.1 Production, Distribution and Sales

According to the Brazilian Quarterly Report of the National Association of Electronic Product Manufacturers (ELETROS) Brazil produced approximately 14 million electric appliances in 2017. The revenue for the same year was R$ 20.5 billion (5.8 billion US dollars) [6].

Home appliances categories manufactured and sold in Brazil range from refrigerators to small appliances. However, 80% of the volume and revenue is concentrated on refrigerators, stoves and washing machines [7]. In 2016 the penetration of refrigerators in Brazilian households was 97.83% (about 66 million products) and washing machines 63% (about 43 million products) [8]. Table 1 presents the penetration of these products in the Brazilian market and other regions [9].

Table 1. Products penetration among Brazil and other different regions.

Region	Refrigerators	Washing machines
Brazil	98%	63%
Western Europe	98%	90%
North America	97%	90%
Australia, New Zealand and Japan	99%	85%

Source: Elaborated by authors

In Brazil, home appliances final consumer sales mainly happen through retails, which can be divided into retails with stores, such as supermarkets, department stores, specialized, wholesalers and independent stores, and retails over the internet (e-tailers), which is basically the e-commerce.

According to a survey conducted by the Boston Consulting Group (BCG) in 2015, online sales account for about 4% of retail purchases in Brazil. Although the figure itself seems to be not significant, the same study shows that more than 50% of all Brazilian retail purchases are influenced by the internet, that is, half of consumers uses internet at some point to make their purchase, either to actually buy or to search prices, stores, models or other information related to the product. For the electronics market, online commerce accounts for 13% of sales, although nine out of ten consumers with internet access have researched the product online before making the purchase [10].

2.2 After Sales

In the legal context, the provision of after-sales services is mandatory in order to support the consumer in case of product failure or dissatisfaction. However, offering quality services, besides being a way to differentiate a product from its competitors, plays an important role in bringing consumer satisfaction [11]. In addition, by increasing consumer satisfaction one can improve consumer retention, loyalty, and hence revenue growth [12].

It is possible to classify the profiles of after-sales services offered by companies into four different strategies [13]:

Product Support. In this more traditional strategy, after-sales service is seen as a "necessary evil." A cost-generating activity designed primarily to manage warranty issues and product failures. The main home appliance companies in Brazil have after-sales services that can be contacted via phone or internet. By contacting the company, the consumer can clear up doubts or schedule a technical visit. The visit, after being scheduled, is carried out by a repair service center, from a network that covers almost the entire national territory.

Revenue Generation. In this view, after-sales services represent an opportunity for revenue generation, through spare parts and accessories sales. It is possible to find, on companies' websites, actions that indicate the generation of revenue through after-sales services, such as the sale of spare parts. Electrolux and Whirlpool Brazilian websites, for example, have a portfolio with thousands of spare parts available for purchase.

New Business Generation. In this approach, after-sales services represent the opportunity to leverage new businesses, such as offering packages of products and services according to specific consumer needs. Some companies are starting the "new business" strategy, such as the partnership sealed in 2018 between the insurance company Assurant and the home appliance manufacturer Whirlpool, in order to sell insurances for used appliances.

In this new business, the insurer sells insurance for products that are no longer covered by the manufacturer's warranty. The consumer can then use the repair service that is offered by the manufacturer company, and performed by authorized repair service centers, with warranty of use genuine parts and certified technicians [13].

Brand Promotion. This approach is similar to the previous one, however the main objective in this case is not profitability with after-sales services, but investment in the long term, to stand out in terms of cost-benefit, quality and functionality, gaining consumer loyalty.

3 CPS Technology

CPS are systems capable of promoting the integration between digital and physical environments, through monitored, controlled, coordinated and integrated operations made by a central computing and communication core. In CPS, embedded systems and communication networks monitor and control physical processes, through sensors and actuators with feedbacks where physical processes are capable of affecting computational processes and vice versa [15]. In the literature, it is possible to find different propositions of CPS architectures. One of them [16] defines an architecture model based on CPS that is divided into three spheres.

The first one is the physical sphere, which comprises physical devices, sensors, actuators and the environment they will modify and control. Devices that make up this sphere must have the ability to collect data from appliances, users, and environment, and transfer it to the control sphere. In addition, they must be able to communicate with the control sphere through their own resources and deliver results to users through actuators.

The second sphere is the support sphere. In this layer, there are elements such as servers and frameworks. These services can be local or located in the cloud and they must be adapted for multiple users.

The third sphere of control is where data is received and monitored. From them, necessary actions that need to be applied in the physical environment are defined. Components of this sphere must have the ability to efficiently and effectively monitor components of the physical sphere, determine the most appropriate action, and send commands in real time, as well as manage physical and service components.

Another architecture for application of CPS is the 5C architecture [17]. In this concept, presented in Fig. 1 the CPS architecture is structured by five layers that guide the system.

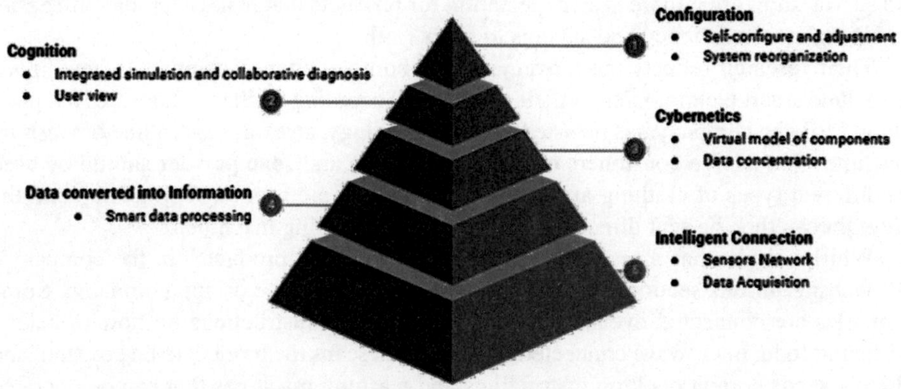

Fig. 1. 5C architecture for CPS. Source: Adapted from Lee et al. (2015).

The first layer (5) is the intelligent connection, which performs data acquisition from the product or the environment around it through sensors network. In the second layer (4), the data collected are converted into information, so that it can be analyzed in order to draw conclusions about the system operation.

The third layer (3), called cybernetics, is responsible for creating a virtual model of the system through the concentration of data and use of algorithms that analyze the system operation. In the fourth layer (2), the cognition process takes place, in which the results analyzed above are formatted for the interpretation of a qualified person.

Finally, the configuration layer (1) is where information and commands from virtual environment return to the physical environment. These commands are defined from data received from previous layers and are able to reorganize the system according to different decisions taken.

By applying these robust trends and computational control codes, CPS can be applied to a wide range of activities, promoting high levels of performance and efficiency. Their applications can make activities faster and more accurate, for example, computer-controlled surgeries, searches and rescues in accidents and disasters, control of large-scale systems such as smart cities, increase human capacities through brain-computer interfaces and body sensors networks, improve life quality through omnipresent monitoring of health indicators, and many other technical and social applications [18].

4 Real Applications of CPS Technology in Home Appliances

Whirlpool Corporation conducted a global survey in 2018 to understand consumer expectations for smart products. More than 2,000 households in four countries (the United States, Brazil, India and France) were interviewed and some of the results released are: 74% are interested in smart products that reduce food waste, 75% have shown interest in products that help water saving, 78% have shown interest in smart products that help money saving, 64% say it is important that products help energy saving, and 69% say they are looking for products that can be personalized into their preferences. Overall, the survey shows that there is an expectation for products that make everyday life easier and bring more efficiency and savings to users [19].

Their research reflects the movement that home appliance companies have made to include smart technologies in their products. According to Electrolux's 2016 annual report [20], the company has invested in this technology, an example is a line of washing machines that advice consumers on which programs and soap powder should be used on different types of clothing and blemishes. Advises are sent through an app and the program can then be sent directly to the connected washing machine.

Whirlpool also has a section dedicated to "connected products" in the company's US website. In this section, you can find products with voice or app command. Some examples are connected ovens, that can receive online instructions on how to bake a particular food, microwave connected to an app that scans the product to be prepared and already sends correct cooking instructions and washing machines that can be operated remotely and have its washing cycle adapted to the user needs [19].

5 Possible CPS Applications in Home Appliances Focused in Customer Services

In order to apply CPS technology in home appliance after-sales services, the entire installed base must be connected to a central system (which can be located in the product itself or externally, such as in the cloud), where all relevant data product operations are processed and stored. The products must contain sensors, which will be responsible for collecting data, as well as actuators, which will be responsible for operating the possible actions sent by the processors [5]. Figure 2 shows the composition of this intelligent product system.

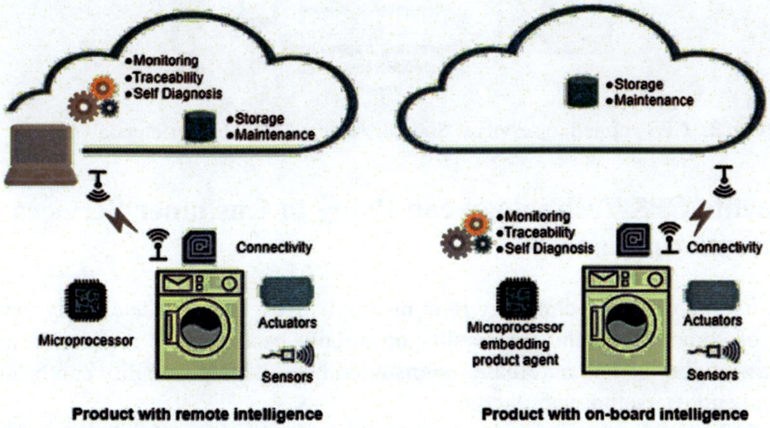

Fig. 2. CPS applied to smart appliances. Source: Adapted from Leitão et al. (2015).

With high volumes of historical data and high analytical capacity, data obtained can be used to improve products and offer new services. In addition, data can also be used to improve existing services. With a network of connected products, companies would be able to locate a defective product in the field and previously identify the problem, taking the right spare part to the technical visit or even repairing it remotely.

In a scenario of high maturity of technological capabilities implemented and high level of products intelligence, products may be able to assess the environment around it, learn from it, self-diagnose and adapt to changes and new needs. This capability becomes more tangible as more products are connected to the network, providing a wealth of data. Figure 3 summarizes these new opportunities for the service business.

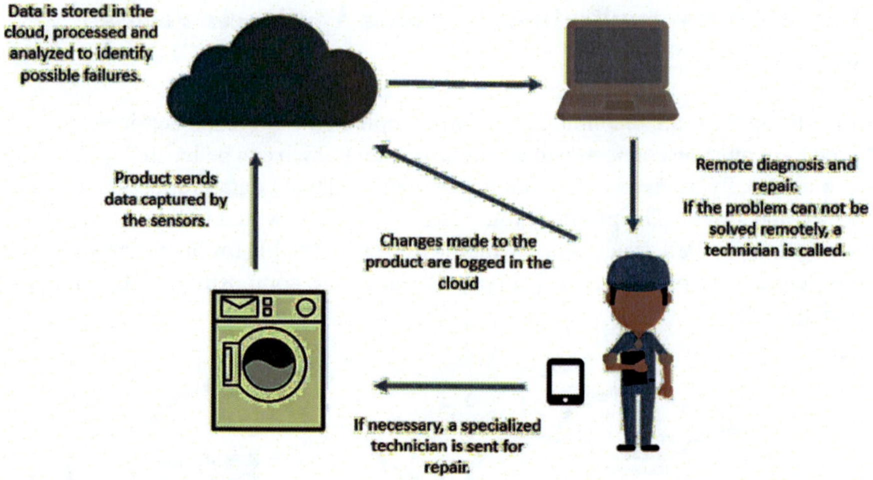

Fig. 3. CPS applied to services. Source: Adapted from Herterich et al. (2015).

6 Benefits CPS Technology can Bring to Customer Services of Home Appliances

The application of CPS technology in home appliances, besides the already mentioned benefits of convenience and practicality on a daily basis, makes it possible to monitor, control, optimize and automate the installed base. This possibility can bring many benefits to end users and manufactures.

The first advantage refers to improvements in new product development and optimization of its operations. Data collected by products can be used to identify opportunities of improvement in their design. From these, new generations of products can be enhanced in order to feature a better performance, from development of more durable components to equipment that consumes less energy [21]. Besides that, data collected not only from products, but also from external agents can be used to optimize products operations. For example, external temperature data could be used to optimize the operation of an air conditioner. Data could also be used to adapt the product performance according to user profile and also to reduce energy consumption.

Another possibility for CPS application is remote product management. With CPS use, it can be possible to remotely increase or decrease functionalities of a product, via software update, according to consumer's needs. For example, rather than selling different models of washing machines with different functionalities, companies could sell only one model of machine. Depending on customer's demands, modular features could be added or withdrawn by subscriptions, which would make the product more responsive to customer needs [21].

CPS can also be used to predict need for technical services and proactive contact of technical support, as well as remote diagnosis and reduction of field services. By collecting product data and identifying usage patterns, it may be possible to predict need for technical intervention before the actual failure. Unlike the existing preventive maintenance actions, which are based on products time of use, predictive actions defined from CPS tend to be more adherent and adapted to each consumer's routine of use. In addition, field data can be used for remote diagnostics in case of failure. So, it would be possible to identify the problem and solve it remotely. Technical visits would be restricted only for cases where remote repair would not be possible, and when applied, would be more efficient since the technician would already know what the problem is and would arrive with correct spare parts and tools needed to fix the problem [21].

Finally, data from products can also be used for other purposes, as new service offerings, according to consumer demand analysis. In addition, data-producing companies can expand their business by selling field data to service providers, who do not have access to it.

7 Challenges

Connected products generate large volume of data. Handling it would demand a lot of energy and might result in an expensive process. Therefore, there is a need to develop lightweight algorithms to process data locally, to send systems only important data. In this context, optimizing data communication in order to reduce it to the minimum without losing performance is a relevant challenge in CPS application, as it can avoid excessive consumption of energy, latency issues and saturation of wireless channels [22].

Still from a technical perspective, consumers are looking for practicality in their everyday life, with products that are easy to use. However, the more functions you want to offer, the more complex the required systems is. Achieving maximum applications with easy usability for final consumers is a major challenge in the development of these systems [23].

Privacy and security are also challenges in the CPS application in home appliances. With large amount of data collected and the information flow, systems must be able to guarantee data protection against hackers or other unauthorized persons, as well as prevent communication failures or interruptions [24]. A survey conducted by Parks Associates in 2018 [25] shows that the percentage of consumers who do not intend to purchase a smart product for their home due to security and privacy issues rose from 21% in the first quarter of 2017 to 32% in the first quarter of 2018. The safety issue covers not only end users but also manufacturing companies, which are interested in having as much data as possible in order to improve its developments, but must be sure these data are not going to be available to competitors, for example.

Therefore, developing robust systems that ensure security and privacy for user's data in a smart and connected world is of utmost importance in CPS. In addition, creating data policies that address customers' privacy concerns and reflect ever-stricter government regulations and transparently defines the type of data collected and how it will be used is an important step to be developed in order to ear customers reliability [26].

8 Conclusions

CPS is an innovative technology that presents many possible applications, including home appliances uses. In addition to the benefits in practicality and usability in final product, CPS use in home appliances also becomes relevant with regard to after-sales services.

Data collected by CPS can provide greater understanding of failures, and consequently improvements in new product development, besides optimization of their operation. Remote monitoring of products and their usage pattern, can provide an outlook of technical intervention needs before the actual failure, through preventive maintenance more aligned with each user's need. In addition, in the event of failure, data analysis may be able to promote remote diagnosis and even remote repair. In a technical visit, this becomes much more assertive and efficient, since the problem has been previously analyzed. All these applications in consumer services are able to improve the quality of services offered and, therefore, increase consumer satisfaction, promote brand appreciation and differentiate it against competitors.

However, the application of CPS brings challenges to be overcome, such as development of solutions that minimize data transfer in order to save energy and make process cheaper, as well as development of robust systems capable of protecting data from hackers and others authorized users in order to make the use of the technology reliable.

Last but not least, even with the existence of challenges to overcome, the application of CPS is clearly relevant and may contribute to the improvement of products, customer service and their value proposition, thus deserving attention and investment in new studies and developments.

References

1. Freitas, A.: Service quality in the context of competitivity. In: Revista Produção, Florianópolis, vol. 5 (2005)
2. Kimita, K., et al.: Framework for analyzing customer involvement in product-service systems. Procedia CIRP **47**, 54–59 (2016)
3. Christopher, M.: O Marketing da Logística: Otimizando processos para aproximar fornecedores e clientes. Futura, São Paulo (1999)
4. Manyka, J., et al.: Disruptive Technologies: Advances that will Transform Life, Business, and the Global Economy. McKinsey Global Institute, London (2013)
5. Leitão, P., et al.: Intelligent products: the grace experience. Control Eng. Pract. **42**, 95–105 (2015)
6. ABINEE: Cenário para a Indústria Eletroeletrônica. Decon - Departamento de Economia, São Paulo (2018)
7. Mascarenhas, H.R.: O Setor de Eletrodomésticos de Linha Branca: Um diagnóstico e a relação Varejo - Indústria. Fundação Getúlio Vargas, São Paulo (2005)
8. IBGE/PNAD, Índices Especiais de Produção Física - Eletrodomésticos (2018)
9. Electrolux, Annual Report, Stockholm (2017)
10. Sonneveld, S., et al.: Varejo no Brasil: A influência do digital sobre o consumo. Boston Consulting Group (BCG), São Paulo (2015)
11. Murali, S., Pugazhendhi, S., Muraldharan, C.: Modelling and investigating the relationship of after sales service quality with customer satisfaction, retention and loyalty - a case study of home appliances business. J. Retail. Consum. Serv. **30**, 67–83 (2016)
12. Possevelt, T., Gerstner, E.: Pre-sale vs. post-sale e-satisfaction: impact on repurchase intention and overall satisfaction. J. Interact. Market. **19**(4), 35–47 (2005)
13. Cavalieri, S., Gaiardelli, P., Ierace, S.: Aligning strategic profiles with operational metrics in after-sales service. Int. J. Prod. Perform. Manage. **56**(5–6), 436–455 (2007)

14. Santos, D.: Assurant inova ao lançar no mercado seguro para equipamentos domésticos usados. Maxpress, São Paulo (2018)
15. Lee, E.: Cyber physical systems: design challenges. In: 11th IEEE International Symposium on Object and Component-Oriented Real-Time Distributed Computing, Orlando, pp. 363–369 (2008)
16. La, H.J., Kim, S.D.: A service-based approach to designing cyber physical systems. In: IEEE/ACIS 9th International Conference on Computer and Information Science, Yamagata, pp. 895–900 (2010)
17. Lee, J., Berahd, B., Kao, H.: A cyber-physical systems architecture for industry 4.0-based manufacturing systems. Manuf. Lett. **3**, 18–23 (2015)
18. Rajkumar, R., et al.: Cyber-physical systems: the next computing revolution. In: Proceedings of the 47th Design Automation Conference, Anaheim (2010)
19. Whirlpool Corporation, The Future Smart Home Needs to Solve Simple Problems to Drive Consumer Adoption: Whirlpool Global Innovation Survey (2018). https://whirlpoolcorp.com/future-smart-home-solve-simple-problems/
20. Electrolux, Annual Report, Stockholm (2016)
21. Herterich, M., Uebernickel, F., Brenner, W.: The next wave of service innovation. In: Series on Research in Information Systems Management and Business Innovation, St. Gallen, vol. 3, p. 52 (2015)
22. Risteska, S.B.L., Trivodaliev, K.V.: A review of Internet of Things for smart home: challenges and solutions. J. Cleaner Prod. **140**, 1454–1464 (2016)
23. Eckl, R., MacWilliams, A.: Smart Home Challenges and Approaches to Solve Them: A Practical Industrial Perspective. Siemens AG, Corporate Technology, Software Architecture, Munich (2009)
24. Probst, et al.: Service Innovation for Smart Industry, Cyber-Physical Systems in the 'Value Network'. Business Innovation Observatory, Luxembourg (2015)
25. O'Dell, C.: Blockchain for Connected Home and Entertainment. Parks Associates, Addison (2018)
26. Porter, M., Heppelmann, J.: How Smart, Connected Products are Transforming Companies. Harvard Business Review, Cambridge (2015)

Check for updates

Improvement in the Method of Assembly of a New Production Process in the PIM

Matheus Oliveira(✉), Karina Cavalcante, Gabriela Veroneze, Marcelo Albuquerque de Oliveira, and Joaquim Maciel da Costa Craveiro

Federal University of Amazon, Manaus, AM 69067-005, Brazil
smatheus_l_oliveira@hotmail.com

Abstract. The present article has an demonstrates the steps of making a connection fastening device that assists in the process of assembling a new production process of a white line company in the industrial hub of Manaus. The process without the presence of this device would generate leakage besides a longer process and a greater probability of error in the assembly. It was decided to build a fastening device to use it in the line, which was built through machining and later followed by surface treatment with zinc. In view of these issues, a series of actions was developed with correct decision-making based on the needs of the project after identifying the problem, its cause and resulting problems. The construction of the device was performed by the supplier, who was chosen based on the best cost and development time at the time. A device capable of solving the problem of leakage, ergonomic improvement and decrease of error variability was obtained.

Keywords: Device · Method · Machining introduction

1 Introduction

With the increasing innovation that we contemplate each day, innovative products are created daily, and with them, new production processes for complex to be assembled process must be created.

Manufacturing companies, in general, use techniques based on new assembly methods, assembly devices, and process automation to reduce error variability and improve their processes.

Hydraulic connections that are used in fluid transfer products have their greatest weakness in the connections that connect them with other objects, because they are at those points where leakage occurs.

This paper searches through machining techniques, decision making, quality tools, and project management, demonstrating the process of developing an assembly device that assists and improves the activities performed in a new production process.

There are basically four decision-making components: belief assessment, value assessment, integration ant metacognition [1].

Collective decision making is a pervasive task for humans. They must decide, where to live, where the next company site will be, or what new product will be developed

L. Pereira et al. (Eds.): IDEAS 2018, SIST 198, pp. 44–51, 2021.
https://doi.org/10.1007/978-3-030-55374-6_5

[2]. There are several different techniques that can be used to improve decision making, either being theoretical like [3–5] or empirical [1, 6–8].

To counterpart these empirical studies [9], mathematical theories have been developed to try and explain how social decisions are made as well as to give a possible guideline on how to make these decisions [3, 10, 11].

Ishikawa diagram was invented by Kaouru Ishikawa, one of the quality pioneers in Japan during the 60s. It is considered one of the seven basic quality tools, it is also known as fishbone diagram, since it reassembles a fishbone structure, as well as cause-and-effect diagram, because it shows possible causes for the observed phenomena [12–14].

2 Development

The development of the new production live was made by using the following steps: a problem identification, using Ishikawa Diagram, passing thru a machining study and finalizing with creating an assembly standard procedure.

The production line that will be discussed below is for a product that comes from a family of beverage making assisted devices, which explains why it is important that no leakage occurs between connections.

2.1 Problem Identification

The hydraulic connections in the product are flanged and threaded, which means that, there is basically a nut that if it's not securely fastened will lead to a leakage. Through an Ishikawa Diagram (see Fig. 1), it was found that there are problems related to labor, which is represented as operational error, material problems, due to the thread of the male or female connections being out of specification leading to an improper fixation by the operator causing leakage, and method due to the manual work used that forced the operator to use movements that are not ergonomic leading to a waste of time.

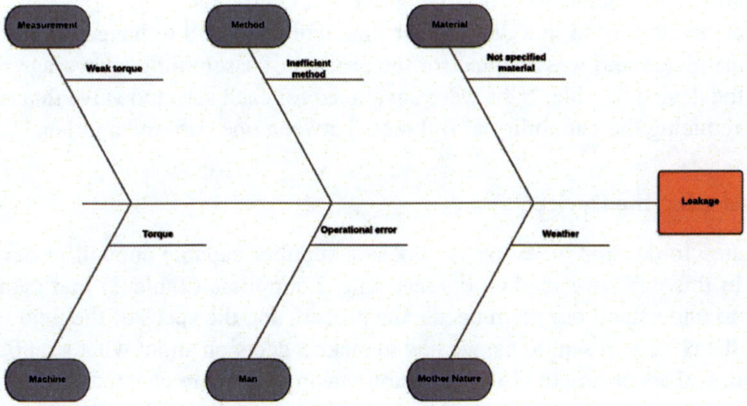

Fig. 1. Ishikawa diagram, Source: Authors

From observing the diagram as well as interviewing peers inside the line it was possible to conclude that the assembly method was inefficient, because the operator use their hands to fix the connections, causing a fragile union between the parts, and was them used as the chosen root-cause

2.2 Starting the Project

For the product assembly process, specific points were selected from the subassemblies of the valves that carry the water to the whole circuit of the beverage machines model A, these valves, like the valve in Fig. 2, require special care, since a leak can easily happen if the fitting of their connections are not well done.

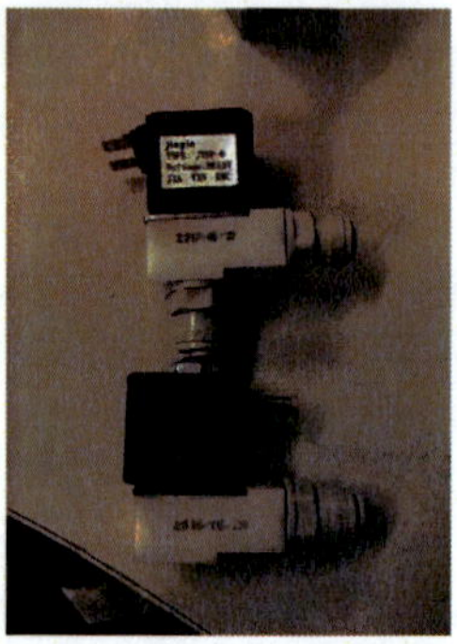

Fig. 2. Solenoid valve used in the model A, every gap is a possible leak point. Source: Authors

As a solution to the leakage problems encountered in the valve connections, assembly devices have been designed so that the connection of these valves is done as accurately as possible. For this, a benchmarking of the previous process was performed, using a device with rails to tighten the male or female connections on the valves, which reduces the possibility of a human error occurring during the process.

This device was used in a dishwasher line, which needed to have a good tightness between the pieces and was adapted for the process of assembling a beverage machine. To make the device flexible, cribs were developed for each selected valve that needed to be fitted, reducing the variability of tightness between one part and another.

2.3 Developing the Device

The first step to develop a device, is choose a supplier capable enough to develop the product. In this case, we made a contact with 2 suppliers (Table 1) that came to the company to understand our main needs, the process and the specs of the final product.

After the price was sent to us, we had to make a decision about what supplier would be the best, and according to [15], a decision making is a more characteristic task of the administrator. However, managers are not the only ones to decide, because the work must be as flexible as the decision-making, but also to make sure that the whole organization, or part of it, takes it as effectively. With this in the management hands, the chosen supplier was supplier 1.

Table 1. Suppliers and their price tag for the service

Supplier	Price (BRL)
Supplier 1	16000
Supplier 2	18000

Source: Authors, 2019

For the winner supplier, a third part contract was drawn, spare parts were ordered as well as a preventive maintenance plan. Here it is important to argument that all the suppliers for this company are in a supplier development program where concurrent engineer is applied [16–18].

2.4 Machining the Device

According to [19], machining operations are those that, at the same time, impart a shape, such as dimensions or finish, or as an element of the chip.

Groover [20] affirms that machining is used as a secondary process. In general, the secondary processes come after basic processes, whose purpose is the initial form of the blank. Machining operations are to transform as initial shapes into the final geometries specified according to the part design. Schematic diagram of the parts is shown in Fig. 3.

The manufacturing process of the devices of the machining device, the metal blocks were used, and the devices were deployed in CNC machines, seeking a greater precision during machining process. In addition, the CNC machines, for small adjustments, could be fulfilled by the supplier itself.

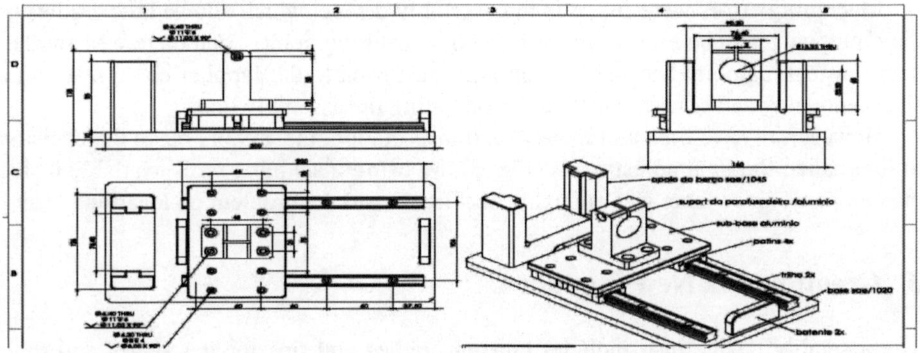

Fig. 3. Design and technical quotes of the assembly device.

After the machining process, as specified, the zinc coated device through the galvanizing process, whose purpose is to prevent corrosion and wear of the material, as it would be used more than 300 times a day inside the assembly line.

2.5 Device's TryOut

With the device implanted in the production line as well as the new process, the combination was tested and it was observed that the valve bore moved downward, which made the screwdriver used to index the part on the valve did not meet the center specified in the drawing.

In the search to find the root cause for the failure mechanism, another Ishikawa Diagram, which is shown below (Fig. 4). From its construction and analysis, it was concluded that the largest possibility of the problem was in the machining error because two devices were ordered from the supplier and only one was problematic.

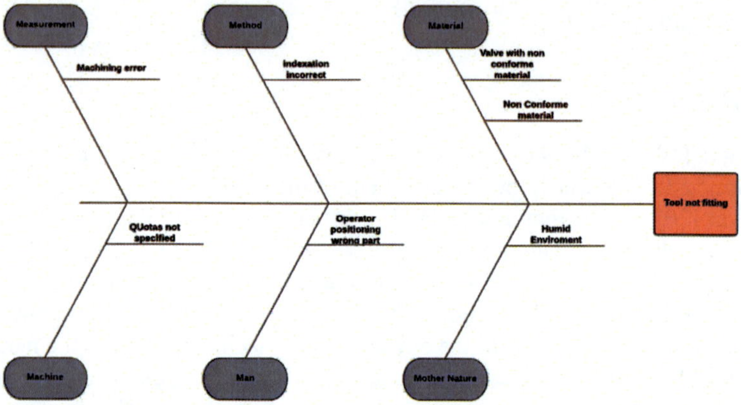

Fig. 4. Ishikawa diagram. Source: Authors

Machining errors were found in the assembly device, which allowed the cradles of the 4 mm valves to be moved upwards from their fitting point. Then there was another decision-making point because returning the part back to the supplier could result in a shipment delay and consequent delay in operating the assembly line.

However, to save time and money on transportation, the authors make the decision to internalize the final adjustments (Fig. 5), by using machining tools available inside the company, since there was enough workforce to aid the problem on location.

3 Creation of a New Standard

The assembly device also includes custom cradles and tips for the valves and their flanged connections, which required setup for mounting of other valves. In doing so, the standard that was previously using KANBAN of ready-made parts, needed to change.

With this information, a redistribution of activities was performed using Yamazumi Board as a tool, which showed the need for a second operator, since the activity time was beyond the specific TAKT time (Fig. 6).

Fig. 5. Machining process. Source: Authors.

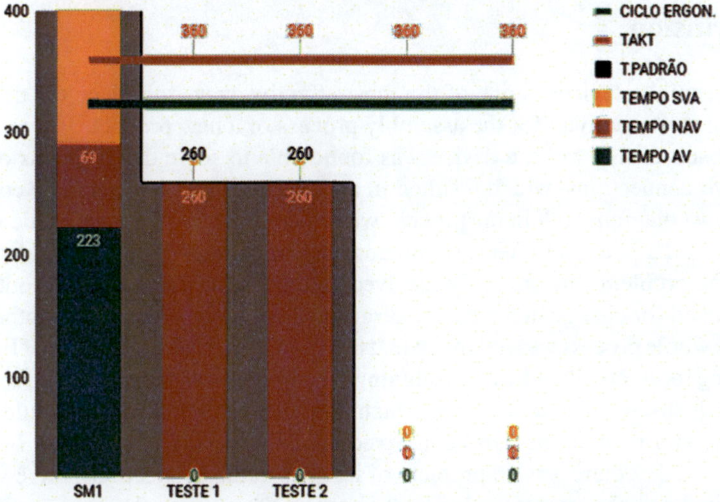

Fig. 6. Yamazumi board after, with the time not appropriate for the process.

With this information, another station was created, with the activities that are independent of the assembly device, which would not make it necessary to buy another one of the same (Fig. 7).

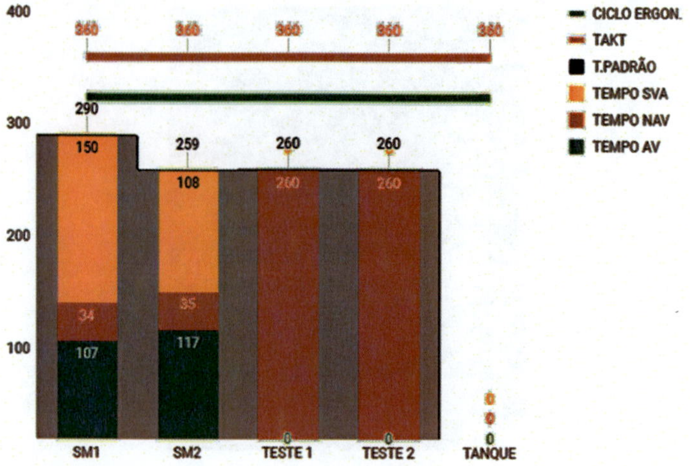

Fig. 7. Yamazumi board after, with the improvement of the new device.

4 Conclusion

Making assembly devices is one of the most effective ways to reduce the difficulty of performing a vital activity for the assembly process of a new product.

As a result of this work, a device was found able to solve the problems of leakage in the valve connections, which resulted in an effective and more accurate connection. The Ishikawa diagram built in the process was able to identify the main mode of failure, its cause and the potential effects on process performance. The decision was found to be ideal, the problem was promptly resolved, and the tests proceeded without showing any other need for change in the device discussed. As a way of keeping another backup device, the whole process was redone so that there was a third party. The use of Production Engineering tools, combined with machining techniques, was very well used to make the device that is discussed in this article, which was vital so that the device could carry out its main activity without the part being fixed outside the device. specified. In addition, it was verified that there was no problem of leakage in the connections, there was also a reduction in the total time required to fix the parts in the process, there was an ergonomic improvement for the operator and also a decrease in the error variability.

References

1. Blacksmith, N., et al.: General mental ability and decision-making competence: theoretically distinct but empirically redundant. Personality Individ. Differ. **138**, 305–311 (2019)
2. Conradt, L., List, C.: Group decisions in humans and animals: a survey. Philos. Trans. R. Soc. B: Biol. Sci. **364**(1518), 719–742 (2008)
3. Chankong, V., Haimes, Y.Y.: Multiobjective Decision Making: Theory and Methodology. Courier Dover Publications, New York (2008)
4. Edwards, W.: The theory of decision making. Psychol. Bull. **51**(4), 380 (1954)
5. Fishburn, P.C.: Utility theory for decision making (1970). Res. analysis corp McLean VA

6. Bharati, P., Chaudhury, A.: An empirical investigation of decision-making satisfaction in web-based decision support systems. Decis. Support Syst. **37**(2), 187–197 (2004)
7. Ford, R.C., Richardson, W.D.: Ethical decision making: a review of the empirical literature. J. Bus. Ethics **13**(3), 205–221 (1994)
8. O'Fallon, M.J., Butterfield, K.D.: A review of the empirical ethical decision-making literature: 1996–2003. J. Bus. Ethics **59**(4), 375–413 (2005)
9. Mann, R.P.: Collective decision making by rational individuals. Proc. Natl. Acad. Sci. **115**(44), E10387–E10396 (2018)
10. Roy, B.: Decision-aid and decision-making. Eur. J. Oper. Res. **45**(2–3), 324–331 (1990)
11. Wierzbicki, A.P.: A mathematical basis for satisficing decision making. Math. Model. **3**(5), 391–405 (1982)
12. Chokkalingam, B., et al.: Investigation of shrinkage defect in castings by quantitative Ishikawa diagram. Arch. Foundry Eng. **17**(1), 174–178 (2017)
13. Stefanovic, S., et al.: Analysis of technological process of cutting logs using Ishikawa diagram. Acta Tech. Corviniensis Bull. Eng. **7**(4), 93 (2014)
14. Wong, K.C.: Using An Ishikawa Diagram As a Tool to Assist Memory and Retrieval of Relevant Medical Cases From The Medical Literature. BioMed Central, London (2011)
15. Chiavenato, I.: Introdução à teoria geral da administração. Elsevier Brasil, Brasil (2003)
16. Chapman, W.: Engineering Modeling and Design. Routledge, New York (2018)
17. Prasad, B.: Concurrent Engineering Fundamentals, vol. 1. Prentice Hall PTR, Upper Saddle River (1996)
18. Sohlenius, G.: Concurrent engineering. CIRP Ann. **41**(2), 645–655 (1992)
19. Ferraresi, D.: Fundamentos da usinagem dos metais. E. Blücher (1970)
20. Groover, M.P.: Introdução aos processos de fabricação. Grupo Gen-LTC, Brazil (2000)

The Strategic Role of MES Systems in the Context of Industry 4.0

José Eduardo da Costa Dias(✉), Fernando Gonçalves de Castro Filho, Alexandre Acácio de Andrade, and Julio Francisco Blumetti Facó

Universidade Federal do ABC, Avenida dos Estados, Bairro Santa Terezinha, Santo André, SP 5001, Brazil
eduardo.dias@ufabc.edu.br

Abstract. Since the early 1990s, Manufacturing Execution Systems (MES) have been playing a pivotal role in managing production activities and integrating business from the top floor to the shop floor, reducing manufacturing costs by increasing overall equipment effectiveness and minimizing unplanned downtime. Cyber-physical systems, the Internet of things and cloud computing are some of the leading technology pillars of the Industry 4.0, also referred as the fourth industrial revolution which is the name given to the current trend of automation and data exchange in manufacturing technologies. In the new context of the Industry 4.0, vertical and horizontal integration of all participants in the production process is required. A smart factory must have collaborative manufacturing systems that can respond in real-time to meet changing demands and conditions in production process, in the supply chain network and its customer needs. Although the concept and application of the main functionalities of MES systems in the manufacturing processes are those prior to Industry 4.0 and therefore they are not considered as one of the technological pillars that represent it, MES systems perform activities which may be considered prerequisites so 4.0 technologies can be implemented efficiently and effectively. This article intends to describe the relation between the main functionalities of the MES with the technological pillars which are the base of the Industry 4.0. in order to position the strategic role of the MES within this new context and business environment.

Keywords: MES · Industry 4.0 · Advanced manufacturing · Meta-analysis

1 Introduction

The concept of Manufacturing Execution Systems (MES) was developed in the United States, in the mid-1990s, when a non-profit organization of manufacturing solutions, MESA (Manufacturing Enterprise Solutions Association), began standardizing these applications in a three-level model of application for an industrial facility: production level, production management level, where the practical application of the MES main functionalities become more visible and the level of enterprise management. This

L. Pereira et al. (Eds.): IDEAS 2018, SIST 198, pp. 52–61, 2021.
https://doi.org/10.1007/978-3-030-55374-6_6

model has been updated and developed over the years; however MES main functionalities were always related to three main process groups: Production, Personnel/Resource Management and Quality [1].

Technological advancements while enabling MES developers to implement meaningful enhancements and better exploit the potential of their features, bring the challenge of keeping MES as an up-to-date tool compatible with new hardware and software systems and with new trends and technologies still under development or maturing process. The companies which implement MES in their processes are challenged as well, since they must keep production chain connected and integrated, not only to the shop floor and top floor levels but to the other levels as well, including the outside world: management, vendors, customers and competitors. MES development companies and their customers must be prepared to follow and to integrate the MES with new technologies, with potential applicability in the manufacturing processes, independently on being them mature technologies already applied to other business processes or disruptive technologies which have the potential to transform and optimize existing manufacturing processes with great innovation potential.

The first three industrial revolutions were characterized by innovative technological disruptions that were notable for transforming the manufacturing processes in their respective epochs and for being identified and named by historians and scholars after their actual events occurred.

The development of the concept and the practical implementation MES Systems in large-scale can be considered as consequences of the Third Industrial Revolution, since its functionalities were the result and could only be implemented only after practical use and development of the technologies pillars of this revolution.

The Fourth Industrial Revolution or Industry 4.0, has been studied and predicted [2] not only through the study of past events and technological leaps, as occurred with its predecessors, but through its component technologies (e.g. Big Data, Artificial Intelligence, Internet of Things), the so-called pillars of Industry 4.0 [3], some of these technologies seem to be more mature and more applicable to manufacturing processes than others which are still under development, but even though they are behind the others in terms of applicability, they have high potential for innovation.

In order to be part of the new environment of Industry 4.0 and to benefit from the competitive advantages provided by its technological pillars, companies must be prepared to integrate their core processes and systems into them. Some technologies which are mature and fully integrated into the processes of these companies will be replaced or modified in their essence; others will be adapted in order not only to be integrated into this new reality but also having its features and benefits enhanced. Thus, systems such as MES, will have to be adapted and enhanced to work together the technological and innovative pillars of Industry 4.0 and also to leverage their resources and benefits in order to remain as critical components in the processes of these companies.

The continuous development and adaptation of manufacturing computer-based systems, such as MES in order to work in harmony with new technologies, even if these technologies have not been specifically created to be used in manufacturing, can become key factors to avoid the obsolescence and consequent replacement of those systems.

The main objective of this article is to present succinctly the research that is being carried out in order to study the strategic role of MES systems in the context of Industry 4.0. The study is being carried out based on a literature review related to the two main themes, MES and Industry 4.0, aiming to verify the feasibility of applying a meta-analysis study to integrate the results of independent studies in order to combine their results obtained and to relate the conclusions to the proposed theme.

2 MES and the Automation Pyramid

Automation means any computer-based system that replaces human labor and aims at quick and cost-practical solutions to meet the complex objectives of industries and services [4]. An automated system contains three essential elements: power to carry out the process and operate the system, an instruction program to direct the process and a control system to execute the instructions [5]. With the use of computer-based systems in manufacturing processes, computers are used not only to diagnose malfunctions but also to take preventive and corrective actions automatically to restore the efficiency of the production to its expected and planned levels. The industrial automation has many functions which are divided into five different levels [4]. These five levels are the components of the Industrial Automation Pyramid as shown in Fig. 1.

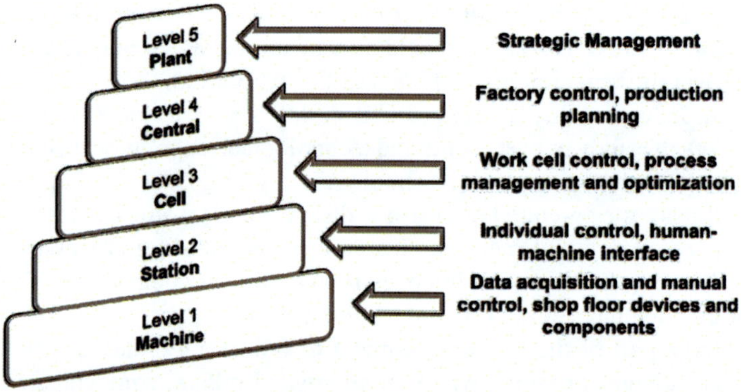

Fig. 1. Levels of industrial automation. Source: Authors based on [6]

Each level of automation there has a set of hardware (machines, equipment and control and measurement devices) and software (specialized systems, databases and operating systems) which interact within and between the respective great requiring human interaction to a greater or lesser degree, either during the operation of the machines and during the manipulation and interpretation of the software application data as well as during the analysis of the data and decision making.

An Enterprise, since adopting in an practical and integrated way the components and applications, among them MES, in each of the five levels, could reach a very high degree of automation and from hence, obtain productivity gains and relevant competitive advantages in comparison to their main direct and indirect competitors, as long as they also did not achieve a degree of automation and integration similar to theirs. MES, under this point of view, could be distinguished as a component/application of high strategic importance in this context, since through its 11 basic functions groups [7] it would execute not only the activities related to one specific level, it is a component in a chain which can integrate the different levels of the pyramid.

3 MES Functionalities and Process Models

MES systems provide the ability to control and monitor manufacturing processes and work in order to ensure data and product integrity and minimize overall manufacturing costs [8]. As a production-oriented solution, as previously described, MES work primarily with Production, Personnel/Resource Management and Quality process groups. Systems Integration and Data Collection are the main features of MES and compound its main base.

Through the data collection, for example, MES is able to capture production data and make it available to management in real-time and, through interfaces, connectors and data replication, MES is able to integrate and interact with other systems in the company like ERP, Personnel and QMS systems on the corporate level and to interact with machines through industrial protocols on the production management level.

MESA [7] divides MES basic functions into 11, as following:

1. Fine planning of workflow: Optimal sequence planning regarding the relevant basic conditions based on the resources available; 2. Resource management with status maintenance: Management and monitoring of staff, machines, tools; 3. Production unit control: Control of the flow of production units based on orders, batches and plan adjustments; 4. Information control: All information relevant to the production process is made accessible to the staff; 5. Operating data logging: Logging of all production-related operating data linked with the production unit; 6. Staff management: Recording and edition of staff working hours; 7. Quality management: Analyses of production-related measurement data in real-time in order to safeguard product quality and be able to identify problems and weak points; 8. Process management: Production process monitoring, including alarm management functions and other features; 9. Maintenance management: Recording the use of operating material and hours of use in order to initiate and support periodic and preventive maintenance tasks; 10. Lot traceability: Recording production-related data across the entire production chain to ensure traceability; 11. Performance analysis: Real-time analysis of the manufactured sizes to downtime, disruptions, piece counter and others in order to allow production efficiency assessments and detection of problems.

Over the years, MES concepts and models had been shaping to the new technologies and processes methodologies employed in the enterprising environment, even though still retaining their core functionalities. c-MES (collaborative MES) term was presented in 2004 in a MESA conference, a new concept applied to the traditional MES, as a fundamental way for competitive advantage obtention [1]. In this new concept, as shown

In Fig. 2, MES assumes a new role as a platform which act as an integrator of the management of the corporation and the shop floor automation, distributing and centralizing information between them.

The adoption of the c-MES model, became important from a strategic point of view in the optimization and support of the decision-making process, through maximizing the potential of obtaining competitive advantages made possible by features that allowed, for example, the obtaining of information in real time and integrated production and distribution planning [9].

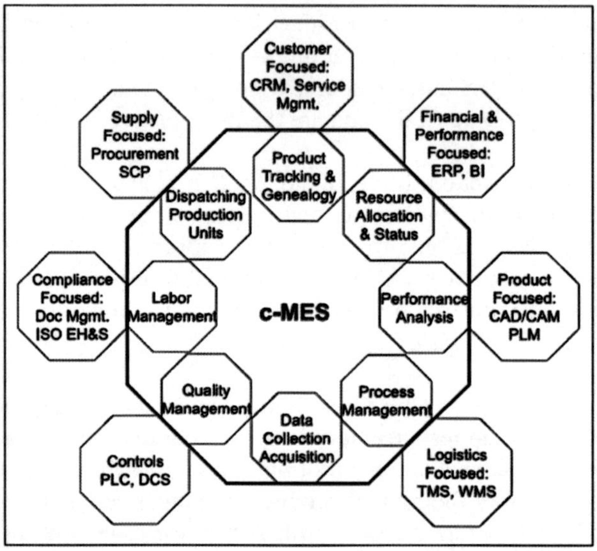

Fig. 2. The function structure of the MESA model [1]

4 Industry 4.0 Pillars

The adoption of new technologies associated with manufacturing processes, whether for execution, control, analysis and planning of these processes is present, enabling a greater or lesser degree of industrial automation, aiming at obtaining competitive advantage with a focus on costs and/or differentiation. The adoption and effectively implementation of new technologies always accompanied the constant evolution of the industry and its processes and technological leaps have provided paradigm shifts and three historical milestones that have been named as Industrial Revolutions [10].

The Fourth Industrial Revolution or Industry 4.0, unlike all others is being predicted before becoming a reality. The previous three industrial revolutions identified as occurring only after the implementation of specific technologies and these implementations were considered as landmarks with occurrence of significant technological leaps [2]. The "forecast" of Industry 4.0 therefore allows companies to take specific actions to implement technologies such as CPS (Cyber-Physical Systems) before it takes place in a broader and more concrete way. In addition to it, when companies integrated CPS into their production processes, logistics and services into current industrial practices would take a step towards transforming their factories today into 4.0 Industry factories [11].

The Internet also plays a essential role in the new manufacturing processes of Industry 4.0, manufacturing processes will use the Internet to achieve the integration of the factory's internal and external networks, the future will be established within an interactive platform based on the Internet and information technology, increasingly integrating factors of production scientifically and making manufacturing processes increasingly automated, connected and intellectualized [12]. Industry 4.0 is based on new technologies which, when effectively applied to industrial production/manufacturing processes, can transform them. The nine technologies which currently are considered as the pillars of Industry 4.0 are [3]: simulation, augmented reality, autonomous robots, the industrial internet of things, cloud computing, cybersecurity, additive manufacturing, horizontal and vertical system integration and Big Data and analytics.

MES, whose standardization of its concepts as a system, occurred in the mid-1990s therefore, after the Third Industrial Revolution and before the Industry 4.0, could be characterized/classified not as a technological leap of this fourth revolution, but rather as a relevant application to be integrated to the new Industry 4.0 pillar technologies, being part of a transition period between the third and fourth industrial revolution. Thus, MES systems could still be considered as relevant agents to obtain advantages, both related to the optimization of production costs and for the differentiation of their products in the market, becoming an actor executing a essential role in this new context, not only being classified as a pre-existent system which is a "survivor" from the pre-industry 4.0 period, but as an agent that is integrated and required to implement and catalyze the disruptive changes that are coming with the Industry 4.0.

In this way, both the companies that develop the MES systems and implement the MES systems and their customers that one's which implement the MES in their manufacturing processes, must be prepared to adapt the functionalities to this new reality, allowing MES to integrate and use technologies like IoT, Big Data, Cloud Computing, to optimize their results and increase their strategic importance in a company's manufacturing and business processes. The proposed research methodology and problem analysis described in the next topic of this paper will deal with the role and strategic importance of MES systems in the context of industry 4.0 as the theme and central problem to be researched.

5 Meta-Analysis Study

According to the subject and the objectives proposed for this research, as well as the characteristics of the literature found on the subject matter of the present study, a meta-analysis study will be adopted based on a review of the literature as a research methodology to be carried out.

The meta-analysis term was coined by Gene Glass in 1976 to give the meaning of analysis of the analyzes. According to Glass, meta-analysis is the statistical analysis of a massive collection of analysis results from individual studies with the purpose of integrating conclusions. Glass applied the idea to a set of results from a series of independent studies that examined the same research questions or similar questions [13], refers to the process of locating, selecting, evaluating and combining information relevant to a research question [14] and is used to improve the understanding, applicability and generalization of a research comparable with several results, not requiring the original data [15].

Even considering that MES systems play a crucial strategic role in the execution, automation and manufacturing control processes in the pre-Industry 4.0 context, the adaptation and transformation of MES core functionalities and its technologies would be necessary to support the companies to implement, in to a greater or lesser extent, the core technologies of Industry 4.0. In order to adapt and integrate the main functionalities of MES to the core technologies of Industry 4.0, the development and implementation costs inherent to these changes will have to be justified, by MES suppliers, being developers and, or implementers and as well as the companies that have already implemented the MES on their manufacturing processes.

In order to make these justifications be considered as necessary and feasible, the understanding of the role and strategic importance of MES systems in the new context of Industry 4.0, where new customer needs are presented to the market, making the competitive environment even more complicated by virtue of the high potential impact of the changes introduced by its technological pillars becomes relevant. The meta-analysis technique will be used, using data researched from existent literature, aiming to improve understanding of the problem.

5.1 Method and Procedures

The following steps will be adopted in order to execute the review of the literature and application of the meta-analysis technique:

Definition of the question to be formulated; research in diverse sources of reliable studies, such as theses and dissertations available in the digital and physical collections of educational institutions, scientific articles, chapters of scientific books and conference papers, addressing the formulated issue or related to the main themes, MES and Industry 4.0; selection of studies and quality evaluation, based on inclusion and exclusion criteria; collection and presentation the data of each study; 5. Evaluation of heterogeneity between studies; calculation and combination of the results of each study; evaluation of the effect of variation of the study validity; interpretation of the results, with evaluation of how much they can be generalized from the review and meta-analysis, according to the characteristics of the material collected during the research.

The procedures to be applied to the meta-analysis study include the following steps [16]:

Goal Setting with the determination of the variables to be searched and cause and effect relationship of variables through an exploratory study; systematization of information with the systematization of the set of data obtained from sources of scientific literature, limitation of bibliographic research in space and development of criteria; data encoding of the publications included to the database according to the selected sources and by experimental objective of each publication; data filtering with the selection of candidate publications for entry into the database according to the objectives of the work, execution of the critical reading of the publications through analysis and systemic interpretation of the information contained therein and insertion of publications into the database; and data analysis including graphics, weights, choice of a statistical model, post-analytical procedures considering the limits, deviations and degree of heterogeneity of the results.

5.2 Literature Review

In order to initiate the literature review with the selection of dissertations, theses and scientific articles related to the subjects to be studied/analyzed (MES and Indus-try 4.0), a base compound of fifteen to twenty academic research specialized websites[1] were preliminarily evaluated and selected, among them Google Scholar, Scielo and Microsoft Academic. Other public databases of thesis and dissertations, will also be used, serving as a complementary basis for conducting the research and execution of the review.

The different criteria of each research method and the type and form of results found in each of them, should also be taken into account, since even when the references found in each site initially would present different results in terms of number and diversity of content, there will be restrictions to be considered, such as occurrence of repetition and crossing results. However the testing phase, with the verification of general results with broader search criteria and data exploration is a prerequisite for the refinement of the data and the creation of inclusion and exclusion criteria, which is basis to execute the meta-analysis study, extracting additional information from the union of the results of the several works found, grouping and standardizing them in order to define the application of the statistical techniques to synthesize the conclusions regarding the subject studied.

6 Preliminary Results and Conclusions

Preliminary research of articles was conducted using Google Scholar search engine, excluding patents. 1,390 results were obtained based on the following search criteria: ("Manufacturing Execution System") AND ("Industry 4.0" OR "Fourth Industrial Revolution" OR "4th Industrial Revolution"). Those results represent all references (e.g. articles, books) where the research terms "Manufacturing Execution System" and "Industry 4.0" were found anywhere in the reference text.

As the main objective of the work is to study the strategic role of MES in the context of Industry 4.0, via meta-analysis based on a review of the literature, it was adopted as an additional search criterion the main functionalities of MES. 11 MES main functionalities were adopted as additional search terms to find references that contained the term Industry 4.0. In order to define clear boundaries to delimitate the research, the search for related publications will be mainly conducted as a structured keyword search [17]. The idea behind this search criteria was to identify in a quantitative manner if those functionalities were identified in the Industry 4.0 related references and for each

[1] Scielo (http://www.scielo.br), World-WideScience (https://worldwidescience.org); Dialnet (https://dialnet.unirioja.es); iSEEK Education (http://education.iseek.com); Microsoft Academic (https://academic.microsoft.com); Google Scholar (https://scholar.google.com); ScienceRe-search.com (http://scienceresearch.com/scienceresearch).

one of the functionalities, the number of references found. Some functionality names have been adapted to allow a more extensive search (e.g. "Operating data logging" was also searched for "Data logging") in the first moment. Table 1 summarizes the results obtained using the search criteria adopted:

Table 1. Google scholar results using MES main functionalities and "Industry 4.0" as search terms.

#	MES functionality	Number of references	% of the total
1	Resource management with status maintenance	4240	28,96%
2	Quality management	2990	20,42%
3	Process management	2600	17,76%
4	Production unit control	1430	9,77%
5	Performance analysis	1430	9,77%
6	Maintenance management	930	6,35%
7	Operating data logging	420	2,87%
8	Information control	321	2,19%
9	Lot traceability	207	1,41%
10	Staff management	57	0,39%
11	Fine planning of workflow	15	0,10%
–	Total	14640	100,00%

The results from Table 1 are not conclusive yet; however, they can be used as primary and potential indicators of which functionalities of MES may be more present in the new context of Industry 4.0 and may indicate, even indirectly, how the MES is inserted in this context.

Starting from the preliminary results and indicators shown in Table 1 and after a search criterion refining, the Literature Review will proceed with the execution of new searches using Google Scholar and other search engines. The plan is to execute a quantitative review using the references selected, categorizing each reference according to the number of references identified for each MES functionality and adopting different weights for each reference based on where the functionality term is found in each reference (e.g. title, abstract, keywords).

After the selection of the references and the tabulation and classification of the data, the meta-analysis study will be performed in order to identify the level of influence of each one of the MES functionalities in the context of Industry 4.0.

References

1. Kletti, J. (ed.): Manufacturing execution system-MES. Springer, Heildelberg (2007)
2. Almada-Lobo, F.: The industry 4.0 revolution and the future of manufacturing execution systems (MES). J. Innov. Manag. **3**(4), 16–21 (2016)

3. Wang, L., Wang, G.: Big data in cyber-physical systems, digital manufacturing and industry 4.0. Int. J. Eng. Manuf. (IJEM) **6**(4), 1–8 (2016)
4. Moraes, C.C., Castrucci, P. L.: Engenharia de Automação Industrial. Rio De Janeiro: LTCn (2007). ISBN 85-216-1532-9
5. Groover, M.P.: Automation, Production Systems, and Computer-Integrated Manufacturing. Prentice Hall Press, New Jersey (2007)
6. Andrade, A.A.: Desenvolvimento de sistema especialista com operacionalidade de aprendizado para operar em tempo real com sistemas industriais automatizados. Ph.D. thesis. Universidade de São Paulo (2007)
7. Meyer, H., Fuchs, F., Thiel, K.: Manufacturing Execution Systems (MES): Optimal Design, Planning, and Deployment. McGraw Hill, Columbus (2009)
8. Popp, S.M.: Manufacturing execution systems (MES). U.S. Patent No 8,491,839 (2013)
9. Younus, M., et al.: MES development and significant applications in manufacturing-a review. In: 2010 2nd International Conference on Education Technology and Computer (ICETC), pp. V5-97–V5-101. IEEE (2010)
10. Lasi, H., et al.: Industry 4.0. Bus. Info. Syst. Eng. **6**(4), 239–242 (2014)
11. Lee, J., Bagheri, B., Kao, H.A.: A cyber-physical systems architecture for industry 4.0-based manufacturing systems. Manuf. Lett. **3**, 18–23 (2015)
12. Cheng, G.J., et al.: Industry 4.0 development and application of intelligent manufacturing. In: 2016 International Conference on Information System and Artificial Intelligence (ISAI), pp. 407–410. IEEE (2016)
13. Cooper, H., Hedges, L.V., Valentine, J.C. (eds.): The Handbook of Research Synthesis and Meta-Analysis. Russell Sage Foundation, New York (2009)
14. Brockwell, S.E., Gordon, I.R.: A comparison of statistical methods for meta-analysis. Stat. Med. **20**(6), 825–840 (2001)
15. Jitpaiboon, T., Rao, S.S.: A meta-analysis of quality measures in manufacturing system. Int. J. Qual. Reliab. Manag. **24**(1), 78–102 (2007)
16. Lovatto, P.A., et al.: Meta-análise em pesquisas científicas-enfoque em metodologias. Revista Brasileira de Zootecnia **36**(2), 285–294 (2007)
17. Seuring, S., Müller, M.: From a literature review to a conceptual framework for sustainable supply chain management. J. Clean. Prod. **16**(15), 1699–1710 (2008)

New Venture Creation, Business Growth and Cases

Locational Dynamics of Academic Spin-Offs: Evidence from Brazil

Filipe Scorsatto[1], Bruno Fischer[1](✉), and Paola Rücker Schaeffer[2]

[1] School of Applied Sciences, University of Campinas, Limeira, Brazil
bfischer@unicamp.br

[2] Institute of Geosciences, University of Campinas, Campinas, Brazil

Abstract. Universities can be perceived as sources of innovation, where academic spin-offs is one of its main vehicles. One of the main points of attention in this debate concerns the pivotal role of universities within entrepreneurial ecosystems. Drawing from this background, this research investigates the locational dynamics of academic spin-offs in the State of São Paulo, Brazil, as well as its association with the respective alma maters of entrepreneurs. 1082 entrepreneurial projects were analyzed covering the period 1998-2017. While research-intensive public universities play a leading role in the generation and retention of entrepreneurs at the local level, the distance from developed markets reduces the concentration of spin-offs in the region. This finding carries implications for the notion that university campuses can leverage levels of regional development, since the causal relationships in this process seem to be tied to elements that are endogenous but also exogenous to the university.

Keywords: Academic spin-offs · Geography of entrepreneurship · Regional development · Universities

1 Introduction

The role of the university in the 21st century goes far beyond the missions related to teaching and training. Today, universities are recognized as institutions that generate and disseminate knowledge, also establishing relationships with business activities and promoting regional development [1]. The research activities of these institutions have become important sources of innovation, with academic spin-offs being one of its main vehicles [2]. In this sense, universities are often seen as central actors of the ecosystems of entrepreneurship and innovation [3]. Such connections present a strategic character in the dynamics of productive systems, given that academic spin-offs have the capacity to generate significant impacts on trajectories of economic growth [4].

Nevertheless, academic spin-offs are heterogeneously distributed in territories, which is due to the heterogeneous availability of complementary resources, institutions and markets [5]. These processes end up generating high levels of spatial concentration in entrepreneurial activity, hindering peripheral regions to tap into its positive impacts [6]. Notwithstanding, important gaps remain in the terms of the aspects that influence the location of these new companies [7].

L. Pereira et al. (Eds.): IDEAS 2018, SIST 198, pp. 65–75, 2021.
https://doi.org/10.1007/978-3-030-55374-6_7

In this context, a critical issue concerns the role of universities as the pivotal elements of local entrepreneurial systems, as areas close to these institutions coincide with higher rates of knowledge-intensive entrepreneurship [8]. For the Brazilian case, issues related to the proximity between academic spin-offs and universities remain as an unexplored field of study. Hence, this chapter focuses on the following research question: Do academic entrepreneurs establish businesses in proximity to their respective *alma mater*?

For our empirical analysis, academic spin-offs have been approached through data from the Small Business Innovation Research Program promoted by the São Paulo Research Foundation (PIPE/FAPESP). A total of 1,082 projects were analyzed covering the period 1998-2017. Findings indicate that while research-intensive universities play a critical role in the generation and retention of entrepreneurs at the local level, the distance from developed markets reduces concentration of spin-offs. This implicates that, on its own, the university campus should not be perceived as a solution for regional development.

2 Entrepreneurial Universities

In addition to traditional research and teaching activities, universities have been establishing strategies to achieve stronger integration with markets through diverse mechanisms. In this context, the entrepreneurial university emerged as the dominant theoretical approach. This view derives from propositions stating that the dynamics of interaction between universities, industry and governments stand for fundamental pillars of innovation systems [9].

Complementarily, the debate on the role of universities includes not only the establishment of relations with the productive sector, but also the creation of companies, technology transfer offices and science parks [10]. Connections between the academic environment and businesses brings benefits to existing firms, but also to the generation of new companies originating from activities carried out in academia [2]. Consequently, universities incorporate the mission of supporting the development of entrepreneurial ecosystems, either through the provision of human resources and technologies for existing enterprises or through the generation of nascent entrepreneurs. However, despite well-known cases of success, the ability of universities to generate new knowledge-intensive firms is highly heterogeneous across institutions [11].

2.1 Academic Spin-Offs

One of the main manifestations of the entrepreneurial university refers to the practice of academic spin-offs [12]. This concept can be defined as the generation of new companies from the university context, with emphasis on the exploitation of results of scientific research [13]. The increase in the relevance of academic spin-offs is associated with the high levels of innovative capacity of these firms [14]. Besides, from a systemic perspective, these firms positively affect economic growth and development in the regions in which they are located [15]. Following this view, universities can be understood as promoters of entrepreneurial capital within the perspective of the ecosystems of innovation and entrepreneurship [10, 16]. However, impacts associated with academic spin-offs are strongly linked to the local environment, which justifies the interest in understanding the locational dynamics of these new ventures.

Location of Academic Spin-Offs. Universities have demonstrated an influential role in the locational decisions of academic spin-offs [17]. Some authors address this issue by referring to "entrepreneurial capital", a concept which can be understood as local-level institutions that promote the emergence of new firms [15]. In this context, geographical proximity is fundamental for the transmission of knowledge, having a significant impact on the generation of knowledge-intensive entrepreneurial activity [18]. The main factors that determine the retention and attraction of academic spin-offs are associated with [19]: (i) ease of collaboration in research, facilitating knowledge flows; (ii) reduction of costs related to these transactions; (iii) dependence on academic R&D activities; (iv) ease to overcome problems that new companies usually face (high costs, lack of equipment, etc.); and, (v) possible social relations between entrepreneurs and universities.

The argument that universities attract and generate new ventures consists in that these firms are embedded in social relations and are prone to agglomerate spatially. In turn, the creation of these firms occurs generally in ecosystems in which universities are present [6, 20]. In a similar vein, literature has identified that new technology-intensive firms are usually established in the same locality where the entrepreneur previously studied [21, 22].

Additional factors that also influence the locational decision of entrepreneurs can be related to levels of economic development of regions, whereas students who graduated in less developed regions tend to have greater geographical mobility [23]. Hence, the location of entrepreneurs seems to be moderated by the socioeconomic context, as urban agglomerations offer easier access to more developed markets, encouraging the retention and attraction of new academic ventures [24].

3 Method

The methodology of this research consisted of two stages. The first stage has a quantitative character and consists of the geographical mapping of academic spin-offs. For this purpose, data from PIPE projects were used to gather a sample of academic entrepreneurship projects. PIPE was created in 1997, inspired by the Small Business Innovation Research (SBIR) Program in the United States and it is managed by the São Paulo Research Foundation (FAPESP).

Although PIPE is not an initiative oriented exclusively to academic spin-offs, it has a bias towards academia, provided that it is funded by an agency that is fundamentally oriented towards supporting research in universities. In any case, we sorted out entrepreneurs that did not have an academic profile in order to track the behavior of academic spin-offs. In order to do so, we sought additional information from individuals in the Lattes database, a platform that stores CVs from Brazilian academics nationwide. Subsequently, we matched companies and information from academic individuals, including the city in which they established their businesses and where they obtained their prior education and training. For this purpose, we considered the last place of academic enrollment of entrepreneurs prior to their respective PIPE grants. The final sample of projects was made of 1,082 companies, covering projects that took place in the period 1998–2017.

The second stage of our empirical approach has a qualitative nature. Four PIPE companies associated with academic entrepreneurship were interviewed. These firms are located in different cities, representing four of the main entrepreneurial ecosystems in the State of São Paulo: Campinas, São Carlos, São José dos Campos and São Paulo. This also offered an in-depth perception of entrepreneurs' connections with the top universities in the analyzed regions. This procedure allowed a closer scrutiny of the locational trends observed in the quantitative exercise. Interviews were recorded and transcribed for analytical purposes[1].

4 Results and Analyses

First, we analyze the distribution of PIPE projects considering the main universities of entrepreneurs' origin (Table 1). As it can be gathered, there is a high concentration of academic spin-offs in five universities, namely: University of São Paulo, University of Campinas, Federal University of São Carlos, State University of São Paulo and the Technological Institute of Aeronautics. These universities account for approximately 80% of the total number of projects. This is in accordance with previous findings [11], which identify that knowledge-intensive entrepreneurial activity is strongly linked to eminent research-oriented universities. Accordingly, these institutions rank among the best schools in Brazil and Latin America. The remaining projects are distributed across 152 different universities.

Table 1. Distribution of PIPE projects.

University	Number of PIPE projects	%
University of São Paulo	450	41.59%
University of Campinas	198	18.30%
Federal University of São Carlos	83	7.67%
State University of São Paulo	72	6.65%
Technological Institute of Aeronautics	53	4.90%
Other universities	226	20.89%
Total	**1082**	**100%**

Source: Elaborated by the authors.

Second, based on data from entrepreneurs' CVs, a predominance of entrepreneurs with a high level of education can be observed, with a particular concentration of academic entrepreneurs with doctoral degrees (Table 2). This coincides with one of the

[1] The research's qualitative phase was previously validated by the Research Ethics Committee of the authors' institution, registered under the process no. 89010418.2.0000.8142.

interviewees, a PhD professor from the Federal University of São Carlos. His company derived from research activities in which scientific knowledge in the field of engineering was the basis for the development of medical equipment. Also, the academic environment provided by the university fostered the creation of relational networks, especially with alumni, who are currently the company's main investors.

Table 2. PIPE Projects according to educational attainment of entrepreneurs.

Educational level	PIPE projects	Retained	% Retained
Bachelor	171	77	45.03%
Masters	243	150	61.73%
Doctorate	668	395	59.13%

Source: Elaborated by the authors.

It should also be noted that entrepreneurs with higher levels of training (masters and doctoral studies) have significantly higher levels of retention in the same locations as where they last attended academic programs. This finding offers interesting insights on matters associated with the strengthening of social networks and how they evolve along the academic trajectory of individuals, making them less prone to geographical mobility over the course of their careers. It also puts emphasis on the strategic role played by highly-qualified individuals in carrying out technology transfer processes through the establishment of knowledge-intensive new ventures that have the potential to fortify local-level ecosystems of entrepreneurship.

Next, PIPE projects are mapped according to their geographic distribution in the State of São Paulo (Fig. 1). The heatmap offers support for previous findings on the concentration trends of knowledge-intensive entrepreneurship in this region [25], identifying a pattern that indicates five dominant ecosystems: São Paulo, Campinas, São Carlos, Ribeirão Preto and São José dos Campos. These municipalities seem to form a single axis of entrepreneurial activity with contiguous ecosystems.

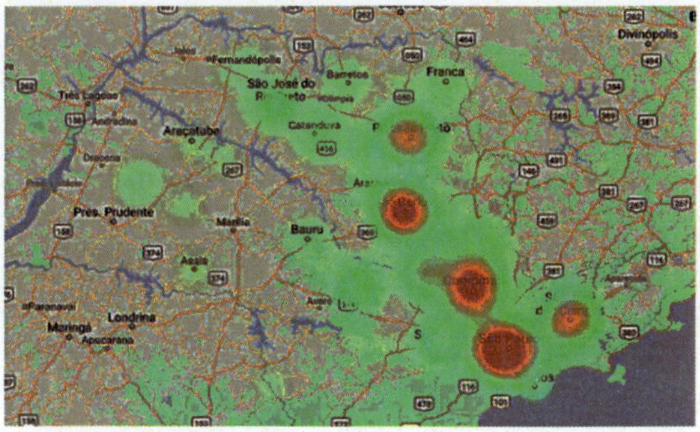

Fig. 1. PIPE Projects Heatmap in the State of São Paulo, Brazil.

The following step in this geographical analysis corresponds to the decomposition of Fig. 1 into two groups: "retained" entrepreneurs (Fig. 2-A) and "footloose" entrepreneurs (Fig. 2-B). While "retained" entrepreneurs (57.7%) have a geographical distribution that clearly delineates the five major ecosystems mentioned above, "footloose" entrepreneurs (42.3%) have a broader spatial reach with a tendency of stronger agglomeration in the Campinas-São Paulo axis. This discrepancy can be associated to the level of development of local markets, as well as their capacity of attracting other types of human capital that can provide companies with enhanced capabilities. In turn, these patterns affect the locational decision of academic spin-offs.

Fig. 2. (A) Heatmap of retained academic entrepreneurs (n = 625); and (B) Heatmap of footloose academic entrepreneurs (n = 457).

In order to further explore these findings, Table 3 connects PIPE projects and the local retention rates per institution. The University of São Paulo, the Federal University of São Carlos and the University of Campinas have strong centralization in their respective main campuses, all located within an axis with a high level of economic development [25]. On the other hand, for the case of the State University of São Paulo, its institutional structure is highly decentralized, covering 24 municipalities in areas with high levels of socioeconomic heterogeneity. Accordingly, this analysis offers additional hints on the complementarity between the retention role of the university and the level of development of local ecosystems.

Table 3. Number of PIPE Projects and share of retained entrepreneurs per university.

University	PIPE Projects	Retained	% Retained
University of São Paulo	450	324	72.00%
Technological Institute of Aeronautics	53	37	69.81%
Federal University of São Carlos	83	56	67.47%
University of Campinas	198	116	58.59%
State University of São Paulo	72	25	34.72%

Source: Elaborated by the authors.

Going deeper into this analysis we verify the microgeography of academic spin-offs in the five main ecosystems of the State of São Paulo. Figure 3 explores the case of São Paulo and Campinas, cities that respectively host the main campuses of the University of São Paulo and the University of Campinas. For the case of the city of São Paulo we notice a strong concentration in the campus area in the western part of the municipality.

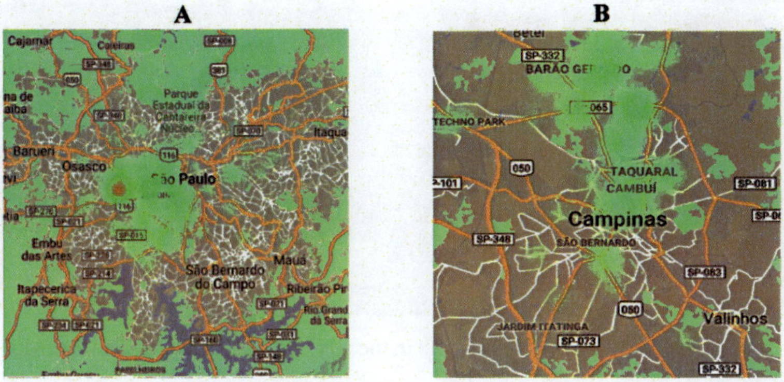

Fig. 3. (A) Distribution of academic spin-offs in the city of São Paulo and (B) in the city of Campinas.

One of the companies interviewed is an example of the importance of the University of São Paulo for the creation and retention of entrepreneurs. Founded at CIETEC, a technology-based incubator from the university, the company was created at the Polytechnic School. The idea behind their technological developments was the subject of a project undertaken by one of the founding partners in the engineering program. It was CIETEC that first helped the company to participate in the PIPE Program, having a faculty member as a co-participant in the R&D proposal. This partnership made it possible to carry out joint research activities and the use of the university's laboratories. This has led to a co-patent between the company and the academic advisor.

In the case of Campinas, a greater dispersion predominates, although the District of Barão Geraldo and its adjacencies represent the main pole of concentration of the spin-offs (northern area of the city). This corresponds to the urban area where the University of Campinas is located, as well as the main campus of the Pontifical Catholic University of Campinas. The company interviewed in Campinas was a spin-off from the Computer Engineering Institute, where the three entrepreneurs connected during their undergraduate studies. Also, it was during their time at the university that these individuals acquired the necessary knowledge for the development of the product marketed by the company. Presently, the company remains related to the university through a continuous flow of qualified human resources and informal relations with faculty members related to mentoring activities.

In the cities of São Carlos, São José dos Campos and Ribeirão Preto (Fig. 4) we can observe a slightly different pattern. Although there are large universities in these municipalities, the main poles of concentration involving entrepreneurial activity are drawn to areas where the main technological parks are located: Parqtec, in the central region of São Carlos, Parque Tecnológico de São José dos Campos, in Via Dutra (the highway that connects the city of São Paulo to Rio de Janeiro), and Supera Parque, in the western part of Ribeirão Preto.

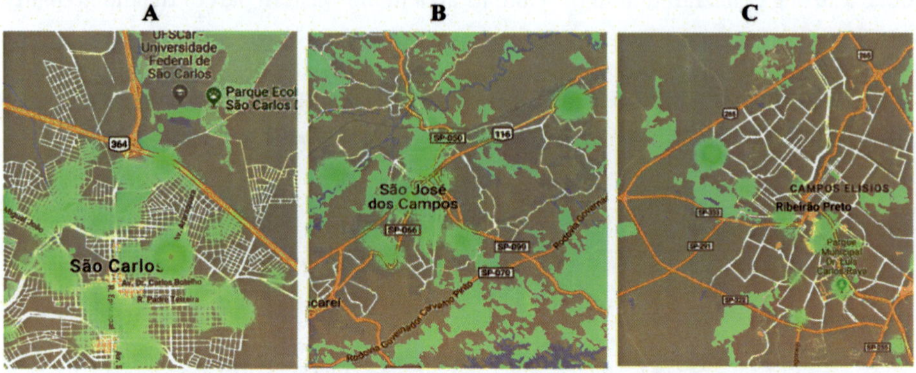

Fig. 4. (A) Distribution of academic spin-offs in the city of São Carlos, (B) São José dos Campos and (C) Ribeirão Preto.

This spatial distribution of companies highlights the beneficial effects of developing dedicated infrastructures for new and established ventures to locate closely. The provision of incubators and science parks can serve the purpose not only of harboring academic spin-offs, but also of connecting them to other agents that are embedded in entrepreneurial ecosystems. Nonetheless, as already highlighted, these structures belong to relatively developed markets and they should not be perceived as 'silver bullets' to engineer functional ecosystems where basic economic conditions are not present.

The company interviewed in São José dos Campos offered support for these perceptions. Currently located in the Technological Park, this firm was initially established in the incubator of the Department of Aerospace Science and Technology (DCTA). Founded by two alumni of the Technological Institute of Aeronautics, it was the university that allowed entrepreneurs to conduct exchanges abroad during their undergraduate studies. These international activities allowed having contact with the technology and the product that would later be improved and developed by their company in Brazil. In addition, emphasis was placed by the interviewees on the teaching and research activities of the university in connection with other agents from the ecosystem: the National Institute for Space Research (INPE), and two other higher education institutions in São José dos Campos, namely FATEC and the Federal Institute.

5 Concluding Remarks

This chapter has addressed the patterns of location of academic spin-offs in the State of São Paulo. As pointed out in the literature, these new companies have a high level of dependence on collaborative arrangements to achieve technological maturity [26]. In this sense, universities with which entrepreneurs have prior relationships can be understood as attraction points for these new companies. The data analyzed suggest relatively high levels of retention of entrepreneurs at the local level. Interviews highlight not only the importance for academic ventures of having facilitated access to human resources, but also the relevance of linkages associated to research activities and shared use of infrastructure.

A first result concerns the formation of an entrepreneurial axis involving five pivotal locations: São Paulo, Campinas, São Carlos, São José dos Campos and Ribeirão Preto. These cities are home to the main universities in the State of São Paulo, and these institutions also respond to intense entrepreneurial activity, with an emphasis on individuals with a doctorate. This result contradicts the widespread perception that a chasm exists between academia and the dynamics of markets. However, although universities are considered anchor institutions of entrepreneurial ecosystems, they cannot meet the market needs of firms, stressing the complementarity between higher education institutions and the productive sector [27]. Accordingly, universities are a necessary condition, but not sufficient to trigger regional economic development [28].

These notions find correspondence with the distribution patterns of academic spin-offs in the State of São Paulo. The São Paulo-Campinas axis, the most developed area in Brazil, presents a high rate of attraction of footloose entrepreneurs, while students from institutions located in economically peripheral regions show stronger mobility patterns. Such distinctions in the localization dynamics of entrepreneurial activity add complexity to the comprehension of entrepreneurial ecosystems processes, given that strategies for the creation of university units in less developed markets may not generate the desired effects in terms of dynamism of the local economy.

For this reason, entrepreneurial ecosystems' approaches have shifted to the adoption of more systemic and relational schemes, aiming at strengthening linkages among agents as a means to achieve success [29]. According to this view, challenges remain for the generation of regional convergence. In the first place, universities' impacts on ecosystems appear to be highly localized [8, 30]. Secondly, based on the endogenous relationship between knowledge-intensive entrepreneurship and economic growth, localities with lower levels of attractiveness for these entrepreneurs tend to achieve worse relative development levels. Ultimately, this situation generates negative feedback loops for the structuring of entrepreneurial ecosystems in laggard localities.

Ethics Declaration.

Conflict of Interest. The authors declare that they have no conflict of interest.

Compliance with Standards Involving Humans as Subjects. All procedures performed in studies involving human participants were in accordance with the ethical standards of the institutional and/or national research committee and with the 1964 Helsinki Declaration and its later amendments or comparable ethical standards. Informed consent was obtained from all individual participants involved in the study.

The research's qualitative phase was previously validated by the Research Ethics Committee of the University of Campinas, Brazil, registered under the process no. 89010418.2.0000.8142.

References

1. Poods, R., Oort, F., Frenken, K.: Innovation, spillovers and university-industry collaboration: an extended knowledge production function approach. J. Econ. Geogr. **10**(2), 231–255 (2010)
2. Collini, S.: What Are Universities for?. Penguin, London (2012)
3. Jiao, H., Zhou, J., Gao, T., Liu, X.: The more interactions the better? The moderating effect of the interaction between local producers and users of knowledge on the relationship between R&D investment and regional innovation systems. Technol. Forecast. Soc. Chang. **110**, 13–20 (2016)
4. Fritsch, M.: How does new business formation affect regional development? Introduction to the special issue. Small Bus. Econ. **30**(1), 1–14 (2008)
5. Stam, E.: Entrepreneurship, Evolution and Geography. [Papers in Evolutionary Economic Geography #09.13]. Utrecht University - Urban & Regional Research Centre (2009)
6. Feldman, M.: The entrepreneurial event revisited: firm formation in a regional context. Ind. Corp. Change **10**(4), 861–881 (2001)
7. Audretsch, D.: Determinants of high-growth entrepreneurship. In: OECD/DBA International Workshop on High-Growth Firms: Local Policies and Local Determinants, Copenhagen (2012)
8. Schaeffer, P., Fischer, B., Queiroz, S.: Beyond education: the role of research universities in innovation ecosystems. Foresight STI Gov. **12**(2), 50–61 (2018)
9. Etzkowitz, H., Leydesdorff, L.: The endless transition: a "triple helix" of university-industry-government relations. Minerva A Rev. Sci. Learn. Policy **36**(3), 271–288 (1998)
10. Audretsch, D.: From the entrepreneurial university to the university for the entrepreneurial society. J. Technol. Transfer **39**(3), 313–321 (2014)
11. Di Gregorio, D., Shane, S.: Why do some universities generate more start-ups than others? Res. Policy **32**(2), 209–227 (2003)
12. Landry, R., Amara, N., Rherrad, I.: Why are some university researchers more likely to create spin-offs than others? Evidence from Canadian Universities. Res. Policy **35**(10), 1599–1615 (2006)
13. Guerrero, M., Urbano, D.: The development of an entrepreneurial university. J. Technol. Transfer **37**(1), 43–74 (2012)
14. Oakey, R., Hare, P., Balazs, K.: Strategies for the exploitation of intelligence capital: evidence from Hungarian research institutes. R&D Manag. **26**(1), 67–82 (1996)
15. Audretsch, D., Keilbach, M., Lehmann, E.: Entrepreneurship and Economic Growth. Oxford University Press, New York (2006)
16. Galan-Muros, V., Davey, T.: The UBC ecosystem: putting together a comprehensive framework for university-business cooperation. J. Technol. Transfer **44**(4), 1311–1346 (2017). https://doi.org/10.1007/s10961-017-9562-3
17. Fischer, B.B., Schaeffer, P.R., Vonortas, N.S., Queiroz, S.: Quality comes first: university-industry collaboration as a source of academic entrepreneurship in a developing country. J. Technol. Transfer **43**(2), 263–284 (2017). https://doi.org/10.1007/s10961-017-9568-x
18. Gilbert, B., Audretsch, D., McDougall, P.: The emergence of entrepreneurship policy. Small Bus. Econ. **22**(3–4), 313–323 (2004)
19. Egeln, J., Gottschalk, S., Rammer, C.: Location decision of spin-offs from public research institutions. Ind. Innov. **11**(3), 207–223 (2004)
20. Alvedalen, J., Boschma, R.: A critical review of entrepreneurial ecosystems research: towards a future research agenda. Eur. Plan. Stud. **25**(6), 887–903 (2017)
21. Baltzopoulos, A., Broström, A.: Attractors of entrepreneurial activity: universities, regions and alumni entrepreneurs. Reg. Stud. **47**(6), 934–949 (2013)

22. Berggren, E., Dahlstand, A.: Creating an entrepreneurial region: two waves of academic spin-offs from Halmstad University. Eur. Plan. Stud. **17**(8), 1171–1189 (2009)
23. Faggian, A., McCann, P.: Human capital, graduate migration and innovation in British regions. Camb. J. Econ. **33**(2), 317–333 (2009)
24. Polonyová, E., Ondos, S., Ely, P.: The location choice of graduate entrepreneurs in the United Kingdom. Miscellanea Geogr. **19**(4), 34–43 (2015)
25. Fischer, B., Queiroz, S., Vonortas, N.: On the location of knowledge-intensive entrepreneurship in developing countries: lessons from São Paulo, Brazil. Entrepreneurship Reg. Dev. **30**(5–6), 612–638 (2018)
26. Lee, K., Lim, G., Tan, S.: Dealing with resource disadvantage: generic strategies for SMEs. Small Bus. Econ. **12**(4), 299–311 (1999)
27. Hayter, C.S.: A trajectory of early-stage spinoff success: the role of knowledge intermediaries within an entrepreneurial university ecosystem. Small Bus. Econ. **47**(3), 633–656 (2016). https://doi.org/10.1007/s11187-016-9756-3
28. Feldman, M.: The new economics of innovation, spillovers and agglomeration: a review of empirical studies. Econ. Innov. New Technol. **8**(1–2), 5–25 (1999)
29. Mason, C., Brown, R.: Entrepreneurial ecosystems and growth oriented entrepreneurship. In: Workshop Entrepreneurial Ecosystems and Growth Oriented Entrepreneurship, OECD LEED Programme. The Hague (2013)
30. Calcagnini, G., Favaretto, I., Giombini, G., Perugini, F., Rombaldoni, R.: The role of universities in the location of innovative start-ups. J. Technol. Transfer **41**(4), 670–693 (2015). https://doi.org/10.1007/s10961-015-9396-9

Check for updates

An Application for Automatic Classification of Unconventional Food Plants

Karla Okada[1(✉)], Eulanda M. Santos[2], José Reginaldo H. Carvalho[2], Carlos Gustavo Nunes-Silva[3], and Max Vasconcelos[1]

[1] Instituto Ambiental e Tecnológico da Amazônia, Rua Tucanos, 49 - Coroado, Manaus, Brazil
karla.okada@iatecam.org.br

[2] Institute of Computer, Federal University of Amazonas, Av. General Rodrigo Octavio Jordão Ramos, 1200 - Coroado I, Manaus, Brazil

[3] Federal University of Amazonas, Av. General Rodrigo Octavio Jordão Ramos, 1200 - Coroado I, Manaus, Brazil

Abstract. Unconventional food plants are eatable species, having one or more parts with nutritional potential, but are not commonly used. These plants have attracted considerable attention, as more and more people have become interested and resorted to these natural resources. In order to be actually consumed, unconventional food plants should be known and disseminated. However, although there are many scientific researches dealing with plants identification in the literature, none of them addresses the automatic identification of these species. This work presents a study focused on the identification of unconventional food plants by means of two different strategies: 1) the classical combination of digital image processing-based feature generation and machine learning; and 2) CNN (Convolutional Neural Network) for feature representation and classification. To do so, the authors generated a database of selected species. The paper also details the process of database collecting, its constitution and representativeness, as well as the experiments performed on comparing the two investigated strategies. In the first strategy, we studied 17 features, including shape and texture features and employed Random Forest as classifier. Since we extracted the features from segmented leaves, the paper also details the segmentation process. Finally, the second strategy applies a CNN pre-trained on ImageNet. The comparative study showed that CNNs achieved lower false positive rates and higher and significant accuracy rates. These results show that computer-aided unconventional food plants identification systems are feasible, which may be important tools to allow non-experts to have access to such a valuable information, since the interested public is large and diverse, ranging from professionals to the general public.

Keywords: Unconventional food plants · Classification · Digital image processing · Deep networks

1 Introduction

Investigations about plants identification have several purposes, such as producing vegetal-based medicines, environmental protection, among others. Recently, with the

L. Pereira et al. (Eds.): IDEAS 2018, SIST 198, pp. 76–85, 2021.
https://doi.org/10.1007/978-3-030-55374-6_8

increasing interest in unconventional food plants, the identification of plants with food potential has become a necessity. This interest is especially due to the possibility of diversifying the daily menu and providing less restricted gastronomy. According to [1], many plants are considered weeds or bushes, just because they grow among cultivated plants or in places where people deem it inappropriate. However, many of these plants are species of great food importance, despite being unknown or neglected by the majority of the population [2].

Unconventional food plants might be defined as plants that have one or more parts that can be used directly in human food, including spices, seasonings and aromatic substances, which are not commonplace or are not common in people's diets in a region, a country or even in the planet. Currently, we have a very homogeneous, monotonous and globalized basic diet.

There are discussion groups in social networks on whether or not one may consume a plant, or any part of it. There are also attempts to identify plant species using mobile applications such as Pl@ntNet[1], GardenAnswers[2], and Plantifier[3], but these applications are not specifically focused on unconventional food plants. Besides, all these mobile applications are collaborative. They take into account the participation of users to enrich the database with images or information about the plants. In this format, there are limitations in the recognition due to the fact that users or a community of specialists are often who define the labels. It may generate an extended waiting time for the response and the labels are subject to a certain degree of uncertainty.

Classification uncertainty might be the key for leading to food benefit instead of intoxication. On the one hand, many plants considered and validated as unconventional food plants have botanical "relatives" sometimes similar, but not considered appropriate for consumption, or that should under no circumstances be consumed. On the other hand, there are plants that are popularly known as toxic, but may be used as human food. Therefore, tools are necessary for providing accurate decisions.

Although there are no study that addresses the automatic identification of unconventional food plants specifically, the literature presents strategies for plants classification. These strategies are usually based on information about leaves, flowers, and fruits, among others. Several authors indicate that leaves present more discriminative patterns when compared to other parts of the plant, because they contain relevant and reliable features [3]. Plant leaves may provide complex information, such as color, morphology and texture, which vary according to growth conditions [4]. Shape information is more often used [5], but information from texture and vein structure can be quite useful. In terms of texture of leaves, textures of the same species may present high variation caused by factors such as maturity levels, acquisition conditions and environmental conditions [3]. It is also important to highlight that, even though there are many approaches for plant identification available in the literature that use information extracted from leaves, there is still need for efficient solutions in terms of storage and processing time, as well higher classification rates.

1 https://identify.plantnet-project.org/.
2 www.gardenanswers.com.
3 https://www.mygarden.org/plants/plantifier.

More recently, some authors have investigated Deep Learning for plants classification, more precisely, Deep CNN (Convolutional Neural Network), or just CNN in this paper. The main advantage of this type of method is the fact that it does not use hand-crafted features, since the network itself performs learning representation and extracts high-level features from the original images. In addition, CNNs have become state of the art in different image classification problems.

Considering this context, this work aims at developing a study on automatic classification of unconventional food plants. Images captured via smartphones are submitted to two pattern recognition processes: 1) a traditional pattern recognition procedure - image segmentation, feature extraction, feature selection and classification; and 2) a CNN employed to represent features and classify images. The obtained results are important to show that developing applicable tools as instrument focused on promoting the dissemination of unconventional food plants is feasible. For instance, this kind of tool may allow a user to take a picture of a plant in any place with the build-in camera of a mobile, and analyze it with an installed recognition application to identify whether it is an edible species. Hence, non-professionals will have access to such valuable information, whose potential is related to economic botany and gastronomy.

2 Related Work

In [6], the authors present a review regarding techniques for extracting plant leaves features. The authors emphasize that the visual features often used in the literature are based on color, vein structure, and texture of the leaf, as well as geometric features such as diameter, perimeter, margin, etc. In terms of features of the leaf vein structure, edge operators such as Sobel, Prewitt and Laplace are used. For texture information, histogram-based methods are used to extract this type of attribute.

The authors in [4] performed the task of classifying leaves using MultiLayer Perceptron (MLP) Artificial Neural Networks and data obtained from the edges of the leaves. Three types of edges were employed: internal veins, isolated leaf edge and leaf edge with overlap/occlusion. MLP has been trained to differentiate leaf edges from internal veins edges. The features were extracted as follows. They have drawn a straight line by crossing the edge perpendicularly along the gradient to generate local features. The line consisted of 3 pixels on both sides of the border, in addition to the center point, i.e. it was a profile of 7 pixels. According to the authors, the 7-pixel profile was sufficient to allow the MLP to distinguish between predictions of leaf edge images and internal veins of leaves. The feature vector was represented in gray level values and normalized between [0, 1]. The edge was classified as a sheet border when more than 60% of the edge pixels were identified as belonging to the edge.

The method proposed in [7] uses color, veins, shape and texture features. The following color features were used (HSV space): mean, standard deviation, asymmetry and kurtosis. In terms of veins, a vector with four values was obtained from a morphological opening operation. Eight shape features were also extracted: Eccentricity, Solidity, Aspect Ratio, Isoperimetric Factor, Elongation, Stochastic Convexity, and Lobedness. Finally, 15 texture attributes were used, including first order attributes: average intensity, average contrast, smoothness, third moment, uniformity and entropy; and second order, extracted from the co-occurrence matrix of gray levels: angular momento, contrast, correlation, entropy, variance, homogeneity, sum of entropy, cluster shade and prominence. The authors used Random Forest and Linear Discriminant Analysis (LDA) as classification methods.

In [3], Local Binary Pattern histograms (LBP) were used as texture descriptors of plant leaves. The texture patterns were then grouped via hierarchical grouping technique, i.e. images of each class (species) were divided into natural groups to express the intra-class variation. Thus, symbolic representations were defined to encode the groups of each class. Finally, kNN was employed for classification. Four databases were investigated, and the classification rates ranged from 79.35 to 97.55.

As mentioned in the introduction, some authors have also investigated the use of deep learning techniques for plant classification. In [8], a CNN was used in a large-scale species identification problem, including animals and plants. The CNN used in this work was InceptionResnetV2, which achieved accuracy below 60%. However, it is important to mention that the database investigated represents a very varied classification problem, since it contains approximately 1.2 million photos of about 20,000 different species of organisms observed in their natural habitat, including plants.

In [9], the CNN model used in the ImageNet Large Scale Visual Recognition Challenge 2012 was specifically applied in the classification of 44 different plant species from leaf images. The authors showed that CNN achieved higher classification rates when compared to traditional methods that use hand-crafted features.

3 Dataset Description

In total, 1,265 digital images of plants were collected from June to September 2017, including six selected species of unconventional food plants (UFP), one species of no food plant very similar visually to one species of UFP and various species of plants grouped in a class called "Not listed". At the end of the collection period, we obtained a database of georeferenced images. These images were collected in different locations distributed in ten different districts of Manaus-Brazil. The selected species, presented in Fig. 1, cover plants that have the leaf as an edible part and are used or represent potential for culinary use, abundance of individuals and diversity of locations all over the city. In addition, taking into account that among plants whose leaves are edible, few can grow in warm and humid regions, the species identified as important in this study stand out for their potential for climate and soil in the urban area of Manaus. The nutritional aspect was also taken into consideration, as described below.

Fig. 1. (1) Bertalha (*Basella alba*); (2) African Spinach (*Celosia argentea*); (3) Jabuti Herb (*Peperomia pellucida*); (4) Taioba (*Xanthosoma taioba*); (5) Capeba (*Piper peltatum*); (6) Indian Spinach (*Asystasia gangetica*).

- Bertalha *(Basella alba):* In addition to the leaves, the tender branches can be consumed in raw salads or stew and soups. It has great nutritional importance for the population, since it is a source of vitamin A and C, calcium and iodine [1, 2].
- African Spinach (*Celosia argentea*): It is considered important source of minerals and vitamins A and C. The main edible part is the leaf, as spinach and cabbage are used. Preference for consumption after cooking.
- Jabuti Herb (*Peperomia pellucida*): It can be consumed raw in salads, such as seasonings and has several nutritional and medicinal properties [1]. Since it is very accessible, it constitutes an important food resource for the population.
- Taioba (*Xanthosoma taioba*): In addition to the leaf, the stem (petiole) and the root are also edible. Since they can easily be mistaken for other species "poisonous tajás", this motivated us to differentiate Taioba from other species of Araceas (botanical family Araceae), here called "False Taioba", which may not be consumed.
- Capeba (*Piper peltatum*): It is better known as a medicinal plant, mainly used in the form of plaster. However, it can be used in stews and stuffed cigars.
- Indian Spinach (*Asystasia gangetica*): It can be used cooked and in salads, including its flower, which is also edible. This species has interesting nutritive properties and it is easy to propagate [1].

Therefore, the database used in our experiments consists of 1,265 instances, divided into 8 classes, as shown in the Table 1 below.

Table 1. Distribution of instances among the classes represented in the database.

Classes	Quantity
Bertalha (*Basella alba*)	101
African Spinach (*Celosia argentea*)	106
Jabuti Herb (*Peperomia pellucida*)	116
Taioba (*Xanthosoma taioba*)	136
Capeba (*Piper peltatum*)	146
Indian Spinach (*Asystasia gangetica*)	153
"False Taioba"	188
"Not listed"	319
Total	1,265

4 Experiments and Results

In this section we describe details of the experimental protocol used, the experiments performed and the results obtained. Initially, we present information about the experimental protocol.

4.1 Experimental Protocol

The experiments were divided into 5 scenarios. In the first and second scenarios, images were classified by the Inception-V3 CNN pre-trained on ImageNet (4,000 iterations). In the first scenario, the original images were used, while in second scenario, images submitted to a manual cropping process were used to train the CNN. Figure 2 shows images obtained from the manual cropping process of the 08 classes.

Fig. 2. Samples of database instances obtained after manual cropping.

The last three experiment scenarios involved using a classic pattern recognition procedure, combining digital image processing (DIP) and machine learning. DIP techniques were used to extract 17 features to represent each image. The procedure assumes that the leaf or branch is in the center of the image, i.e. the central pixel is inside the leaf. In order to take advantage of the feature shape of an ellipsoidal envelope, a polar ellipse scan was implemented, the ratio between the major axis and the minor axis being proportional to the same aspect ratio of the image. Each pixel $p\ (i, j)$ of the image is indexed by $p\ (\rho, \theta)$, where ρ is the length of segment l that joins the pixel to the center of the ellipse and θ the angle between l and the major axis of the ellipse. The pixel neighborhood, in which the operators and filters are applied, is defined as all pixels belonging to the same track and sectors, and the image was divided into 10 tracks and 12 sectors, each with 30° of aperture.

The automatic segmentation algorithm compares the pixels of the image with those of its neighborhood, discarding those whose intensity value in each of the channels R, G, B and H, S, V are outside the range defined by the mean and standard deviation. The resulting image is framed and several morphological and texture features are extracted, as described below. The first 09 are features of shape, while the remaining (08) are of texture. Parentheses indicate the number of values of each feature.

- Centroid (2): The coordinates (x, y) of the center of mass of the segmented area;
- Eccentricity (1): Eccentricity of the ellipse that has the same 2nd moment of the segmented region;
- MajorAxisLength (1): Length in pixels of the major axis of the ellipse that has the same 2nd point as the segmented region;
- MinorAxisLength (1): Length in pixels of the minor axis of the ellipse that has the same 2nd moment of the segmented region;
- Orientation (1): Angle (between 90 and −90) formed between the x-axis of the image and the major axis of the ellipse that has the same 2nd moment of the segmented region.
- FilledArea (1): Area in pixels of the segmented region, assuming that all pixels within the perimeter have been filled;
- EquivDiameter (1): Diameter of the circle that would have the same area of the image (4 * Area/PI);
- Solidity (1): Defined by the ratio Area/ConvexArea;
- Extent (1): Defined by the following ratio Area/(BoundingBox Area);
- Equivalent grayscale image intensity on tracks (10): arithmetic mean of pixel intensity on each of the 10 concentric ellipsoidal tracks of the grayscale image, from the innermost to the outermost;
- Standard deviation of the grayscale intensity equalized in the tracks (10);
- Average image intensity formed by the entropy of a 9×9 neighborhood on the graytone image per track (10): arithmetic mean of the pixels in each of the 10 concentric ellipsoidal trails of the image resulting from the entropy filter of a 9×9 neighborhood applied over the grayscale image, from the innermost to the outermost;
- Standard deviation of the image intensity formed by the entropy of a 9×9 neighborhood on the grayscale image per track (10);
- Average H-channel intensity of the image per track (10): the arithmetic mean of the pixels in each of the 10 concentric ellipsoidal H-channel tracks of the image, from the innermost to the outermost;
- Standard deviation of the average channel H intensity per track (10);
- Average image intensity formed by the entropy of a 9×9 neighborhood on the H-channel of the image per track (10): arithmetic mean of the pixels in each of the 10 concentric ellipsoidal trails of the image resulting from the applied 9×9 neighborhood entropy filter on the H-channel of the image, from the innermost to the outermost;
- Standard deviation of the average entropy of a 9×9 neighborhood on the H channel, per track (10).

The Random Forest machine learning technique was used for the classification task of the species in the last three series of experiments. In the third scenario, the algorithm was trained and tested with all 17 features. In the fourth scenario, only the 09 shape features were used, while in the last scenario only the texture features were applied. In all experiment scenarios, the same base partition was used. In this way, a direct comparison between the results obtained in the 5 scenarios was possible. The database was divided into 3 partitions using the holdout validation strategy. The distribution of the instances between the partitions is: 888 for training, 252 for validation and 125 for test. The results obtained in the scenarios investigated are presented and discussed in the next section.

4.2 Results

The 5 scenarios investigated are compared in Table 2 in terms of accuracy and false positive rate.

Table 2. Results of the scenarios investigated

Scenario	Accuracy	False positive rate
Inception-V3 – Original data	83,7%	7,14%
Inception-V3 – Manual Crop	83,7%	6,74%
Random Forest – All attributes	64,68%	21,82%
Random Forest – Shape attributes	34,52%	37,27%
Random Forest – Texture attributes	58,33%	22,73%

This table shows that the results obtained by the CNN (in both scenarios) were significantly higher than the results obtained by the Random Forest using all the shape and texture features. When we compare the results obtained by varying the groups of features, it is possible to observe that the attributes of texture are more relevant than those of shape. However, the combination of the two groups of attributes is superior to the use of the groups separately.

Although the recognition rates obtained by the CNN in scenarios 1 and 2 were similar, the errors made in each scenario presented some differences. To illustrate these differences well, Table 2 highlights the false positive rate produced by each method. This rate was calculated considering the False Taioba and Not Listed classes only, thus, whenever an image of one of these two classes was incorrectly classified as belonging to an edible class, an error was considered. Therefore, when images of edible classes are incorrectly classified as belonging to other edible classes, this error was not considered as false positive. This rate shows that the manual crop helped to decrease the f positive rate. On the other hand, it caused distortions in classes of edible species. It is possible to observe that, the manual cropping action produced a slight increase in the accuracy rates of the False Taioba and Not Listed classes. However, classes of edible species such as Indian Spinach were more confused.

Figure 3 shows some images of the False Taioba and Not Listed classes that were incorrectly classified in scenario 1 and scenario 2 of the experiments performed. It can be emphasized that CNN made more errors in the classes that have smaller leaves, because crop manual introduced distortions of shape and size in leaf images.

Therefore, these results indicate that the use of pre-trained CNNs, with transfer learning, can capture more adequately the regularities and specificities of each class investigated in this work. However, the false positive rate obtained still needs to be reduced in order to allow the system to be used in real applications. In addition, new features need to be tested, especially features related to leaf veins.

Fig. 3. Some Scenario 1 and Scenario 2 classification errors

5 Conclusions and Future Work

This work presented a study on classification of unconventional food plants, considering 06 species of edible species, a class of species not listed and a class of inedible species. The work involved the creation of a database, generated from images acquired mainly via smartphones, the process of extraction of shape and texture features - obtained from the leaves of the plants automatically segmented, and classification. In addition, convolutional neural networks were used to classify the raw images, i.e. without using the hand-crafted features.

The results show that the segmentation process was able to extract the leaves of the plants in the images. However, although we used texture and shape features, the feature extraction process did not produce enough discriminant attributes, since the classification rates obtained through a traditional pattern recognition method did not exceed the results achieved by the deep network. As future work, it is intended to generate new features, especially related to leaf veins, since the literature indicates that this type of attribute is highly relevant. Furthermore, a principal components analysis (PCA) might also be targeted for future work, as the amount of possible features demands the identification of the subset of representative ones.

In terms of CNN, the results are quite interesting, even the database being considered small for this type of approach, which makes it promising for the problem investigated. In addition, as the images were generated by different devices and in different lighting conditions, the classification task is significantly complex. However, false positive rates can be considered high. Therefore, it is notable that the results of CNN can be improved, by increasing the database size using data augmentation, for instance. Finally, the results show a great potential for using an automatic classification system in a final application to identify unconventional food plants or similar problems, which may be important to allow these plants to be known and disseminated.

References

1. Kinupp, V.F., Lorenzi, H.: Plantas Alimentícias não Convencionais (PANC) no Brasil. Guia de identificação, aspectos nutricionais e receitas ilustradas. Instituto Plantarum de Estudos da Flora LTDA, São Paulo (2014)
2. Cardoso, M.O.: Hortaliças não convencionais da Amazônia. Brasília: EMBRAPA-SPI-Manaus: EMBRAPA-CPAA (1997)
3. Naresh, Y., Nagendraswamy, H.: Classification of medicinal plants: an approach using modified LBP with symbolic representation. Neurocomputing **173**, 1789–1797 (2016)
4. Xia, C., Lee, J.-M., Li, Y., Song, Y.-H., Chung, B.-K., Chon, T.-S.: Plant leaf detection using modified active shape models. Biosys. Eng. **116**(1), 23–35 (2013)
5. Silva, P.F.B., Marçal, A.R.S., Da Silva, R.M.A.: Evaluation of features for leaf discrimination. In: Image Analysis and Recognition (ICIAR 2013). LNCS, vol. 7950, pp. 197–204. Springer, Heidelberg (2013)
6. Ab Jabal, M.F., Hamid, S., Shuib, S., Ahmad, I.: Leaf features extraction and recognition approaches to classify plant. J. Comput. Sci. **9**(10), 1295–1304 (2013)
7. Elhariri, E., El-Bendary, N., Hassanien, A.E.: Plant classification system based on leaf features. In: 9th International Conference on Computer Engineering Systems (ICCES 2014), pp. 271–276 (2014)
8. Mo, J., Frank. E., Vetrova, V.: Large-scale automatic species identification. In: Australasian Joint Conference on Artificial Intelligence 2017, Advances in Artificial Intelligence, pp. 301–312 (2017)
9. Lee, S.H., Chan, S.C., Wilkin, P.: Deep-plant: plant identification with convolutional neural networks. In: IEEE International Conference on Image Processing (ICIP 2015), pp. 452–456 (2015)

Tactical Design: Understanding, Experimenting, and Learning from Design to Organizational Growth

Bruno Raphael de Carvalho Santos(✉), Luana Bittencourt Saraiva, and Claudete Barbosa Ruschival

Universidade Federal do Amazonas, Manaus, Amazonas 69080-900, Brazil
brunuph@gmail.com, bitluana@gmail.com, claudete@ufam.edu.br

Abstract. The design is an area of study that has multiple approaches. When applied as a tool to achieve a competitive advantage in a business environment, it becomes Design Management. However, to obtain advantages and results, it is necessary to make understandable Design concepts and its principles among internal collaborators. Thus, the present study used the exploratory bibliographical research, which starts with the formation of the theoretical fundaments and the dialogue between different authors to understand the conceptual basis of the subject. The literature review was elaborated with theses, base authors, dissertations, and articles to understand the possibilities tested of how the process of integration of Design Management in companies occurs. As a result, to a competitive advantage in the market, the implementation of design at the operational, tactical, and strategic levels are essentials. The tactical level connects the strategic vision with the operational level and therefore, must disseminate the actions and concepts of design through the corp. There are several ways to spread the strategic vision such as graphic design applied in internal and external communication, improvements in the work environment by interior design, product design for new product concepts, among others. Finally, the team's learning about the importance of the application of design in the various sectors of the company is a contributing factor to innovation and organizational growth.

Keywords: Design management · Tactical design · Organizational growth

1 Introduction

Traditionally, the design is an area of knowledge that seeks to solve problems for the most diverse human needs, and it uses a design process that includes the phases of planning, design, and execution for the improvement of the physical systems. These developed by the design of products. Intending to optimize communication between people, institutions, and companies, the design emerges to enables the most diverse forms and tools of visual communication using signals, symbols, images, and printed matter. With the technological advances driven mainly by human needs, the range of the

L. Pereira et al. (Eds.): IDEAS 2018, SIST 198, pp. 86–94, 2021.
https://doi.org/10.1007/978-3-030-55374-6_9

large design area has expanded, giving rise to new qualifications, such as web design, interaction design, social design, interior design, packaging design.

In the business context, the design has a broad potential to be explored. Much more than granting an visual appeal to products and services, the design can be used from a strategic point of view, giving companies a competitive advantage through innovation in products or management models, as communicates with the public of interest, in the positioning of the brand and in the higher value perceived by the customers. Making design a tool to achieve a competitive advantage within the company consists of a strategic decision of the highest level of the corporation. This set of strategic actions using the design to obtain a competitive differential is called Design Management.

However, for the organization to experience all the advantages of this management with tangible results, it is necessary to disseminate knowledge among the employees, responsible for implementing this new vision about the importance of Design for the development of organizational strategies. In this sense, [1] the internal public becomes, in fact, the first customer to be considered by Design.

Therefore, this article aims to verify how to build a shared understanding of design principles so that everyone in the company has the chance to experiment, learn, and grow as a team. For reach this goal, it is explained what the design is, showing its scope and limits. Then, how this area behaves in the business context, that is Design Management. Next, due to the importance of disseminating the knowledge about Design as an innovative tool in the company, the knowledge management was studied, as well as how the concept propagated and integrated into the organization. Finally, it displays how this area can be experienced by the employees, for their assimilation, learning and consequently the growth of the team.

This research is qualitative, exploratory, and textual to understand the possibilities tested of how the process of integration of Design Management in companies occurs. Based on the reading and interpretation of classic books on the subject, in addition to theses, dissertations and articles published in periodicals about design management of the last ten years, located in national databases in Portuguese language using the key-words: design management, strategic design, tactical design, design communication, and functional design. In this way, it uses the phenomenological method [2] to qualitatively investigate the different approaches defended by the authors of the area, promoting a dialogue between them, to understand the conceptual basis of the subject, in order to interpret it from multiple perspectives.

2 Design: Concepts and Principles

Everywhere one can find and interact with the design, when opening a door, walking down the street, while sitting down, lying down, eating, working, anyhow, the design is in our daily experience [3]. Also, according to the author, there are other contacts with the design that are intangible, such as how a team is managed, in the transportation of products, in the work processes. This scope may leave doubts about the limits of design, so it is necessary to define the concepts that approach this study.

There are several descriptions of what design is. It is difficult to explain the term since that in Portuguese, there is not adequate meaning for the English word of origin [4]. It is a profession that works with concepts, design, with information managed efficiently and with visual appeal. Design is a process of problem-solving focused on people [3] while the authors [5] depart from the etymological analysis of the word. The word comes from the Latin language "designare" translated as "designate" and "draw" the same translation maintained for the English "design" [5]. So the analysis presents the idea of "designate" as the intention, a project, that is the design applied to the development process of something and "draw" as a visual representation, a model, or design in the final product. So, the design can be in the whole process of creation, in the project, in the intention, in the execution, in the final product.

Thus, the professional that works with design does not do it alone because it relates to other disciplines [3]. This flexibility promoted its entry into various fields of knowledge, in which Graphic, Product, Interior, Audiovisual Design, and was developed Design Management, as mentioned in the introduction of this study.

As for the management, it is a term of difficult delimitation; however, generically it is understood as a process that occurs within the organizations. The difficulty of a more precise definition is because management is associated with different activities as well as a design [6]. According to the authors, the function of management is organizing, integrating, directing, and maintaining productive processes. This concept started after the production process became fragmented.

The entry of design into the management of the corporate world was from the shift from a linear Taylor model to a more informal, flexible organizational model [5]. Then, in the next session will be presented the Design Management as a strategy to guide the company's vision in its different levels: strategic, tactical, and operational [7].

3 Design Management

The most innovative companies in the world, such as Apple, Nike, Starbucks, Target, Walt Disney, Whirpool, use design as an integrative resource to innovate efficiently and successfully [8]. These design-led companies maintained a competitive edge of 228% compared to others in the stock market [8].

Even with this data, many corporate cultures do not understand the need to invest in design. This fact is due to the difficulty in finding methods to measure the impact of the use of the design in the leadership of the company [8]. The designers themselves who are responsible for clarifying the contributions they can make to the company proactively do not, so other employees also do not value them [9]. In other words, they do not know how to sell their service, which further contributes to the devaluation of the profession [3].

However, understanding the benefits that design can bring to help the organization achieve its goals is critical to the implementation of Design Management, as it takes care of the implementation of design as a formal program of activities within the corporation as part of its philosophy [5].

Therefore, the insertion of the design towards the growth of the company and according to its ideals and purposes it can execute in three levels: corporate, business and operational [3] or strategic, tactical and operational level [5]. Each of them has its strategy, average, and a measure of performance in the company, concluding that they are as requirements for the improvement of productive operations [3].

Detailing the levels at which design can interact, the operational level work with sales, marketing, and production departments, and is at the end of the process in product generation [10]. Generally, the designer starts at this level of performance in companies, executing actions, and daily operations such as the delivery of products or services [3]. The decisions taken at the strategic and tactical levels take shape at this stage, generating actions such as the creation of flowcharts, infographics, maps etc. [1].

At the tactical level, also called functional or business level, the goal of the design is to reach the purposes of the company's vision as defined at the strategic level through the agreement between the strategic, and the operational level [11]. It has the perspective of design as a process, and it is at this stage that the company culture is disseminated in plans for each department as a new product line, for example [3]. The sectors of a company involved are usually human resources, financial, and research and development [10].

Design at the strategic or corporate level deals with the organization's general management, where decisions with the most significant impact for the company occur [10]. At this point, the design ensures its insertion into the company's culture, since this hierarchy guides other strategies, aligns the vision and mission of the company with the objectives, critical success factors, and key performance indicators, generating value for the product and the customer [3]. In this way, the impact of design at the strategic level in the company is the possibility of innovation [1]. Based on this, two examples of the underlying assumptions can be disseminated throughout the corporation to be understood by corporate workers, based on the analysis done by the Design Management Institute and the Motive agency: (i) the use of design methods to understand customer needs; (ii) reformulating complex problems to get new ideas [8].

However, in most cases, the design is not understood as a strategy of differentiation and innovation in the management of the company, only being operational and sometimes reaching the tactical [9]. However, the tactical level is responsible for the link between the operational and the strategic [5] and can generate communication between all stages of the organization through People Management, or practices that disseminate the design by the company [1]. Therefore, it is necessary to discuss how this propagation of the knowledge about the design within the company can happen to obtain the results and advantages inherent in this area of activity.

4 Knowledge Management and Shared Understanding of Design

In a society where the dynamism in relationships and consumption is the rule information is shared globally in a matter of seconds and new products are generated daily, while others become obsolete, and in that frenzy, the surviving companies are those who hold knowledge and uses it as a competitive advantage to produce innovation in its products, processes, and services. Therefore, greater competitiveness obtains organizations that stand out for being the creator and disseminator of knowledge.

In this sense, many Japanese companies invest in knowledge [12]. For these authors the company is not a machine, but a living organism, which can have a sense of identity and fundamental purpose, being equivalent to organizational self-knowledge - the shared understanding of what the company stands for, where it is going, what type of the world wants to live and, most importantly, how to make this world a reality. Now, an organization has professionals from different areas acting in a collaboratively, and the interaction of the different formations and points of view are opportunities for the creation, sharing, and dissemination of knowledge.

Organizational learning occurs when a double looping learning is implemented in the organization, through a virtuous circle in which new information is used to challenge ideas and concepts already accepted and well established, thus developing new prospects for the future [13].

In this multifaceted context, knows the design, which, as explained in the previous topic, has its concepts, principles, and possibilities of contribution for companies. However, there are still challenges regarding the practical implementation of design knowledge in companies, their relationship with innovation, and how it can add value to corporations, requiring a long process of individual and collective maturation [14]. The fact that there are few reliable, professional statistics, as well as the lack of design management courses and research policies with long-term design missions, helps to explain the difficulties that the area for their implementation in companies [5]. Managers need firm milestones, reliable information, and the assurance that they can finance the design safely.

To make design understandable to those who do not specialize in the area, it should first describe the nature of the profession, the different areas that practice the design and the various methods that design professionals use in their work [5]. After that, the authors recommend evaluating the relevance of design to the science of management and discovering what can be learned from the creative process. Finally, they explain that it is necessary to evaluate the impact of design on corporate performance to determine what managers can gain from it.

Design must be introduced in an organization in a gradual, responsible, and deliberate way to be effective [5], graded from integration into stages, through a series of successive projects. Responsible for even starting with a single project, design integration requires the support of senior managers to demonstrate the strategic character of design and to leave aside the idea that it is difficult to manage. Lastly, deliberate the reason the design must be conducted at all levels, based on the corporate values, supported in all divisions of the company, and having fluid communication with senior management.

Regarding communication, intermediate managers (tactical level) are at the intersection of vertical and horizontal information flows of the company and serve as a bridge between the visionary ideals of top management (strategic level) and the frontline of business (operational level) [12]. Managers at the tactical level mediate "what is" and "what should be" and remake reality according to the company's vision. The design manager must be present throughout the organization and should always reiterate that design can help any department and in discussions of themes that have a philosophical basis such as ethics and how it relates to sustainable development, corporate values, customer value, aesthetics [5].

Summarizing the author's approach, the organization can disseminate and integrate design knowledge in the following ways:

- Training, contact, listening and communication procedures;
- Coordination with human resources management, because it can help locate the tools needed to support and understand design, first among design partners, then among the rest of the company's employees;

- Communication policy referring to the design. Using the newsletter of the company or developing a newsletter or website of the design department that everybody can access throughout the company by the Intranet;
- Internal communication around the design to involve all of the company, such as an information display in the hall or printed about its activities.

Therefore, based on the studied framework, it was verified that, although it is not a simple process, it is possible to perform a shared analysis of Design in an organization, being an activity of the design manager of the tactical level. After making evident the role of this area of knowledge within the company, whether, through meetings, training or internal communication, it is necessary to disseminate this knowledge that is mainly given by experimentation, explained below.

5 Experimenting Design Concepts and Team Growth

There is the argument that some people's sector is not of design. So, the author takes the warehouse sector to ask the following questions: Is it well signaled? Is it easy to find the stored materials? Are the documents well designed? Also the foodservice? Are there any menus? Is the company brand printed on plates and napkins? Therefore, it concludes that design can have a relationship with all departments of the company, regardless of its impact on the final product or service [9].

In this perspective, all organizations are systems of forms and, therefore, can be classified according to design disciplines, considering [5]:

- Design of environments: workspaces, reception areas, factories, exhibition spaces;
- Product design: machinery, commercial products;
- Packaging design: commercial products, promotional material;
- Graphic design: letterhead, notices, invoices, files, reports, computer screens, advertising, signs, trade names, and technical documentation.

There are, therefore, innumerable ways of experimenting design within an organization, such as the product design to create the most appropriate furniture for the execution of the work activities, as well as the development of any tools with ergonomic aspects that can facilitate the execution of the work [5]. The authors also talk about information design as fundamental to synthesize and design information that is readable and with perceived meaning, materialized in the form of statistics, diagrams, and organizational charts. As for web design, they say that this area is another way for employees to experiment design, which may be part of a seasonal campaign of internal marketing, but also to other situations that require planning, control, communication, training, e-mail, and corporate social networking sites and other mechanisms of digital interaction between company and employees.

Design is also seen as a People Management sector, because it enables the internalization of strategies, including the company's identity in the organizational culture [1]. The result of this interaction, according to the author, corresponds to the Tactical Design that, in the communication aspect, aligns the identity of the company with the internal public, programming actions that communicate the essence of the company to employees. About culture question, it stimulates the pride of belonging, through internal actions that favor the participatory culture (design culture, design). Besides, it leverages the engagement between the internal public and the company's philosophy, resulting, finally, the propagation of values, mission, and vision that make sense for people and emotionally link them to the organization.

As an example of these actions used in the internal organizational environment, Tactical Design is vital in the process of monitoring and reviewing practices (audits), training (conferences, workshops, courses), dissemination and exchange of information (meetings, panels, discussions, forums) and various corporate events, because without the involvement of the internal public in actions like these, the Tactical Design becomes unfeasible [1]. This vision corroborates the statement of Tomás Maldonado when he says that the designer is never alone; he never works alone. Therefore, he is never a whole. That is, without the commitment and participation of all with the specified design strategies, it is difficult to maintain the value they attribute to the organization.

Therefore, it is necessary to value and stimulate teamwork, since they play a central role in the knowledge-creating enterprise, by providing a shared context in which individuals can interact with each other and engage in the constant dialogue, which depends on useful reflection, creating new points of view and integrating individual and collective perspectives [12].

In this way, to favor the evolution of the organization through design and obtain valid results, the designer must interact with other areas, integrate the systematization of processes with freedom of creation, fundamental for the efficiency of processes and the sedimentation of knowledge in organizations [15]. Communication is the guiding thread between areas that are, sometimes, understood as disjointed. Therefore, exercise is fundamental to the practices of designers [16]. Finally, the connection between the concepts of Tactical Design and the internalization of strategies can explain the preponderance of the Tactical level among the opportunities of design contribution to the internalization of organizational strategies [1]. The Fig. 1 below summarizes the topics studied:

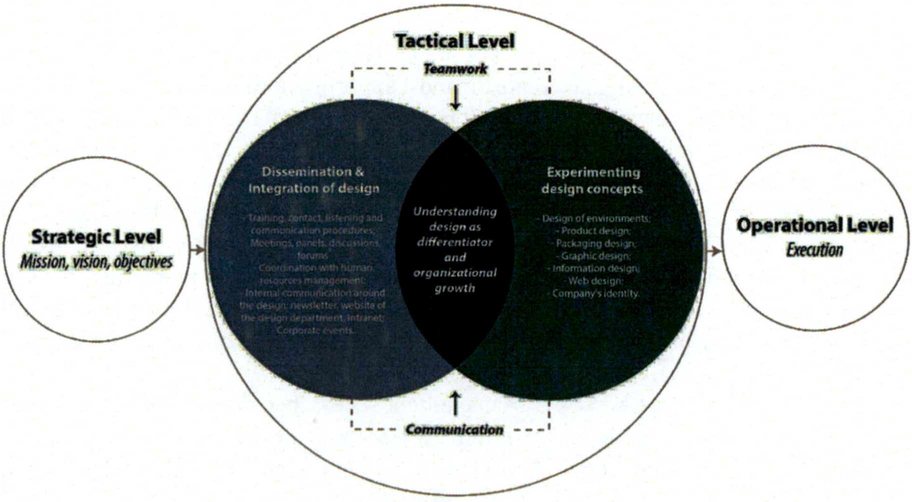

Fig. 1. Synthesis of dissemination, experimentation and organizational growth. Source: Adapted from [5] and [1].

After reflecting on the several authors cited, it is possible to make considerations about the shared understanding of design principles, as presented below.

6 Final Considerations

Design is a multidisciplinary field that each area and project contributes with its vision to the solution of problems, always taking as base the needs of the user, client, and the general public. In the world of business, design transposed the concept of executor and reached the vision and mission of the company at its highest level, as was concluded by the authors studied. As a result, companies that have adhered to this vision have achieved a differentiating advantage in the market.

The differential that the design provides is not necessarily attached to the figure of the designer, because even with several professionals of the area meeting, the company may not make the proper use of their skills, for just not understanding their competitive edge. For this reason, it is fundamental to make the entire team of internal collaborators aware of the dissemination of the strategic perception of design in the various sectors of the company, which competes directly at the tactical level.

This diffusion of the knowledge of design as a competitive differential for the company can be made, as previously explained, in a gradual manner and through several internal communication projects, such as workshops and training. This awareness promotes the experimentation of design in several departments, either in the improvement of the internal and external environment, internal communication methods, document and process standardization, risk reduction, production cost decrease, continuous improvement from the business.

Thus, this research exposed the contribution of design management with a focus on tactical design. It is recommended applying these improvements as a case study for experimentation beyond the conceptual but practical study.

References

1. da Silva Júnior, J.E.: The contribution of design applied to organizational strategies. 150 p. Dissertation (Master in Design e Graphic Expression) – Graduate Program in Graphic Design and Expression, Federal University of Santa Catarina, Florianópolis (2015)
2. Tesch, R.: Qualitative Research: Analysis Types and Software Tools. The Falmer Press, Basingstoke (1990)
3. Best, K.: The fundamentals of design management. Bookman, Porto Alegre (2012)
4. Strunck, G.: Design Living. 2AB, Rio de Janeiro (2004)
5. de Mozota, B.B., Klöpsch, C., da Costa, F.C.X.: Design Management: Using Design to Build Brand Value and Design Corporate Innovation. Bookman, Porto Alegre (2011)
6. Palmieri, A.R., de Figueiredo, L.F.G.: Proper and relevant role of design Management. DaPesquisa em Artes **7**(9) (2012)
7. Moreira, B.R., Bernardes, M.M.S., Almendra, R.A.: Design management in practice: discussion of human and procedural factors involved. Manag. Proj. Tecnol. **13**(1), 59–74 (2018)

8. Westcott, M.: Design-driven companies outperform S&P by 228% over ten years – The 'DMI Design Value Index'. DMI: Design Management Institute (2014). https://www.dmi.org/blogpost/1093220/182956/Design-Driven-Companies-Outperform-SP-by-228-Over-Ten-Years–The-DMI-Design-Value-Index. Accessed 13 Dec 2018
9. Phillips, P.L.: Briefing: The Project Management of Design. Blucher, São Paulo (2007)
10. Sousa, B.: Design Management in Portuguese: Design Management or Holistic Design?. Esad arte + design, Portugal (2012)
11. Tanure, R.L.Z., Kistmann, V.B.: The appropriation of design management: a case study in the clothing sector. Des. Tecnol. (2010)
12. Takeuchi, H., Nonaka, I.: Knowledge Management. Bookman, Porto Alegre (2008)
13. Motta, F.C.P., de Vasconcelos, I.F.G.: General Theory of Administration. Cengage Learning, São Paulo (2010)
14. Moreira, B.R., Bernardes, M.M.S., Van Der Linden, J.C.S.: Simultaneous implementation of design practices in product developers. Study Des. **24**(2), 44–65 (2016)
15. Demarchi, A.P.P., Fornasier, C.B.R., de Martins, R.F.F.: Organizational model of self-management applied in a craft production and evaluated by knowledge management based on design thinking. Estudy Des., 1–15 (2011)
16. Moreira, B.R.: Design management in practice: framework for enterprise implementation. (Doutorado em Design) – Graduate Program in Design, Federal University of Rio Grande do Sul, Porto Alegre (2016)

State-of-the-Art on Furniture Design: A Visual Review

Ana Carolina Correa de Medeiros, Roger Pamponet da Fonseca(✉), and Augusto César Barreto Rocha

Universidade Federal do Amazonas, Manaus, Amazonas 69080-900, Brazil
rogerpamponet@ufam.edu.br

Abstract. The furniture design is a crucial element to propose smarter houses. For that matter, the research and development of new furniture pieces depend on examining the design trends and the future tech that will make such objects real. By using the Visual Data Mining process, this paper aims to investigate the scientific production from 2013 to 2018, to understand the current design trends and the future tendency for furnishing. The systematic survey follows five phases: domain knowledge; pre-processing of data; pattern extraction; post-processing of data; and to use of knowledge. This search protocol provided three categories of analysis: (I) Production + Framework + Customization + Internet of things; (II) Algorithm + Shape Editing + Flexible Furniture + Modular Furniture; (III) Human-Centered Design + Smart Furniture. Each one of these categories resulted in different sections, with the subsequent visual analysis of furniture design trends. The 4th part studies the innovations of furniture production. Part 5 explores the relation of mathematics and design, mainly the use of algorithms to design flexible furniture. Section 6 reviews the use of smart furniture on mini-living houses, exclusively in Brazil. The conclusion provides an exam of furniture trends, regarding the Internet of Things, smart furniture and flexible furniture. The furniture design is an essential field for the study of human needs. The Data Mining Survey emphasizes the importance of design to enhance the life quality of consumers.

Keywords: Innovation · Furniture design · Data mining

1 Introduction

Research on the State of Art, in any area, is essential for anyone looking to analyze the direction innovation will take. It is necessary to proceed with technological surveillance as a way to detect the trends, especially the ones that relate to Design.

The innovative vocation of organizations results from continuous stimuli for technological development [1]. Technical surveillance relates to perceiving how some technologies prove themselves in the market as a commodity [1]. The patterns are, therefore, the anticipation of human behavior. New trends will always emerge every year, so it is essential to research its fundamentals to understand how they will influence the prospecting of current and fresh products [2].

This article aims to demonstrate the importance of the perception of innovative trends in furniture design to discuss the main possibilities for the future in furnishings. By using

L. Pereira et al. (Eds.): IDEAS 2018, SIST 198, pp. 95–104, 2021.
https://doi.org/10.1007/978-3-030-55374-6_10

a Visual Data Mining approach, this article objectives to analyze the written scientific publications and point out the innovation path in such areas.

2 Methodology: A Visual Data Mining Approach

Design is a multidisciplinary field, and as such, it may be challenging to find a suitable approach to accomplish the systematic review of scientific research. As Prodanov and Freitas say, scientific research has to be systematic and critical, aiming to substantiate the analysis of a subject [3]. Flick remark, the innovation and scientific evolution have changed the very own scientific methodology of research, especially the qualitative research practice [4].

For that matter, the present exploratory research seeks to offer more excellent proximity to the problem presented, through literature research on the subject, through systematic reading of papers, journals, and scientific literature, especially those included in the period from 2013 until 2019. The methodology applied to this systematic review, as Blum, Merino, and Merino explain, is a visual method for systematic review in design based on concepts of Data Mining [5].

This methodology has five phases: (I) domain knowledge; (II) pre-processing of data; (III) pattern extraction; (IV) post-processing of data; (V) the use of knowledge [5]. Blum, Merino, and Merino adduce that the identification of the problem, followed by the pre-processing of data and by the transformation of the said data into knowledge is the primary path to follow through Data Mining as a scientific method to Design research.

3 Innovation in Furniture Design: A Visual Survey

Using the methodology proposed by Blum, Merino, and Merino [5], this visual survey followed the phases mentioned in the methodology. The first phase, the Domain Knowledge, relates to the database selection (the databases selected were World-WideScience.Org, Science Direct, and Capes Periódicos - Brazil). The second phase is the Pre-processing of Data, with the selection of a period (2013–2019). The patter extraction is the combination of descriptors. The descriptors selected: (Furniture) OR (Furnishing) AND (Innovation), to establish the domain knowledge (phase one). The systematic examination presented the following results (Table 1):

Table 1. Results of bibliometric research using the keywords "Furniture OR Furnishing AND Innovation" (2013–2019)

Period	Scientific sites	Number of scientific publications
2013–2019	WorldWideScience.Org	**939 papers categorized as** article (710); periodical (57); report (25); thesis (13); peer-reviewed article (5)
2013–2019	Science Direct	**3820 papers categorized as** Review articles (220), Research articles (2,849), Encyclopedia (50), Book chapters (546), Conference abstracts (17), Other (124)
2013–2019	Capes Periódicos (Brazil)	**97884 articles categorized as** Peer-reviewed articles (42,039), articles with no peer-revision (55,845)

Source: Elaborated by authors.

After finishing the three first phases (Domain Knowledge, Pre-processing of Data e Pattern Extraction), it is possible to explore the data found by post-processing (Phase IV), recognizing the most used keywords among the articles found on the Patter Extraction phase (Phase III). Table 2 illustrates these results.

Table 2. Bibliometric research adding keywords to furniture OR furnishing AND innovation, using the science direct platform (2013–2019)

Keywords added	Number of scientific works published		
	Science Direct	WorldWide Science.Org	Capes Periódicos
Production	2,786	651	5,278
Algorithm	2,879	754	4,869
Framework	2,037	342	4,261
Internet of things	1,010	409	4,145
Flexible furniture	1,614	252	2,099
Shape editing	877	439	2,038
Customization	1,880	551	884
Modular furniture	763	286	2,022
Smart furniture	746	216	1,525
Human-centered design	1,296	53	911

Source: Elaborated by authors.

For the post-processing of data (Phase IV), there was a selection based upon on the number of citations of each article, the relation of the topics of each article, the direct relationship between the themes and the similarity of the scientific approach of the theme, aiming to indicate the relevance of each article. To go further on the visual Data Mining method, the use of knowledge (Phase V) helps to select the articles referenced in the next part of this paper. This selection resulted on a visual collection, separated in three categories: (1) Production + Framework + Customization + Internet of things; (2) Algorithm + Shape Editing + Flexible Furniture + Modular Furniture; (3) Human-Centered Design + Smart Furniture.

4 Furniture Production Now: Innovation, Trans-Disciplinary Design, and Internet of Things

This study's first category is Production + Framework + Customization + Internet of things. By the post-processing of data, it was possible to find articles that would be relevant to pinpoint the innovation trends for Furniture Design. The next table (Table 3) will demonstrate those results.

The study of human behavior and societies is a task for sociology and many other social sciences; however, the design theory has become progressively more interdisciplinary [9]. The ever-increasing density of everyday dilemmas makes Design important

Table 3. Articles selected after the Visual Data Mining process. **First category: Production + Framework + Customization + Internet of things**

Year	Theme	Authors	Relevance
2017	Smart sensory furniture	Bleda et al. [6]	**146 citations.** Related to IoT, Sensors, AAL, Ambient Intelligence, and Ubiquitous Systems
2017	Updatable PSS	Marques et al. [1]	**40 citations.** Related to services and product design
2014	Design for longevity	Hebrok [7]	**5 citations.** Related to product lifecycle
2019	Internet of things	Mahieu et al. [8]	**3 citations.** Related to the Internet of Things and Personalization
2014	Trends for furniture	Zamoner et al. [2]	**2 citations.** Related to furniture design trends
2018	Trans-disciplinarily design	Vieira [9]	**1 citation.** Related to design theory

Source: Elaborated by authors.

as a way to create innovative solutions for these *Wicked Problems* - that are complex issues solved via Design propositions [1, 2, 9].

Vieira [9] points out that Design correlates with several disciplines, both in the practical sphere, as well as in the scientific aspects and in the academic environment. Design, therefore, serves to solve wicked problems through the interaction between several areas, to identify invariants and specific characteristics of the product [9]. Zamoner *et al.* [2] will comment on the importance of examining and forecasting trends in Design, especially in Brazil. The innovation of everyday processes can appear to be slower in comparison to other countries where cutting-edge technology seems more welcomed by consumers.

As Marques *et al.* [1] mention, "Product-services are becoming increasingly complex and diverse with advances related to technology and customers' needs" (p. 3). Those needs are complex, and consumers press the industry for greener solutions, making reconfigurable and updatable product-services a necessity for enterprises to keep up with innovation [1]. This kind of innovation demands that organizations move from product design to system design [1], changing the whole production process and moving to a scenario where design is a trans-disciplinal effort [9].

The innovative process of Design also points to the Internet of Things (IoT) as a methodology that can change the furniture production process [6–8]. For Bleda *et al.* [6] the use of sensors to make systems more intelligent is a prospect for the future of furniture technology. There are sensors in numerous methods, from hospital beds to office chairs. Measuring the variables of use via the network of sensor intercommunication serves to create systems much more attentive to humans.

Thus, embedded devices, wireless sensor networks, human-computer interactions, artificial intelligence, ubiquitous computing, intelligent sensory furniture, and Ambient Assisted Living (ALL) have become technologies to anticipate users' needs without being intrusive to them [6]. For Mahieu *et al.* [8] the Internet of things not only refers to computer systems and robotics but can also link to environments and objects in those environments. The robots can assist in many aspects (e.g., Nutrition monitoring, Homework assistance, Therapy monitoring) and a complete system can interpret raw data provided by IoT sensors using the context of a domain knowledge [8]. The sensory furniture can use the same method to put up with modifiability aspects that can predict other scenarios on ALL.

The life cycle analysis (LCA) of home objects, is also an important subject [7]. People "tame" their things when they are brought close (to their homes); items are redeemed when they are re-rendered after spending too much time near users; and when things go out of everyday use and are removed from the house, passed on or thrown away, being *dis-domesticated* [7]. Thus, Hebrok [7] analyzes that products related to life, death, and love (or the affection) of its users. The big challenge is how to make a piece of furniture that uses emotions and attachments, that transcribes values, meanings, and symbols, and align itself with utility. For this to become a reality, it is imperative to understand that design is a trans-disciplinal effort [9], to apply other scientific knowledge to furniture production (e.g., psychology to understand human relations, computer systems to use IoT solutions).

5 Algorithms: From Mathematics to Furniture

Following the visual Data Mining methodology, the second group is Algorithm + Shape Editing + Flexible Furniture + Modular Furniture. The next table (Table 4) will demonstrate the results of the exploratory research of those keywords.

Table 4. Articles selected after Visual Data Mining. **Second category: Algorithm + Shape Editing + Flexible Furniture + Modular Furniture**

Year	Theme	Authors	Relevance
2017	Wood manufacturing system	Zhao et al. [10]	**171 citations.** Related to flexible manufacturing systems using algorithms systems
2017	Furniture made of roombots modules	Nigolian et al. [11]	**44 citations.** Related to Robotics and Control systems for reconfigurable surfaces
2018	Convertible furniture design	Zhou and Chen [12]	**4 citations.** Related to using algorithms to make convertible furniture
2018	Shape editing	Ibrahim and Yan [13]	**1 citation.** Related to using algorithms to make convertible furniture

Source: Elaborated by authors.

The application of algorithms to design is not something new. Mathematics relates to several other sciences as a method to provide analysis of different possibilities, even for furniture production. Nigolian *et al.* [11] demonstrates the use of "Auto-reconfigurable modular robot interface using virtual reality: furniture arrangement made of roombots modules" that allows the creation of pieces of furniture through immersion in augmented reality in which the interface of the program operates the movement of the robots and in the spatial configuration of the said objects.

The study of algorithms for reconfiguration of modular furniture using design and computer science is a matter of innovation. Authors such as Ibrahim and Yan [13]; Zhao *et al.* [10]; and Zhou and Chen [12], write about the use of algorithms as an auxiliary method in the design of flexible furniture.

There is an immediate need for all furniture to become convertible so that spaces are better occupied [13]; however, it creates a problem of geometric order: how to fold these elements so that there is no collision or breaks in the structure of the furniture? Zhou and Chen [12] indicate the application of computer graphics to prevent collisions. In the same way, they consider using an algorithmic tree of junction possibilities to manipulate the creation of the mobile more automatically and efficiently [12]. Therefore, Zhou and Chen [12] propose an algorithm that solves a three-dimensional problem: to make a mobile compact and expansive enough to guarantee its *multifunctionality*. Following a similar path, Ibrahim and Yan [13] have perfected a structure for folding furniture based on the analysis of movements, joints, and smart structures. This path has the following steps: start the action, then fold the structure, and at last the segmenting of the object. In this way, Ibrahim and Yan [13] seek to perfect the folding order and the energy required to perform the object's transformation.

Zhao *et al.* [10] study a flexible manufacturing model that can be adapted to any production system using the *Object-Oriented Timed Colored Petri Net* and GASA (Genetic Algorithm). Thus, the calculation algorithm serves to create a tree of possibilities for the manufacture of objects, based on the genetic variance of a living organism. Therefore, the study by Zhao *et al.* [10] contributes to the formulation of multifunctional objects that adhere to the industry's production methodology.

6 Smaller Homes, Smarter Furniture

The third category of this research is Human-Centered Design + Smart Furniture. The Data Mining process found the results shown in Table 5. Those results relate to Brazilian's cases hence, Human-Centered Design (HCD) must connect to the reality in which it will be inserted [14].

The Human-Centered Design methodology was proposed by IDEO in 2015 [14], and idealized by Tim Brown. Brown [22] states that to apply the HCD, the designer needs the design thinking phases. So, the design thinking phases for design creation are - empathize (with users); define (what are the users' needs, what is their problem, and the designer's insights); ideate (by challenging assumptions and innovative solutions); prototype (to start creating the solutions); and the final phase that is testing the solutions [22]. According to Scherer *et al.* [19], this methodology is an excellent way to serve users by having in their actions such as: listening, that is, searching directly within the target audience; create solutions, usually as a team; and implement, through prototypes and direct testing with clients.

Table 5. Articles selected after Visual Data Mining. **Third category: Human Centered Design + Smart Furniture**

Year	Theme	Authors	Relevance
2018	Evolution of the built environment	Ludovico and Brandão [15]	**9 citations**. Related to the architecture of small living spaces
2018	A human-centered design toolkit to small living	Pezzini, Schulenburg, and Ely [16]	**6 citations**. Related to furniture design for small living spaces
2015	Ergonomics of the built environment	Oliveira and Mont'alvão [17]	**5 citations**. Related to furniture design for small living spaces
2018	Framework for human-centered furniture design	Araujo and Vergara [18]	**4 citations**. Related to Human-Centered design
2017	Human-centered design in furniture projects	Scherer et al. [19]	**4 citations**. Related to Human-Centered design
2015	Human-centered design for small houses in Brazil	Siqueira and Costa Filho [20]	**1 citation.** Related to furniture design for small living spaces
2018	Human-centered design for small houses in Brazil	Frossard and Pessôa [21]	**1 citation**. Related to furniture design for small living spaces

Source: Elaborated by authors.

Siqueira and Costa Filho [20] analyze that the criteria for establishing human needs for the furniture are more subjective and should lean towards to humanize residential spaces. There is a tendency for the furniture design is to provide a participative space in which to listen, to create and, finally, to deliver, constitute a continuous activity throughout the project [16, 18–22].

The HCD also relates to the minimal spaces for living found in urban areas. It is challenging to buy bulky residences in the urban areas of the new metropolises. Housing spaces have become smaller and smaller, and for this reason, research on furniture that meets the demand of people living in minimal spaces has become a growing niche within the furniture market [23]. This trend is not unique to Brazil, as Zhou and Chen [12] write, in all large cities around the globe, every inch is vital. Following the same idea, Ibrahim and Yan [13] analyze that buying a dwelling in today's cities is quite expensive. Besides, social housing programs, such as the "*Minha Casa, Minha Vida*" Program, have reshaped Brazilian cities due to mass-produced residences in urban spaces. These residences usually have limited room spaces, which require special attention to the production of specific furniture for this current market gap [19, 21].

The smart furniture conception also depends on the observation of users (potential or real ones). Oliveira and Mont'Alvão [17] analyze the interactions of the human-environment interface and the displacement from one point to the other to perform a specific activity. The user's cognitive analysis of environmental perception serves to generate an ergonomic diagnosis of the environment to establish a systemic view of the built space [17]. However, Araújo and Vergara [18] point out that this analysis cannot be automatic; it should consider user experience, to be consistent, and to transcribe a hierarchy between activities (motivated by needs), actions (subordinated to goals), and operations (determined by objective conditions of action). Human needs, therefore, are evolutionary and individualized. The use of Post Occupancy Evaluation (POE) can help in generating greater flexibility of the space arrangement [15, 17].

Araújo and Vergara [18] also point out that housing is evolutionary, insofar as a change is part of an evolutionary process of system use; the user experience shapes the house to the needs of the residents. The long-term experience in which there is a motivating need (*why*) in which the planned or subconscious operations (*what*) wherein the fulfillment of conditions of action (*how*) [18]. Similarly, Scherer *et al.* [19] realize that furniture should be versatile, multifunctional and reconfigurable (modular, aerial, adjustable and expandable furniture) since living spaces are becoming very restricted and this furniture should be easy to assemble and have high durability.

7 Conclusions

The furniture design is essential to enhance the life quality within the postmodern houses. Since the homes are becoming even smaller, it is crucial to understand the trends for innovation in furniture design.

Life quality depends on several indicators; the qualitative order, other of quantitative order. It is possible to relate those indicators to the manufacture of products and produce goods that attend consumers' needs. The psychological aspects plus the sensory aspects connects to the psychology of products [23]. The product's objectives describe what the product intends to be its requirements and attributes [24]. The physiology of the product relates to the materials used and to the manufacturing processes applied [23, 24]. All the connections between design and consumers need innovative approaches, and this spreads to furniture design.

The Visual Data Mining methodology is a tool to enhance Design scientific research, aiding in the exam of trends for furniture design. This research points to innovation in furniture production especially regards ALL and IoF. The customization of products is a route to make furniture even more flexible and smart. Design can relate to many knowledge fields such as computer science and psychology, aiming to create pieces of furnishing that will suit people needs.

To foster the adoption of technologies can be challenging. This paper provides a visual survey on tendencies in Design, aiming to help researchers finding the technology routes to follow through innovation in furniture design. There is still a vast field for research on furniture, expressly regarding the sustainable raw materials for furniture production and embedded smart systems due to IoT and other solutions of Industry 4.0.

Acknowledgments. We want to share our appreciation to *Universidade Federal do Amazonas PPGD* technical staff for their practical assistance and *FAPEAM*, for partially funding this research project.

References

1. Marques, M., Zacharewicz, G., Agostinho, C., Jardim-Gonçalves, R.: Reconfigurable and updatable product-service systems: the path for sustainability and personalization. In: D'Ambrogio, A. et al. (eds.) Society for Modeling & Simulation International (SCS). Proceedings of the Symposium on Model-driven Approaches for Simulation Engineering, v. Society, 23–26 April. POD Publ: Curran Associates, Inc. (2018)
2. Zamoner, M.T.S.C., Razera, D., Heemann, A., Barauna, D.: Trend research: benchmarking for furniture design. P&D Des. **1**(4), 1–11 (2014)
3. Prodanov, C.C., Freitas, E.C.: Methodology of Scientific Work: Methods and Techniques of Academic Research and Work. Feevale, Novo Hamburgo (2013)
4. Flick, U.: Introduction to Research Methodology: A Beginners Guide. Penso, Porto Alegre (2013)
5. Blum, A., Merino, E.A.D., Merino, G.S.A.D.: Visual method for systematic design review based on data mining concepts. DA-Pesquisa **11**(16), 124–139 (2016). https://doi.org/10.5965/1808312911162016124
6. Bleda, A.L., Fernandéz, F.J., Rosa, A., Zapata, J., Maestre, R.: Smart sensory furniture based on WSN for ambient assisted living. IEEE Sens. J. **17**(17), 5626–5636 (2017)
7. Hebrok, M.: Design for longevity: taking both the material and social aspects of product-life into account. J. Des. Res. **12**(3), 204–220 (2014)
8. Mahieu, C., Ongnae, F., Backere, F., Bonte, P., Turck, F.: Semantics-based platform for context-aware and personalized robot interaction in the internet of robotic things. J. Syst. Softw. **149,** 138–157 (2019)
9. Vieira, S.L.S.: Design transdisciplinarity: distinct reality levels. Gestão & Tecnologia de Projetos **13**(1), 101–114 (2018)
10. Zhao, Y., Liping, S., Lijing, W., Xinyu, F., Jiawei, Z., Yanchun, X.: The study on reconfigurable algorithm of the wood flexible manufacturing system based on OOTCPN-GASA. In: IEEE Xplore, EI. 2nd International Conference on Control and Robotics Engineering, ICCRE 2017, pp. 68–72 (2017)
11. Nigolian, V., Mutlu, M., Hauser, S., Bernadino, A., Ijspeert, A.: Self-reconfigurable modular robot interface using virtual reality: arrangement of furniture made out of roombots modules. In: RO-MAN 2017 - 26th IEEE International Symposium on Robot and Human Interactive Communication, January 2017, pp. 772–778 (2017)
12. Zhou, J., Chen, X.: Convertible furniture design. Comput. Graph. (Pergamon) **70**, 165–175 (2018)
13. Ibrahim, M., Yan, D.M.: Fold and fit: space conserving shape editing. Comput. Graph. (Pergamon) **70**, 316–326 (2018)
14. Ideo: Field Guide to Human-Centered Design, 1st ed. Ideo, Boston (2015)
15. Ludovico, S.S.A., Brandão, D.Q.: Characterization of the morphological identity of the architectural space of an evolutionary dwelling. Gestão e Tecnologia de Projetos **13**(1), 39–58 (2018)
16. Pezzini, M., Schulenburg, R., Ely, V.H.M.B.: A human-centered design toolkit to small living. Des. Tecnologia UFRGS **15**(1), 23–50 (2018)

17. De Oliveira, G.R., Mont'alvão, C.R.: Methodologies used in the studies of Ergonomics of the Built Environment and a modeling proposal for Interior Design projects. Estudos em Design **23**(3), 150–165 (2015)
18. De Araujo, G.O., Vergara, L.G.L.: Activity theory and affordances as a framework for approaching user experience. Estudos em Design **26**(1), 113–131 (2018)
19. Scherer, F.V., Azolin, B.R., Guimarães, F.C., Parolin, G.: Development of a furniture line through a user-centric design methodology. Des. Tecnologia UFRGS **7**(14), 135–146 (2017)
20. De Siqueira, C.N., Costa Filho, L.L., Users' needs in residential spaces, in the perception of architects and interior designers. Estudos em Design **23**(3), 36–45 (2015)
21. Frossard, G.C., Pessôa, S.S.M.V.: Design of Environments and the Metropolises: a reading of the contemporary context. In: Blucher Proceedings, Colóquio Internacional de Design 2017, vol. 4, no. 3, p. 805 (2018)
22. Brown, T.: Design Thinking. Campus, Rio de Janeiro (2010)
23. Lefteri, C.; Design Materials. Translation: Henrique Eise Toma - São Paulo: Blucher - Brasil (2017)
24. Lefteri, C.: How to Design: 92 Manufacturing Techniques for Product Design. Alves, M.A.L. (trans.), 2 ed. Blucher, São Paulo (2013)

Study on the Relationship Between Design Management and Business Project Management

Úlima Santos, Alef Santos, and Claudete Barbosa Ruschival(✉)

Amazonas Federal University, Av. General Rodrigo Octavio, 1200 Coroado I, Manaus, AM 69067-005, Brazil
ulima.usds@gmail.com, alef.vernon@hotmail.com, b_claudete@hotmail.com

Abstract. Design management is a process that aims to guarantee the company survival from the correct application of the design resources inside the corporation. Business management has the same intention of ensuring the company's success. However, both practices use different methods. We carried out a qualitative research in order to collect bibliographic data about design management and business management's own methods. To this end, we used techniques of a systematic review of the literature according to Sampaio and Mancini (2007). From the data analysis, we could observe the existence of three methods that can be employed simultaneously in a company: Design Thinking, Scrum, and PMBOK. We noticed that the methods of Design Thinking could be applied together with Scrum or PMBOK. However, design management methods are exclusive to this area and bring the user closer to the product or services that come from this process, while business management methods engage all the company's resources, not only those from design.

Keywords: Design management · Business project management · Methods · Project management · Scrum · PMBOK · Design Thinking

1 Introduction

New technologies have developed new forms of communication and relationships, influenced mainly by the interaction between mobile telephony and the dynamics of the exchanging information in social networks. This fact has definitively changed how products and services are consumed, requiring companies to make radical changes in their management methods. In order to face these changes, some authors understand there is an urgent need to insert a new managing vision in companies, whose pillars must be on a multidisciplinary structure that favors creativity, innovation, learning, and adaptation. Consequently, companies will be better able to respond quickly to the market and ensure their survival.

This scenario demands innovative and differentiated solutions. In this context, the design stands as a viable alternative for its multidisciplinary and strategic character, since its methodology ends in the search for the kind of solutions mentioned above.

L. Pereira et al. (Eds.): IDEAS 2018, SIST 198, pp. 105–114, 2021.
https://doi.org/10.1007/978-3-030-55374-6_11

It is necessary to go beyond the project aspect for the company to benefit from the values given by design. The designer has to be inserted in the organizational structure as a decision-maker of strategies, participant, and acting in the corporate management model adopted. In this case, the designer would have the function of proposing design strategies in all areas, inoculating the design in the strategic process of the corporation, that is, performing design management.

Design management proposes to plan and assist the development of new products and services from their embryonic phase to their life cycle. It can also aid in the choice of materials and manufacturing techniques best suited to the product's functionalities. Besides, design tools and methods are feasible to support the various stages of business management, such as planning the most appropriate products and services and meeting the expectations and wishes of the company or the public to which it aims to cater.

When it comes to projects, we questioned the methods of business project management in order to provide a strategic framework for the designer's performance. At the same time, we searched for the design management method more conducive to its realization according to the methods of the company, to then meet the current dynamics and complexity of the consumer market. We assumed that interaction between design management methods and project management is possible, in a way that innovation and creativity would be the main aspects of the company's design culture. The purpose of this study is to understand how this interaction could occur from the interaction between existing methods.

This study's contribution is in the interpretative response that the interaction between these methods can produce or offer, for example, in the construction of values of products, processes, and services related to usability, satisfaction, safety, comfort, customization and others.

2 Theoretical Referential: Design Management and Business Management

In order to relate design management and business management, both sectors needed a conceptual survey. The purpose of the following subtopics is to contemplate this survey.

2.1 Design Management

Design is a creative activity that aims to answer needs, with the sense of planning, designing, and creating [1]. Knowing that design exerts a specific and integrated function in the processes of the company [2], it is vital to apply it as a formal program of activities within an organization, to help it achieve its goals. This practice consists of the successful management of people and processes that enable the creation of results that are part of the consumer routine [1]. Design management is understood in a multidisciplinary way [3], involving the management of relationships between disciplines, also including different roles [1].

Thus, design management is essential within organizations, as it can define and execute cost reduction, plans, and goals as well as coordinate the development of innovative products in the market [4]. To design to be effective must be inserted in a gradual, responsible, and deliberate way, needing to be managed at all hierarchical levels of the company, not only in design projects [2]. This process has three levels, according to [2]: (i) operational, where the focus is on the execution of projects; (ii) tactical, which consists on the organization and assignment of functions, and; (iii) strategic, where the implementation of the design to standardize the vision and mission of the company occurs [5].

All of these levels encompass the decision-making of the manager, who is responsible for: (i) Understanding the clients' profile; (ii) Involving strategic levels of the company to think about alternatives which serves the customers' needs; (iii) Involving of everyone in the company in problems solution, making them appreciate the effort and feel responsible for the success of the project, which helps in the sales pitch made to the customer. In this way, the design manager directly influences the services and products that a company offers [2].

2.2 Business Management

The concept of management involves the efficient use of tangible and intangible resources of the company to achieve organizational goals [6]. For that, there are necessary activities that the manager must carry out. According to [6, 7] and [8], these functions are: (i) Planning: determining goals and decision activities; (ii) Organization: appointing tasks and activities to people and departments and allocating resources; (iii) Direction: influence, motivation, communication and development of people; (iv) Control: measuring performance according to pre-established goals.

The managers are classified into different levels within management [6]: (i) senior management (board), where the manager establishes the objectives, policies, and strategies of the company; (ii) average administration, where administrative tasks are implemented, involving coordination and conflict resolution; (iii) operational management. Where the manager supervises and directs the operational work of the production processes, this last phase, according to the author, is the one that most relates to the execution of the routine tasks of the organization.

For [7] and [8], modern managers need to take a multidisciplinary view of management, seeking to promote creativity, learning, adaptation, and innovation. [2] adds that this model is susceptible to the design process, since it favors a simple structure, stimulates action and experimentation, and centralizes the essential values of the management activity.

As for project planning and execution methods, each area, both design management and business management, uses specific methods for the development and control of its activities. Design management, for example, has Design Thinking, while business management has Scrum, PMBOK, Prince, Canvas, among others that companies can apply according to the aim of the project.

3 Methodology

This research had a qualitative approach and bibliographical and documentary procedures. We carried a qualitative survey out of theses, dissertations, and scientific articles dealing with project management related to design management and business management. For this, we used techniques of systematic literature review that, for [9], are means of research based on data sources of the literature on a specific topic and that allow a summary of the highest possible number of studies on a particular problem. We used this technique in order to identify the most commonly used design and project management methods cited by researchers.

Using these techniques, [3] created a framework to facilitate the organization of research steps, which we adapted to this article, namely: 1. Define the questions: "How can enterprise project management methods provide a strategic framework for the designer's performance?" and "Which design management method is most conducive to its realization in order to better meet the current dynamics and complexity of the consumer market?" 2. Identify the databases we were going to consult, keywords and search strategies: The platforms for the research were Google Academic, online journals of Design Studies, DAPesquisa, Design, Technology and Management, and Project Technology, considering the publications of the last five years. The keywords used in the search were: design management; business management; management; management of design methods; business management methods; management methods. 3. Establish criteria for the selection of the article: it must deal with design management or business management, and may or may not include the two topics in the same article. 4. Search in the chosen databases based on the strategy defined: we conducted it over fourteen days, selecting the sources by title. 5. Compare the search and set the initial selection of articles: we saved a total of 58 files for analysis. After reading the abstract, we selected 25 papers for this research. 6. Apply the criteria in the selection of articles and justify possible exclusions: We excluded all of the works that did not deal with design management or business management as the main topic. 7. Analyze critically and evaluate all studies included in the review: We read and summarized the information in the topics Theoretical Referential, Results and Discussion of this article. 8. Prepare a critical summary, summarizing the information provided by the articles included in the review: We provided a critical summary with information referring to enterprise management and design methods applied in companies, including the results obtained by the authors. 9. To present a conclusion, informing the evidence about the effects of the intervention: From the reading and analysis of all the selected papers, we could verify how the interaction between design management and project management methods can occur and what benefits the company could get from such interaction. Through this steps, it was possible to identify the processes most used by the authors. We listed this processes in Table 1 below.

From the reading of the selected material, we noticed that the method of design management most approached by the authors was Design Thinking, while the most approached methods of project management in the enterprise scope were Scrum and PMBoK.

1. Design Thinking is divided into four phases [29]: (1) Immersion: when the project team studies the context of the problem, from the point of view of the company and their client, in order to define the problem and identify the needs and opportunities that will guide the creation of solutions in the next phase. (2) Analysis and synthesis: the team organizes the data obtained in the previous item in order to generate

Table 1. Primary methods used for design management and business management

References	Design management methods	Management of business projects methods
[10–15]	Design thinking	–
[16–20]	–	Scrum
[20–25]	–	PMBoK
[26]	–	CDS Model
[27]	Design management integrator disk	–
[28]	Design management framework	–
[27]	–	CMMI

Source: The authors.

standards that will turn into tools such as Personas, Blueprints, Conceptual Maps, among others, which will participate in the creation of solutions. (3) Ideation: it aims to create innovative ideas for the project theme through the synthesis tools created in the previous phase to stimulate creativity and create solutions that are following the context of the subject. (4): Prototyping: it is an auxiliary function of the validation of generated ideas and can occur throughout the project in parallel with Immersion and Ideation. We can contemplate this phase from the perspective of the project team, which consists of shaping the idea by elaborating it with as much detail as possible, increasing the level of solution fidelity from the user's point of view. The user, when interacting with the model created and in different environmental contexts, can provide inputs for the project's improvement and adequacy;

2. Scrum, is an agile process focused mainly on software development and is divided by [15] into seven steps, defined by the authors as (A) Product Backlog: list of priorities related to product functionalities that the team must consider during the project. (B) Sprint Planning: a meeting between the Scrum Master and his team to define the goals of each Sprint and to prioritize the Backlog items. (C) Sprint: cycles of up to one month in which, at the end of each, the company must deliver a tangible value result to the customer. (D) Daily Scrum: fast meeting of up to fifteen minutes, ideally at the same time each day, to discuss how the Sprint progress is occurring. (E) Definition of Done: It is a document (DoD) that certifies that the team has finalized the work of Sprint and it is necessary that every member of the team is aware of how that accomplishment occurred, guaranteeing transparency. (F) Sprint Review: a meeting between the Scrum team, its master, and the clients for discussion and inspection of the results in order to identify which adaptations they will make in the backlog. (H) Sprint Retrospective: A meeting between team members to evaluate the performance of the previous Sprint, identifying which items were successful, and what improvements they will make for the next Sprint.
3. PMBOK is a process divided into five phases [16]: (I) Initiation: where the team identifies a need that the problem will solve. At this stage, they define the project manager, the stakeholders, and the objectives of the project. (II) Planning: where

they detail the objectives, and plan the actions necessary to achieve them. This phase specifies the activities, resources, costs, and everything that will be carried out by the project in order to avoid the appearance of unforeseen events. (III) Execution: the activities that they planned are carried out, according to the requirements initially established. (IV) Monitoring and Control: the two previous phases occur in parallel and focus on controlling everything that is being executed by the project, always comparing with the forecast and making corrections when necessary. (V) Closure: consists of the evaluation of the project results, as well as the discussions on the positive and negative aspects that occurred during its procedure.

From the studies of these methods, we proposed interactions between them, with Design Thinking as the design management model more conducive to interacting with the other methods.

4 Results

To synthesize the content of the results of the study, we proposed an interaction between the design management method and project management, seeking to start a dialogue between them.

Scrum is a framework that has: 1. Three responsibility groups, the Product Owner, the Scrum Master, and the Scrum Team; 2. Three activity moments, Sprint Planning, Sprint Review, and Daily Scrum, and; 3. Three artifacts, the Product Backlog, the Sprint Backlog, and the Burndown. The interactions of the activities performed bases the development of this method. Scrum works cyclically with repetitions of tasks until all items requested by the client are approved. It also works with a list of mandatory items, that is, those that add value to the final product, and the Daily Scrum meetings decides the entire development process.

Design Thinking focuses on the user, on their journey and their actual problems to find innovative and viable solutions collectively and collaboratively. Through multidisciplinary meetings, it looks for ways to innovate. It stem from a reflection and an abductive thinking to immerse, both in the needs of the company and the needs of end-user and thus propose ideas that solve the problem. Uses prototyping to test the product and know if it meets the needs of the end-user.

The PMBOK, being the guide with the best practices on project management, guides the processes and efforts that must be undertaken to achieve the success of the project, as well as its evolution and development. For this, it defines as working system as an initial term of the project, the creation of a management plan and also when and how to close each phase of the project.

We concluded, comparing the three forms of management, that it is possible to carry out projects integrating the abductive thinking of the designer thinker in the Scrum method, as we can see in the scheme of Fig. 1 below.

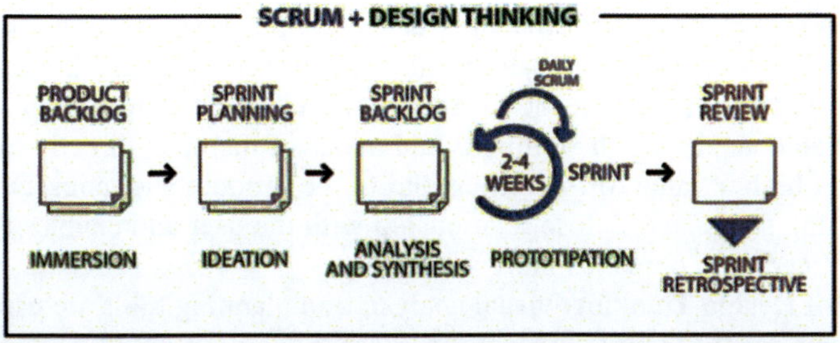

Fig. 1. Interaction between the phases of Design Thinking and Scrum.

As we can see in Fig. 1, the Immersion stage, which defines the problem, needs innovation opportunities for the solution of the project and could be connected to the Product Owner's activities since it defines the Product Backlog items. Thus, it would include the needs of the user as essential in the project. It could also be done in the PMBOK Initiation stage, which establishes the premises and requirements of the project and the monitoring of customer satisfaction, as we show in Fig. 2.

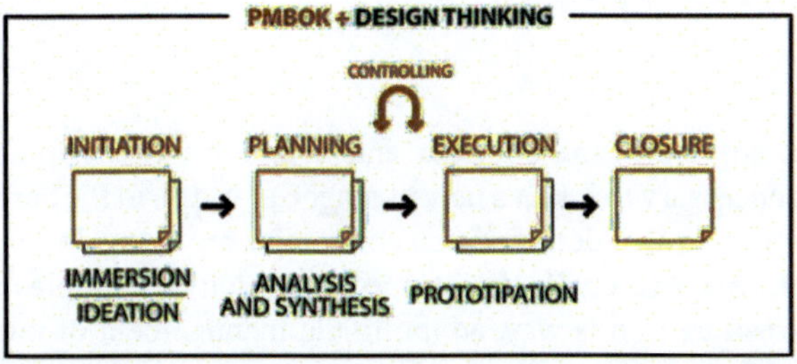

Fig. 2. Interaction between the phases of Design Thinking and PMBOK.

After these brief considerations and the schemes shown, we could discuss how each phase of Design Thinking can integrate the two other methods one by one and we wrote it in the topic below.

5 Discussion

Step 1 of Design Thinking, corresponding to Immersion, interacts easily with stage I of the PMBOK, the initiation, since both seek to understand in detail the problem they seek to solve. However, Scrum's Step A, Product Backlog, despite doing this contextualization, takes into account only the functional aspects of the product that it intends to design. It does not, at this stage, cover the needs of the user of the product in question. At this point, the immersion suggested by Design Thinking can bring benefits to Scrum if it is employed in its initial phase, because focusing on the user will reduce the chances of rejection when the product is tested, for example.

The same is true when comparing Phase 4 of Design Thinking with Scrum phases C, E and F, and PMBOK III and IV. Prototyping in Design Thinking is performed not only on the perspective of the project manager or the company in question but also on the perspective of the user, who participates in and influences the process. Such participation would be convenient in the execution, revision, and monitoring phases of the other methods. When considering the opinions and needs of the user, directly identified, we can attribute a higher value of the differential to the product and, consequently, to the company. Doing so creates a close relationship with the user and enables the company to meet their needs and demands more assertively.

In step 2 of Design Thinking, visual analysis and planning tools are used to prepare the project team for the solution generation phase. Such tools could be implemented in Scrum's phase B and PMBOK's phase II to provide a broader view of what the team needs to do and how they can do it. Likewise, the tools used in Phase 3 of Design Thinking can be adapted to Scrum phases C and D and phase III of the PMBOK, since it seeks to stimulate creativity from not only the tools used but also from a team with varied abilities, which allows the emergence of innovative ideas.

Concerning Scrum's phase G and PMBOK's phase V, we could not find any correspondent in the context of Design Thinking. This process can be completed more fully if the project team aligns it to one of the two other methods studied here, since both offer suggestions for closing projects once the team already solved the problem in question.

6 Conclusion

For a company to survive in the complex and dynamic market, modified by the new technologies, it is necessary to insert a new managerial vision in the company. According to the literature review, it was identified that companies are using design tools more often to solve internal and external challenges, since it is a relatively simple methodology with a high impact on results. In this way, adopting the management of integrated design to the management of the company allows a greater reach to success, considering both have the same objective.

Also, the application of design management in conjunction with business project management methods would bring more creative and consequently more innovative results, stimulating companies to use creativity as a tool to develop a more dynamic and competitive design culture.

Design Thinking, in addition to other project management methods, would contribute to building a new vision for project teams, providing collaborative work and different skills. This way, both people directly involved with the project itself (the developers), and people at whom results are aimed (customers and users) contribute to the final product or service. This type of participation would provide a more assertive scope in the results, since the projected products would be the result of the participation of a multidisciplinary team, composed of members like designers, engineers, managers, and suppliers involved, bringing the perception of the end-user as value and competitive differential.

Given the importance of the theme to ensure the correct use of available resources and the success of a company, future studies should be carried out on it. We recommend an applied and in-depth research with the simultaneous use of a business management method and a design management method. It is also essential to analyze how the company will manage the human resources that involve the multidisciplinarity of professionals within the same team. The results and conflicts that could arise from this relationship would be the subject of a new study.

References

1. Best, K.: Design Management Foundation. Bookman, Porto Alegre (2012)
2. de Mozota, B.B.: Design Management. Bookman, Porto Alegre (2011)
3. de S Libanio, C., Amaral, F.G.: Design management aspects addressed in dissertations and theses in brazil: a systematic review. Eletron. Sci. Mag. Prod. Eng. **11**(2), 565–594 (2011)
4. Vieira, S., Perassi, R.: Design management - branding image perception. Des. Stud. **21**, 1–21 (2013)
5. Fernandes, R.Q.K.: Design management in the toy industry (2015)
6. Silva, R.O.: Administration Theories. Pearson, São Paulo (2008)
7. Carvalho, J.E.: Business Management - Core Principles, 4th edn. Edições Sílabo, Lisboa (2016)
8. Daft, R.L.: Administration, 3rd edn. Cengage Learning, São Paulo (2018)
9. Sampaio, R.F., Mancini, M.C.: Systematic review studies: a guide for judicious synthesis of scientific evidence. Braz. Mag. Physiotherapy **11**(1), 83–89 (2007)
10. Brown, T.: A Powerful Methodology for Decreeing the End of Old Design Thinking Ideas. Elsevier, Rio de Janeiro (2010)
11. Steinbeck, R.: Design thinking as a creativity strategy in the distance. Educommunication Sci. Mag. **19**(37), 27–35 (2011)
12. de Araújo, S.M.M.., da S Júnior, J.E., de Figueiredo, L.F.G., de Sousa, R.P.L., Merino, A.D., Merino, Gi.S.A.D.: Design thinking as a tool for the internal business audience. vol. 8, pp. 273–285 (2013)
13. Buchanan, R.: Worlds in the making: design, management, and the reform of organizational culture. She Ji J. Des. Econ. Innov. **1**(1), 5–21 (2015)
14. Demarchi, A.P.P., Fornasier, C.B.R., de F Martins, R.F.: Design thinking and its visuals codes in the design management for the innovation: incremental innovation by sidme model. DAP Mag. **11**, 191–211 (2016)
15. Demarchi, A.P.P., Fornazier, C.B.R., Martins, R.F.D.F.: Design management humanized by design thinking from conceptual relationships. Projetica Mag. **2**(1), 19–36 (2011)
16. Schwaber, K., Sutherland, J.: Scrum guide - a definitive guide to scrum: the rules of the game (2017). https://www.scrumguides.org/index.html. Accessed 10 Dec 2018
17. Junior, A.F., do Prado, A.E., e Oliveira, A.C.M.D., Perucci, C.C., Junior, A.C.P.: Scrum practice analysis used in a multinational Brazilian company in the information technology sector. Iberoam. J. Proj. Manag. **7**(1), pp. 37–46 (2016)
18. Junior, W.A.M., de S Rodrigues, M.J., de M Souza, P.A., Nogueira, R.F.G.: Inventory control: process management using the Kanban tool supported by Scrum agile methodology. Res. Soc. Dev., **8**, 1–21 (2018)
19. Silva, L.H.F.P., Soares, L.S.: Evaluation of scrum adoption by software companies that market enterprise management systems. Braz. Mag. Manag. Eng. **7**, 137–165 (2018)
20. de Sousa, J.C.A.: Comparative Study of Agile Methodologies and PMBOK. Polytechnic Institute of Viseu (2018)
21. Alvarez, K.L.E.: Implementation of the PMBOK® methodology of the project management institute to improve productivity in the execution of MG trading SAC projects. Lince (2018)
22. de F Freitas, V.I.G.L.: Environmental management in the areas of knowledge and life cycle in projects as proposed for PMBOK: a review (2017)

23. Munk, R.: Project management practice: a case study of the adherence and dissonance of a Brazilian company system to PMBOK. Getúlio Vargas Foundation (2016)
24. Pinto, E.B., Vasconcelos, A.M., Lezana, Á.G.R.: PMBOK and CMMI approaches on successful software projects. Proj. Manag. Mag. GeP **5**, 55–70 (2014)
25. da S Silva, B.O., Boente, K.P., Boente, R.M.P.: Project Management Knowledge Areas: A Look at PMBOK. 5th Edn. EDU.TEC Scientific Digital Magazine of Faetec, vol. 7, No. 1 (2017)
26. Cuneo, M., et al.: Design management : diagnosis based on competitiveness, differentiation and sustainability in an association of artisans in Southern Brazil. In: IFDP 2016: Systems & Design: Beyond Processes and Thinking, pp. 593–612 (2016)
27. de F Martins, R.F.: Design Management as an Organizational Strategy - A Model for Integrating Design in Organizations. Federal University of Santa Catarina (2004)
28. Moreira, B.R.: Design management in practice: framework for enterprise implementation (2016)
29. Vianna, M., Vianna, Y., Adler, I.K., Lucena, B., Russo, B.: Design Thinking Business Innovation. MJV Press, Rio de Janeiro (2011)

Leadership in Social Business: A Needed Context

Ricardo Pereira(✉), Daniela de Oliveira Massad, Gertrudes Aparecida Dandolini, Gregory Falavigna, João Arthur de Souza, and Édis Mafra Lapolli

Federal University of Santa Catarina, Florianópolis, SC, Brazil
rikardop@gmail.com

Abstract. The definition of leadership and the search for an ideal leader model are recurrent concerns of organizations and the common theme in academic research. The present study is an example of this when analyzing the role of the leader in social affairs. These organizations are complex structures that demand more studies, since they have as a priority a social mission and an altruistic action, on the other hand, they seek the creation of economic value as a necessary condition to maintain financial sustainability. The analysis of the literature showed that the concept of social business is still poorly defined, and its limits to other fields of study remain obscure. Leadership in social business is evolving, lacking clarity of definition and appropriate tools to characterize, develop, and test it.

Keywords: Social enterprise · Social business · Leadership

1 Introduction

The world that we will meet in a few decades may not be the same. Social chaos, hunger, war, carelessness with the environment, poverty, depletion of natural resources, among many other ills, are current problems and need resolution, aiming at the preservation of the human species and the world that or fence. In this sense, one must seek to achieve economic, social, and environmental sustainability. In this context, the phenomenon of entrepreneurial practices motivated by social, environmental, cultural, and political purposes began to emerge. It has grown on a world scale and gained prominence to the point of beginning to arouse media interest, academic, and government [1].

The organizations originated from these entrepreneurial practices are social businesses (social enterprises, social enterprises, enterprises 2.5, inclusive businesses or impact businesses), and maybe hybrid organizations, which combine characteristics of non-profit organizations with the financial sustainability of traditional companies. The complexity of these organizations demands unique attributes for their managers, and the role of the leader becomes vital to the achievement of the mission by these social businesses.

The present study aims to identify the approach to the role of the leader in social business. Leadership in social business is a poorly studied subject, lacking theoretical

L. Pereira et al. (Eds.): IDEAS 2018, SIST 198, pp. 115–125, 2021.
https://doi.org/10.1007/978-3-030-55374-6_12

and empirical studies for the development of models and instruments of analysis. Thus, this paper identifies, in the literature, studies that address leadership and social business aiming to analyze the role of leadership in the achievement of the social business mission and which leadership profile has been appropriate for these organizations, from the literature found.

2 Theoretical Basis

2.1 Social Business

In recent years, organizations have developed complex structures. Social businesses (social enterprises or social enterprises) emerge as new organizational models that need further study. These organizations, also called "businesses with social impact" or "impact businesses" have the social mission as a priority, with the creation of economic value as a necessary condition to maintain financial sustainability. The non-profit, in most cases, and their altruistic performance characterize the peculiarity of these organizations.

Bengali economist and banker Muhammad Yunus won the Nobel Peace Prize in 2006, and his work in the field of microcredit in Bangladesh has become a benchmark for business models that create social impact as a mission.

A social enterprise is an "enterprise designed to solve a social problem" (p. 10), but which must be financially sustainable, not with the intention of generating profits, but with creating social value for the community and (customers, employees, suppliers, investors and the whole of society) and cover at least their operational costs [2].

While in a traditional business model, the main objective is to generate economic value for shareholders or owners, in a social business the focus is to create social value, the creation of economic value necessary only to maintain the financial sustainability of the enterprise [3] and push it further to achieve its social mission.

Table 1. The spectrum of social enterprise.

	Nonprofit organizations	Social enterprises	Traditional companies
Activity	Purely philanthropic	Mixed	Purely commercial
Method	Driven by mission	Driven by mission and market	Driven by the market
Goal	Create social value	Mixed value	Create economic value

Source: Adapted from Dees [4]

Table 1 presents the spectrum of social enterprise compared to nonprofits and traditional business and shows how social enterprises combine a mix of social values and goals with business practices. The activities of the social enterprise may be economical or not, but the mission is social change and development [5].

Some authors consider that social business can also make a profit; however, there are doubts about this, due to the tensions that arise while seeking to generate economic value, gain scale and productivity for the business and still achieve the desired social impact. A business performance approach can serve as an ally in supporting social initiatives to achieve financial results while remaining focused on the social mission of the business [6].

As for the social impact generated by social enterprises, from a static perspective, it is related to the efficiency of the solutions, goods, and services that social entrepreneurs offer. Under a systemic and dynamic approach to social impact, innovation is the central contribution of social entrepreneurs [7].

"Social entrepreneurs are agents of change who institute new patterns of value creation that other actors may adopt, ultimately realizing an even higher static impact. The impact of social entrepreneurship is thus, above all, a transformative one" [7] (p. 236).

Considering the peculiarity of social business and the contribution of the social entrepreneur to society, the role of the leader becomes essential for these organizations.

2.2 Leadership

"Leadership is the process of influencing others to understand and agree on what needs to be done and how to do it" [8] (p. 8); involves influencing a group of individuals who have a common purpose. It can be a small group of tasks, a group of communities, or a large group that covers an entire organization. Leadership is about an individual who influences a group of others to achieve common goals. Some researchers conceptualize leadership as a trait or a behavior, while others see leadership from a processing perspective or relational point of view [9].

Leadership takes into account characteristics, behaviors, influence, patterns of interaction, role relationships, and occupation of an administrative position [8]. In considering leadership, it is valuable to distinguish between person, position, and processes [10].

Research on the person as a leader tends to focus on skills, personality, engagement styles, gender differences, and behaviors of individual leaders. The second view of leadership occurs when the focus is on the formal position occupied by the leader in an organization. A third approach is to consider leadership as a set of processes or dynamics that occur among individuals, groups, and organizations. In this approach, leadership is a set of activities or processes related to motivating and influencing people and shaping and achieving results [10].

Over time, researchers have proposed many different theories and styles of leadership, as there is no particular theory or style that can be considered universal.

The great man theory, trait theories, contingency theories of leadership, situational theory, behavioral leadership theories, participatory leadership theories, affiliated leadership, authentic leadership, transactional leadership, transformational leadership, charismatic leadership, service leadership, ethics leadership, entrepreneurial leadership, distributed leadership, direct and indirect leadership, LMX theory, among others, are examples of theories and styles that attempt to explain the leadership phenomenon [11].

This diversity of theories means that situations, contexts, culture, work environment, new laws and regulations, information overload, organizational complexities, and psychosocial developments, notably impact the concept of leadership, making it compatible with changing organizational dynamics [12].

"There are almost as many different definitions of leadership as there are people who have tried to define it" [13] (p. 7). The definition of leadership and the search for an ideal leader model are recurrent concerns of organizations and the common theme in academic research. For example, in a survey (TITLE ("leadership")) in the Scopus database, we find about forty-five thousand articles that deal with the subject.

When it comes to the study of leadership and the role of the leader in social affairs, one notices the paucity of research that deals with the interconnection between the two themes or their inadequate exploration of the role of leaders in training, growth, effectiveness, and decline of these organizations.

While there are several similarities between leaders in the field of social enterprise and leaders in the private, public, and nonprofit sectors, the levels of complexity, ambiguity, and lack of established theoretical and practical knowledge base make creating leadership in the business sector much more challenging [14].

It is with this instigation that we have the direction of the present study, which is to verify in the corresponding literature the interconnection of the topic of leadership with social business and to ascertain the role of the leader in the achievement of his organizational objectives.

3 Methodology

The present article conducts exploratory research to evidence the studies carried out on leadership in social business, using a theoretical-conceptual methodology. The purpose of this methodology is to construct a panorama on the subject, providing a basis for future studies, that seek the improvement of the available concepts [15].

As for the data source, the study will be of the bibliographic type; a mode of study that uses documents such as: books, periodicals, encyclopedias, critical essays, dictionaries and scientific articles. This type of study has as differentiating characteristic the fact of appealing to scientific sources, and not directly to the facts of the empirical reality [16].

In this sense, in the first moment, we define the search strategy, the descriptors, and the database, then we establish the inclusion and exclusion criteria, calling this step of data collection. In the next step, called data selection, we read summaries, keywords, and titles of publications, organize related studies, and identify the selected studies. We conclude this stage by categorizing the selected studies.

Finally, we present the analysis, interpretation, and discussion of the results, the revision and synthesis of the acquired knowledge, proposing future studies.

3.1 Data Collection

Between January and February 2019, we selected the articles to compose the basis of analysis. We divide this stage into two phases; the first one was for (a) pre-selection of articles and the second for (b) filtering of those articles.

The pre-selection stage of articles consists of three activities: defining the keywords, defining the database, and fetching the articles in the database with the keywords. The pre-selection of the articles begins with the contextualization of the theme, describing how the researchers understand it. This characterization of the theme creates a singularity because it delimits and particularizes its content. The ambiguity of the understanding of the research topic may provide a deviation from the focus of the research.

The present study aims to answer the following question: What is the role of the leader in social business?

The keywords chosen for the present study were: leader*, "social business", "social entrepreneur", and "social entrepreneur*". Through the combinations of keywords were searched in the Scopus database and The Leadership Quarterly, using the following Boolean equation: ("leader *" AND "social enterprise" OR "social enterprise" or "social entrepreneur *").

The definition for the chosen database considered its relevance in the area of administration and social sciences, also justified by the multidisciplinarity of the publications stored. Besides we decided to do separate research in The Leadership Quarterly as the leading magazine on leadership.

3.2 Selection of Data

At this stage, we searched for articles that incorporated the title, abstract, and keywords of the fields, without temporal delimitation, we obtain 329 scientific articles. However, we found that the articles had no direct relationship between leadership and social business. In a continuous process, we opted for the search only by title, obtaining a table of publications, according to Table 2, which brought the interconnection between the two researched topics.

Table 2. Articles selected by database.

Database	Number of articles
Scopus	36
The leadership quarterly	1
Total	37

Source: Authors of Research (2019)

Selection of articles in the database Scopus considered the number of quotes and reading their summaries. Some documents had the same number of citations since they had not been cited previously, retrieving 15 articles from the Scopus database and an article from The Leadership Quarterly. Table 3 shows the articles selected for analysis.

Table 3. Articles of the bibliographic Portfolio.

Year	Author	Title/Source	Cited by
1999	Prabhu	Social entrepreneurial leadership. Career Development International	79
2012	Smith; Besharov, Wessels, Chertok	A paradoxical leadership model for social entrepreneurs: Challenges, leadership skills, and pedagogical tools for managing social and commercial demands. Academy of Management, Learning & Education	60
2013	Felicio, Gonçalvez, Gonçalvez	Social value and organizational performance in non-profit social organizations: Social entrepreneurship, leadership, and socioeconomic context effects. Journal of business research	53
2010	Ruvio, Rosenblatt, Lazarowitz	Entrepreneurial leadership vision in nonprofit vs. for-profit organizations. The Leadership Quarterly.	43
2010	Litzky, Godshalk, Bongers	Social entrepreneurship and community leadership: A service-learning model for management education, Journal of Management Education	20
2012	Maak, Stoetter	Social Entrepreneurs as Responsible Leaders: 'Fundación Paraguaya' and the Case of Martin Burt, Journal of Business Ethics	13
2008	London	Leadership and Advocacy: Dual Roles for Corporate Social Responsibility and Social Entrepreneurship, Organizational Dynamics	11
2018	Thorgren, Omorede	Passionate Leaders in Social Entrepreneurship: Exploring an African Context. Business & Society	7
2018	Muralidharan, Pathak	Sustainability, transformational leadership, and social entrepreneurship. Sustainability	7
2014	Mass, Seferiadis, Bunders, Zweekhorst	Social entrepreneurial leadership: Creating opportunities for autonomy	3

(*continued*)

Table 3. (*continued*)

Year	Author	Title/Source	Cited by
2014	Tyan, Smith	Entrepreneurial leadership of social enterprises: Challenges and skills for embracing paradoxes. Journal of Leadership Studies	3
2018	Newman, Neesham, Manville, Tse	Examining the influence of servant and entrepreneurial leadership on the work outcomes of employees in social enterprises. The International Journal of Human Resource Management	2
2019	Kelly, Lee	Cultural leadership ideals and social entrepreneurship: an international study. Journal of Social Entrepreneurship	0
2019	Erpf, Ripper, Castignetti	Understanding Social Entrepreneurship Based on Self-Evaluations of Organizational Leaders–Insights from an International Survey. Journal of Social Entrepreneurship	0
2018	Pasricha, Singh, Verma	Ethical Leadership, Organic Organizational Cultures and Corporate Social Responsibility: An Empirical Study in Social Enterprises. Journal of Business Ethics pp. 941–958, (2018)	0
2018	Pasricha, Raom	The effect of ethical leadership on employee social innovation tendency in social enterprises: Mediating role of perceived social capital. Creat Innov Manag.	0

Source: Survey data (2019)

4 Analysis and Discussion of Results

This section shows the results we get from research in terms of the knowledge built in the researcher to understand and identify research opportunities through a selection of articles.

Preliminary to the analysis of selected articles, it is essential to make explicit some findings. Of the articles selected, at least 30% of them have been published in the last two years. These data indicate that although leadership studies are exhaustive and numerous, their application to entrepreneurship/social affairs is still laconic [17] and recent.

The characterization of social business presents some distortions. Numerous are the conceptual influences, understandings, and nomenclatures, characteristics of a new and evolving field. We have identified various denominations related to social businesses, such as: social enterprises, 2.5 companies, pyramid base companies, impact businesses, or inclusive businesses. Despite the terminological divergence, one of the points of consensus is that there are social enterprises to solve socio-environmental problems.

What sets ordinary entrepreneurship apart from social entrepreneurship is simply motivation – money stimulates the first group; altruism stimulates the second. An excellent social leader must play essential roles and have well defined strategies to mobilize people because their actions make the difference to obtain positive results in the desired objectives [18].

Social enterprises that apply business practices to achieve social goals and social entrepreneurs are considered agents of change. They bring about social change, with the mission of finding solutions to social, environmental, and educational problems. The purpose of social enterprise is not to generate profit, but to provide quality of life [19]. Social entrepreneurs organize entrepreneurial social networks in their communities and represent the leading role in creating social and economic value [20].

The studies analyzed can be identified from two perspectives: on the one hand, a set of articles that relates some style of leadership to social affairs [22–25]; other articles attempt to identify the ideal traits [5, 26, 27] and the role of entrepreneurial social leaders in the formation, growth, functioning, effectiveness, decline, and closure of entrepreneurial social organizations [5].

Also the creator of entrepreneurial social organizations can exhibit many of the characteristics and behaviors of the classical business leader in the process of creating and managing their organizations [5]. The challenges and associated skills to effectively manage the emerging stresses of social mission and business outcomes require a unique set of skills beyond those necessary to achieve an organization's business goals.

The importance of the role of the leader in organizations and the role played by social entrepreneurs in social enterprises is emphasized, communicating the mission of the organization effectively, to achieve its social objectives.

Other authors, however, point out that passionate leadership is related to transformational leadership, and this, in turn, has an essential influence on entrepreneurial action to fulfill social mission [25]. This passion can strengthen the organizational power of social entrepreneurship through more significant resource mobilization, commitment among members of the organization and legitimacy in society; contribute to creating higher social value, anchoring the social mission among the target public; and emphasize the entrepreneur's own identity, motivating and encouraging him to remain engaged in the social business [26].

In this sense, we have identified some leadership theories, such as the transformational one, which states that the leader's ability to motivate subordinates and achieve results higher than expected causes the members of the organization to achieve higher levels of motivation and morality by interacting with others [27, 28], responsible leadership [24] and ethical leadership, where leaders should be an example to be followed, their actions should always be guided by ethics [21].

Besides ethical leadership attitudes influence the behavior of the organization's employees. Ethics is essential for organizations to have transparency and credibility, bringing more positive results in decision making, providing a competitive differential [29].

Finally, we can see in the analysis the miscegenation of the concepts of leadership and entrepreneurship. The will of some authors to coin a new theory of leadership, by some denominated as entrepreneurial social leadership, takes advantage of the characteristics of the social entrepreneur and characterizes him as a leader.

In conclusion, some authors indicate that there are business opportunities when the market is not in equilibrium, and entrepreneurs discover and act on them, bringing the market back to equilibrium [30]. Although this statement was four decades ago, the context of social, environmental and economic imbalance makes it more current than ever, highlighting the emergence of social business and the importance of the leader in this context.

5 Conclusions

The concept of social business is still poorly defined, and its limits to other fields of study remain obscure. It may be a problem for some, for others, it can be seen as a unique opportunity for researchers from different fields and disciplines.

Leadership in social business is evolving, lacking clarity of definition and appropriate tools to characterize, develop, and test it.

This paper aimed to identify the role of the leader in social business studies. The analysis showed that the social leader needs to have well-defined strategies to mobilize people and achieve desired goals, and his role as a social advocate is vital in communicating the organization's mission effectively. Entrepreneurial action is influenced by the leader's passion for fulfilling the social mission. The social leader can motivate subordinates and becomes an example to be followed, through their actions guided by ethics and transparency.

It is evident from the analysis that leadership is particularly valuable in companies where the individual is essential and where the environment is characterized by constant change. In this context, not only the leader but also the followers need to be involved in exploring and taking advantage of the opportunities and in achieving the organizational goals.

The analysis identifies a convergence between the fields of entrepreneurship and leadership. For a more in-depth analysis of both constructs, we suggest empirical deepening through case studies that demonstrate leadership styles and leadership profiles that fit into these new organizational models.

References

1. André, K., Pache, A.: From caring entrepreneur to caring enterprise: addressing the ethical challenges of scaling up social enterprises. J. Bus. Ethics **133**(4), 659–675 (2016)
2. Yunus, M., Moingeon, B., Lehmann-Ortega, L.: Building social business models: lessons from the grameen experience. Long Range Plan. **43**(2–3), 308–325 (2010)
3. Mair, J., Marti, I.: Social entrepreneurship research: a source of explanation, prediction, and delight. J. World Bus. **41**(1), 36–44 (2006)
4. Dees, J.G.: Enterprising nonprofits. Harvard Bus. Rev. **76**, 54–69 (1998)
5. Prabhu, G.: Social entrepreneurial leadership. Career Dev. Int. **4**(3), 140–145 (1999). https://doi.org/10.1108/13620439910262796

6. Mair, J., Sharma, S.: Performance Measurement and Social Entrepreneurship. In: Volkmann, C., Tokarski, K., Ernst, K. (eds.) Social Entrepreneurship and Social Business. Gabler Verlag, pp. 175–189 (2012)
7. Beckmann, M.: The impact of social entrepreneurship on societies. In: Volkmann, C., Tokarski, K., Ernst, K. (eds.) Social Entrepreneurship and Social Business. Gabler Verlag, pp. 235–254 (2012)
8. Yukl, G.: Leadership in Organizations, 7th edn. Prentice Hall, Upper Saddle River (2010)
9. Northouse, P.: Leadership: Theory and Practice, 3rd edn. Sage Publications, Thousand Oaks (2004)
10. Hartley, J., Hiksman, B.: Leadership Development: A Systematic Review of the Literature. NHS Leadership Centre, London (2003)
11. Amanchukwu, R.N., Stanley, G.J., Ololube, N.P.: A review of leadership theories, principles and styles and their relevance to educational management. Management **5**(1), 6–14 (2015)
12. Amabile, T., Schatzel, E., Moneta, G., Kramer, S.: Leader behaviors and the work environment for creativity: perceived leader support. Leadersh. Q. **15**, 5–32 (2004)
13. Stogdill, R.M.: Handbook of Leadership A survey of the Literature. New York Free Press (1974)
14. Jackson, B., Nicoll, M., Roy, M.: The distinctive challenges and opportunities for creating leadership within social enterprises. Soc. Enterp. J. **14**(1), 71–91 (2018)
15. Forza, C.: Survey research in operations management: a process-based perspective. Int. J. Oper. Prod. Manag. **22**(2), 152–194 (2002). https://doi.org/10.1108/01443570210414310
16. Oliveira, M.: Como fazer Pesquisa Qualitativa. Petrópolis, Rio de Janeiro: Vozes (2007)
17. Brunelli, M.: Empreendedorismo social e liderança: revisão da literatura de 2006 a 2017. Cadernos de gestão e empreendedorismo (2018)
18. Kelly, L., Lee, B.: Cultural leadership ideals and social entrepreneurship: an international study. J. Soc. Entrepreneurship **10**(1), 108–128 (2019)
19. Erpf, P., Ripper, M., Castignetti, M.: Understanding social entrepreneurship based on self-evaluations of organizational leaders-insights from an international survey. J. Soc. Entrepreneurship **10**(1), 108–128 (2019)
20. Mass, J., Seferiadis, A, Bunders, J., Zweekhorst, M.: Social entrepreneurial leadership: creating opportunities for autonomy. Johns Hopkins Research Series on Social Entrepreneurship, pp. 223–255 (2014)
21. Pasricha, P., Singh, B., Verma, P.: Ethical leadership, organic organizational cultures and corporate social responsibility: an empirical study in social enterprises. J. Bus. Ethics **151**, 941–958 (2018)
22. Smith, W., Besharov, M., Wessels, A., Chertok, M.: A paradoxical leadership model for social entrepreneurs: challenges, leadership skills, and pedagogical tools for managing social and commercial demands. Acad. Manag. Learn. Educ. **2**(3), 463–478 (2012)
23. Tyan, Y., Smith, W.: Entrepreneurial leadership of social enterprises: Challenges and skills for embracing paradoxes. J. Leadersh. Stud. **8**(3), 42–45 (2014)
24. London, M.: Leadership and advocacy: dual roles for corporate social responsibility and social entrepreneurship. Organ. Dyn. **37**(4), 313–326 (2008)
25. Muralidharan, E., Pathak, S.: Sustainability, transformational leadership, and social entrepreneurship. Sustainability **10**(1), 567–589 (2018)
26. Thorgren, S., Omorede, A.: Passionate leaders in social entrepreneurship: exploring an african context. Bus. Soc. **57**(3), 481–524 (2018)
27. Litzky, B., Godshalk, V., Bongers, C.: Social entrepreneurship and community leadership: a service-learning model for management education. J. Manage. Educ. **34**(1), 142–162 (2010)
28. Felicio, J., Gonçalvez, H., Gonçalvez, V.: Social value and organizational performance in non-profit social organizations: social entrepreneurship, leadership, and socioeconomic context effects. J. Bus. Res. **66**(10), 2139–2146 (2013)

29. Pasricha, P., Raom, P.: The effect of ethical leadership on employee social innovation tendency in social enterprises: mediating role of perceived social capital. Creat. Innov. Manag. **27**(3), 270–280 (2018)
30. Kirzner, I.: Competition and Entrepreneurship. University of Chicago Press, Chicago (1973)

The Positive Impacts of Using the Service Design Approach for Expanding the Innovative Potential on Business

Giovanna Silva(✉), Ricardo Gaspar, Júlio Francisco Blumetti Facó, and Alexandre Acácio de Andrade

Federal University of ABC, Santo André, SP, Brazil
gdebelis@ufabc.edu.br

Abstract. With the increasing influence of the Experience Economy, the conceptual differentiation of the terms Goods and Services has lost relevance in the context of generating value for the final customer. Culture, structure and strategy centered on the client (internal and external), regardless of their size and time of life, stand out. It is in this scenario that the Service Design approach stands out as a strategy for the development of innovative solutions and initiatives in companies of all sizes, due to their low cost and complexity of implementation. The present study seeks to structure, based on bibliographic review (academic and market) and case studies, a deeper understanding of a framework proposed based on design approaches to help companies understand the potential positive impacts it can have on the business strategy, sustainability and innovative potential.

Keywords: Service design · Experience economy · Small business · Innovation

1 Introduction

"When you have two coffee shops, side by side on a street, both selling the same coffee, for the same price, the service design is the factor that leads the decision about which one are you going to come into." [1]

With the continuous growth in technology and information, the speed in which the market changes is faster than ever. We are now on an era of the economy where not only the distinction between goods and services lost its value, the extension of their deliveries is no longer enough to foster economic growth, create and maintain jobs and keep the economic prosperity [2]. To keep growing and evolving their reach to meet customer needs, companies must evolve on the capabilities that allow them to stage experiences and create value on new and innovative ways.

A complementary aspect that reinforces the importance of this discussion is the context of entrepreneurship in Brazil. While micro and small companies hold 70% of the jobs [3], only 27% of the GDP (Gross Domestic Product) comes from this same group [4]. In addition to those numbers, about 20% of all small business started close in the first two years [5]. Some aspects that may influence that scenario can be related

L. Pereira et al. (Eds.): IDEAS 2018, SIST 198, pp. 126–135, 2021.
https://doi.org/10.1007/978-3-030-55374-6_13

to the reasons that often cause those companies to fail. When talking about mortality in micro and small companies, based on researches conducted by Sebrae (the national institute that supports and provides data about those companies), some of the main reasons appointed by the report [5] are:

- The previous status of the entrepreneur (employed or unemployed);
- The previous experience of the entrepreneur on the company's sector;
- The motivation to start a business;
- The previous planning before starting the business;
- The quality of the management.

As for medium and big companies, the incorporation of concepts from design and services can come as a new way to look at both the trends of the market and the needs from all parts involved in performing a better service or solution to a specific need, evolving the company's potential to innovate. This evolution in strategy can be crucial for long time competitiveness.

On this context, the service design approach comes as a complement for the traditional tools and fields on business management and design, bringing new optics to the strategic design of companies and services. The fundamental concept on the approach is to seek and evolve on a deeper understanding of the parts involved on the delivery of a service, and to completely design the experience with a process based on iteration and co-creation with all parts involved [1].

The purpose of this work is to stimulate companies to incorporate this new design approach by connecting its principles and concepts with the potential positive impact on the business's strategy, competitiveness and sustainability.

2 Literature Review

2.1 Service Design

The service design approach is an interdisciplinary field, where the objective is to help designing services that create maximum value to the people involved on the process and the society. In order to do so, it connects and explores several aspects of the organization, the environment, the economy and a deep understanding of costumer's, employees' and other stakeholder's needs [6].

An important highlight here is that, in service design, the focus of the co-creation is expanded from a customer centric to a human centric approach. It does so by considering all real stakeholders on the process, and designing better solutions for each one of them. That is relevant because when services and solutions are created based on people's needs, it is easier to create real value [6].

"The biggest missed opportunity in the development is that organizations don't think about their customers as valuable, productive assets in the delivery of a service, but as anonymous consumers of products" [6].

The practical side of the service design approach makes use of design frameworks and tools for the development of better solutions, and, in doing so, it presents synergy to the design thinking methodology on the ideations phase of the solution, suggesting a sequence of steps to be executed in order to better co-create with costumer and other stakeholders.

Other relevant principles when talking about service design are [6]:

- The Service Framing
  To analyze the company's value creation as a whole and treat it as a service and a full experience.
- The Human Centricity
  The service design approach drives the innovation focus from the customer to all stakeholders involved on any level with providing the service.
- The Qualitative Approach
  In design, the quantitative approach is important to provide paths in the beginning of projects and context analyses. But once the main challenges are uncovered, the qualitative research takes place to allow deeper understanding of needs of the people involved in the process and provide richer insights.
- The Ecosystem Approach
  To better design services and experiences, it is fundamental to understand all the connections between the designed service and the environment, as well as the economic aspects of it.

As we discuss the impact of the service design approach on organizations and their innovative capabilities, the concepts shown above are some of the key insights that can change the way business models and value propositions are developed.

2.2 Design Thinking

As we talk about design, it is relevant to bring up some context about design thinking. Design thinking is an approach from the design field that proposes a structured methodology for solving complex problems [7, 8].

When put in practice, the design thinking approach derives in a methodology to enable a structured (but flexible) path to the design and co-creation of creative solutions.

The methodology is usually set out as a double diamond shaped framework, where phases of "diversion" (moments where the project team is stimulated to think of different solutions and ideas, amplifying the scope of the research) are intercalated with phases of "conversion" (when the project team is encouraged to reduce the number of alternatives and select one way to pursue).

Each of the phases has specific objectives and deliverables, guiding the group to the solution while engaging collaboration, empathy and co-creation.

3 Discussion

As discussed earlier, when working with service design, we bring tools, methods and methodologies from de design field, and adapt them, in order to better serve business needs.

That way, the present framework was designed based on the knowledge available on several studies available on both service design and design thinking fields, in order to provide a baseline for the discussion on innovation, competitivity and sustainability of business.

Based on a design approach, the framework is divided into six sequential and iterative phases, each one with a different deliverable, in terms of understanding and creating an innovative solution. To better analyze and discuss the impacts of the approach, each step is followed by a strategic analysis of the exercise proposed.

The six phases on applying the service design approach to an organization are:

1. Understanding the Challenge
2. Understanding the Need
3. Defining Design Principles
4. Ideating Solutions
5. Prototyping & Testing
6. Implementing Changes

It is important to notice that, just like in design thinking, the idea behind the phases is that the group working on the framework should have completed each phase before moving on to the next one. That does not mean finished sessions cannot be revisited, it only establishes touch points with deliverables in which is possible to analyze the evolution of the work and assure the assertiveness of the next phases.

3.1 Phase 1: Understanding the Challenge

For the first step on the development of new initiatives, it is required to fully understand the context to which the company wishes to develop the solution for. This means knowing the value the company has to offer, knowing the history of the market and understanding the parts involved on the problem and their role on it.

3.1.1 Outcomes

After the conclusion of the studies on phase 1, the company should be able to:

- Understand the market, the opportunity gaps existing, and indicate what market need they want to address when designing the solution.
- Understand the historic context of the market and identify future tendencies.
- Have a map of the stakeholders involved on the context and identify their role and relevance on it.

3.1.2 Strategical Impacts

Strategic changes in approach to projects may already be noted in the first step. The approach of starting the project with a desk research can provide the company with a more forward-looking and market-oriented vision and can act as an essential element in the decision-making process on the path to be taken throughout the initiative.

In addition, the development of the stakeholder map is an essential strategic artifact to support decision making and analysis of possible partnerships and new value propositions that contribute to the ecosystem of the company.

It is important to note that this phase can also generate potential innovation gains by broadening benchmarking research to similar scenarios, providing relevant market insights and trends.

3.2 Phase 2: Understanding the Need

After obtaining a clear view of the context and the opportunities they would like to address, the group should focus on understanding how the stakeholders interact in this scope, what are their particular needs and what they consider valuable or challenging in the current situation of the selected problem.

When talking about services and experiences, it is essential to understand the human aspects of the delivery, such as behavioral science principles responsible for underlying human interactions [9].

3.2.1 Outcomes

After the conclusion of the studies on phase 2, the company should be able to:

- Understand the customers needs.
- Understand other stakeholders needs.
- Understand what is perceived as valuable or challenging in the context (opportunities).

3.2.2 Strategical Impacts

Phase 2 represents one of the highest gain potentials in the short, medium and long term.

One of its main characteristics corresponds to the greater inclusion of human factors in the project, and the extension of the scope of the research with users (end users, users involved in the process, or service providers).

This phase usually yields gains beyond the results of research and insights, such as increased employee awareness and engagement, a sense of customer appreciation, and greater prospects for opportunities in the ecosystem beyond the usual value propositions. In this way, it can represent strong gains in brand and change management.

3.3 Phase 3: Defining Design Principles

The third phase is one of the most important on the framework, because it establishes an important turning point between understanding and creating. It is in this moment that all the research made starts making connections and establishing the delimitations to the creation phases.

Design principles are predefinitions about what has to be considered when designing the solutions. Based on the research made, this phase links the discoveries about the market needs and tendencies, and people's needs. Thinking from service design optics, it is essential to consider the particular needs of all stakeholders directly involved in the process – customers, employees, society, and so on.

At this point, the group should also be able to think about ways to measure the success of the solution according to the challenge selected and the expectations of the stakeholders. The metrics selected here will also be used as a support for prioritization of the solutions.

3.3.1 Outcomes

After the conclusion of the studies on phase 3, the company should be able to:

- Specify design principles from the point of view of the different stakeholders previously mapped.
- Understand the design principles as guides to the ideation of possible solutions.
- Select metrics to prioritize the initiatives and measure the success of the solutions.

3.3.2 Strategical Impacts

The third step proposed by the framework can be considered an essentially strategic element. The objective of the phase is to align objectives among the internal stakeholders of the process, and the metrification of results and gains from its evolution.

In terms of strategic artifacts, design principles at that time can represent a strong element of composition in tactical and strategic plans, facilitating the alignment of efforts and objectives throughout the organization, and interacting with strategic design discipline (application of design thinking for the development of business strategies).

3.4 Phase 4: Ideating Solutions

On phase four, the group can start thinking about ideas to better answer the mapped needs. The design principles previously defined on phase three are the main source for guidance on the ideation processes.

A key point to this phase is to allow the group to ideate on a free environment. The only important things to keep as rules are the design principles. Other than that, the team can think about simple, complex, quick or long solutions.

At the end of this phase, the group should select an idea to be prototyped and tested. The selection of the idea should be aligned to the metrics defined on phase 3, this way, the idea selected should be able to bring the company as closer as possible to the goal of better solving the problem pursued.

3.4.1 Outcomes

After the conclusion of the studies on phase 4, the company should be able to:

- Have a list of ideas on how to address the problem selected that respect the delimitations given by the design principles.
- Select an idea (or a combination of ideas) based on its capacity to bring results on the metrics selected on phase four.

3.4.2 Strategical Impacts

After developing design principles aligned with company strategy and project insights, the fourth step ensures the involvement of the project team in the design of solutions for greater value generation.

The way this step is conducted can enable greater strategic and creative participation of different levels of teams in the solution of a need of the organization, besides having the potential of dissemination of mindset, culture and strategy, making possible innumerable gains of satisfaction and recognition.

3.5 Phase 5: Prototyping and Testing

On phase five, the goal is to test the market with the idea selected, in order to check the route. To do so, it is necessary to develop a sample of the solution (prototype), so that it is possible to identify the consumer's responses to it.

After the development of a prototype, it should be tested separately on each group of stakeholders, to guarantee the evaluation of all design principles.

3.5.1 Outcomes

After the conclusion of the studies on phase 5, the company should be able to:

- Capture feedback from the public participant on the tests.
- Measure the success of the solution (the one that has been prototyped and tested) according to the metrics previously defined.
- Make adjustments on the solution, if needed, based on the feedback received.

3.5.2 Strategical Impacts

The fifth step is strategically relevant for providing learning and enabling solution testing in a structured and secure manner. When it comes to capacity to innovate, this is one of the most learning-intensive stages for the entire organization.

The way in which the projects are conducted and developed in teams that aim at prototyping provides an environment focused on the development of the solution and dedicated to the real and empirical evaluation of its applicability. For small businesses or organizations with high risk aversion, innovation can pose a major challenge in terms of balance between risk appetite and development of new and deferred products and services. The culture of prototyping and testing can present a healthy path for the inclusion of a sustainable approach to internal innovation.

3.6 Phase 6: Implementing Changes

The idea on phase six is to provide ways in which the company can integrate the initiatives that had good acceptance on the market to their business, with the ability to be sensitive to feedback from the customers and employees about its success and further needs to adjustments.

In addition to the implementation of the solution, another important deliverable on this phase is the feedback from the execution of the framework to the company. It is important that the process executed on those six phases be gradually internalized on the way the entrepreneurs and companies operate on a day-to-day basis. That is comprehended in several aspects over the strategic, tactical and operational levels of the organization, for example, culture, strategic planning, project management, and so on.

3.6.1 Outcomes

After the conclusion of the studies on phase 6, the company should be able to:

- Implement the solution tested, if desired, or go back to the framework to test new solutions (iterate).
- Start (or evolve on) the process of internalizing the main strategic exercises on the framework (start with understanding the problem, design solutions considering all stakeholders and their needs, and other important tools).
- Disseminate the service design and design thinking mindsets over the organization through leadership, culture, objectives and aligned incentives.

3.6.2 Strategical Impacts

Among the main objectives of the sixth and last stage, we can mention:

1. The evolution of initiatives from projects to operation.
2. The crucial moment of change management and cultural change within companies.

The way the final stages of initiatives are conducted within a company (whether small or large) influences the adherence of mindset and behavior that will be disseminated in the form of culture in the organization.

Thus, it is important that all factors listed here as strategic impacts are understood and internalized by managers, so that this relevance is passed on and understood throughout all levels of employees.

It is also important to note that team involvement in projects reinforces the meaning of teamwork, collaboration, co-creation and internalizes the sense of purpose and responsibility in their day to day activities, representing one of the main strategic and sustainability gains organizations.

4 Conclusions

With constantly changing technology and its consequent impact on substantial market changes and consumer behavior, it becomes increasingly important for companies to turn their strategy to tapping the human potential available to their employees and society. The service design approach offers visions, tools and methods to facilitate and optimize the application of this new mindset, quickly, strategically, inclusively and at low associated cost.

The present work aimed to provide inputs for strategic decision making in organizations that stimulate the application of this new approach internally in their businesses, in a structured, systematic and sustainable way, aiming to provide subsidies for the substantial expansion of their competitiveness and capacity to innovate sustainably.

As demonstrated in the discussion session, the service design approach is deeply connected to business strategy, having the potential to have impact several dimensions of a business model - increasing and consolidating the customer base, improving relationship opportunities with its clients and stakeholders, building and expanding strategic partnerships, developing new channels, and more.

For future work on the subject, we suggest going further to the definition of tools and methods for each phase, testing the methodology on entrepreneurs and companies and reporting the process, and mapping more deeply and with metrics its success and opportunities. The main objective of the project is to provide entrepreneurs tools and strategic understanding of how to use them to the development of more innovative, sustainable and competitive business models.

Acknowledgments. *I would like to thank my family and friends for the unconditional support I received at every stage of my life and career and in this work.*

Especially my mother, first supporter and defiant and eternal mentor and coach.

To my father and grandmother, for the opportunity and gift to discover and work from an early age on my passion and purpose.

And to my teachers and co-authors, for the opportunity given, and for all the support, shared knowledge and words of encouragement that led to the development of this work and many others to come.

References

1. Stickdorn, M., Schneider, J., Andrews, K., Lawrence, A.: This is Service Design Thinking: Basics, Tools, Cases, 1st edn. Wiley, Hoboken (2011)
2. Pine, B.J., Gilmore, J.H.: The Experience Economy. Harvard Business Press, Boston (2011)
3. Estadão. (22 de Fevereiro de 2019). Pequena empresa responde por 70% dos empregos. Fonte: Journal Estadão. https://economia.estadao.com.br/noticias/geral,pequena-empresa-responde-por-70-dos-empregos,70001963654
4. Sebrae. (22 de Fevereiro de 2019). Micro e pequenas empresas geram 27% do PIB do Brasil. Fonte: Sebrae. http://www.sebrae.com.br/sites/PortalSebrae/ufs/mt/noticias/micro-e-pequenas-empresas-geram-27-do-pib-do-brasil,ad0fc70646467410VgnVCM2000003c74010aRCRD
5. Sebrae.: Sobrevivência das Empresas no Brasil. Sebrae, Brasília (2016)
6. Polaine, A., Løvlie, L., Reason, B.: Service Design: From Insight to Inspiration. Rosenfeld Media, New York (2013)
7. Brown, T., Katz, B.: Change by design. J. Prod. Innov. Manag. **28**(3), 381–383 (2011)
8. Cook, L.S., Bowen, D.E., Chase, R.B., Dasu, S., Stewart, D.M., Tansik, D.A.: Human issues in service design. J. Oper. Manag. **20**, 159–174 (2002)
9. Goldstein, S.M., Johnston, R., Duffy, J., Rao, J.: The service concept: the missing link in service design research? J. Oper. Manag **20**, 121–134 (2002)
10. Benson, P.G., Saraph, J.V., Schroeder, R.G.: The effects of organizational context on quality management: an empirical investigation. Manag. Sci. **37**, 1107–1124 (1991)
11. Teixeira, J., Patrício, L., Nunes, N.J., Nóbrega, L., Fisk, R.P., Constantine, L.: Customer experience modeling: from customer experience to service design. J. Serv. Manag. **23**, 362–376 (2012)

12. Hill, A.V., Collier, D.A., Froehle, C.M., Goodale, J.C., Metters, R.D., Verma, R.: Research opportunities in service process design. J. Oper. Manag. **20**, 189–202 (2002)
13. Sangiorgi, D.: Building a framework for service design research. In: 8th European Academy of Design Conference, pp. 415–420. GBR (2009)

Low Cost Solution for Home Brewing and Small Brewing Business Using Raspberry Pi

Alexandre Acácio de Andrade[1](✉), Álvaro Batista Dietrich[2], Júlio Francisco Blumetti Facó[1], and Ricardo Reolon Jorge[1]

[1] Universidade Federal do ABC, Bairro Jardim Antares, Rua São Paulo, s/n, São Bernardo do Campo, São Paulo, Brazil
{aacacio,julio.faco,reolon.ricardo}@ufabc.edu.br
[2] Universidade Federal do ABC, Bairro Santa Terezinha, Avenida dos Estados, Santo André, São Paulo 5001, Brazil
alvaro.dietrich@ufabc.edu.br

Abstract. In several countries artisanal beer producers are common and this tendency can be observed also in Brazil where a significant growth of the sector in recent years took place. This work explores an automation option directed to the production of small batches of beer where the important process variables such as times and temperatures are controlled. The implementation of the proposed automation makes use of low cost hardware, the Raspberry Pi 3, along with free software, the CraftBeerPi which easily allow the improvement of the quality and uniformity of brewing process. Since the hardware used has network communication capabilities, it is possible to remote monitoring the process, through "IoT" technology, which allows real-time checking of variables and alerts on all stages of the production process.

A production line capable of producing 20 L of beer per batch was set up to perform the tests and prove the concepts, making it simple and effective to follow up and control the process variables by a mobile phone. The conclusions of this work are that the presented automation option has low cost and high flexibility, which makes it an innovative and attractive solution for artisans and small brewing business who seek to reproduce and compose beer recipes that would previously only be possible with the use of sophisticated and expensive equipment.

Keywords: Brewing process · Low-cost automation · Raspberry Pi · CraftBeerPi

1 Introduction

Beer is the most consumed alcoholic beverage in the world and has been following man for more than 10,000 years. Over time, the beverage was modified according to the regions of production and customs of each era that it crossed, receiving new ingredients and undergoing changes in its manufacturing process leading to the great diversity of methods and styles of production, which continues to grow today.

One of the factors in the recent history of beer that has led to this rise in diversity is the home brewing. This culture gained momentum in the early 1980s in the United

L. Pereira et al. (Eds.): IDEAS 2018, SIST 198, pp. 136–145, 2021.
https://doi.org/10.1007/978-3-030-55374-6_14

States of America and today is a very popular hobby in that country, since the brewing culture in the United States goes back to the beginning of its colonization by the English. The English school of brewing is one of the most important in the world (together with the German and Belgian schools).

Automation of manufacturing processes is essential for industrial competitiveness in order to reach minimal cost, product quality assurance and flexibility in production. It is clear that the advantages of automation also extend to small and micro-producers who, usually, have very limited financial resources.

Alternative and low-cost technologies are the most affordable considering the financial reality of micro and small business owners.

A great example of such technology is the Raspberry Pi which is a low-cost microcomputer which can be used in many different applications. A good example is that it can monitor and control various factors of a production process such as pressure, humidity, temperature and others [2]. Several industrial sectors make use of the control and monitoring of variables such as temperature and pressure in order to maintain assure product quality and process safety. Raspberry Pi can replace in some cases a Programmable Logic Controller (PLC), but at a much lower cost. Naturally, Raspberry Pi would not be suitable for controlling brewing process in large scale due to compatibility, flexibility and reliability levels required by industrial standards, but can be a great low-cost solution [3] for the small entrepreneur.

Automation is essential for those who want to compete in the beer market, due to their characteristic of homogenization of production which results in quality product, but the usual solutions for industrial automation are not always accessible to small entrepreneurs. Therefore, small entrepreneurs need automation at a reduced cost, but capable to meet the requirements of safety, quality and production capacity.

The greatest difficulty for the production process of craft beers is the various peculiar requirements for each stage of the manufacturing process, such as time and temperature control. Such specificities guarantee the quality of the beer at the end of the whole process of brewing and fermentation.

From this research problem was elaborated and tested an automation solution to meet small entrepreneurs requirements, keeping overall costs as low as possible.

There are different devices that have been developed to meet the demand for low cost automation, and in this work we chose to use Raspberry Pi which is one of the lowest cost solutions in the market. The Raspberry draws attention due to its ease of use and its versatility to suit the needs of each problem. The focus of this work was in two main aspects of the brewing process: time control and temperature control.

2 Beer Manufacturing

In Brazil, the Decree 6871/09 from 2009 states the following definition: "Beer is the beverage obtained by the alcoholic fermentation of brewer's wort from barley malt and potable water, by the action of yeast, with the addition of hops". There is no definition for the concept of artisanal beer in Brazilian legislation. However, there is the bill PL 5191/2013 which is still waiting approval under legislative process with the aim of creating specific legislation for producers of craft beer. The project will define as an artisan producer the one that produces up to thirty thousand liters per year. On the other hand in the USA the Brewers Association (BA), the association of American Artisan Brewers, defines artisanal beer by the volume of production, the production process and the fact that the owner belongs to a local community [4].

Another point of relevance for the Brazilian bill PL 5191/2013 concerns the variety of different types of beer available in the artisanal beer market, which is not limited to the Pilsen type, which is the most widely produced and consumed beer in Brazil [5, 6]. Therefore, the automation process for craft beer must be flexible in order to produce several different beer recipes.

2.1 Brewing Process - Bottlenecks for Micro Entrepreneurs

In order to specify the automation requirements for brewing process, the following steps can be summarized for beer production: extracting sugars from malt, separating the malt from the sweet liquid, boiling the liquid to sterilize it and adding the hops, cooling and letting the fermentation occur after the addition of yeast. In general, for the same bier type the process is similar, independent of the manufacturer. Therefore, what can differentiate the final product are the ingredients and the execution and control of each step of the manufacturing process. In Brazil there is a group of obligatory ingredients for brewing, and a brewer can innovate with any ingredient as long as he respects the obligatory ingredients [7, 8].

2.1.1 Grinding of Malt

The purpose of grinding malt is to expose its endosperm, the white starch-rich part that we readily recognize in rice for instance, but without destroying or grinding the bark of the grains. Such care is very important for both the enzymatic action and for the filtering of the wort. The smaller the pieces, the more efficient the degradation of the starch and, consequently, the extraction of sugars. However, excessive grinding also increases the breakage of the bark, thus becoming clarifying the wort more complicated [9].

2.1.2 Mosturation

Mosturation is the name given to the transformation of molecules present in malts for brewing. In this process is generated a sweet liquid, full of aromas and of varied color, to which we give the name of wort.

For the implementation of the project, a technique developed in England called "Single-Infusion" is used, that is, the cooking of the malt in a single temperature range. With this technique the batching is totally geared to the saccharification of the wort, in which the conversion of the starch into simpler sugars occurs. The optimal selection of temperature causes both the enzymes that convert the malt starch present in the malt (α-amylase and β-amylase) into sugars to proceed properly.

In this step we can choose some characteristics of the final product. A milder temperature batching, in the range of 63 °C, produces a wort with higher proportion of fermentable sugars relative to the non-fermentable ones, because β-amylase activity is favored, resulting in a drier beer with less body. On the other hand, if the temperature is between 68 to 70 °C, α-amylase activity is favored, leading to the production of a wort with higher content of non-fermentable sugars, resulting in a thicker and dense beer [10].

2.1.3 Boiling and Hops Addition

Many fundamental transformations for the manufacture of a high quality beer occur during boiling, which must be intense and vigorous. We can mention among the main transformations the sterilization, the concentration of the wort, the evaporation of volatile compounds produced during the saccharification and the mosturation, as well as the isomerization of the alpha acids of hops used in the wort. This latter phenomenon, for example, promotes the solubilization of alpha-acids, which provides its characteristic bitterness. The slight darkening of the beer through caramelization and the Maillard reaction, and the coagulation of proteins, which contribute to the production of a more crystalline beer, are results of boiling process.

For different characteristics of beer, the hop must be added in different parts of the boiling process. If the intention is to provide only bitterness, the hops must be added at the beginning of the boiling process, which increases protein coagulation and evaporation, consequently increasing the wort density resulting in the desired bitterness. If the intention is just to provide flavors, the hops must be added to the pot in the final 30 min of boiling. These late additions allow part of the essential oils of hops to be retained in the wort, increasing its contribution to the final aroma.

2.1.4 Yeasts

Yeast should not be added to the boiling wort; otherwise it will become unusable. Thus, the wort must be cooled down to the temperature range around 20–25 °C, varying the optimum temperature of growth for the yeasts, depending on the type of beer to be produced.

2.1.5 Fermentation and Cold Sedimentation

At this stage the yeast is added to ferment the hopped wort. Fermentation generates carbon dioxide that needs to be eliminated from the fermenter. This stage lasts approximately seven days and the measurement of the final density of the wort determines the alcoholic strength.

After this period, known as primary fermentation, the fermenter is cooled down in the range of 2 to 4 °C, usually within a refrigerator, for another seven days at least for cold sedimentation of coagulated proteins and flocculation of yeasts, resulting in a more translucent beer, with improved and stable flavor and aroma.

2.1.6 Packaging and Priming

Ready-to-eat beer is a delicate and fragile product, easily contaminated. Thus, vessel transfers should be done in a controlled environment. Any material that may come into contact with the liquid, including bottles and lids, should be sterilized with peracetic acid, commonly used in industry for this purpose.

Beer gas is necessarily lost during fermentation. To obtain a carbonated beer, a procedure called priming is performed, which consists in adding a small amount of sugar to the beer, so that light fermentation occurs in the bottle, leaving it carbonated and allowing the formation of foam. The proportion of the priming to be added in each bottle is of vital importance, because in the latter phase a new fermentation will occur. In two weeks the beer must be carbonated and a good part of the remaining yeast should be settled in the bottom of the bottle.

3 Low Cost Automation Solutions and Figures

Due to the fact that micro-entrepreneurs usually face budget constraints, solutions were searched that met the needs, but without the high costs of standardized industrial equipment. After researching several options, the following equipment/solutions were decided upon.

3.1 Raspberry Pi

Raspberry Pi is a mini microcomputer with size comparable to a credit card and has very low cost. The small device has a processor, graphics processor, USB (Universal Serial Bus) and HDMI (High-Definition Multimedia Interface), RAM (Random Access Memory), power input and expansion bus, that is, a complete computer of reduced size and cost [11, 12].

The Raspberry Pi project was developed by the Raspberry Pi Foundation whose main goal was to develop a low-cost device that would be accessible to any school so that it would be possible to spread knowledge about technology [2].

Currently Raspberry Pi is not only used for educational bias, it is a great alternative to solve other problems like low cost automation. You can use the device to manage spreadsheets, control robots, and process control sensor data. The device supports different programming language like C++, Python, Assembly, Java etc. [13].

3.2 Operating System (OS) Raspberry Pi

There are several operating system options for Raspberry Pi and you should consider what features you want in Raspberry and then choose what best suits your needs. Some OS options are [14]:

- Raspbian - It is the OS based on the Debian operating system (Linux). Raspbian is the most widely used OS in Raspberry when it comes to using the device as a computer with limited functionality, yet very light and easy to learn. The OS is complete and has most of the features of a standard computer
- Ubuntu- Is a better-known Linux OS and is used in Raspberry when the interest is to simulate a full-featured computer. Ubuntu is a bit more difficult to use, but it has more tools and turns raspberry into a traditional computer.
- Open Source Media Center (OSMC) - Is an operating system used for Raspberry to be used to play high-quality media (video and audio). For users interested in using the device as a media center.

- Recalbox- Is an operating system used to transform Raspberry into a game emulator. For users who want to use the gaming device.

In the proposed solution Raspbian SO was chosen due to the ease of learning and for being the most suitable for beginners, besides being considered standard by the Raspberry Foundation and also be the most used. Raspbian is a Linux distribution derived from Debian, a fairly complete OS that allows full control of the device. It is possible to use the internet, Libreoffice and several other applications. A good example of an available application, which will be used in the proposed solution, is CraftBeer, used precisely in the management of production of craft beers.

Debian OS was developed by the association called Debian Project whose goal was to create an open source operating system. Debian uses the Linux kernel which is the central computer code that performs the basic operations and ensures the operation of the other programs. Debian has more than 50,000 precompiled and easy-to-install software, plus a manager that enables the use of all these features. Raspbian is a Debian optimization developed for Raspberry Pi hardware and most of the software available for Debian is also available to run on Raspbian optimally [15].

3.3 Raspberry Pi Model Used in This Work

There are different options for Raspberry Pi settings that can be used. In the proposed solution it was decided to use the Raspberry Pi 3 Model B. The device has a dimension of 85.6 mm × 56 mm × 21 mm, 1 GB RAM, 1.2 GHz Quad Core processor and various peripheral connectors. (see Fig. 1). Figure 1 shows the schema of the device [16].

- A - Processor: The device has a Quad Core 1.2 GHz Broadcom BCM2837 64bit CPU (Central Processing Unit)
- B - SD (Secure Digital) Memory Card Slot: The memory card slot on the device serves as a hard disk for Raspberry.
- C - USB port: The device has two USB ports
- D - Ethernet port: This is a standard RJ45 Ethernet port
- E - HDMI Input: This type of port provides digital audio and video outputs.
- F - DSI (Display Serial Interface): Used to connect an LCD or OLED screen with a 15-pin flat ribbon cable.
- G - Analog Audio Output: This is a standard analog connector used to conduct high impedance loads (Amplified Speakers).
- H - GPIO (General Purpose Input/Output): These are programmable data input and output ports. There are 40 pins.
- I - CSI (Camera Serial Interface): This connection allows a camera to be connected directly to the card.

3.3.1 Accessory Modules

In order to control the temperature and time of each stage of the beer production process, a set of accessory modules, designed to be compatible with the Raspberry Pi project and

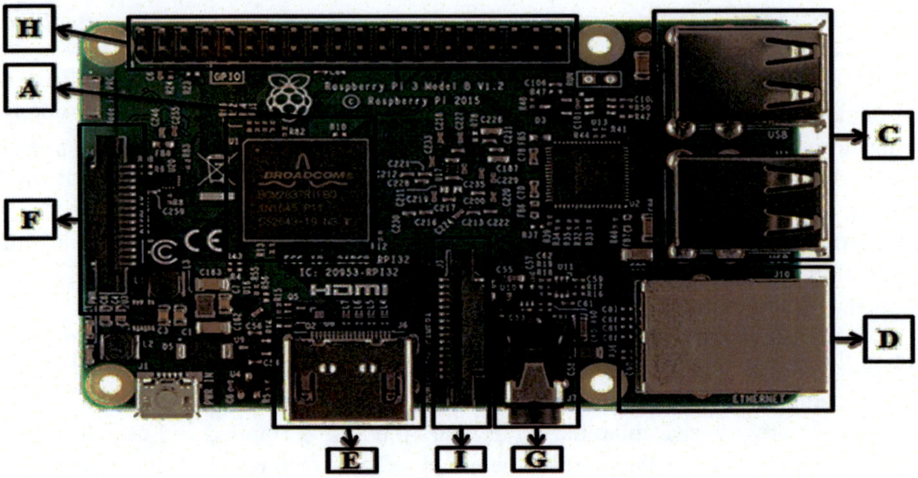

Fig. 1. Top of Raspberry Pi. Source: Autors

with the CraftBeerPi software, will be presented being integral part of the proposed solution.

The project was designed to be "plug and play" and modular, allowing installation facilities and flexibility of adaptation in virtually any provision of small-scale beer production equipment, ranging from single-vessel equipment to tri-block kitchens, adding the control functionality.

The project has no involvement in the development of software that we will use as a graphical interface, neither in the Raspbian operating system nor in the Raspberry Pi project. The accessory modules are only intended to be a compatible and integrative solution to these projects. The set consists of 3 modules:

- Control Module; – Power Module; and – Relay Module.

3.3.2 Software

The project was developed to use CraftBeerPi software, software created by Manuel Fritsch, a German from the northwestern city of Padeborn. Distributed for free on his blog, Manuel teaches how to connect his software in a variety of ways to Raspberry Pi [17, 18].

4 Methodology

Initially a research was done to familiarize with the brewing process. This led to highlight the craft beers which directed the research in order to identify automation opportunities within their manufacturing process. There were possibilities to automate the whole process of manufacturing artisan beers and to do that, low cost solutions were researched [19].

After searching for some solutions, such as Arduino and Low Cost PLCs, we chose to use Raspberry Pi, because in addition to meeting the low cost criteria, it is also scalable if we need to increase the operation. It was decided to use free and open-source software well known among artisan brewers, the CraftBeerPi. CraftBeerPi software was configured in Raspberry Pi and with the help of the set of designed accessory modules, which includes controller, power module and relays, and monitored and controlled the variables relevant to the manufacturing process as temperature and time. In fact keeping temperature ramps is the biggest justification of automation

4.1 Configuration Raspberry Pi and Craftbeerpi

In order to test and validate the proposed low-cost solution, an artisanal brewing process was carried out in an experimental set up, where the Raspbian operating system was installed on Raspberry Pi using a SD card with the OS downloaded from the official page of the Raspberry Pi Project. The OS was easily installed and configured using the step-by-step procedure provided by the Raspberry Pi Project.

CraftBeerPi software was then installed, following the step-by-step provided by the official software site. Installation was performed via command prompt.

5 Tests and Results

The test setup comprises the following modules and components:

- Raspberry Pi/SD memory card 8 GB/water pump
- Control module/power module/Relay module
- Freezer/50 L caldron/Temperature sensors DS18B20

5.1 CraftBeerPi PID Configurations

The system start begins with the PID configuration (Preprogrammed Control Algorithm) so that the system quickly meets the temperature ramps requested by the CraftBeerPi software.

In this test, a parameter definition list made available by the CraftBeerPi was used as parameters "PIDArduinoLogic", a control code available for free. For the PID adjustment, only the Proportional and Derivative controls were used [11], considering that in the case of a ramp temperature process, Integral control helps the controller to eliminate the steady state error and keep the set point.

With these parameters defined, test steps for production simulation were created with four temperature ramps: 25 °C to 65 °C, 65 °C to 68 °C, 68 °C to 75 °C and 75 °C to 80 °C, since for beer production process usually increasing temperature ramps are required. To create the steps, the parameters were selected using the Craftbeerpi task command. Due to the large temperature difference between the initial temperature of the water (approximately 25 °C) to the first set point, it took almost 2 h to reach the requested temperature.

After this ramp, the controller remained stable all the time, presenting good response curve. These preliminary temperature ramp tests presented satisfactory results for the production process. After the preliminary tests, two beer recipes of the type (*witbier*) were elaborated and the results were verified.

6 Conclusions/Final Considerations

A low-cost, highly flexible and functional automation solution, easily replicable, has been successfully crafted and tested, thus serving micro-entrepreneurs who want to work in the craft beer business.

The sharing mechanisms of programs and experiences, available in online forums, also offer to the interested parties' diverse possibilities of personalization of content.

The Raspberry pi system has advantages over traditional PLC-based automation solutions, as its acquisition cost is below USD 100.00 and with very similar process performance (of course the application described is not critical or costly in case of failure, which would lead to the need to use a PLC).

References

1. Andrade, A.A., et al.: ELABORAÇÃO DE BANCADAS DIDÁTICAS PARA AUTOMAÇÃO INDUSTRIAL BASEADAS EM CLPs E FREIOS DE FOUCAULT. Revista de Ensino de Engenharia 35.2 (2017)
2. RASPBERRYPI. Raspberry Pi Foundation. https://www.raspberrypi.org/about/. Accessed 01 Mar 2019
3. RASPBIAN. About Raspbian. http://raspbian.org/RaspbianAbout. Accessed 01 Mar 2019
4. BREWERS Association CRAFT BREWER DEFINED. https://www.brewersassociation.org/statistics/craft-brewer-defined/. Accessed 01 Mar 2019
5. Brunelli, L.T., Mansano, A.R., Venturini Filho, W.G.: Physicochemical characterization of beer produced with honey. Braz. J. Food Technol. **17**(1), 19–27 (2014)
6. Padilha, A.C.M, et al.: Capacidade absortiva na produção artesanal de cervejas. Espacios **37**(35) (2016)
7. Borges, P.F.O.: Concentração do mercado de cerveja no Brasil e a participação das microcervejarias (2015)
8. SEBRAE. Cervejas Artesanais: Um Mercado Para Ficar de Olho. http://startupsebraeminas.com.br/cervejas-artesanais-um-mercado-para-ficar-de-olho/. Accessed 01 Mar 2019
9. Henson, Cynthia A., Duke, S.H.: A comparison of standard and nonstandard measures of malt quality. J. Am. Soc. Brew. Chem. **66**(1), 11–19 (2008)
10. Mathias, B.D., et al.: An identity perspective on coopetition in the craft beer industry. Strateg. Manag. J. **39**(12), 3086–3115 (2018)
11. Garrett, F.: Como funciona o Raspberry Pi? Entenda a tecnologia e sua aplicabilidade. http://www.techtudo.com.br/noticias/noticia/2014/11/como-funciona-o-raspberry-pi-entenda-tecnologia-e-sua-aplicabilidade.html/. Accessed 01 Mar 2019
12. Paiva, O.A., Moreira, R.O.: Raspberry Pi: dispositivo de 35 dólares para visualização de imagens DICOM. http://www.scielo.br/pdf/rb/v47n2/pt_0100-3984-rb-47-02-99.pdf. Accessed 01 Mar 2019
13. Thomsen, A.: Primeiros passos com o Raspberry Pi. https://www.filipeflop.com/blog/tutorial-raspberry-pi-linux/. Accessed 01 Mar 2019

14. Delaney, S.F.: RealPi-a real time operating system on the Raspberry Pi. Diss. University of Nevada, Reno (2018)
15. Lenz, M.: Distributing debian packages. In: Python Continuous Integration and Delivery, pp. 77–85. Apress, Berkeley (2019)
16. ROBOCORE. Raspberry Pi 3 - Model B. https://www.robocore.net/loja/produtos/raspberry-pi-3-model-b.html#inf_tecnicas. Accessed 01 Mar 2019
17. CRAFTBEERPI. Craftbeerpi Installation&Doc. http://web.craftbeerpi.com/. Accessed 01 Mar 2019
18. Fritsch, M.: Craftbeerpi Github Manual. https://github.com/Manuel83/craftbeerpi/. Accessed 01 Mar 2019
19. Quintino, L.F., et al.: Kit didático de baixo custo para práticas interdisciplinares em cursos de Engenharia Eletrônica. Revista Principia **1**(34), 124–131 (2017)

Design, and Systems Development with Data Analytics Management in the Digital Age

Relating Design Management and Project Management: Application of the PM Mind Map Tool in the Creative Design Process

Bruno Perdigão Pacheco[1](✉), Mateus da Silva Bento[2](✉), Almir de Souza Pacheco[2](✉), and Magnólia Grangeiro Quirino[2](✉)

[1] Federal Institute of Education, Science and Technology of Amazonas, Manaus, Brazil
brperdigao@gmail.com
[2] Federal University of Amazonas, Manaus, Brazil
mateusbentorp@gmail.com, almirdesigner@gmail.com, mquirino@ufam.edu.br

Abstract. This article aims to relate the design management and project management during the creative design process through the PM Mind Map tool for the development of the name and brand for a company of customized services for selling fruits products. In this sense, it is a study of a qualitative nature, using the technical procedures of bibliographical and applied research. The elements and perspectives involved in the creation of the brand were structured in a collaborative way according to the visual model. The PM Mind Map reveals itself as a project management tool that has not yet been widely disseminated and deepened in the academic-scientific area. However, when linking its application with organizational performance based on the concepts of Design Management, it was noticed a clarification of the project stages, greater control over deadlines and costs, more explicit norms that facilitated the submission to requirements and constraints, delimitation of tasks' division more understandable, monitoring and control of risks through discussed premises. In this way, it is recommended to deepen the relation between them in future research.

Keywords: Design Management · Project Management · PM Mind Map

1 Introduction

The term Design Management originated in Great Britain in the 1960's [1], which emerged as the management of relations between design agencies and their clients. The look at design ceased to have an exclusively aesthetic character and became strategic [2]. The Design Management is necessary for solving organizational problems by directing, planning, executing and controlling activities and managing resources efficiently.

Design management encompasses continuous processes, business decisions, and strategies that enable innovation [3]. In the search for resource management, the design management directs elements present in the strategic management of the organization

L. Pereira et al. (Eds.): IDEAS 2018, SIST 198, pp. 149–159, 2021.
https://doi.org/10.1007/978-3-030-55374-6_15

- among them: vision, mission and values, based on the levels strategic, tactical and operational.

In the last decades there has been an increase in the search for savings of financial, technological and human resources by the organizations during the realization of their projects. The PMBOK guide defines a project as "a temporary effort undertaken to create a unique product, service or result" [4]. A project consists of four phases: conception, planning, execution and closure [5].

In this context, project management emerges, which integrates efforts, skills and tools to achieve the goals defined by the organization. The Project Management must be applied in an efficient and collaborative way, based on tools that contribute to its effectiveness and control of results. In this case, PM Mind Map is a project management tool that ensures integration between the levels of design management and project development.

The questioning that guides this study goes through the search for an increase in the collaborative virtue of the creative process of design - how to favor the participation of the collaborators during project management? The objective of the article is to relate the design management and project management during the creative design process through the PM Mind Map tool for the development of the name and brand for a company of customized services for selling fruits products. Methodologically, it is a qualitative study, using the technical procedures of the bibliographic research and the application of the PM Mind Map tool during the creative design process.

2 Theoretical Reference

2.1 Design Management

Design management is what identifies a company, the difference of its competitors and defines its clients and investors [1]. It is the competent management of people, projects, processes and procedures that idealize and execute products, services and experiences [6]. Gorb says that design management corresponds to the correct use of the design resources available in the corporation during the execution of projects, and the manager has a fundamental role in the importance of design in this process [7].

Based on this, it becomes intrinsic to design management, a strategic planning directed to the view of the design, having it as a guide and fulfilling its stages closely. Campomar describes planning stages, but he warns that not necessarily they should be applied in an unrestricted way due to the time spent and cost to carry out such strategic actions: situation analysis; goals; action plan; implementation of the action plan; and control of results [8].

The element that distinguishes design management is the constant communication of design to, strategically, contribute for the choice of the best paths to be adopted by the company, using the following resources: physical or material, financial, human, market and administrative, being tangible or intangible [9]. This implies in the systemic and effective participation of the design in the three organizational levels: strategic, tactical and operational. Fialkowski and Kistmann point out the objectives of each level. The strategic level aims to position the company strategically, with its objectives and creating a competitive advantage in the market. The tactical level aims to plan activities and provide resources to generate actions that position the company in the strategic way stipulated. The operational level aims to develop actions and solutions, realizing in a tangible way the position established in the strategy[1] [10].

2.2 Project Management

Project management is defined as "the application of knowledge, skills, tools and techniques to project activities in order to attend their requirements" [11]. Kerzner complements by treating project management as "planning, scheduling and controlling a series of integrated tasks in order to achieve their goals successfully for the benefit of the project participants" [12].

The integrative nature of project management to solve a problem of the organization is observed, based on the collaboration of the people involved and the joint application of skills and techniques. Faced with increased interest in the area, Eiras points out that new approaches to project management have emerged and the issues addressed have diversified beyond the traditional schedule, budget and scope [13].

Among the tools for project management is the PM Mind Map, whose model of approach favors the integrated visualization of the phases that permeate project development (initiation, planning, execution, control) and provides information for its closure [14].

2.3 PM Mind Map

The PM Mind Map is a project management tool developed by Paulo Mei which is based on the concept of mind maps and applying project management principles of Project Management Institute (PMI). Based on aspects related to neuroscience and neuroleading, the tool has among its principles: stimulus to simplicity with content; providing a visual and intuitive structure for project management; providing value to the organizational strategy and stakeholders; speed in execution and response to changes in a planned form; leadership and participation [14].

The PM Mind Map is composed of 15 elements, distributed in 6 perspectives: business (elements: value proposal, objective and business results); product (elements: product and requirement); influences (elements: interested and other external influences); conditions (elements: restrictions and risks); execution (elements: resources, deliveries, time and cost scheduling, total cost) and control (elements: result and estimates)[2].

3 Application of the PM Mind Map Tool in the Creative Design Process

The project's development to create the name and brand of the company considered all the perspectives and elements present in the PM Mind Map project management tool.

---

[1] Fialkowski and Kistmann synthesized levels of design management according to the variables: forces, deadline, results and involved. For more information, consult original work.

[2] For more information about the meaning of each element, consult Mei (2015).

For this purpose, the visual model of the tool in size A1 was used, with fixation of post-its in a collaborative way. The title of the project "Creation of name and brand for company selling fruit products" was defined and project manager: João da Silva[3]. Figure 1 presents the PM Mind Map project for name and brand creation for the company:

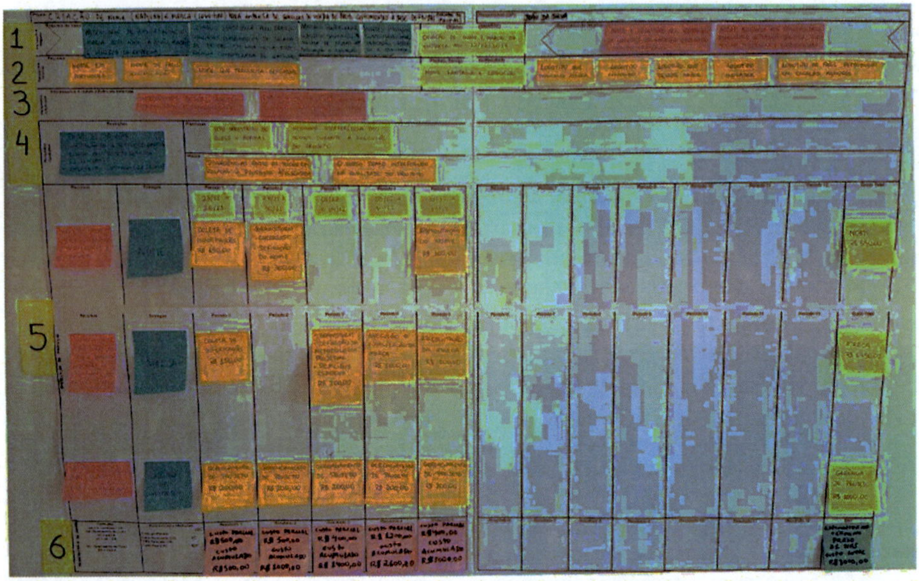

Fig. 1. PM Mind Map project name and brand creation for the company. Source: Elaborated by authors

Line 1 - In the Business Perspective, the value proposition, the objective and the desired results were defined. In this perspective, the strategic elements of the company were considered, by knowing: vision - to distribute through Brazil various combinations of fruit salad as a new customized experience of consumption; mission - to provide healthy consumption of fruit salad with dedication and innovation; values - sustainability, dedication, innovation, freedom and quality. The elements defined in the business perspective in the PM Mind Map of the project were:

- Value proposal: the need to have a name and a brand that is linked to the vision, mission and values of the company.
- Objective: Creation of the company's name and brand until 12/12/2018[4].
- Results: name and logo that express the company values and have a positive impact; generate the empathy of the stakeholders.

[3] All proper nouns in this article are fictitious names and have no relation to the names of the authors.

[4] All dates in this article are written in the US mode (month/day/year).

Line 2 - In the Product Perspective, the product, service or result and its requirements were identified. In the project there are two results (fancy name and brand). For each result, the requirements were indicated:

- Requirements for the name of the company: being in Portuguese language, easy to assimilate and convey the feeling.
- Requirements for the brand of the company: transmit joy, being colorful, innovative, that denote flavor and that is easy to reproduce on a reduced scale.

Line 3 - In the Influences Perspective, the interested people and the external influences for the project are pointed out:

- Interested people: partners, possible customers and competitors.
- External influences: graphics, software and computers, in addition to the short term.

Line 4 - Under the Conditions Perspective, the constraints, premises and risks for the project were identified, by knowing:

- Restrictions: deadline of 20 days, limitation of graphic software (Corel Draw), project team reduced (2 people) and limited budget to R$ 3,000.00[5].
- Premises: unrestricted use of colors and shapes and no interference from company partners during project execution.
- Risks: divergence between the partners of the company in relation to the proposal presented and the possible short-term interference in the quality of the project.

Line 5 - In the Execution Perspective, the resources, the deliveries, the periods for delivery of packages with the required deadline and budget and the total costs are indicated. For the project, there are two deliveries:

- Name
- Brand

For the delivery of the company's name, it is necessary as resources João da Silva (designer) and José de Souza (human resources), office (furniture and materials) and Internet. The packages for the delivery of the name are divided into:

- Collection of information in the period from 11/23 to 11/26/18. Cost R$ 150,00;
- Brainstorm, checklist, discussion and definition of the name in the period from 11/27 to 11/30/2018. Cost R$ 300,00;
- Presentation of the name in the period from 12/11 to 12/13/2018. Cost R$ 100,00.
- The total cost of delivery of the name corresponds to R$ 550.00.

The company logo is delivered by João da Silva and José de Souza, office (furniture and materials), Internet and graphic design software. The delivery is divided into four packages, by knowing:

[5] All monetary values mentioned in this article are in Real (Brazilian currency).

- Collection of information in the period from 11/23 to 11/26/2018. Cost R$ 150,00;
- Brainstorm, definition of design methodology and first sketches from 12/01 to 12/04/2018;
- Execution and finalization of the brand from 12/05 to 12/10/2018. Cost R$ 1.000,00;
- Present logo in the period from 12/11 to 12/13/2018. Cost of R$ 100,00.
- The total cost of delivering the logo is equivalent to R$ 1.450,00.

The Execution Perspective also features on the project team formed by João da Silva (manager) and José de Souza (collaborator), project manager for major deliveries and project management:

- Each period was equivalent to R$ 200,00 for all five delivery packages.
- Total cost of project management is R$ 1.000,00.

Line 6 - The Control Perspective consists of 5 results and their respective costs, in addition to the estimate at the end of the project, as described below:

- Result 1 (period from 11/23 to 11/23/2018): partial cost - R$ 500,00. Accumulated cost - R$ 500,00.
- Result 2 (period from 11/27 to 11/30/2018): partial cost - R$ 500,00. Accumulated cost - R$ 1.000,00.
- Result 3 (period from 12/01 to 12/04/2018): partial cost - R$ 400,00. Accumulated cost - R$ 1.400,00.
- Result 4 (period from 12/05 to 12/10/2018): partial cost - R$ 1.200,00. Accumulated cost - R$ 2.600,00.
- Result 5 (period from 12/11 to 12/13/2018): partial cost - R$ 400,00. Accumulated cost - R$ 3.000,00.
- The estimated project completion is 21 days and has a total cost of R$ 3.000,00.

Through the management of the project, it was possible to create a name and brand proposal for the company. The PM Mind Map tool allowed the development of the project in order to consider the strategic management of the company.

3.1 The Creative Design Process Embedded in the PM Mind Map Tool

The creative design process must go beyond the production of visual output because design is embedded in many areas of decision-making within the organization [15]. Cooper and Press point to a problem of undervaluation and under-utilization of design in companies and affirm that the answer lies in the management of design as a process that brings the universe of design closer to the universe of management [16].

As a process, design management favors collaboration, since it occurs through the infusion, an informal process with the goal that all the employees of an organization become aware of and become involved with the developed design projects [17].

In order to create a company's name and brand, Gavin Ambrose and Paul Harris's Design Thinking methodology was used, which is subdivided into seven steps: problem definition; research; generation of ideas; proposal elaboration; selection of proposals; implementation of the project; and learning [18].

The collaborative levels of the creative design process were accentuated when inserted into the PM Mind Map tool. Table 1 shows the association between the steps of the design methodology and the constituent perspectives of the PM Mind Map tool:

Table 1. Association between design methodology and PM Mind Map perspectives

Design thinking	PM Mind Map perspectives					
	Business	Product	Influences	Conditions	Execution	Control
Problem definition	Value offer; goal	–	–	–	–	–
Research phase		Requirements	Stakeholders and external influences	Restrictions; premises; scratches	Period 1	–
Generation of ideas	–	–	–	–	Resources; deliveries; Period 2; period 3	–
Proposal elaboration	–	–	–	–	Resources; deliveries; period 3	–
Selection of proposals	–	–	–	–	Resources; deliveries; period 4	–
Project implementation	–	–	–	–	Resources; deliveries; period 5	–
Phase of learning	Business results	–	–	–	–	Results 1 to 5; total cost; ENT

Source: Elaborated by authors

The association between design thinking and the PM Mind Map tool resulted in the implementation of the name and brand project for the company of customized services for selling fruits products, according to the Fig. 2 below:

Fig. 2. Name and brand suggested for the company. Source: Elaborated by authors

4 Relations Between Design Management and the PM Mind Map Management Project Tool

From the PM Mind Map visual model developed for the company, it is possible to gauge some relationships between design management and project management. According to Soares, while design management is focused on the macro activity of design strategies, with decision-making power in an organization, structured to shape the company profile, products and/or services, the management of the project coordinates tasks, time and resources, through techniques and tools, creating conditions for the project development, without losing sight the organizational strategies [19].

A relevant factor in design management refers to the management of resources at all levels of the organization, indicated by Mozota as a purpose of the activities of the design management implemented through the communication of the organizational objectives [15]. In the PM Mind Map developed for the project of creation of the brand and name of the company, physical, material, financial, human, market and administrative resources were considered. The resources and the respective items used in the project are exposed below:

- Physical and material: notebooks, office (furniture and material), internet, software of graphic design;
- Financial: available budget from the company (R$ 3.000,00);
- Human: João da Silva (designer) and José de Souza (public relations);
- Market: name and logo created in view of the company's position in the market;
- Administrative: they are not listed in the project.

After reviewing the available resources for the development of the project, the stages of strategic planning in design were related to the elements presented in the PM Mind Map project management tool. The strategic planning stages are formed by analysis of the company's situation, definition of objectives and goals, preparation of action plan, execution of the plan and control of results. The company's positioning strategy is the differentiation, pointed out by Kotler and Keller, as the act of developing a set of significant differences to distinguish the company's offer from what its competitors are offering [20]. Thus, the idea of differentiation reached by the company permeated all phases of project development. Table 2 presents the relationship between strategic planning and PM Mind Map tool.

Table 2. Relationship between strategic planning and PM Mind Map tool

Organizational objective: Creation of company name and brand		
Stage of the strategic planning	Description	Elements of PM Mind Map
Analysis of situation	The company has the differentiation as strategy. There are few companies that trade fruit salad focusing on customization	Interested and other external influences; Value offer; and Requirements
Goals and objectives	Objectives: To build a brand with solid concepts; Satisfying the expectations of the strategic public (stakeholders); Consolidation in the market; Conquer the national market; Be recognized for environmental responsibility; Be recognized for excellence in service Goals: Obtain 20% of the fruit salad market in the metropolitan area of Belém in the first 3 years; Open branches or franchises in all five regions of the country within 10 years; Allocate at least 95% of the waste generated for recycling or composting	Value offer; and Business results
Elaboration of action plan	It was done according to elements presented in the PM Mind Map tool	Goal; Business results; Requirements; Interested and other external influences; Restrictions; Premises; Resources; and Deliveries
Execution of the plan	It was done according to elements present in the PM Mind Map tool	Resources; Deliveries; Periods; and Results
Control of results	It was done according to elements present in the PM Mind Map tool	Periods; Results; and Total Cost Control

Source: Elaborated by authors

Finally, the PM Mind Map project management tool was classified with the three levels of design management (strategic, tactical and operational), indicating the elements that connect from one level to another, as shown below.

- Strategic: value proposal; goal; business results; interests and other external influences; and constraints (the latter connects the strategic and tactical levels);
- Tactical: restrictions; requirements; products; services or results; premises; risks; resources; and deliveries (the latter connects the tactical and operational levels);
- Operational: deliveries; periods; results; and total cost.

The relations between design management and project management are verified, with PM Mind Map as a parameter. It has been observed that design management permeates all the constituents of project management and the results of these contribute to the survival of the organization in the market.

5 Final Considerations

Project management tools help teams and organizations decisively as facilitators in executing control. Design management in this context must seek to solve problems immersed in each stage of strategic planning and at all levels - strategic, tactical and operational - of a company. Paulo Mei's PM Mind Map reveals itself to be a project management tool that has not yet been widely disseminated and deepened in the academic-scientific area; however, this article has sought to relate its application to organizational performance based on the concepts of design management, in order to collaborate with dissemination of the tool mentioned above. The result of this tooling application of the project relating to the management of the design was the generation of name and brand for a company selling fruit-based products.

It was realized a clarification of the stages of the project, a greater control over deadlines and costs, more explicit norms that facilitated the submission to requirements and restrictions, delimitation of division of tasks more understandable, and monitoring and control of risks through debated premises. The use of the simple or complex project management tool within the perspective of design management has proved to be very useful, so it is recommended to deepen the relationship between them in future research.

References

1. Mozota, B., Klopsch, C., Campelo, F.: Gestão do design. ARTMED, Porto Alegre (2011)
2. Bahiana, C.: A importância do Design para sua empresa. CNI-COMPI-SENAI/DR-RJ, Brasília (1998)
3. Design Management Institute (DMI): What is design management? Accessed 12 Dec 2018
4. Project Management Institute: Um Guia do Conhecimento em Gerenciamento de Projetos (Guia PMBOK), 6th edn. Project Management Institute, Newtown Square (2017)
5. Menezes, L.: Gestão de Projetos, 3rd edn. Atlas, São Paulo (2009)
6. Best, K.: Fundamentos da Gestão do Design. Bookman, Porto Alegre (2012)
7. Gorb, P.: Design Management. Van Nostrand Reinhold, New York (2001)
8. Campomar, M.C.: O plano de marketing. Revista Marketing **51**, 54–58 (1977)
9. Mozota, B.: Design Management. Éditions d'Organization, Paris (2002)
10. Fialkowski, V., Kistmann, V.: Gestão de design e inovação incremental guiada pelo significado. Estudos em design **26**(2), 28–53 (2018)

11. Project Management Institute: Um Guia do Conhecimento em Gerenciamento de Projetos (Guia PMBOK), 6th edn. Project Management Institute, Newtown Square, p. 10 (2017)
12. Kerzner, H.: Gestão de projetos: as melhores práticas, 2nd edn, pp. 15–16. Bookman, Porto Alegre (2006)
13. Eiras, F.: Evolução das pesquisas de gestão de projetos: um estudo bibliométrico do International Journal of Project Management. GEPROS-Gestão da Produção, Operações e Sistemas **12**(1), 211–234 (2017)
14. Mei, P.: PM Mind Map®: a gestão descomplicada de projetos. Brasport, Rio de Janeiro (2015)
15. Mozota, B.: Design Management: Using Design to Build Value and Corporate Innovation. Allworth, New York (2003)
16. Cooper, R., Press, M.: El diseño como experiencia: el papel del diseño y los diseñadores en el siglo XXI. Editorial Gustavo Gili, Barcelona (2009)
17. Dumas, A., Mintzberg, H.: Managing design, designing management. Des. Manag. J. **1**(1), 37–43 (1989)
18. Ambrose, G., Harris, P.: Design Thinking. Bookman, Porto Alegre (2011)
19. Soares, V.: Disciplina Projeto de Produto. Curso de Desenho Industrial. Escola de Belas Artes, Rio de janeiro (2002)
20. Kotler, P., Keller, K.: Administração de marketing, 12th edn. Pearson Prentice Hall, São Paulo (2006)

Check for updates

Development of the Innovative Design of an Automatic Equipment to Aid in Physical Rehabilitation

Roberta Goergen[1,2](✉), Marianna Gioppo de Souza[2], Maurício Oberdörfer[3], Matias Alles Hubert[2], Jocarly Patrocínio de Souza[3], Luiz Antônio Rasia[2], and Antonio Carlos Valdiero[4]

[1] Federal Institute of Education, Science and Technology Farroupilha, Campus Panambi, R. Erechim, 860 - Planalto, Panambi, RS 98280-000, Brazil
betinhamtm@gmail.com

[2] Regional University of the Northwest of the State of Rio Grande do Sul, R. Pref. Rudi A. Franke, 540 - Arco-Íris, Panambi, RS 98280-000, Brazil

[3] University of Passo Fundo, BR 285, São José, Passo Fundo, RS 99052-900, Brazil

[4] Horizontina Faculty, Avenida dos Ipês, 565, Horizontina, RS 98920-000, Brazil

Abstract. This paper addresses the development of an innovative design of an automatic equipment to aid in physical rehabilitation. Physical rehabilitation aims to help people with disabilities or about to get disabilities to interact with their social environment. During rehabilitation, the physiotherapist judges the training in the patient's experience. For rehabilitation to be beneficial, the interaction between physiotherapist and patient is important. Several robotic devices have been developed by companies or institutions but with a high cost. The aim of this work is to present a robotic equipment for the physical rehabilitation of lower and upper limbs in patients who suffered muscle injury, stroke or surgery. A man-machine interface was developed to control the robot and manage the physical rehabilitation activity. Therefore, the physiotherapist begins to operate the system by entering the patient information and the type of exercise. The patient's reactions will be computed during the exercises and detected by position and force sensors. Thus, from this innovation, patients can perform repetitive movements, exercising the affected limb, during the rehabilitation period, always following the recommendations of the physiotherapy professional. The project team is multidisciplinary and composed of engineers, mathematician, programmer and physiotherapists. So, security must be ensured by software and hardware in the system. Therefore, the manipulator robot can perform all active and passive exercises, as well as learn specific movements of exercises and execute them with or without the physical therapist through the human-machine interface.

Keywords: Rehabilitation · Robotic · Interface man-machine

1 Introduction

The theme of robotics for rehabilitation is of great importance in the present times. In the last 10 years, we emphasized several studies in automated physical rehabilitation.

L. Pereira et al. (Eds.): IDEAS 2018, SIST 198, pp. 160–167, 2021.
https://doi.org/10.1007/978-3-030-55374-6_16

Physical rehabilitation aims to bring the capabilities that have been lost in patients who have suffered some type of illness, muscle injury or accident. Moreover, they cannot be treated with medicines. The number of patients who need physical rehabilitation increases every day.

Rehabilitation is a physical and psychological recovery for the process of the disabled person in order to achieve social reintegration. Different robotic architectures were developed and applied in the rehabilitation of human upper and lower limbs [1].

This robotic equipment enables the patient to perform repetitive activities. Increasing the motor skill of the development of functional activities. Aiming at the recovery of patients with motor disabilities. Training your muscles and your movements gradually. Utilizing robotic technology to support and increase the productivity and efficiency of therapists.

Most robotic assistive devices are exoskeletons that aid or increase the movement of the upper and lower limbs. In the SPEXOR project the objective is to present a new exoskeleton of the spine to prevent back pain in physically fit workers and support workers with low back pain [2, 3].

Research on the development of exoskeleton for rehabilitation lower and upper limbs represent a considerable part of studies in robotics. A systematic survey of the design and development of multiple exoskeletons for lower limbs was reviewed by [4].

This article presents the development of an innovative equipment of a modular robotic device for rehabilitation of lower and upper limbs. A new pneumatic modular robotic solution for rehabilitation that is low cost and simple design. The patient's reactions during rehabilitation are computed by position and force sensors. Then evaluated by the physiotherapist.

This article is organized as follows: a brief review on rehabilitation is reported in Sect. 2, and is done the design methodology in Sect. 3. To illustrate the results is presented the innovative design of an automatic equipment of physical rehabilitation in Sect. 4. In the Sect. 5 contains conclusions.

2 Review of Rehabilitation

Several researches show that in rehabilitation the repetitive movements of injured human limbs may help the patient regain function in the injured limb. Robots, cable manipulators and exoskeletons are the three types of mechanical systems used for movement rehabilitation [5, 6].

In the last decade, several robotic devices for rehabilitation have been developed to restore the mobility of affected limbs. These systems can be grouped according to the rehabilitation principle that follow in Fig. 1.

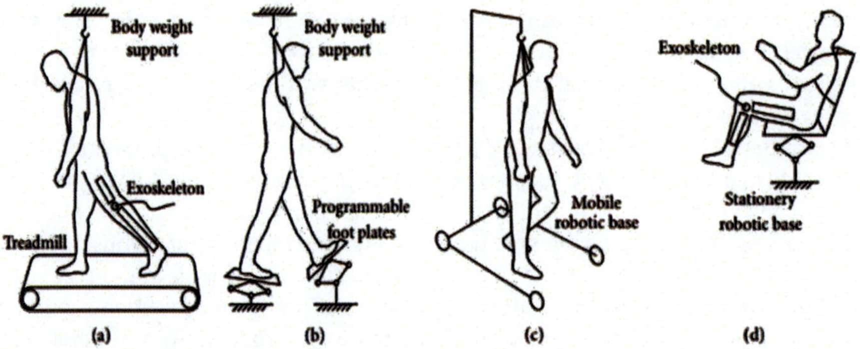

Fig. 1. Robotic system types for rehabilitation: (a) treadmill gait trainers, (b) foot-plate-based gait trainers, (c) overground gait trainers, (d) stationary gait and ankle trainers. Source: Adapted [7]

Figure 2 shows the main movements of the shoulder. The use of robots for upper and lower limb rehabilitation has increased considerably. Robot-based rehabilitation provides an accurate assessment of motor recovery and automates simple tasks that overwhelm health care professionals such as physical therapists and occupational therapists.

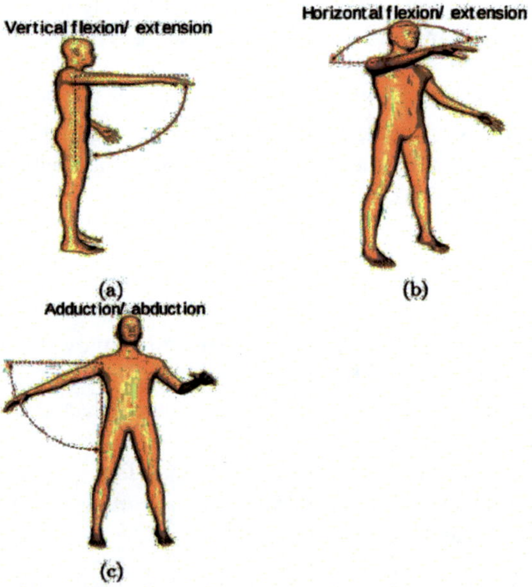

Fig. 2. Upper-limb rehabilitation movements: (a) Vertical flexion-extension, (b) Horizontal flexion-extension, (c) Adduction-abduction. Source: Adapted [8]

The basic rehabilitation movements of the flexion/extension lower limbs are shown in Fig. 3.

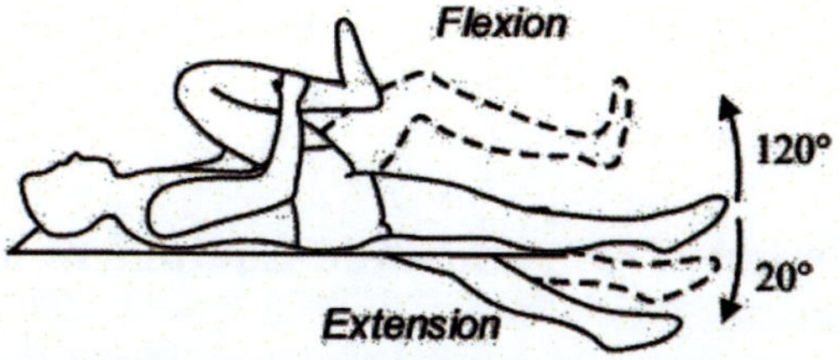

Fig. 3. Lower limbs basic movements. Source: Adapted [9]

This proposed study involves the search for knowledge in large areas, such as engineering, programming, mathematical modeling, robotics and physiotherapy. This multidisciplinary team aims to develop innovative design of an automatic equipment to aid in physical rehabilitation of lower and upper limbs. Pneumatic actuators, with controllable, stiffness are used to avoid injury to the patient and damage to the equipment.

According to [10], pneumatic systems are becoming increasingly used because of their high speed and strength capacity as well as their relatively low strength and cost. A composite robot with pneumatically actuated actuators has low cost advantages, high power-to-size ratio, installation flexibility, are clean and do not pollute the environment [11–13].

3 Design Methodology

For the development of innovative automatic physical rehabilitation equipment, we performed a survey of data and information on the needs and functionalities that the project would require. The methodology used in the project follows the steps proposed by [14], illustrated in Fig. 4.

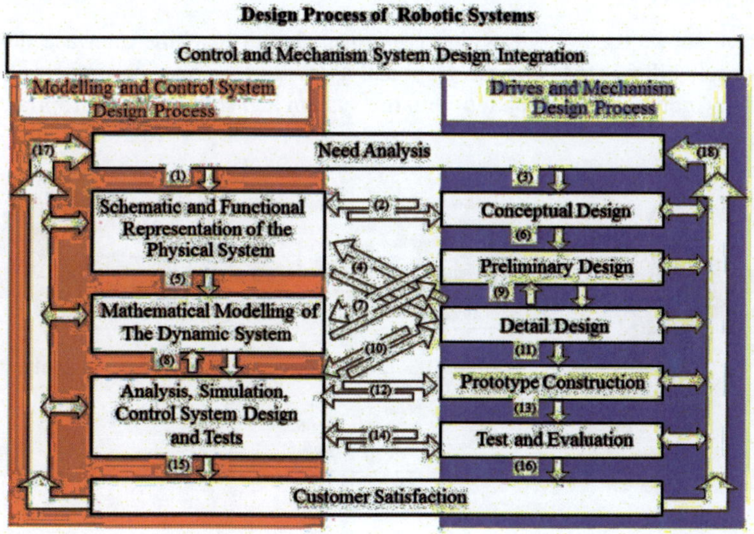

Fig. 4. Design methodology of smart machines and robots. Source: Adapted [14]

Thus the design process, in main phases, can be translated as Needs Analysis, Conceptual Design, Preliminary Design, Detail Design, Prototype Construction, Test and Evaluation and Final Documentation.

4 Innovative Design of an Automatic Equipment of Physical Rehabilitation

In the design, see Fig. 5, an exoskeleton to work the physical rehabilitation of lower and upper limbs will be designed. The proposed robotic mechanism presents two independent movements (flexion/extension of the knee and flexion/extension of the arm).

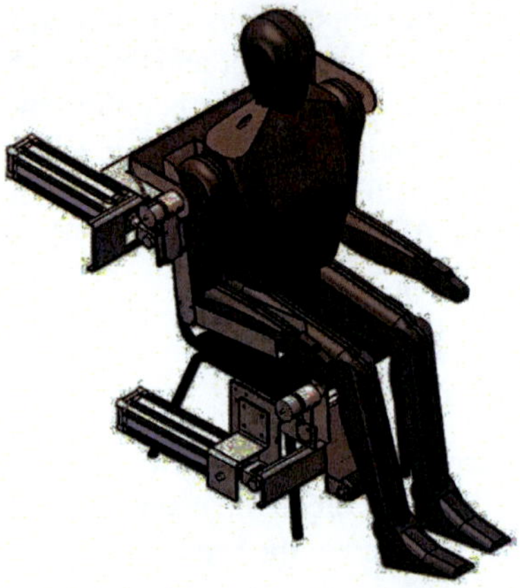

Fig. 5. A conceptual design of the pneumatically driven robotic workbench for rehabilitation of lower and upper limbs

Figure 6 shows the system, which comprises a man-machine interface design and a robot for rehabilitation. Using the graphical user interface, the physiotherapist inserts information (input) about the patient into the system and selects the appropriate type of exercise.

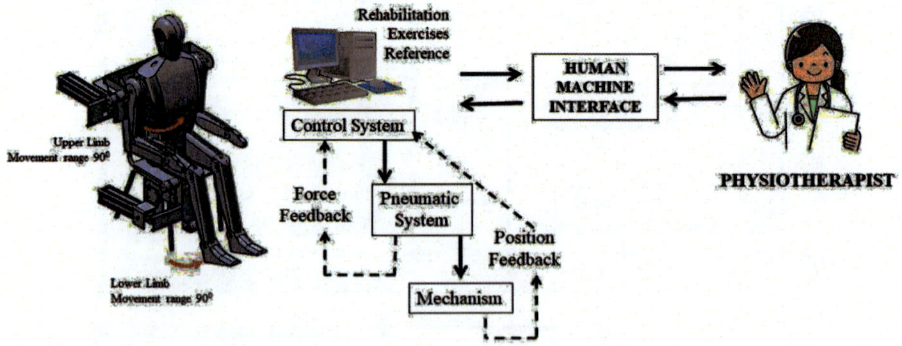

Fig. 6. The pneumatically driven robotic workbench for rehabilitation of lower limb

There are several communication methods to provide the interface between a man and a machine. The interface for the communication between the user and the machine is programmed in the desktop computer, as well as the signal processing software that sets the control commands to the robot.

The program can interact with the robot that is also connected to the same WiFi network, receiving information such as position, orientation or sending commands.

The experimental prototype is being developed at the Innovation Center for Automatic Machines and Servo Systems (NIMASS/UNIJUÍ Campus Panambi), which has adequate computational and experimental infrastructure for the construction of a workbench of tests to verify and to validate the performance of the modelling and force control on pneumatic actuators.

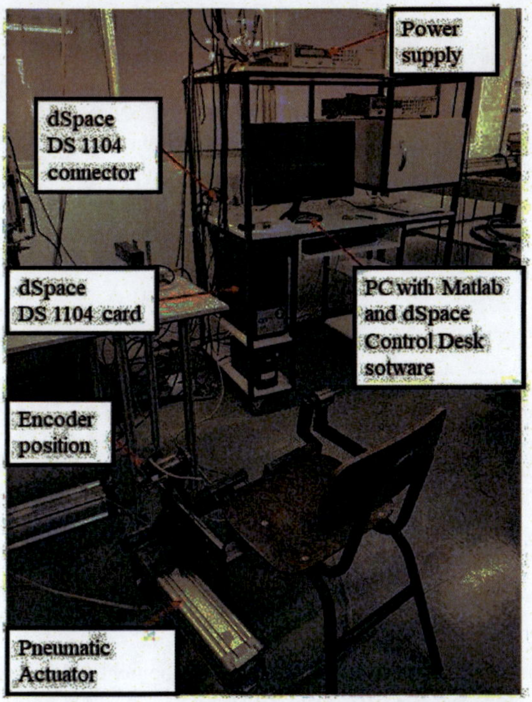

Fig. 7. Experimental workbench for rehabilitation

Figure 7 shows our experimental bench for physical rehabilitation is being developed and tested. Initially we are working with experimental tests for lower limb movement. The simulation results indicate that the new robotic mechanism produces satisfactory movements for the lower limbs.

The robot for rehabilitation can perform all active and passive exercises, as well as specific movements of exercises and execute them through the man-machine interface. Thus, it can serve as an aid to the physiotherapist. Implementing movements that are difficult to perform for the physiotherapist.

5 Conclusion

This article details the development of a new and innovative pneumatic robotic modular solution for lower and upper limbs rehabilitation that is low-cost and simple design. A contribution of this article is the application of pneumatic robots to assist in the physical rehabilitation of patients. We present the human-machine interface for to control the robot and manage the physical rehabilitation activity.

This research helped in the development of innovative equipment for the health area, more specifically for rehabilitation. It was also possible to identify the importance of multidisciplinarity for the generation of innovative ideas, based on knowledge, methodologies and tools of engineers, programmers, mathematicians and physiotherapists. The results of this article could guide the development and optimization of other similar equipment.

Acknowledgments. This study was financed in part by the Coordenação de Aperfeiçoamento de Pessoal de Nível Superior - Brasil (CAPES) - Finance Code 001. This work has financial support from Fapergs (project number 17/2551-0001014-0, Edictal 02/2017). The authors also would like to thank CNPq (National Council for Scientific and Technological Development) and Finep (Funding Authority for Studies and Projects) by financial support at the Innovation Center for Automatic Machines and Servo Systems (NIMASS) in UNIJUÍ University. And Federal Institute Farroupilha - IFFAR.

References

1. Gonçalves, R.S., Carvalho, J.C.M., Ribeiro, J.F., Salim, V.V.: Cable-driven robot for upper and lower limbs rehabilitation. In: Habib, M.K. (ed.) Handbook of Research on Advancements in Robotics and Mechatronics, pp. 284–315. IGI Global, Hershey (2015)
2. Babič, J., Mombaur, K., Lefeber, D., et al.: SPEXOR: spinal exoskeletal robot for low back pain prevention and vocational reintegration. In: González-Vargas, J., Ibáñez, J., Contreras-Vidal, J.L., et al. (Orgs.) Wearable Robotics: Challenges and Trends [s.l.], pp. 311–315. Springer, Cham (2017)
3. Babič, J., Petrič, T., Mombaur, K., et al.: SPEXOR: design and development of passive spinal exoskeletal robot for low back pain prevention and vocational reintegration. SN Appl. Sci. **1**(3), 262–266 (2019)
4. Aliman, N., Ramli, R., Haris, S.M.: Design and development of lower limb exoskeletons: a survey. Robot. Auton. Syst. **95**, 102–116 (2017)
5. Gonçalves, R.S., Carvalho, J.C.M.: Robot modeling for physical rehabilitation. In: Robotics: Concepts, Methodologies, Tools, and Applications, pp. 1212–1232. IGI Global (2014)
6. Barbosa, A.M., Carvalho, J.C.M., Gonçalves, R.S.: Cable-driven lower limb rehabilitation robot. J. Braz. Soc. Mech. Sci. Eng. **40**(5), 1–11 (2018)
7. Díaz, I., Gil, J.J., Sánchez, E.: Lower-limb robotic rehabilitation: literature review and challenges. J. Robot. (2011). Article ID 759764, 11 pages
8. Hernandez, E., Valdez, S.I., Carbone, G., Ceccarelli, M.: Design optimization of a cable-driven parallel robot in upper arm training-rehabilitation processes. In: International Symposium on Multibody Systems and Mechatronics, pp. 413–423. Springer, Cham (2017)
9. Valdivia, C.H.G., Ortega, A.B., Salazar, M.A.O., Escobedo, J.L.C.: Design and analysis of a new robotic mechanism for lower limbs rehabilitation. In: 2013 International Conference on Mechatronics, Electronics and Automotive Engineering, pp. 15–20. IEEE (2013)

10. Todorov, E., Hu, C., Simpkins, A., Movellan, J.: Identification and control of a pneumatic robot. In: 2010 3rd IEEE RAS & EMBS International Conference on Biomedical Robotics and Biomechatronics, pp. 373–380. IEEE (2010)
11. Pradipta, J., Klünder, M., Weickgenannt, M., Sawodny, O.: Development of a pneumatically driven flight simulator Stewart platform using motion and force control. In: 2013 IEEE/ASME International Conference on Advanced Intelligent Mechatronics, pp. 158–163. IEEE (2013)
12. Abry, F., Brun, X., Sesmat, S., Bideaux, E.: Non-linear position control of a pneumatic actuator with closed-loop stiffness and damping tuning. In: 2013 European Control Conference (ECC), pp. 1089–1094. IEEE (2013)
13. Riachy, S., Ghanes, M.: A nonlinear controller for pneumatic servo systems: design and experimental tests. IEEE/ASME Trans. Mechatron. **19**(4), 1363–1373 (2014)
14. Valdiero, A.C., Rasia, L.A.: Gestão de projetos de pesquisa e desenvolvimento de produtos mecatrônicos. In: Desafios em engenharia industrial, pp. 89–106. Unijuí, Ijuí (2016)

Stuttering and the Use of Facebook as a Tool for Interaction Between People Who Stutter: A Content Analysis

Gabriela Costa Arouck de Souza(✉) and Franciane da Silva Falcão

Department of Design and Graphic Expression, Federal University of Amazonas, Av. General Rodrigo Octávio, n. 3000, CEP: 69.077-000, Manaus, AM, Brazil
gabrielaarouck@ufam.edu.br

Abstract. Stuttering is a speech disorder that negatively impacts most daily activities that require communication, primarily oral, of people who have disfluency. The daily communicational difficulties are more evident for people who directly follow the life of a person with this disfluency, and affect the professional and social life as a whole of the person with stuttering. Given the possibilities of technological solutions available for communicational issues, this study aimed to understand the social situations of more significant communication difficulties in the view of the stutterer, and point out the contents and interactions most sought by this social group that could be made available with the aid of a specific technological resource for the person with stuttering. For that, the technique of content analysis of the posting of group members about the stuttering of the social network of Facebook, of Brazilian nationality, was applied. Among the contents/information interactions identified as a great help to overcome problems faced daily by people with stuttering, are: contacts and scheduling service with speech therapists specialized in this area, auxiliary platform of daily exercises that make up the treatment of this access to didactic information and practices to cope with difficult day-to-day communication situations, support network.

Keywords: Stuttering · Interaction · Facebook

1 Introduction

Disfluency is characterized by any waves that occur during the speech flow of anyone, whether stuttering or not. The manifestation of stuttering occurs in the exaggerated presence of disfluencies [1]. Stuttering is a disorder of fluency that is characterized by repetitions, ruptures, and prolongations that happen involuntarily in a spontaneous speech that directly affects the communication of the individual. According to the Brazilian Institute of Fluency [2], the incidence of stuttering in the country is 5% of the population. This means that 10 million people have some level stuttering. The permanence of the disorder is 1%, about 2 million Brazilians stutter chronically.

Stuttering is accompanied by negative feelings, usually such as frustration, shame, anxiety, and inferiority in the other [3]. Any communication situation can pose a threat

L. Pereira et al. (Eds.): IDEAS 2018, SIST 198, pp. 168–178, 2021.
https://doi.org/10.1007/978-3-030-55374-6_17

to the stutterer and consequently arouse these negative emotions. Most of the time, the communication disorder occurs on a wide range of activities at school, at work, and also at home [4–8]. Some students and teachers stereotype stuttering negatively. A comparison between students who have fluent speech and students who stutter, it is possible to note the higher risk of rejection to stuttering students, increasing the possibility of being targets of bullying [9–11].

Blood [12], for example, investigated experiences of self-esteem, life satisfaction, and bullying among 54 adolescents with stuttering and 54 adolescents without stuttering. Adolescents with stuttering compared to those who did not have the disorder reported significantly more victimization among peers and low self-esteem. In another study, it was shown that the risk of bullying is significantly higher in stuttering adolescents than adolescents without the disorder [13].

Like children and adolescents, adults who stutter also experience daily difficulties. Klompas and Ross [14] studied the quality of life in adults who stutter and found negative impacts of stuttering on emotion, self-esteem, self-image, work performance, academic and peer and teacher relationships. When interviewing 232 stutterers, Klein & Hood [15] found that 40% of stutterers agreed that job choice and earnings were negatively affected by the disorder and the other 70% believed that the chances of being hired or promoted were reduced, according to declarations. It is remarkable the harmful influence that this disorder has on people's lives, whether in social life, in the practical relationships between parents and children, colleagues, partners in love, in homes, in schools or work environments.

A few years ago, mobile health (m-Health), defined as the use of mobile computing and communication technologies for health care and public health, has been growing steadily. The target audience for this type of service are doctors, nurses, patients, and even healthy people [16]. The use of mobile technology can support diagnosis and compliance with treatment guidelines, as well as increase administrative efficiency and keep necessary information of patients [17, 18].

With the impact of stuttering on the different dimensions of the life of the patient with disfluencies, this study aimed to understand the social situations of greater difficulty in communication in the stutter's vision, and to point out the information and content interactions most sought by this group that could be made available with the aid of a specific technological resource for the person with stuttering.

2 Materials and Methods

This article is exploratory research that is based on an identification of reports of problematic situations experienced by people who have stuttering, through the application of the technique of content analysis in postings of groups in social network Facebook.

2.1 Study Object

From a survey conducted on the social network Facebook to raise the number of groups focused on the subject on the disorder, a search was done on the tool "search" with the word "gagueira" (Portuguese from the word stuttering) and "stuttering", and the search was filtered by selecting the "groups" option, located on the same page. Then there are 29 national groups and 44 international groups. The groups were analyzed some members and status, categorizing them into active or non-moving. The absence of movement of the group was determined due to the restriction of information from the closed groups.

The object of study of this work were the reports published in 4 groups of Facebook on the theme stuttering, all of the Brazilian origin, whose date of posting was in the period from September 25, 2018, to October 25, 2018. Groups that fit in the previously established criteria, were selected for the analysis of the publications, following the methodology of content analysis.

2.2 Study Environment

The study environment of this article is the social network of Facebook. According to the Internet Steering Committee in Brazil [19], about 107 million Brazilians are Internet users, which corresponds to 61% of people ten years of age or older. The use of online social networks is 78% among the most popular activities carried out by its users. According to Folha de São Paulo [20], Brazil reached the mark of 127 million monthly active users in the first quarter of 2018, is one of the five largest markets of the company. For this reason, as the study environment for analyzing user reports on the subject of stuttering, Facebook was chosen. In addition to these data, we opted for the analysis of the publications on Facebook instead of interviews because of the difficulty that the people who have the disorder would face to responding orally to the questions made by the author.

2.3 Method of Data Analysis

For the right understanding of the interaction of members of Facebook groups, a content analysis was selected, according to Bardin [21], which can include social network analysis. The content analysis analyzes the communications, what was spoken in interviews, or what is verified by the researcher. The analysis of material aims to classify themes or categories that help the more profound understanding of discourses [22]. The stages of the techniques, developed by Bardin [23], are organized in three phases: pre-analysis, material exploration, and treatment of results, inference, and interpretation.

The first phase is characterized by the systematization of the initial ideas placed by the theoretical framework and by the establishment of indicators for the interpretation of the acquired information. It is understood in floating reading, the choice of the documents, formulation of the hypothesis, and lastly the elaboration of the indicators, with the purpose to interpret the material collected.

The second phase consists of the exploration of the material, where the construction of the coding operations is done, which considers the cuttings of the texts in units of registers, the definition of rules of counting and classification and aggregation of the information in categories, be they symbolic or thematic. The first category is grouped according to the related themes and offer to the initial categories. Then, the initial categories are grouped according to the theme for which intermediate categories arise, and the process is done again for the creation of the final categories [24].

The third and last phase developed by Bardin [23] comprises the treatment of results, inference, and interpretation, the understanding of the manifest and latent contents inserted in all the collection material. A comparative analysis is carried out through the juxtaposition of the categories listed in each analysis, highlighting the similar and different aspects.

3 Results and Discussion

Next, the results of the initial, intermediate and final categories, according to the method of content analysis, selected in the social network Facebook, will be displayed, based on the inclusion criteria established by the authors.

3.1 Results of Content Categories: Initial Category

The initial categories are characterized as the first impressions about the reality being studied. In the coding process, a total of 52 categories were produced. Each category consists of words, taken from the selected posts according to the inclusion criteria created by the author. It is important to note that there were no requirements for the naming of the categories as well as the quantity of the categories. The following table illustrates the naming given to each starting category (Table 1):

Table 1. Initial category

Initial category		
1. Sessions with a specialist speech therapist	19. Live on Instagram on the subject of stuttering	37. Job search/difficulty
2. Professional help	20. Lose memory and cure stuttering	38. Improvement in verbal fluency
3. Disclosure of specialist in Neuro-Linguistic Programming	21. Case of cure	39. Stuttering Decreased Attempt
4. International stuttering Day	22. Prejudgement	40. Inclusion in society
5. Stuttering is not funny, has treatment	23. Acceptance at work	41. Difficulty expressing yourself
6. Brazilian meeting of people who stutter	24. Difficulty finding a place to practice therapy exercises	42. Low self-esteem
7. Event DIAG 2018/RJ	25. Personal fight	43. Presentation of work and frustration

(continued)

Table 1. (*continued*)

Initial category		
8. Postgraduate in Fluency and Language	26. Rejection in employment/job interviews	44. Finish sentences and be interrupted
9. Training Course - Fluency Workshop	27. Loss of opportunity	45. Lead a normal life
10. Results of therapies	28. Training/chances of not stuttering	46. Acceptance of defects
11. Stuttering Device	29. Failures, afflictions, annoyance, and nervousness	47. Do not give up
12. Methods of overcoming	30. Experience of a stuttering teacher	48. Personal Acceptance
13. Visualization Exercise	31. the difficulty of interacting with unknown people	49. Do not be ashamed
14. The benefit of being placed at the same height as a child	32. Lack of confidence and much effort	50. Patient commitment
15. Global Therapy Questioning	33. Bliss/sadness	51. Trust for being able to express yourself
16. Doubts about stuttering	34. Stuttering worsening	52. Do not miss good opportunities
17. Study on anticipation of stuttering	35. Withdrawal due to stuttering	
18. Understanding Stuttering	36. Physical exercise	

Source: Elaborated by authors

For the refinement of the data analysis, it is necessary to group the initial categories, which results in the intermediate categories, presented in the later section.

3.2 Results of Content Categories: Intermediate Category

After the presentation and discussion of the initial categories, the intermediate categories were elaborated, totalizing into six categories. Due to the page limit, only three initial categories were presented in each table as an example for the formation of the intermediate categories, in order to help the understanding of the results of the same. The agglutination of the first three initial categories originated the first intermediate category, named professionals of the area, presented in the following Table 2:

Table 2. Intermediate category I – Area professionals

Initial category	Guiding concept	Intermediate category
Sessions with a specialist speech therapist	One member of the group questions the effect of expert sessions	**I) Area professionals**
Professional help	Reflective text on the search for professional help	
Disclosure of specialist in Neuro-Linguistic Programming	Disclosure of a specialist recommended by one of the group members	

Source: Elaborated by authors

The second intermediate category, disclosure, is carried out the surveyed of events and courses that are released by the participants of the analyzed Facebook groups. In Table 3, the second intermediate category is displayed.

Table 3. Intermediate category II – Events and courses

Initial category	Guiding concept	Intermediate category
International stuttering Day	Disclosure of graphic material.	**II) Events and courses**
Stuttering is not funny, has treatment	The slogan used in campaign awareness materials	
Brazilian meeting of people who stutter	Disclosure of event artwork	

Source: Elaborated by authors

Then, Table 4 presents the third intermediate category, termed Therapies, Exercises, and Treatments. This category aims to group the subjects of therapies, exercises, and treatments shared by the members of the Facebook groups.

Table 4. Intermediate category III – Therapies, exercises, and treatments

Initial category	Guiding concept	Intermediate category
Results of therapies	Reflective text on the results of therapies	**III) Therapies, exercises, and treatments**
Stuttering Device	Questioning the use of the SpeechEasy device to the members of the group improves	
Methods of overcoming	Dissemination of methods of overcoming in video	

Source: Elaborated by authors

The fourth intermediate category is focused on issues about stuttering, discussed by users of Facebook groups. Table 5 illustrates the process of forming this intermediate category.

Table 5. Intermediate category IV – Stuttering

Initial category	Guiding concept	Intermediate category
Doubts about stuttering	Sharing videos of people who stutter, professionals	**IV) Stuttering**
Study on anticipation of stuttering	Research on anticipation of stuttering disclosed in the group	
Understanding Stuttering	Video shared by one member about stuttering	

Source: Elaborated by authors

The fifth intermediate category, personal experiences, highlights the sharing of people experiences, positive and negative feelings caused by the disturbance. Table 6 presents the fifth category:

Table 6. Intermediate category V – Personal experiences

Initial category	Guiding concept	Intermediate category
Prejudgement	Prejudices suffered by members of the group shared in text form	**V) Personal experiences**
Acceptance at work	The slogan used in campaign awareness materials	
Difficulty finding a place to practice therapy exercises	Questioned by a member of the group	

Source: Elaborated by authors

The formation of the sixth and last intermediate category, called motivational, is based on the messages of motivation written or shared by the members of the groups analyzed, presented in the following Table 7:

Table 7. Intermediate category VI – Motivational

Initial category	Guiding concept	Intermediate category
Lead a normal life	Text about personal experience plus motivational message	**VI) Motivational**
Acceptance of defects	Shared video about personal acceptance, dealing with defects	
Do not give up	Personal experience followed by a motivational message	

Source: Elaborated by authors

3.3 Results of Content Categories: Final Category

The final constitution consists of three categories called "Disclosure," "Practical information," and "Self-help." The final categories represent the synthesis of the apparatus of significations, raised in the course of the data collected. Table 8 shows the formation of the first final category:

Table 8. Final category I – Disclosure

Intermediate category	Guiding concept	Final category
I) Area professionals	In addition to sharing the names of specialized professionals, there are also questions about the results obtained in the care of speech-language pathologists	**I - Divulgation**
II) Events and courses	Posting of disruption awareness campaign material, lectures, events, graduate and courses	

Source: Elaborated by authors

Table 9 presents the second final category, entitled, Practical Information, as follows:

Table 9. Final category II – Practical information

Intermediate category	Guiding concept	Final category
III) Therapies, exercises, and treatments	Information, questions, and suggestions of therapies, exercises on the disorder	**II - Practical information**
IV) Stuttering	Information about stuttering, video links, discussions started by members	

Source: Elaborated by authors

Finally, the last final category, called self-help, contributes to the finalization in the process of interpretations that will collaborate in the understanding of the project (Table 10).

Table 10. Final category III – Self-help

Intermediate category	Guiding concept	Final category
V) Personal experiences	Experiences experienced by group members	**III - Self-help**
VI) Motivations	Motivational messages that support	

Source: Elaborated by authors

The content analysis, elaborated methodology Bardin [23], aided in the investigation of the interaction between the members of the Facebook groups. We have a final result, three categories called "Disclosure," "Practical information," and "Self-help." It is noted that reports of personal experiences are published more frequently, as well as issues directly related to stuttering. Personal experience as an oral presentation before a broad audience, job interviews, and difficulty interacting socially with other people are some examples of the publications found. It is observed that these are common difficulties, which cause embarrassment when doing simple tasks, such as talking on the phone or just asking someone for information. These people have found a way to expose their problems, obtain·information about the disorder, and also have the social support of the members. The exchange of information and experiences fuel the existence of the group, making Facebook a channel of communication between people who stutter. The answer obtained in these analyses contributed to the understanding, as well as to the development of this work.

4 Conclusion

It is remarkable, that stuttering can affect several aspects of the life and daily life of a stutterer. Through this work, we can see some difficulties reported through the analysis made by the social network Facebook. Difficulties involving oral communication, social interrelationship, practical situations as a simple phone call, already cause discomfort to the people who have the disorder. It is worth mentioning that the method of analysis used to achieve the objective of this article was of such importance since it was possible to extract with more precision the situations experienced by members of the social network groups, in a more free and spontaneous way. Interviews and other techniques of inquiry could cause constraints because of the communicational difficulty that this public has.

We then look at technological resource solutions that can facilitate some of the day-to-day tasks. The development of an m-health application could help in aid of treatments, as well as the accompaniment of professionals specialized in the field. An example would be the development of a stuttering system that offers to the schedule of consultations by the application with specialized professionals, without the stutter needing to talk on the telephone, applications that help in the treatment, the exercises that the stutter needs to practice daily. Creating a system that discusses the problem, questions, and schedules queries would be a solution. It is hoped that this study will contribute as a content aid to the development of specific technological resources for people with stuttering.

References

1. Oliveira, C.M.C., Cunha, D., Santos, A.C.S.: Risk factors for stuttering in disfluent children with familial recurrence. Rev. Audiol. Commun. Res. **18**(1), 43–49 (2013)
2. Brazilian Institute of Fluency
3. Dias, T., Alves, C., Vandenberghe, L.: The treatment of stuttering and the therapist-client relationship: a case study. Acta Comportamentalia **22**(3), 352–364 (2014)
4. Bloodstein, O., Bernstein Ratner, N.: A Handbook on Stuttering, 6th edn. Thomson, Delmar Learning, Clifton Park (2008)
5. Shafiee, B., Mehralian, Z.: Disfluency Disorders in Children's Speech, 11th edn. Isfahan University of Medical Sciences, Isfahan (2003)
6. Train, Y., Blumgart, E., Craig, A.: Subjective distress associated with chronic stuttering. J. Fluency Disord. **36**(1), 17–26 (2011)
7. Yaruss, J.S., Quesal, R.W.: Overall Assessment of the Speaker's Experience of Stuttering (OASES): documenting multiple outcomes in stuttering treatment. J. Fluency Disord. **31**(2), 90–115 (2006)
8. Craig, A.: An investigation into the relationship between anxiety and stuttering. J. Speech Hear. Disord. **55**(2), 290–294 (1990)
9. Dorsey, M., Guenther, K.: Attitudes of professors and students toward college students who stutter. J. Fluency Disord. **25**, 77–83 (2000)
10. Turnbull, J.: Promoting greater understanding in peers of children who stammer. Emot. Behav. Diffic. **11**, 237–247 (2007)
11. Davis, S., Howell, P., Cooke, F.: Sociodynamic relationships between children who stutter and their non-stuttering classmates. J. Child Psychol. Psychiatry **43**, 939–947 (2007)
12. Blood, G.W., Blood, I.M., Tramontana, G.M., Sylvia, A.J., Boyle, M.P., Motzko, G.R.: Self-reported experience of bullying of students who stutter: relations with life satisfaction, life orientation, and self-esteem. Percept. Mot. Skills **113**(2), 353–364 (2011)
13. Blood, G.W., Blood, I.M.: Bullying in adolescents who stutter: communicative competence and self-esteem. Contemp. Issues Commun. Sci. Disord. **31**, 69–79 (2004)
14. Klompas, M., Ross, E.: Life experiences of people who stutter, and the perceived impact of stuttering on quality of life: personal accounts of South African individuals. J. Fluency Disord. **29**, 275–305 (2004)
15. Klein, J.F., Hood, S.B.: The impact of stuttering on employment opportunities and job performance. J. Fluency Disord. **29**(4), 255–273 (2004)
16. Free, C., Phillips, G., Felix, L., Galli, L., Patel, V., Edwards, P.: The effectiveness of M-health technologies for improving health and health services: a systematic review protocol. BMC Res. Notes **5**(14), 2–7 (2010)
17. Sherry, J.M., Ratzan, S.C.: Measurement and evaluation outcomes for mHealth communication: don't we have an app for that? J. Health Commun. Washington **1**(17), 1–3 (2012)
18. Vázquez, M.Y.G., Sexto, C.F., Rocha, Á., Aguilera, A.: Mobile phones and psychosocial therapies with vulnerable people: a first state of the art. J. Med. Syst. **40**(6), 1–12 (2016)
19. Internet Management Committee in Brazil. Research on the use of information and communication technologies in Brazil: TIC Domiciles 2016. São Paulo (2017)

20. Folha de São Paulo: Facebook reaches 127 million monthly users in Brazil. https://www1.folha.uol.com.br/tec/2018/07/facebook-chega-a-127-milhoes-de-usuarios-mensais-no-brasil.shtml. Accessed 19 Sept 2018
21. Bardin, L.L.: Content Analysis. Presses Universitaires de France, Paris (1977)
22. Silva, A.H., Fossá, M.I.T.: Content Analysis: example of the application of the technique for qualitative data analysis. Qualitas Revista Eletrônica **17**(1) (2015)
23. Bardin, L.: Content Analysis. Edições 70th, São Paulo (2011)
24. Fossá, M.I.T.: Proposition of a construct to analyze the culture of devotion in family and visionary companies. Thesis (Doctorate in Administration) – The Federal University of Rio Grande do Sul, Porto Alegre (2003)

Exploratory Study on the Behavior of the Brazilian Financial Market Using Google Trends

Fernando Gonçalves de Castro Filho[1(✉)], José Eduardo da Costa Dias[1], Alexandre Acácio de Andrade[2], and Julio Francisco Blumetti Facó[2]

[1] Universidade Federal do ABC, Avenida dos Estados, 5001, Bairro Santa Terezinha, Santo André, São Paulo, Brazil
fernandocastro23@yahoo.com.br, eduardo.dias@ufabc.edu.br

[2] Universidade Federal do ABC, Rua São Paulo, s/n, Bairro Jardim Antares, São Bernardo do Campo, São Paulo, Brazil
{aacacio,julio.faco}@ufabc.edu.br

Abstract. Search engines changed the way people find relevant information on the internet. We suggest that investors utilize these internet tools during their investment decision process and that the datasets generated by search engines are related to stock market movements. The purpose of this study is to analyze the relationship between this social phenomenon and moves on the Brazilian stock market. Following previous studies, we analyze the correlation between the Google query volumes for terms related to a specific index of the Brazilian stock market (Ibovespa) and stock market metrics (Open, Closing, High and Low prices). Our results presented a positive correlation between the investors' attention, represented by the Google search volumes, and the market prices. These results suggest that the Google Trends data are more positively correlated to the highs of the financial data, suggesting that investors tend to search more on the web when the market is reaching its peak or vice versa. This article highlights the potential that this source of information has on the understanding of the Brazilian stock market.

Keywords: Google Trends · Investor behavior · Brazilian stock market

1 Introduction

Prices fluctuate at all times, this is the result of the interaction between the participants of the market that is offering the product of interest. These participants can range from companies with significant capital to small individual investors. The thing in common here is that all of them buy or sell their products according to their expectations about the future of the market.

In normal conditions, what is expected of the consumer – in the case of this article, we can substitute consumer by the investor – is that they will always seek a way to increase their satisfaction by reducing any uncertainty involved in the process of choosing a product, or an investment. Therefore, consumers and investors tend to take a more active

L. Pereira et al. (Eds.): IDEAS 2018, SIST 198, pp. 179–188, 2021.
https://doi.org/10.1007/978-3-030-55374-6_18

behavior when searching for information related to products or investments, which today, thanks to technological advances and greater accessibility, Internet searches became an essential part of this process. When a consumer searches for products online, this usually leads to the purchase of this product [1].

According to Simon [2], the investor's decision-making process begins with a stage of gathering information. As previously stated, the internet has facilitated the process of collecting information over time, and search engines can be considered the superior technology for this purpose. This type of technology helps when web browsing and, in most cases, users use it as an entry point to the World Wide Web [3].

In 2006, Google made it possible for anyone to access information related to the frequency of specific search terms that were being entering their search engine. This tool, called Google Trends – since its data originate from Google users – has been used as a way to explain some social phenomena; a simple example is the detection of influenza epidemics according to the search for words related to the symptoms of the disease [4]. This type of data has applications to a wide range of subjects, from forecasts of the unemployment level of the United States [5] to house and apartment price forecasts [6].

Specifically, for the financial market, this type of analysis has been used to predict the price of different assets. According to Da [7], the volume of Internet searches related to stock market index names can be interpreted as a measure of investors' attention on the market. When this investor verifies a growing demand for information about a particular market index, it has a high chance of using search engines as a source of information.

Given this scenario, the present study seeks to understand if it is possible to use web-generated search data as a source of information to predict the behavior of the Brazilian financial market. According to the recent literature [8–10], on the United States financial market, this type of data is used, but what we propose in this article is to explore these same properties in a market that is smaller and is placed inside a developing country and by that, more dependent on foreign capital.

This work is organized as follows: the presentation of the bibliographic review pertinent to the subject. Right after, the scientific methodology used is detailed in each of its steps. Following, we will have the exposure of the results and our considerations on the topic.

2 Literature Review

The literature review of this study is divided into four major parts. The first one aims to explain the origin of search engines and how their data are being used to make estimates of social phenomena related to Financial markets. The second part is responsible for detailing the functioning of the financial market and the particularities of the capital market of Brazil, compared to the United States. Finally, we try to understand how the investor behaves in these markets, explaining the possible reasons for the adoption of certain types of behavior.

2.1 Search Engines and Finance

The entire search process begins with the need for information. The user is looking for something: from this, he formulates the query in its verbal form, choosing a keyword. The system of the search engine then returns the documents that, within a collection of documents stored in its database, corresponds to the query [3].

Search engines are tools designed to retrieve information from the World Wide Web (WWW). Due to the volume of data processed, every search engine has as its main competencies - in addition to searching and returning relevant results - the administration of large databases [3].

Every time a user accesses a search engine and performs a query, data such as time, search term, and the region where the computer is located are saved into the databases. This data can be classified as web search data. These "digital traits" that we leave can provide insights into a broad range of disciplines [11] and can be applied in virtually any marketplace where Internet searches precede the operation, which holds true for a large share of the world economy [6].

In today's world, there is a wide variety of data sources that refer to the digital traits that consumers leave on the Internet in their daily lives; likes; tweets; location; searches; among others. Each of these data can tell us something about the behavior of users on the smallest possible scale and therefore assist in predicting behavior, movements and even prices of markets.

According to Wu [6], markets may vary in the horizon of predictability, depending on the discrepancy between the digital search process that precedes the transaction and the transaction itself. Therefore, the stock market, when observed in short periods, is more susceptible to manipulations and, with that, requires a higher quality of data.

For the Brazilian financial market, Ramos [10] found a significant relationship between Google's research data and the market itself. The authors used as keywords; stock tickers, index names and words related to the fixed income market to explain the volatility, trading volume and future returns over the analyzed period (from 2007 to 2014).

Given the particular characteristics of the Brazilian financial market, which will be addressed in greater depth during the next chapter, the work of Ramos [10] result shows that searches originated both in Brazil and abroad have a positive correlation to financial returns and volatility when related to individual companies and indexes. The same does not occur for terms related to the fixed income market.

2.2 Brazilian Financial Market

For Andrezo [12], the financial market can be classified as a set of institutions, markets, and assets that enable the transfer of resources between borrowers (who needs capital) and investors (who lends the capital). In this article, we will only deal with a segment of the capital market; the stock market. This specific type of market allows investors to buy or sell public traded companies and companies to access capital by transferring a part of the ownership.

In Brazil, there are four basic types of investors: individuals; legal persons; external investors and institutional investors. Institutional investors tend to invest compulsorily by following with the rules of the composition and diversification of their investment portfolios. This group has high importance for the financial market as a whole: given their weight; these guarantee a certain level of stability to the market. As an example of representatives of this group we have; pension funds, investment funds, insurance, and capitalization companies and investment companies [13].

The group of individual investors, individual or legal, are represented by "ordinary" persons or companies that participate in the market, buying or selling assets individually. Any individual or group of persons, whether physical or legal, investment funds or any other type of investment entity with a residence, head office or domicile abroad is considered to be an external investor [13].

An interesting characteristic of the Brazilian market is the distribution of types of investors and their respective market share. It's noticeable in Fig. 1 that, currently half the Brazilian market is made up of foreign investors. These showed strong growth between the years 2010 to 2014, and have maintained steady growth since then. Only 17.7% of the market share is held by individual investors [14].

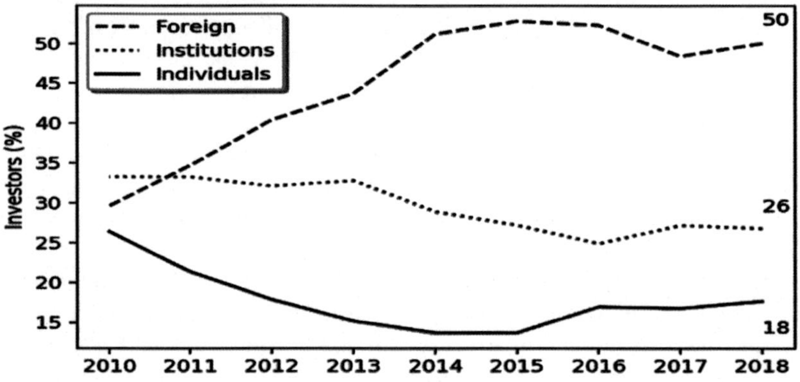

Fig. 1. Evolution of Investors' participation in the Brazilian financial market. Source: Elaborated by authors.

When evaluating the performance of investments involving stocks, the Bovespa index (Ibovespa) is always indicated as the reference for such a comparison, considered the primary benchmark for the variable income market. The Ibovespa is revalued every four months and is the result of a theoretical portfolio of assets. Its main objective is to be the indicator of the average performance of asset prices with the highest marketability and representativeness of the Brazilian stock market (approximately 80% of the number of trades and the financial volume of the entire market), being composed only of shares and units of companies which meet the criteria of its methodology [15].

Currently, the index is composed of 64 companies representing 25 sectors of the Brazilian economy. Among the areas, the Financial Intermediaries sector (28.9%) is the most representative, consisting of four of the largest commercial banks in the country. Next, we have the Oil, Gas and Biofuels sector (15.7%), which is strongly represented by Petrobras shares (PTR3 and PTR4) and in third place, we see the Mining sector (12.2%), with Vale as a strong company (VALE3) (B3, 2018). The combination of these three large sectors accounts for about 60% of Ibovespa's composition, which shows an intense concentration of the Brazilian stock market.

2.3 Behavioral Finance

According to Fama [16], an efficient market is a market in which security prices always entirely reflect the available information. The question of how fast information reaches the participants and affects their trading decisions has generated one of the pillars of the Efficient Market Hypothesis (EMH). The prices of this type of market are the reflex of how investors to allocate their capital, by assuming that they are behaving rationally.

On the other hand, behavioral finance has emerged to challenge the EMH and emphasizes the importance of human emotion, sentiment, and mood on the financial decision-making process. This new way of viewing the market does not expect it to be efficient, at the most general level, behavioral finance is the study of human fallibility in competitive markets, bringing the attention to some rational actions of the individuals, but that in fact could be being developed unconsciously [17].

It's common sense that making right investments decisions can become a challenge of controlling one's own emotions, because pursuing immediate pleasures are not always the best decision when it comes to investments, meaning a sub-optimal choice. According to the Prospect Theory [18], decision weights are not probabilities as they do not obey the probability axioms and they should not be interpreted as measures of degree or belief. This theory seeks to describe how mental structures interfere in the decision process involving risk.

Individuals, in general, value gains and losses against a reference level and do not think about the final state of wealth (the reflection effect), this means that among investors, there is a greater concern with price oscillations than in knowing whether the final results that can be achieved address their objectives. Also, investors tend to retain losing investments over a long period, while making gains quickly with winning positions (disposition effect): this bias generally limits gains and amplifies losses. Another exciting discovery of the Prospect Theory is that individuals tend to prefer a small decline over a short probability of a substantial loss (certainty effect). For example, people prefer an 80% safe gain of 3,000 to a chance of 80% gain of 4,000. Thus, the decision will be in favor of the alternative of lower risk even if this brings less expected benefit [18].

It is difficult to sustain the case that people in general, and investors in particular, are entirely rational [17]. Overestimating our technical knowledge can lead to the illusion of a skilled investor – even if the good results come from the inertia of following a trend or from chance – and make ourselves excessively confident, leading to worse performance on investments [19].

In the investment process, it is common to find difficulties in filtering which information is relevant and which are not in determining the price of an asset. In general, while the individual investor is vulnerable to recent news reports with greater exposure to the media, the professional investor tends to be susceptible to the opinion of his professional community.

There is psychological evidence that shows that people tend to deviate from rationality in the same way, and not randomly. If we consider that the average investor forms their investment portfolios more based on emotions than on facts, thus, buying and selling would be highly correlated among investors. They would tend to buy the same assets, the ones that are getting the most attention on media [17].

Large financial movements, and ultimately crises, are often driven by collective phenomena such as the herd effect. Such an effect can be described as when a set of investors position their investments in the same or similar directions, based almost exclusively on the fact that many others are performing that same positioning. The fear of losing an idea of profitable investment is often the driving force behind this kind of behavior [20].

3 Methodological Procedures

This study presents an exploratory character since it aims to provide an overview of a subject, until then, little explored for the Brazilian market. Descriptive and empirical, because it presents characteristics of a phenomenon and establishes a relationship between the variables utilizing data [21]. The following subsections will expose the data characteristics and the statistical method utilized.

3.1 Data

Google Trends is a service that analyzes a random part of Google's searches to calculate how many searches were done for specific terms. The tool is based on two variables: time and location. Each Google Trends data point is calculated by dividing the total of searches for a particular geographic region and the period that total searches cover, resulting in a measure of relative popularity. The resulting numbers are scaled in a range of 0 to 100 based on the ratio of a topic to all searches in all topics. For example, a value of 100 represents the peak popularity of a term during the determined period [22].

The service does not provide search information for daily periods, except for frequent search terms, which is not our case. Therefore, we conduct our analyses for monthly periods. Data were also obtained using region filters, both for Brazil and Global, what we want to validate here is whether there is any difference between the research relationships of the different regions, since the Brazilian financial market has a strong influence of foreign investors.

There are some limitations to using this service, in addition to providing only a normalized index of interest – not the total volume of searches –, the region where the search is filtered may also lead to different results. In areas like China and Russia for example, Google's search engine is not the most commonly used tool, which directly affects results. In the case of Brazil, the dominance of the Google search engine is clear. In the country, for the year of 2017, the mechanism had 97% of the search engine market, against a shy 1.7% for the second place [23].

The financial information was extracted from the Yahoo Finance portal. The service is provided free of charge and, it receives the information from the data company called ICE Data Services and passes it on in the original format received with a delay of 15 min (in the case of Brazil).

According to Elder [24], the opening price generally reflects the opinion of amateur investors about the value. Opening prices, most of the time, are closer to the minimum or maximum of that period, and during the day, the price of the asset tends to recover and reach a more moderate value. Still, according to the author, the closing price tends to reflect the opinions of professional investors, characterizing them as more active at the end of trading periods.

3.2 Statistical Analysis

The statistical analysis initiates by data exploration. The purpose of this step is to better understand the data to be studied and to describe the datasets utilizing descriptive tools such as tables, graphs, histograms, averages, among others. The principle here is to first analyze the variables separately so that we can then check if there is any relation between them [25].

To check if there is a linear relation between the variables, we are utilizing the Pearson correlation coefficient (PCC) that measures the direction and intensity of the linear relationship between two quantitative variables. The PCC has a value that varies between $+1$ and -1; where 1 is the total positive linear correlation (meaning that when the score of one variable raise; we expect the score of the other observation to raise as well, and vice versa). 0 is no linear correlation, and -1 is the total negative linear correlation (the opposite movement is expected, if a score of one variable goes up, the other variable goes down, and vice versa) [25]. The PCC is a standard statistical tool, and it was utilized on two other papers [7, 26] related to the topic.

There are a variety of ways of performing statistical checks, for example; in its work, Ramos [10] used the Autoregressive Vector Model (VAR) and Granger causality tests to verify the relationships between the research volumes and the rates of return, volatility and volume of trading conducted for stock market companies and domestic fixed income market.

In Rao's work [26], as its first step, the author calculates the PCC and cross-correlation between the time series. The second step consists of applying a Granger causality test to find if there is causality between the two variables. On the third and last step, Rao uses the Expert Model Mining System (EMMS) – which incorporates a set of competing methods – to verify if the features proposed in the earlier sections can be useful to predict prices of a financial instrument. In this work, we will perform the data exploration and PCC to verify what information can be extracted from the data.

4 Results and Considerations

In this section, results of the data exploration and the Pearson correlation coefficient are presented. Figure 2 shows a line plot of the Google Trends data, with no region filter, and a line plot of the closing price of the Ibovespa. This figure illustrates the behavior of the two variables, and it can be perceived that it may have a positive correlation overall, but on specific periods of time (i.e. between the year 2008 and 2009), the correlation appears to be negative.

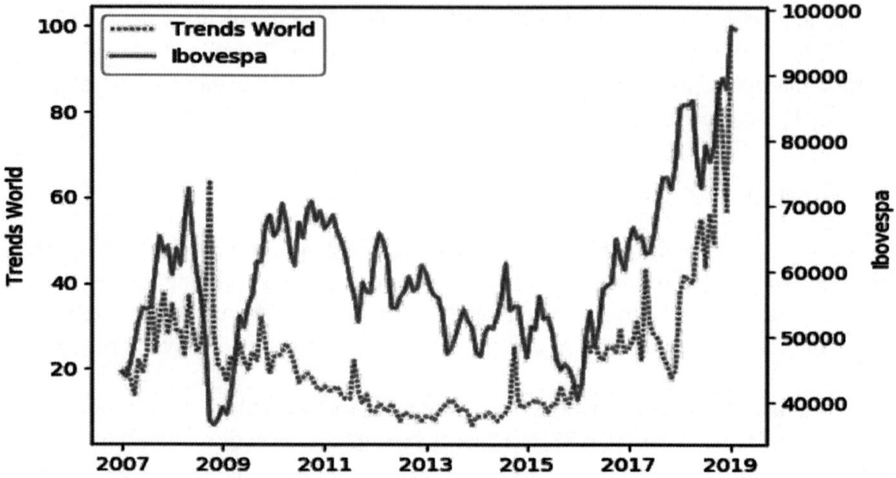

Fig. 2. Line plot of the Google Trends data and Ibovespa. Source: Elaborated by authors.

Search data for the term Ibovespa, with no region filter, presented a positive and significant correlation with different financial metrics of the index for the period between 2007 and the beginning of 2019. As we can observe in Table 1, the most relevant correlation is between the peak of the price on the period observed (high) and the trends data. This corroborates with the hypothesis that the Brazilian financial market is influenced by foreign investors, but not in a significant way because the difference between the filtered and non-filtered trends data are irrelevant. What can be inferred from the correlations table (Table 1) is that the Google Trends data are more positively correlated to the highs on the financial data, suggesting that investors tend to search more on the web when the market is reaching its peak or vice versa. It's also worth mentioning that we do not observe any overall negative correlation on the period.

Table 1. Correlations table.

Correlation	Open	High	Low	Close
Trends World	0.6109	0.6403	0.5458	0.6157
Trends Brazil	0.5588	0.5910	0.4897	0.5686

Source: Elaborated by authors

When isolating and separating the "High" financial variable (the most correlated) and the "Trends" variable with no region filter in periods of one complete year, we can check in which year the investors interest on the web for the term "Ibovespa" are more correlated to price movements on the index itself. Table 2 displays the correlations between these two variables separated by year.

Table 2. Correlations table between High and Trends variables.

High	2007	2008	2009	2010	2011	2012
Trends	0.7009	−0.1106	0.4765	−0.1120	0.1234	0.6009
High	2013	2014	2015	2016	2017	2018
Trends	−0.4938	0.3828	−0.5795	0.7428	−0.5256	0.2566

Source: Elaborated by authors

We can observe within three (2008, 2010, 2011), out of the twelve periods, a neutral correlation. It is observed a negative correlation on the periods of 2013, 2015 and 2017, also accounting for 25% of the periods. And in 50% of the periods, it is observed a positive correlation between the two variables, peaking at 0.7428 in 2016.

This work is still in progress but it's relevant to remember that correlation does not imply causation, but it's clear that there is a significant correlation between searches registered by search engines and the investor's attention, represented by the Google search volumes, to market movements. To establish a causation relationship between these variables a further statistical analysis must be applied.

In future works, we plan to add a broader range of keywords related to the Brazilian Financial Market and compare its results of relevance with random words. Another plan is to combine different companies from different sector and check if there is any correlation between those companies, and their respective keywords, to the trends data.

References

1. Shim, S., Eastlick, M.A., Lotz, S.L., Warrington, P.: An online prepurchase intentions model. J. Retail. **77**, 397–416 (2001)
2. Simon, H.A.: A behavioral model of rational choice. Q. J. Econ. **69**, 99 (1955)
3. Lieberam-Schmidt, S.: Analyzing and Influencing Search Engine Results Business and Technology Impacts on Web Information Retrieval. Gabler (2010)
4. Ginsberg, J., Mohebbi, M.H., Patel, R.S., Brammer, L., Smolinski, M.S., Brilliant, L.: Detecting influenza epidemics using search engine query data. Nature **457**, 1012–1014 (2009)
5. Ettredge, M., Gerdes, J., Karuga, G.: Using web-based search data to predict macroeconomic statistics. Commun. ACM **48**, 87–92 (2005)
6. Wu, L., Brynjolfsson, E.: The future of prediction: how Google searches foreshadow housing prices and sales. SSRN Electron. J. (2009)
7. Da, Z., Engelberg, J., Gao, P.: In search of attention. J. Finance **66**, 1461–1499 (2011)
8. Preis, T., Moat, H.S., Stanley, H.E.: Quantifying trading behavior in financial markets using Google Trends. Sci. Rep. **3**, 1684 (2013)
9. Curme, C., Preis, T., Stanley, H.E., Moat, H.S.: Quantifying the semantics of search behavior before stock market moves. Proc. Natl. Acad. Sci. **111**, 11600–11605 (2014)
10. Ramos, H.P., Ribeiro, K.K.M., Perlin, M.S.: The forecasting power of internet search queries in the Brazilian financial market. RAM. Revista de Administração Mackenzie **18**, 184–210 (2017)
11. Bordino, I., Battiston, S., Caldarelli, G., Cristelli, M., Ukkonen, A., Weber, I.: Web search queries can predict stock market volumes. PLoS ONE **7**, e40014 (2012)

12. Andrezo, A.F., Lima, I.S.: Mercado financeiro aspectos conceituais e históricos. Atlas (2007)
13. Fortuna, E.: Mercado Financeiro: Produtos e serviços. Qualitymark (2013)
14. Dados de mercado. http://www.bmfbovespa.com.br/pt_br/servicos/market-data/consultas/dados-de-mercado/
15. Ibovespa | B3. http://www.b3.com.br/pt_br/market-data-e-indices/indices/indices-amplos/ibovespa.htm
16. Fama, E.F.: Efficient capital markets: a review of theory and empirical work. J. Finance **25**, 383 (1970)
17. Shleifer, A.: Inefficient Markets: An Introduction to Behavioral Finance. Oxford University Press, Oxford (2000)
18. Kahneman, D., Tversky, A.: Prospect theory: an analysis of decision under risk. Econometrica **47**, 263 (1979)
19. Grinblatt, M., Keloharju, M.: What makes investors trade? J. Finance **56**, 589–616 (2001)
20. Bouchaud, J.-P.: The (unfortunate) complexity of the economy. Phys. World **22**, 28–32 (2009)
21. Gil, A.C.: Métodos e técnicas de pesquisa social. Atlas (2014)
22. Trends Help. https://support.google.com/trends#topic=6248052
23. Search Engine Market Share Brazil. http://gs.statcounter.com/search-engine-market-share/all/brazil#quarterly-200901-201702
24. Elder, A.: Como se transformar em um operador e investidor de sucesso: Entenda a psicologia do mercado financeiro, técnicas poderosas de negociação, gestão lucrativa de investimentos. Elsevier Brasil (2004)
25. Moore, D.S., Pessoa, C.F.C., Carneiro, P.D.G.: A estatística básica e sua prática: D.S. Moore. Livros Técnicos e Científicos (2005)
26. Rao, T., Srivastava, S.: Modeling movements in oil, gold, forex and market indices using search volume index and Twitter sentiments. In: Proceedings of the 5th Annual ACM Web Science Conference on - WebSci 2013 (2013)

Systematic Review of Data that Uses Project E in the Development of Digital Products Focusing on User Experience

Paula Silva(✉) and Franciane Falcão

Universidade Federal do Amazonas, Manaus, Amazonas, Brazil
paulacostapdas@gmail.com

Abstract. Project E is a methodology for design projects created in 2010 by Meurer and Szabluk. The purpose of this article is to evaluate how Project E has been applied in projects of digital product development focusing on user experience. This study is a systematic review of data that uses the Project E method. We paid attention especially to which phases were fulfilled and which techniques were used in each one of them, as well as the result of the whole work. A total of 26 articles were analyzed, being separated into 3 different application structures of Project E. From these identified structures, we then analyzed the importance of the techniques applied in each of the steps of generating an innovative design result.

Keywords: Project E · Design · Development of digital products · User experience · Product design methods · Systematic review

1 Introduction

According to ISO 9241-11 of 1998, usability is defined as "the ability of a product to be used by specific users to achieve specific objectives with effectiveness, efficiency and satisfaction, in a specific context of use." User experience is defined as "the experience a product creates for people who use it in the real world" (James Garret 2011).

Given the difficulty of understanding and applying these concepts, Jesse James Garret developed a theory present in the book "The Elements of User Experience" (2011). In this book, Garrett isolates the elements in surface, skeleton, structure, scope, and strategy.

These elements correspond to layers that provide user experience when using a product, whether it is physical, web or mobile. When all these layers are produced according to the needs of users and designers goals, a satisfactory experience is created. Garret's theory also proposes that a design project should be generated starting by strategy plan and ending by the surface plan. Thus, the development of no plan should end before the conclusion of the previous plan, making them interconnect and have coherence, leading to better results.

Based on this theory, Meurer and Szabluk (2010) published a methodology called Project E, in which each layer of user experience is examined separately and based on Garret's ideas.

L. Pereira et al. (Eds.): IDEAS 2018, SIST 198, pp. 189–196, 2021.
https://doi.org/10.1007/978-3-030-55374-6_19

To better understand Project E as well as to study its application, we've decided to analyze articles that applied this methodology by creating a systematic review that will help to develop our final project.

2 Fundamentals

Meurer and Szabluk (2010) define Project E as a design methodology to guide and optimize the development of Friendly Graphical Interfaces (FGI) for different systems and interactive virtual-digital products that are recommended for medium to high complexity projects. This methodology is composed of 6 steps named by the 6 user experience elements listed by Garret, being: Strategy, Scope, Structure, Skeleton, Aesthetics, and Execution.

The authors state that Project E does not necessarily have to be developed sequentially, allowing the designer to change previous steps or skip initial steps if they have already been defined by previous projects. The following is a brief description of each of the steps and some recommended application techniques:

2.1 Defining the Actors

It is necessary to define the assignments of the team that will develop the project. They must be divided into two main groups: visual programming and computer programming, as well as the project manager, who organizes and validates the process.

2.2 Strategy

During the development of Strategy, the project context is identified through some steps recommended by the authors. These are design issues, scenario definition, initial and final situation, equalization of project factors and taxonomy.

In addition to these steps, some analyses of similar apps are also recommended. These analyses can be linguistic, diachronic, synchronic, tangential, design analysis, heuristic analysis and a list of project requirements and constraints.

2.3 Scope

In this stage, content must be organized into modules, sub-modules, and categories through different techniques. Definition of functionalities, tools and task scenarios should also start, being the first step to generate alternatives.

2.4 Structure

At this point design aspect is predominant, being necessary to work on organizational charts of the website or app as well as flowcharts of the tasks that can be performed in it. It is important to design this step keeping in mind the requirements listed in the Strategy step.

2.5 Skeleton

In this level, wireframes are defined as a part of the process to organize the final interface. It is recommended that elements present in several screens are positioned in the same place (e.g.: logo, search area, global and local navigation, and content areas).

It is also advised that wireframes are organized into task flowcharts, making it easier to correct any possible usability issues.

2.6 Aesthetics

Aesthetics is the final stage of FGI's development. This stage is fundamental to promote the usability of the product since the user will be in direct contact with the interface. Some aspects that deserve special attention are study and definition of grids, composition, diagramming and visual graphic identity.

2.7 Execution

The Execution stage defines interactive models to simulate the main functionalities of the product, so they can be tested and final programming can start.

3 Methodology

For a qualitative survey of Project E's application, we performed a systematic review following the model proposed by Brereton et al. (2007). This is to identify, evaluate and interpret relevant research related to a specific problem, topic, area or phenomenon of interest. Component steps of the systematic review are hereafter explained.

3.1 Research Question

The question posed by the present study is: "How Project E has been applied in interface design projects?"

3.2 Development of a Search Protocol

As suggested by Brereton (2007), we began with a preliminary search that aimed to evaluate the number of articles relevant to this research.

We started by using the Capes Platform, one of the biggest virtual libraries in Brazil that contain national and international scientific production. At first, we used the advanced search method and required terms "Project E" and "design" to be present in the content. There were 769 results. From the initial reading, it was possible to notice that many of these did not use the Project E method, confusing the term with the words.

Then we began to section the found content, selecting articles reviewed peer journals, obtaining 374 results. The following filter included only articles published since 2010, the year in which Project E was launched, making total results decrease to 305.

The next step was to select only publications in the design topic, decreasing results to 10. Following initial reading articles' titles, 4 of them were removed for having no relation with Project E and the design of web interfaces. After reading the abstract of the six remaining articles, it was concluded that none of them used the Project E method, therefore none of the researches on Capes Platform brought results on the subject.

Thereafter we looked for another platform and found the "Project In Action" website, created by authors who developed Project E. It was created for teachers to apply Project E with their students and publish the results. Succeeding the search, a total of 67 projects were found, published since the year 2014, all of which had the keywords "Project E" and "Design".

3.3 Criteria for Selecting Articles

The evaluation of articles was done through a two-step screening process, as described by Brereton *et al.* (2007). For the first step, we used date-of-publication criteria, selecting the 15 oldest articles, published between 2015 and 2014, and the 15 most recent, published in 2018. To have a general view, we've decided to use only 30 articles, considering reading time and processing of information as a factor. This way, we were able to compare initial articles that used Project E methodology with state-of-the-art techniques.

In the second stage, all 30 selected articles were read and articles that didn't fit the inclusion criteria were excluded. By doing this, 4 more articles that didn't apply Project E in a project of design of interface were removed (Table 1).

Table 1. Search protocol – September 2018.

How Project E has been applied in interface design projects?	Database:
	Project in action
Search string: "Project E AND design"	67
Only articles published since 2010	67
Only articles in design topic	67
Total	67
Selection of articles, 1st screening	30
Selection of articles, 2nd screening	26

Source: Elaborated by authors

4 Results

When analyzing the 26 articles selected in Project. In Action platform, we noticed that there were three main ways of applying the stages proposed by the method, being them presented below.

- Apply project E in four parts: Contextualization, Deconstruction and verification, Reconstruction and Identity and differentiation (24 articles);
- Divide into four parts, which are: First stage of activities, Second stage of activities, First phase of the semester project and Second phase of the semester project (1 article);
- Apply method in 5 steps: Perception, Target, Configuration, Sketch and Refining (1 article).

Most of the articles presented the first form of application, containing Contextualization, Deconstruction and verification, Reconstruction and Equity and Differentiation. Of all articles, the most frequent products are apps, accounting for 65.4% of the total, followed by games (11.5%), sites (11.5%), platforms (3.8%) and digital magazines (3.8%).

During the initial classification, we also considered the level of innovation presented in each project. For this, we were based in the degree of novelty proposed by Davila et al. (2008). According to the authors, innovations can be classified as i) incremental: improvement of existing products and processes; ii) semi-radical: involves a significant change in the business model or technology used by the company, which would not be achieved with incremental innovation; and iii) radical: supply of products and processes in a totally new way (for the world or the company), causing changes in the industrial sector they belong. From this concept, we classified each of the projects developed according to their results and in which category it would fit the most.

After the evaluation, we considered that 15.4% of articles resulted in radical innovations, 34.6% semi-radical innovations and 50% in incremental innovations.

We concluded that authors Pantaleão et al. (2014), Menna et al. (2017), Reis et al. (2017) and Silva et al. (2018) were the ones that obtained results with better innovative quality.

4.1 Rating Parameters

By comparing these steps with those found in Project E, we were able to separate which techniques are most used in each step of the method and create the following graphic. After analyzing it, we concluded that the techniques most used in the Strategy stage are: creation of an infographic presenting the chosen theme, design questions (answering what, why, how and for whom), definition of personas, study of similar and references, structural analysis, morphological and heuristic analysis, visual identity analysis, terms pertinent to the project and checklist.

In the Scope stage, a few techniques were presented and used, being recognized only requirements of the project and morphological matrix.

In the Structure stage, the techniques of creation of scenarios for the digital product and definition of tools, functionalities, and contents were found.

In the Skeleton stage, the techniques of development of wireframes, wireflows, and use cases and the creation of grids and meshes were present.

In the Aesthetics stage, the techniques of development of visual signature, definition of typographic fonts, definition of the imagery, and definition of the chromatic matrix were used.

And in the Execution, only the presentation of the final screens was encountered (Fig. 1).

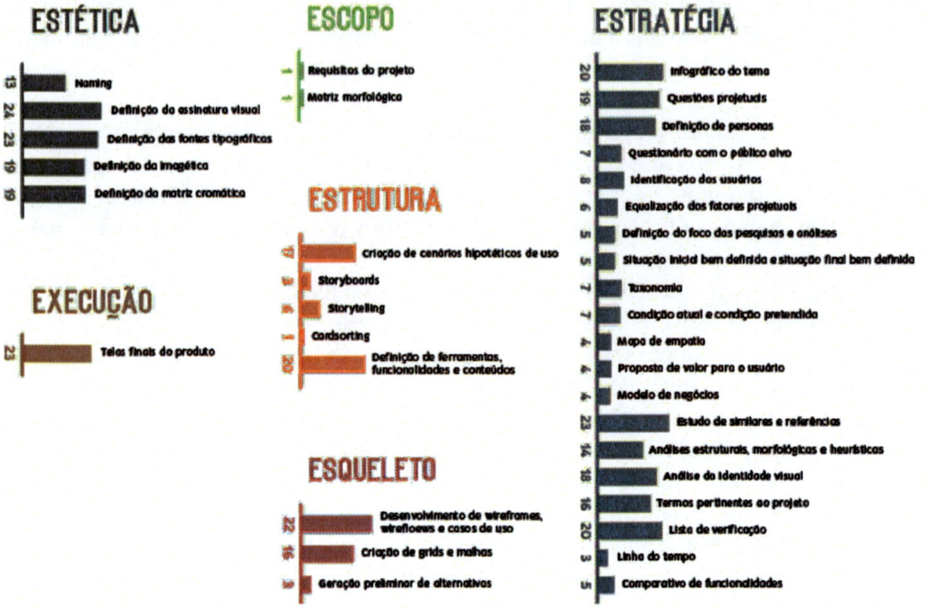

Fig. 1. Infographic presenting general results. Source: Elaborated by authors

Regarding the highest innovative quality articles, the most used techniques in Scope are infographics of the theme, creation of personas, the study of similar and references, structural, morphological and heuristic analyzes and checklist. In the Structure stage: the creation of hypothetical use scenarios, storyboards, storytelling and definition of tools, functionalities, and contents. In the Skeleton stage, there is the development of wireframes, wireflows and use cases and the creation of grids and meshes. In Aesthetics, we encountered the techniques of development of visual signature, definition of typographic fonts, definition of imagery and definition of the chromatic matrix. And in Execution, there is only the presentation of final screens. None of these papers presented techniques attributed to the Scope phase (Fig. 2).

5 Discussion

As we have seen previously, Project E is usually not applied with each of its steps being performed separately, although they can be found scattered throughout the text.

In Strategy stage, is notable that one of the most used techniques is the creation of an infographic about the chosen theme, as a way of visualizing project data in a more dynamic and informative way. Also, there are the structural, morphological, heuristic and visual identity analyses of similars and referents. These analyses help the author understand the current market and its needs.

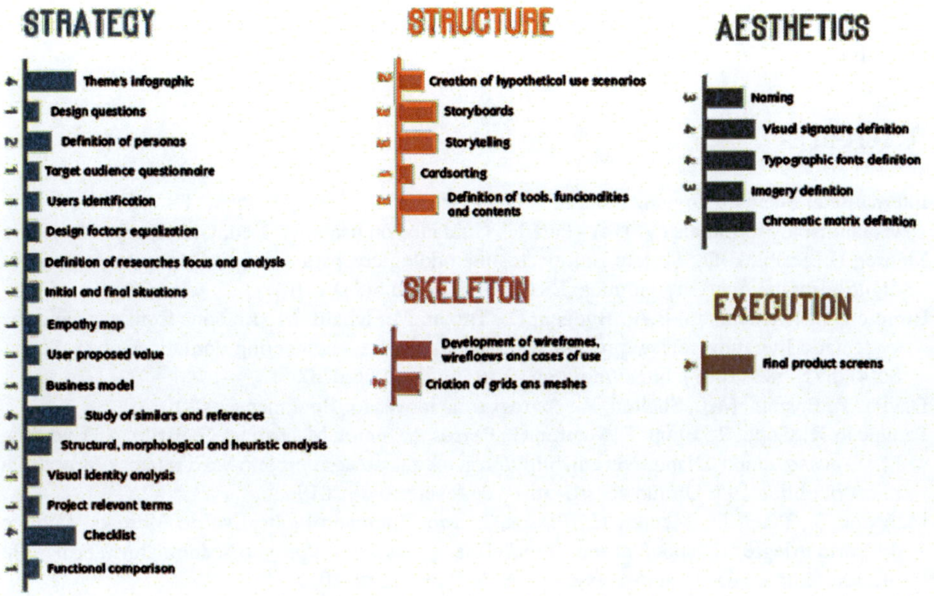

Fig. 2. Infographic presenting results of articles with the highest innovative quality. Source: Elaborated by authors

Following the analyses of articles that presented the greatest degree of innovation, it is noticeable that the Structure stage requires special attention, with the creation of hypothetical scenarios of use, storyboards, storytelling, and definition of tools, functionalities, and contents. With the application of these techniques, authors were able to create projects that fulfill their customer's desires and adapt to the environments they use, granting a radical innovation in their field.

6 Conclusions

Project E contributes to digital product projects by having a strong focus on development aiming user experience. A great number of techniques used in its development have exactly this purpose, such as those of creations of personas, identification of target public, analysis of similar and references and creation of scenarios of use with storytelling and/or storyboards.

Techniques less frequent in projects studied were those addressed to the Scope stage, which focuses on the definition of the system and its differentiation. With a greater application of this stage, perhaps we could have found a bigger number of projects with a high level of innovation.

We conclude that the use of Project E can assist the development of digital products and increase their usability. Despite the presence of different levels of innovation, we noticed that projects that stood out the most presented a great presence of techniques to improve the user experience in the Structure stage, which is a fundamental step.

Therefore, the most important techniques in these works are those that require users consultation.

References

International Standard Organization. ISO 9241-11: Ergonomic requirements for office work with visual display terminals (VDTs) - Part 11: Guidance on usability. 1 ed. Genebra, p. 22 (1998)

Meurer, H., Szabluk, D.: Projeto E: aspectos metodológicos para o desenvolvimento de projetos dígito-virtuais. Ação Ergonômica: ERGODESIGN II **5**(2) (2010)

Brereton, P., Kitchenham, B.A., Budgen, D., Turner, M., Khalil, M.: Lessons from applying the systematic literature review process within the software engineering domain. J. Syst. Softw. **80**(4), 571–583 (2007). https://doi.org/10.1016/j.jss.2006.07.009

Davila, T., Epstein, M.J., Shelton, R.: As regras da inovação. Bookman (2008)

Pantaleão, R., Zupo, R., Bopp, T., Meurer, H., Passos, J., Junior, M., Zaffari, G., Brito, A., Everling, M.: O nosso tempo. Disponível em. http://projetoemacao.com/projetoAcao/relatorio/projetoPublicadoPublico/195. Último acesso em 19 de fevereiro de 2019

Menegon, J., Patel, L., Meurer, H., Passos, J.: Meu Mercado (aplicativo do Mercado Público de Porto Alegre). Disponível em. http://projetoemacao.com/projetoAcao/relatorio/projetoPublicadoPublico/365. Último acesso em 19 de fevereiro de 2019

Menna, K., Jacques, M., Meurer, H.: Respeita as mina. Disponível em. http://projetoemacao.com/projetoAcao/relatorio/projetoPublicadoPublico/664. Último acesso em 19 de fevereiro de 2019

Reis, E., Lucas, L., Meurer, H.: Tudo já existe. Disponível em. http://projetoemacao.com/projetoAcao/relatorio/projetoPublicadoPublico/659. Último acesso em 19 de fevereiro de 2019

Silva, C., Magnus, F., Nunes, J., Barcellos, L., Meurer, H., Capra, A., Menezes, L.: Mapa de curiosidades sobrenaturais. Disponível em. http://projetoemacao.com/projetoAcao/relatorio/projetoPublicadoPublico/840. Último acesso em 19 de fevereiro de 2019

Garret, J.J.: The Elements of User Experience. New Riders (2011)

Affordance Based Design: The Affordance Digital Stimuli Tool to Stimulate Creativity During Product Design

Karuliny Marques(✉), André Ogliari, Rodrigo Bastos Fernandes, and Matheus Gomes

Federal University of Santa Catarina, Trindade. C. P. 476, Florianópolis, Brazil
karulinycristie@gmail.com

Abstract. Decisions made in the early stages of product development influence considerably on the product life-cycle costs. Thus, it is necessary to search for creative solutions in the conceptual phase, subject to creativity barriers during the ideation process, such as the functional fixedness. In this regard, the present research proposes the Affordance Digital Stimuli (ADS) to assist designers in finding creative solutions for natural and easy-to-use products. Were conducted brainstorming sessions supported by the ADS with twenty engineers to evaluate the proposed stimuli. The application of the ADS was assessed using eye-tracking equipment, which measured the flow and focal points of the user. The evaluation of the stimuli was also performed through a questionnaire considering issues related to the stimuli configuration and its' contribution to the ideation of solutions. The results evidenced that the proposed ADS were effective to inspire the ideation of solutions with affordance characteristics related to aesthetics, safety, and usability attributes and, therefore, is effective in mitigating the functional fixedness and stimulating creativity during ideation sessions.

Keywords: Affordance design · Creativity · Analogy · Conceptual design · Product design

1 Introduction

To generate solutions during the conceptual design phase of product development brainstorming sessions are usually conducted as a creativity method. Recent researches have shown that the brainstorming method, without the aid of stimuli while contributing to the generation of many ideas [1], do not always result in creative ones [2].

In this perspective, the development of methods to help designers generate creative solutions has become the goal of research groups to increase the chances of success in PDP (Product Development Process) [3]. As examples, the Biostimulators – that aggregate concepts of Bionics [2], the Udins – that concern the idea of Universal Design [4], and the Trends of Evolution (TEs) of TRIZ applied to product planning [5], all aim to stimulate designers' creativity during the ideation process, particularly during brainstorming sessions.

L. Pereira et al. (Eds.): IDEAS 2018, SIST 198, pp. 197–207, 2021.
https://doi.org/10.1007/978-3-030-55374-6_20

Among the concepts and the variety of stimuli already instrumented, the present research identified the potentiality on the use of affordance concepts [6] adapted to the field of Engineering [7] and Design [8], combined with aesthetic, usability and safety ideas, in order to achieve intuitive, easy to use and safe products. In this context, is presented the instrument entitled Affordance Digital Stimuli (ADS).

2 Affordance in the Field of Product Design

The affordance concept proposed by psychologist James J. Gibson [6] in the field of Ecological Psychology to study animal-environment interactions. Gibson [6] defined affordance as "[…] something abstract and mediating the relationship between the (animal) subject and environment."

In the field of engineering, the affordance concept has the potential to contribute to the understanding of user-product relationship through two approaches [7], considering object-user or object-object interaction. Given these approaches, affordance is established as an idea that "[…] connects the structure of the environment (object) with the capacity of the human user to determine which behaviors are possible to be connected" [9]. Therefore, "[…] affordance indicates which behaviors are possible" [9].

The affordance concepts have been applied in different situations such as: in architecture [10]; robotics [11]; and the modern information system for physical and digital use [12]. For the present research, the affordance concept is associated with the basic attributes of product: aesthetics, safety, and usability, aiming at the development of intuitive, easier to use, and safer products.

The aesthetic attributes are related to the appearance of the product perceived in symbols, shapes, interaction, movement, material, texture, name, and figures [13]. These are properties that potentiate possibilities of action, that is, the detection of affordances [14], and may influence the decision to buy the product.

The usability attributes comprise aspects related to the use of the product, in other words, its' usability, which regard "[…] the extent to which specific users can use a product, to achieve specific objectives, with effectiveness, efficiency, and satisfaction, in a context of specific use" [15]. And, the safety attributes are related to human factors in man-machine interaction. In this sense, whether through training or product limitations, man can make mistakes in the use of the product [16]. Therefore, the safety attributes, along with the affordance concepts, are essential to the establishment of undesirable actions to avoid them through a better product design [17].

To apply the affordance concepts with the discussed attributes to product design, two product usage levels must be considered: the operational level, which corresponds to a user-product interaction (e.g. characteristics such as colors, sound, and smell); and the functional level, related to the user-product communication through the structure and layout of the product (e.g. shapes and textures). According to the standard, the affordance concept and attributes can promote a new understanding of the relations that occur between the product and the user during the use of the product [18]. Thus, the present research proposes the ADS to assist design teams during the ideation of new products.

3 The Affordance Digital Stimuli (ADS)

The ADS were elaborated to operationalize through a web-based database (See Fig. 1) stimuli to the design team to enhance the generation of solutions during brainstorming sessions.

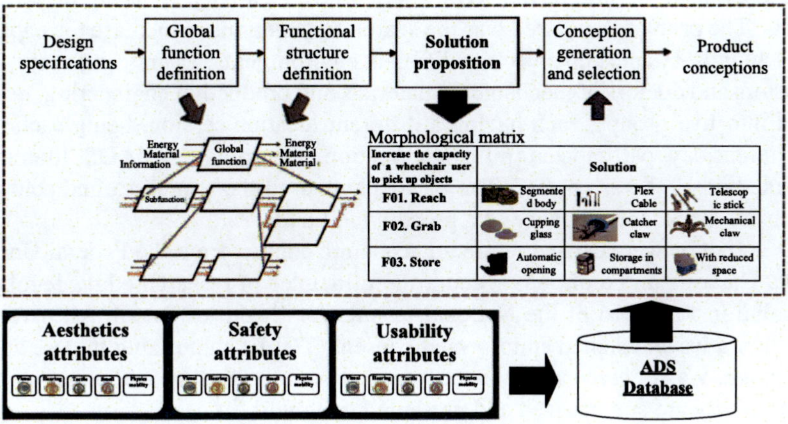

Fig. 1. Overview of ADS content used in the conceptual design phase. Source: Authors

The ADS was created and implemented in a web-based system to make the stimuli more available for design teams' usage. Web platforms, when compared to physical cards, facilitate future expansions of the stimuli base and the search for context-based stimuli. Each affordance digital stimulus developed, took into account human ways of interaction, such as vision, hearing, touch, and physical-motor skill. Each stimulus has a textual description of the attribute associated with affordance concepts, representative abstract illustrations, a suggestion on how to proceed with the abstraction of content, and also some application examples of the ideas in existing products.

In the WEB environment the ADS were also grouped according to the relation with technical functions of products (e.g., reach, grab, store) and with potential user's tasks (e.g., check, trigger, control, save). At the initial page of the proposed instrument are the search fields (See Fig. 2): technical functions and the user tasks. Through these filters, the ADS search is carried out based on the problem-solving context.

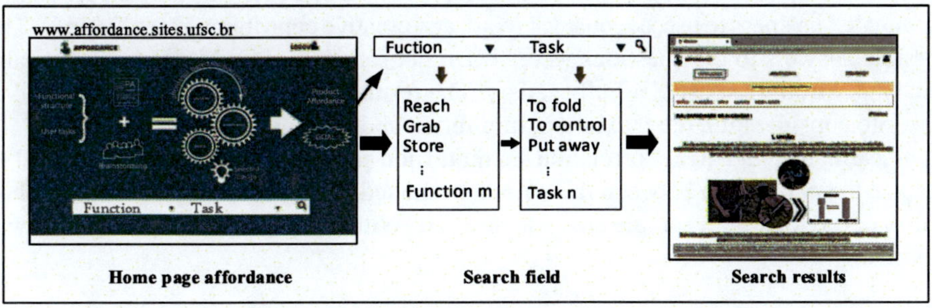

Fig. 2. Web environment with ADS. Source: Authors, 2019

4 Evaluation of the ADS

4.1 Materials and Methods

This work is the result of a master's degree thesis from the corresponding author (first author), which had the second author as advisor. Five brainstorming sessions was conducted to evaluate the use of the ADS as stimuli to creativity, focusing on affordance concepts. The problem proposed was the search for solutions to increase the capacity of a wheelchair user to pick up objects in different environments. Twenty engineers, including masters and doctors of mechanical, materials and production engineering, randomly divided into five groups, each having a different ideation session, being each session in a different day, participated in this application. To evaluate the ADS, the research-action method [19] was applied. The participants used the stimuli to find solutions to the problem, under the guidance of a process facilitator.

The activities were carried out in an academic context from the Federal University of Santa Catarina and took into account the principles of research ethics regulated by the resolution 510/2016 of the National Council of Health of Brazil. All participants were voluntaries and signed an informed consent (TCLE) authorizing the use and right of the image. Wasn't given any monetary compensation. The time and instruments used in each activity are summarized and presented in Table 1.

Table 1. Activities for the ADS application

Activity	Time (min)	Instruments/Guidance
1. Oral and graphical presentation of the problem	10	(i) Form with a description of the problem and functional synthesis
2. Performance of the affordance website and its ADS tool	15	(ii) ADS for creative stimulation
3. Generating the solution	45	(iii) Forms, for records of ideas generated; (iv) Eye tracking equipment
4. Application of the evaluation questionnaire	5	(v) Questionnaires for the survey on the use of the ADS by the participants

Source: (Authors, 2019)

The criteria to ADS's evaluated by the users [20], through the questionnaire took into account its' easiness to use, adequacy, layout, and creative contribution (see Table 2). The possible answers to the questions were: 1. Do not satisfy, 2. Poorly satisfy, 3. Partially satisfy, 4. Strongly satisfy, 5. Fully satisfy. The results of the questionnaire analyzed to take into consideration the maximum, minimum, and average scores.

Spearman's statistical correlation method (non-parametric data) [21] was used to analyze the correlation between the criteria evaluated. The level of significance applied was $\alpha = 0.05$, and the sample size was $n = 20$. With the p-value, the correlation was reported.

The eye-tracking equipment prescribed in activity 3, was used as a quantitative test tool to evaluate the systems' usability, parameterized by the efficacy, efficiency, and satisfaction metrics [15]. The most common eye-tracking parameters are based on fixations and, or saccades, and among the most widely applied fixation-based parameters one can find: Heat maps, Hit Ratio, Revisits, Average Fixation Duration, and First fixation sequence. These metrics can be used to define the observer's involvement (in this case, the evaluators in the use of the ADS) [22].

The purpose of this tool is to provide information about the attention of the user (the ADS evaluators), by capturing the eye movements related to specific ADS areas of interest. An Area of Interest (AOI) is a region of the ADS selected to identify metrics specifically for those regions (See Fig. 4).

Heat maps indicate visually the fixation points (i.e., interface areas that attracted more attention), or even the group main focus of attention.

Revisits provide information about how often the participants returned his gaze to a specific point, defined by an AOI. This parameter allows the researcher to examine which areas repeatedly attracted the participant attention, which may be associated with his motivation and concentration.

Average Fixation informs how long the average fixation lasted. If an image leads to a long period of fixation, longer than other image proposed, then it may be worthy exploring the reasons.

Another measure is the saccades or saccadic movements, characterized by non-linearized eye movements continuously (as in a line-by-line reading of a text), but it jumps from one attachment to another [23] as evidenced in the distribution of the circular points on AOI. The metric of sequence can also define it. The increased time of fixations in the same informational position can be proportional to the increase of the cognitive load to process the information [22]. In this sense, the eye-tracking data analysis software maps the fixations on an affordance concept with the defined AOIs (See Fig. 3).

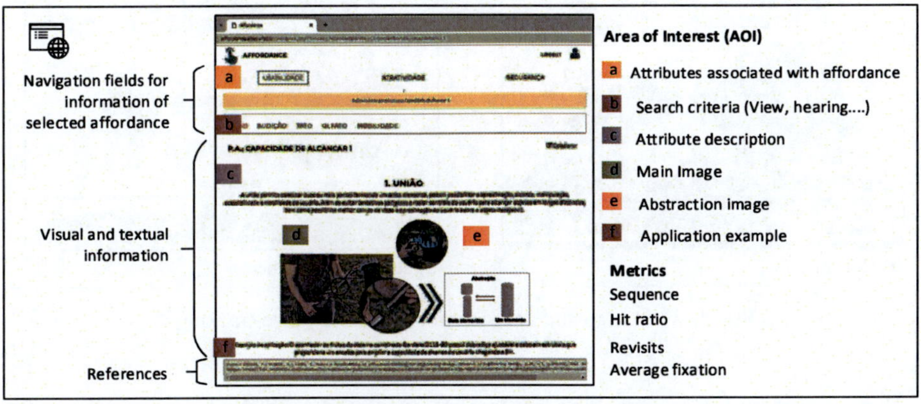

Fig. 3. Metrics for each AOI by eye-tracking

4.2 Results and Discussion

Figure 4 shows participants during brainstorming using the ADS. All participants were voluntaries and signed an informed consent (TCLE) authorizing the use and right of the image issued through the Federal University of Santa Catarina.

Fig. 4. Examples using the ADS in the brainstorming session.

According to the results of the heat mapping metric analyzed for each AOI, with the use of the eye-tracking equipment can be drawn the attention distribution (See Fig. 5). The AOIs defined for evaluation were: the main image, the attributes description, and the application example. The Fig. 5a show the highest intensity of attention areas is those that approach the red color. Figure 5b shows the lines connecting the attention focus of the evaluator in the defined AOIs.

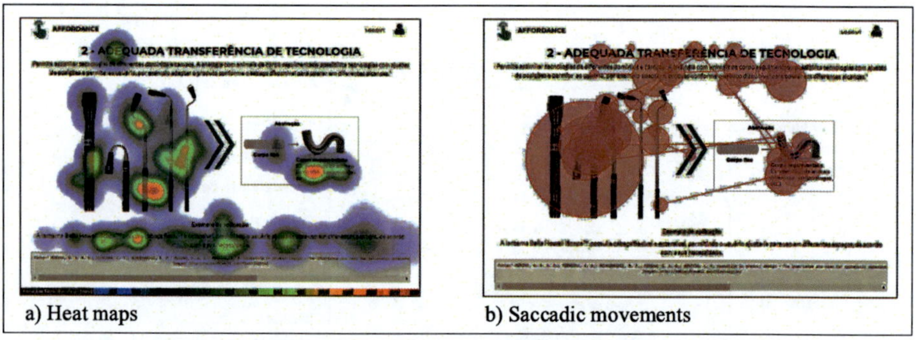

Fig. 5. Heat Map and Saccadic movements by Eye-tracking

The report by the eye-tracking equipment generated with the value middle of the parameters evaluated of the five users that used the equipment in eleven stimulators, show that: the AOI that presented the highest average of Hit Ratio value was the main image (100%), followed by the image of abstraction (75%), the description of the attribute (75%) and finally the application example (50%). The AOI that presented the highest value of Revisits was the main image (11.8), followed by the application example (11), the attribute description (6.3) and the abstraction image (1.7). The Revisits value is understood as a comparison value between AOIs. The more revisits in an AOI, the higher the interest of the evaluator at it. The AOI that presented the highest Average fixation value was the main image (241.1 ms), followed by the attribute description (232.8 ms), the abstraction image (199.5 ms) and the application example (108.1 ms). As for the fixation sequence, the order of attention of the evaluators was first the main image, followed by the description of the attribute, the example of application, and at last the image of abstraction. By the comparison of the ideas' generated records with the AOIs it was possible to observe that the main image presented a more significant potential for cognitive mobilization in relation to the textual AOIs, since the concepts present through an image were transferred to the solution by means of analogy making, as shown in the figure (see Fig. 6).

The ADS retrieved shows a telescopic pen (see Fig. 6a) that illustrates the generic affordance concept of *Form* [13] associated with the usability attribute and user control [24, 25]. The affordance concept considered in the example ADS consisted of a functional affordance, that is, the form of the structure allows that the user intuitively expands or

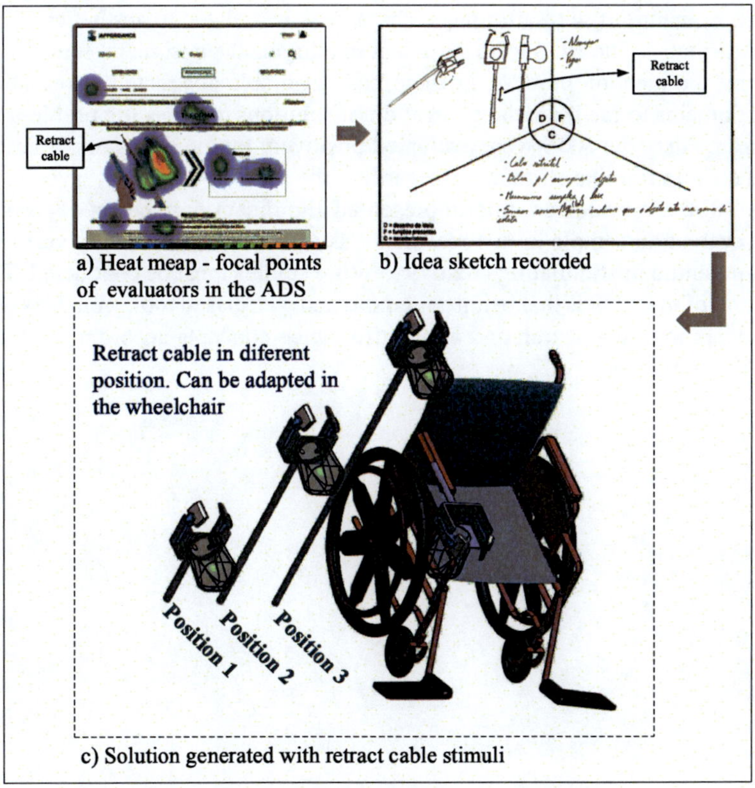

Fig. 6. Example of ADS content transferred to a solution

reduce the rush by adjusting it to the space needed to pick up objects. The evaluators generated solutions like show the example in Fig. 6b using the ADS.

The results of the qualitative evaluation of the use of the ADS, based on the questionnaire applied to the participants with maximum values allowed equal 5, showed an average value above 4. That is, the scores for the ADS development and evaluation criteria demonstrated adequate elaboration and definition of the ADS, confirming the potential of the proposed stimuli to support creativity and its contribution to mitigate the functional fixation. Other result was through analysis made by with Spearman correlation, according to Table 2.

The Spearman correlation test [21] (See Table 2) showed that the degree of correlation for all criterion is above the moderate (i.e., from 0.50 to 0.70), and the p-values were way lower than the significance (α value of 0.05). This result highlighted the existence of influence of one variable on the other, emphasizing that all criterions pointed must be considered to the elaboration of ADS with the same degree of importance.

5 Final Considerations

The research results evidence that the proposed stimuli were effective to inspire the generation of solutions with affordance characteristics related to aesthetics, safety, and usability attributes. The ADS were sufficient to mitigate the designers' functional fixedness during the ideation process, as they make available abstracted information from different domains to facilitate the generation of solutions to a specific problem by analogy making. Also, the ADS were evaluated positively by the engineers, pointing your adequate elaboration like stimulators regarding the facility, adequacy, layout, and contribution. All these evaluated criteria presented significant correlation degrees. As the data and evaluation sample in this research was small, as a preliminary verification of the ADS potential in stimulating creativity, future studies must be conducted. The main contribution of the ADS is that being a modular and web-based tool, it enables designers and engineers to easily search and apply affordance concepts according to the design problem requirements.

Table 2. Correlations of the evaluation criteria

		Easiness to use	Adequacy			Layout	Creative contribution	
		Readability	Content organization	Amount of information	Sequencing of the information	Layout	Contribution to a generation of SP	Contribution of abstraction images
Easiness to use	Readability	1						
Adequacy	Content organization	0.898 (p = 0.000)	1					
	Amount of information	0.898 (p = 0.000)	1 (p = 0.000)	1				
	Sequencing of the information	0.811 (p = 0.000)	0.761 (p = 0.000)	0.760 (p = 0.000)	1			
Layout	Layout	0.672 (p = 0.001)	0.568 (p = 0.009)	0.567 (p = 0.009)	0.620 (p = 0.004)	1		
Creative contribution	Contribution to a generation of SP	0.750 (p = 0.000)	0.689 (p = 0.001)	0.685 (p = 0.001)	0.912 (p = 0.000)	0.587 (p = 0.006)	1	
	Contribution of abstraction images	0.757 (p = 0.000)	0.661 (p = 0.002)	0.661 (p = 0.002)	0.845 (p = 0.000)	0.730 (p = 0.000)	0.893 (p = 0.000)	1

Source: (Authors, 2019)

Ethics declaration
Conflict of interest
The authors declare that they have no conflict of interest.
Compliance with standards involving humans as subjects
All procedures performed in studies involving human participants were in accordance with the ethical standards of the institutional committee and with the 1964 Helsinki Declaration and its later amendments or comparable ethical standards. Informed consent was obtained from all individual participants involved in the study.

References

1. Linsey, J.S., Clauss, E.F., Kurtoglu, T., Murphy, J.T., Wood, K.L., Markman, A.B.: An experimental study of group idea generation techniques: understanding the roles of idea representation and viewing methods. J. Mech. Des. **133**, 01–15 (2011)
2. Fernandes, R.B., Ogliari, A.: Enhancing creativity through Biological Stimuli during new products ideation. Int. J. Innov. Educ. Res. **6**(10), 332–350 (2018)
3. Pahl, G., Beitz, W., Feldhusen, J., Grote, K.H.: Engineering Design: A Systematic Approach. (Translation from the German Language edition: Konstruktionslehre), 3rd edn. Springer, London (2007)
4. Messerschmidt, P.H.Z.: Sistemática para planejamento de produtos orientado por princípios de projeto universal. Dissertation (master's degree in mechanical engineering), Universidade Federal De Santa Catarina, Florianópolis, p. 223 (2018)
5. Abdala, L.N., Fernandes, R.B., Ogliari, A., Löwer, M., Feldhusen, J.: Creative contribuitions of the methods of inventive principles of Triz and Biotriz to problem solving. J. Mech. Des. **139**(8), 082001 (2017). https://doi.org/10.1115/1.4036566
6. Gibson, J.J.: The Ecological Approach to Visual Perception. Lawrence Erlbaum Associates, Inc., Hillsdale (1986). (Original work published 1979)
7. Maier, J.R.A., Fadel, G.M.: Affordance: the fundamental concept in engineering design. In: Proceedings of ASME Engineering Conference, American Society of Mechanical Engineers. International Design Engineering Technical Conferences and Computers and Information (2001)
8. Norman, D.A.: The Design of Everyday Things-Basic. New York (1988)
9. Maier, J.R.A., Fadel, G.M., Battisto, D.G.: An affordance-based approach to architectural theory, design, and practice. Des. Stud. **30**, 393–414 (2009)
10. Salingaros, N.: Why we need to grasp our surroundings: object affordance and prehension in architecture. J. Arch. Urbanism **41**(3), 163–169 (2017)
11. Yamanobe, N., et al.: A brief review of affordance in robotic manipulation research. Adv. Rob. **31**(19–20), 1086–1101 (2017)
12. Benbunan-Fich, R.: An affordance lens for wearable information systems. Eur. J. Inf. Syst. **28**(3), 1–16 (2018)
13. Hekkert, P., Cila, N.: Handle with care! why and how designers make use of product metaphors. Des. Stud. **40**, 196–217 (2015)
14. Xenakis, I., Arnellos, A.: The relation between interaction aesthetics and affordances. Des. Stud. **34**, 57–73 (2013)
15. Merino, G.S.A.D., Teixeira, C.S., Schoenardie, R.P., Merino, E.A.D., Gontijo, L.A.: Usability in product design - the importance and need for systematic assessment models in product development – USA-design model (U-D). Work **41**, 1045–1052 (2012)
16. Guber, N.D.: Responsabilidade no projeto do produto: uma contribuição para a melhoria da segurança do produto industrial. Dissertation (Masters in Production Engineering). Universidade Federal de Santa Catarina, Florianópolis, p. 74 (1998)

17. Maier, J.R.A., Fadel, G.M.: Identifying affordances. In: International Conference on Engineering Design, Paris, France, Paper No. ICED'07/591 (2007)
18. Galvão, A.B., Sato, K.: Affordances in product architecture: linking technical functions and users' tasks. In: ASME IDETC/CIE, DETC2005–84525, Long Beach, CA, 24–28 September 2015 (2005)
19. Gil, A.: Como Elaborar Projetos de Pesquisa, 4th edn. Atlas, São Paulo (2002)
20. Vernadat, F.B.: Enterprise Modeling and Integration: Principles and Applications. Chapman & Hall, London (1996)
21. Mukaka, M.M.: A Guide to appropriate use of correlation coefficient in edical research. Malawi Med. J. **24**(3), 69–71 (2012)
22. Gobbi, A.G., Merino, E.A.D., Merino, G.S.A.D., Gontijo, L.A.: Uso do eye tracking para obtenção de medidas quantitativas em testes de usabilidade: Um estudo focado na medida da satisfação. idt Alves Díaz Merino, Leila Amaral Gontijo HFD **6**(11), 106–125 (2017)
23. Bylinskii, Z., Borkin, M.A., Kim, N.W., Pfister, H., Oliva, A.: Eye fixation metrics for large scale evaluation and comparison of information visualizations. In: ETVIS Workshop on Eye Tracking and Visualization (2015), pp. 235–255. Springer, Cham (2017). https://doi.org/10.1007/978-3-319-47024-5_14
24. Iida, I.; Buarque, L. Ergonomia: Projeto e Produção. São Paulo, (2016)
25. Cormier, P., Olewnik, A., Lewis, K.: Toward a formalization of affordance modeling for engineering design. Res. Eng. Des. **25**(3), 259–277 (2014)

Check for updates

Efficient Methods for System Design Analytics

Petter Krus(✉)

Linkoping University, Linkoping, Sweden
petter.krus@liu.se

Abstract. In the design of complex products systematic methods for managing requirements and system architecture are used. Different analysis methods, such as system simulation, can be used to predict the behaviour and properties of a system for a point design. However, there is also a need for methods to establish relations in a design through quantitative analysis, where different aspect can be presented and visualized in such a way that the designer can understand the design space and the trade-off between requirements, as a support for design decisions. In this paper an overview of a range of computationally efficient methods that can be used on underlying models of the system. These include well known methods such as system optimization and sensitivity analysis. Here is also introduced methods for visualizing functional correlation. The methods are demonstrated on an small electric commuter aircraft.

Keywords: House of quality · Sensitivity analysis · Functional correlation · Systems engineering

1 Introduction

When designing products and systems the trace-ability of requirements and implementation [1] is of high importance. In reality, however, this link is static at best. This means that there is little room to negotiate requirements efficiently at a subsystem level. In [4] a suit of methods were put together in order to get an overview of relations in a design. Here some of these concepts will be discussed more in depth. Figure 1 shows the different domains of customer (or stakeholders) requirements, the functional characteristics, and the design parameters. The design problem in a more narrow sense is about the mapping of functional requirements to design parameters. This is identified in Axiomatic design [7] as the design matrix. Here this is analytically by differentiation from an under laying model. New technology breeds new products, which in turn suggests new direction of development for technology. Regardless of the mechanism, however, the product needs to be matched to some customer needs. A method that acknowledges this is the QFD, see Hauser and Clausing [2]. QFD is a technique for translating customer/stakeholders requirements into engineering characteristics (or simply, functional characteristics). The relationship between

L. Pereira et al. (Eds.): IDEAS 2018, SIST 198, pp. 208–218, 2021.
https://doi.org/10.1007/978-3-030-55374-6_21

the customer requirements, the engineering characteristics and design parameters is shown in Fig. 1.

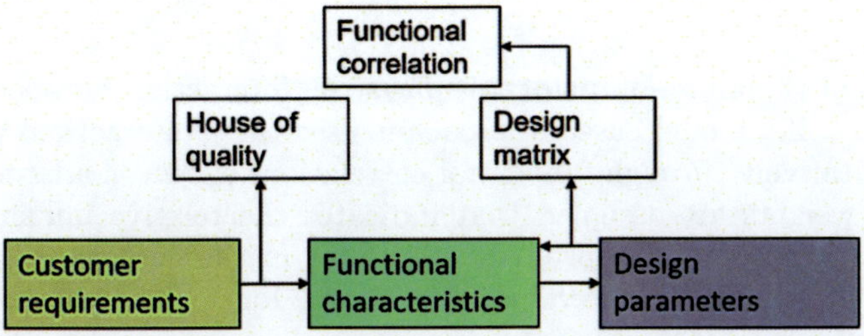

Fig. 1. The relationship between the customer (stakeholders) requirements that are translated into engineering characteristics that should be realized by the design parameters.

1.1 Definition of Requirements

The house of quality is a method used in QFD to map the relation between customer requirements and engineering characteristics (hereafter called functional characteristics). Here, the house of quality is used to provide trace-ability in a quantitative way between individual design parameters and customer requirements. Here priorities of the functional characteristics can are calculated as (Fig. 2):

Functional Characteristics	Functional correlations			
fc1		c_{12}	c_{13}	c_{14}
fc2			c_{23}	c_{24}
fc3				c_{25}
fc4				

Customer requirements	Functional Characteristics f_{c1}	f_{c2}	f_{c3}	f_{c4}	Customer req. priorities
Cr1	K_{C11}	K_{C12}	K_{C13}	K_{C14}	P_{c1}
Cr2	K_{C21}	K_{C22}	K_{C23}	K_{C24}	P_{c2}
Cr3	K_{C31}	K_{C32}	K_{C33}	K_{C34}	P_{c3}
Cr4	K_{C41}	K_{C42}	K_{C43}	K_{C44}	P_{c4}
Sign	ϕ_1	ϕ_2	ϕ_3	ϕ_4	
Demand or whish (D or W)	D/W	D/W	D/W	D/W	
Target values	v_1	v_2	v_3	v_4	
Functional char. priorities	1.00	1.00	1.00	1.00	
Normalized functional char. priorities	1.00	1.00	1.00	1.00	

Fig. 2. The "House of quality" (or QFD-matrix). The roof is tilted in order to be able to be presented in a spread sheet.

$$p_{fi} = \sum k_{c,ij} p_{cj} \tag{1}$$

which can also be written in matrix form as:

$$\mathbf{p_f} = \mathbf{K_c^T} \mathbf{p_c} \tag{2}$$

The roof of the house-of-quality displays the interaction between functional characteristics. If a model based approach is used these interactions can be evaluated quantitatively through functional correlations shown in a later section. In the matrix, weights are assigned that indicates the relative influence of functional characteristics on the product attributes in the customer requirements. These are usually assigned the values *0, 1, 3* or *9* for no, small, medium or large influence.

Furthermore, the customer priorities are set to the right of the table. Using Eq. (1) the functional characteristics priorities can be calculated. This indicates the relative importance of different functional characteristics. The target values for the functional characteristics are displayed below the table.

The requirements on functional characteristics can then be listed. In addition to the information given by the QFD-matrix, also information whether the requirements are demands *"D"* or just wishes *"W"*. Some requirements are determined by regulations and cannot be negotiated. They should therefore be demands. In fact they could even be omitted from the QFD-matrix in order not to make it unnecessarily cluttered. Not all requirements are meaningful to have target values on. In some cases it is just a requirement of "existence" or "absence" In some cases a property is rather subjective, the definition might be ambiguous, or just difficult to enumerate, such as handling and feel, and here the value of "1" indicate that it is equally good as a certain reference product. In the model based QFD, the requirements are linked to models that can be manipulated by design parameters, to produce the "best" design. Furthermore it can be linked to system optimization for optimization with respect to the system requirements.

1.2 Formulation of Objective Function

The design optimization problem is in general a constraint multi-objective optimization problem. Here, the Functional characteristics constitutes the objective that are to be formulated into an aggregate objective function. One often used objective function is:

$$f_o = \sum_{i=1}^{n} (\frac{y_i}{y_{i0}})^{\varphi_i \gamma_i} \tag{3}$$

Here

$$\varphi_i = \begin{cases} 1 \text{ for minimization of } y_i \\ -1 \text{ for maximization of } y_i \end{cases} \tag{4}$$

Furthermore, the exponent γ_i can be established from the priorities. E.g

$$\gamma_i = 1 + p_{f,i} \tag{5}$$

2 Sensitivity Analysis

Sensitivity analysis is an useful tool to study relationships between system parameters and functional characteristics. Sensitivity analysis can quickly give an overview over what parts of the design that are of importance for the desired behaviour [4], it also gives an overview of the different couplings in the designs that is a key property in Axiomatic Design [7]. Furthermore, it can be used to study the influence of disturbances and uncertainties in parameters and constants. Sensitivity analysis is the primary tool for studying the degree of robustness in a system. Assuming the system:

$$\mathbf{y} = \mathbf{f}(\mathbf{x}) \tag{6}$$

where $\mathbf{f}$ is a non-linear function. However, using linearization around a nominal point, this can be written as

$$\mathbf{y_0} + \mathbf{\Delta y_0} = \mathbf{\Delta x J} + \mathbf{f}(\mathbf{x_0}) \tag{7}$$

where $\mathbf{J}$ is the Jacobian, where the elements are defined as:

$$k_{ij} = \frac{\partial y_i}{\partial x_j} \tag{8}$$

Example: Electric Light Vehicle. An electric vehicle has functional requirements, range (at constant speed 70 km/h) and acceleration time (0–70/h) and design parameters of battery size m_b, and engine power P_m).

The range can be calculated from first principles as (considering just air resistance):

$$R = \frac{2k_b m_b \eta}{C_d A_0 \rho v^2} \tag{9}$$

Here: k_b is the battery energy density, m_b is the mass of the battery. η is the combined efficiency of battery and motor. C_d is the aerodynamic drag coefficient. A_0 is the frontal area and v is the vehicle speed.

The acceleration time can be calculated as

$$t_a = \frac{mv^2}{2P\eta_a} \tag{10}$$

Where the total weight is:

$$m = m_0 + m_b \tag{11}$$

The design matrix is

$$\begin{pmatrix} R \\ t_a \end{pmatrix} = K \times \begin{pmatrix} m_b \\ P_m \end{pmatrix} \tag{12}$$

The sensitivity matrix K can be calculated as:

$$K = \begin{pmatrix} \frac{\partial R}{\partial m_b} & \frac{\partial R}{\partial P_m} \\ \frac{\partial t_a}{\partial m_b} & \frac{\partial t_a}{\partial P_m} \end{pmatrix} \tag{13}$$

This yelds:

$$K = \begin{pmatrix} \frac{2k_b\eta}{C_d A\rho v^2} & 0 \\ \frac{v^2}{2P\eta_a} & -\frac{(m_0+m_b)v^2}{2P^2\eta_a} \end{pmatrix} \tag{14}$$

This expression shows that Range is only dependent on the battery size, at least with the simplified model used here. The acceleration is on the other hand dependent on both motor and battery. The influence of battery comes from the fact that the battery brings a penalty in the form of weight. Hence the system is "decoupled" in the nomenclature of axiomatic design.

2.1 Normalized Sensitivities

If the system is complex and the sensitivity matrix large, it may be difficult to get an overview of the system since the different parameters may have values of different orders of magnitude. The functional characteristics are normally also of different orders of magnitude. In order to make it easier to get an overview of the sensitivities some kind of normalized dimensionless sensitivities are needed. The first approach to normalize the sensitivities is to employ the following definition from [10].

$$k^0_{ij} = \frac{x_j}{y_i}\frac{\partial y_i}{\partial x_j} \tag{15}$$

In this way a non-dimensional value is obtained, that indicates how many percent a certain functional characteristic is changed when a system parameter is changed one percent. In this way it is much easier to assess the relative importance of the different system parameters.

Example: Normalized Sensitivities of the Electric Light Vehicle. For the electric light vehicle example the following sensitivity matrix is obtained.

$$K^0 = \begin{pmatrix} \frac{m_b}{R}\frac{\partial R}{\partial m_b} & \frac{P_m}{R}\frac{\partial R}{\partial P_m} \\ \frac{m_b}{t_a}\frac{\partial t_a}{\partial m_b} & \frac{P_m}{t_a}\frac{\partial t_a}{\partial P_m} \end{pmatrix} \tag{16}$$

With (9) and (10) it can also be written as:

$$K^0 = \begin{pmatrix} \frac{m_b}{\frac{2k_b m_b \eta}{C_d A_0 \rho v^2}}\frac{\partial R}{\partial m_b} & \frac{P_m}{\frac{2k_b m_b \eta}{C_d A_0 \rho v^2}}\frac{\partial R}{\partial P_m} \\ \frac{m_b}{\frac{mv^2}{2P\eta_a}}\frac{\partial t_a}{\partial m_b} & \frac{P_m}{\frac{mv^2}{2P\eta_a}}\frac{\partial t_a}{\partial P_m} \end{pmatrix} \tag{17}$$

With (14) this leads to the following normalized sensitivity matrix:

$$K^0 = \begin{pmatrix} 1 & 0 \\ \frac{m_b}{m_0+m_b} & -1 \end{pmatrix} \tag{18}$$

Clearly this representation is much easier to asses. Many of the elements gets trivial. It says that increasing the battery mass will have a direct proportional effect on the range. Furthermore acceleration time is inversely proportional on motor size. If the battery m_b is very small compared to the rest of the vehicle m_0 the first element on the second row also becomes zero.

One general observation that can be made is that if a function can be written as:

$$y = ax^b \tag{19}$$

Then the normalized sensitivity becomes

$$k^0 = \frac{x}{y}\frac{\partial y}{\partial x} = \frac{x}{ax^b}abx^{(b-1)} = b \tag{20}$$

That is, the normalized sensitivity just becomes the exponent of the function.

3 Functional Correlation

An important aspect of the functional characteristics is whether they are conflicting or coherent. This kind of information is important when negotiating requirements since it can highlight areas of conflicting requirements where compromises have to be made, and also areas where there is not meaningful to reduce requirements.

A measure of dependencies is the correlation of the functional characteristics, as presented in [4] and in [9]. It is limited to the interval [−1, 1], and it is a symmetric matrix, so there is no information regarding the dominant direction of dependency. The correlation is a measure of the angle (cosine) between two row vectors in the sensitivity matrix. If the correlation is one, they are completely aligned. If it is zero they are orthogonal and if it is minus one, they are pointing in the opposite direction

$$c_{ik} = \frac{\frac{1}{n}\sum_{j=1}^{n} k_{ij}^0 k_{kj}^0}{s_i s_k} \tag{21}$$

The *adjusted functional characteristics correlation* can be written as:

$$c_{a,ik} = \varphi_i \varphi_k \frac{\frac{1}{n}\sum_{j=1}^{n} k_{ij}^0 k_{kj}^0}{s_i s_k} \tag{22}$$

whereas before $\varphi = 1$ if a large value is desirable, or required, and $\varphi = -1$ if a small value is desirable, or required. Here the standard deviations in the sensitivities are:

$$s_i = \sqrt{\frac{1}{n}\sum_{j=1}^{n} (k_{ij}^0)^2} \tag{23}$$

3.1 Example: Functional Correlation of an Electric Light Vehicle

As an example the functional correlation matrix is calculated for the electric light vehicle. It is assumed that:

$$m_b = m_0/4 \tag{24}$$

The functional correlation matrix is calculated from:

$$c_{a,ik} = \varphi_i \varphi_k \frac{\frac{1}{n}\sum_{j=1}^{n} k_{ij}^0 k_{kj}^0}{s_i s_k} \tag{25}$$

where

$$s_i = \sqrt{\frac{1}{n}\sum_{j=1}^{n} (k_{ij}^0)^2} \tag{26}$$

We had that for the light vehicle the normalized sensitivity matrix is:

$$K^0 = \begin{pmatrix} 1 & 0 \\ \frac{m_b}{m_0+m_b} & -1 \end{pmatrix} \tag{27}$$

$$s_1 = 1/\sqrt{2} = 0.707 \tag{28}$$

$$s_2 = \sqrt{\frac{1}{2}\left(\left(\frac{m_b}{m_0 + m_b}\right)^2 + (-1)^2\right)} \tag{29}$$

$$s_2 = \sqrt{\frac{1}{2}\left(\left(\frac{1}{5}\right)^2 + (-1)^2\right)} = 0.721 \tag{30}$$

$$c_{a,12} = \varphi_i \varphi_k \frac{\frac{1}{n}\sum_{j=1}^{n} k_{ij}^0 k_{kj}^0}{s_i s_k} = 1(-1)\frac{\frac{1}{2}\left(\frac{1}{5}\right)}{0.707 \times 0.721} = -0.196 \tag{31}$$

This yields the correlation matrix as:

$$C_a = \begin{pmatrix} 1 & -0.196 \\ -0.196 & 1 \end{pmatrix} \tag{32}$$

This shows that the requirements are conflicting, although at a low magnitude, since a larger battery required for a larger range will increase the weight and hence have a negative influence on the acceleration performance. However, the smaller the battery is compared to the total weight, the smaller this influence is.

4 Example

As an example a small electric transport aircraft is studied. Electric power is considered as a low emission alternative for aircraft, [3]. However, the low energy density of batteries place severe restrictions on their range. Here we can use design analytics to study the relations in the design. The aircraft has a capacity of ten passengers. First the customer/stakeholders requirements are established with their priorities, then a QFD matrix is defined that maps the customer/stakeholders requirements onto the functional characteristics. This is shown in Fig. 4. Here target values are given for the functional requirements as well as if there are hard requirements, demands D, or desirable or wishes W (Fig. 3).

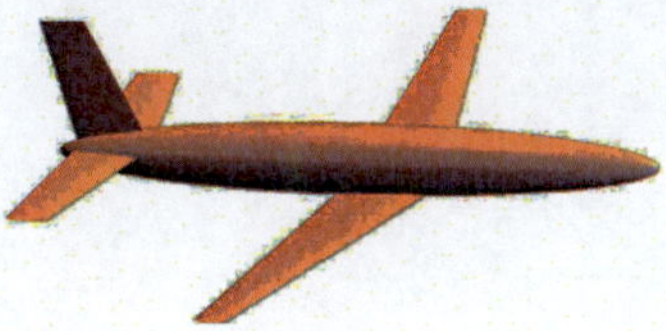

Fig. 3. Basic aircraft geometry assumed in the example.

Customer/Stakeholders requirements \ Functional Characteristics	Range	Payload	Liftoff distance	Landing distance	Takeoff weight	Required weight quotient	Buffeting speed quotient	Optimum cruise speed	Landing speed	Liftoff speed	Stall speed cruise	Consumtion	Rotation	Stability	Cost	Customer req. priorities
Airline	9	9	3	3					3	3		9			9	1.57
Crew			3	3					3	3			9	9		0.71
Environment												9				0.29
Maintenance org.					3											0.86
Manufacturer					3										9	1.14
Passenger	9	3														1.00
Regulatory bodies			3	3		9			3	3			9	9		2.00
Society																0.43
Sign	1	1	-1	-1	-1	-1	-1	1	-1	-1	-1	-1	1	1	-1	
Demand or whish	D	D	D	D	W	D	D	W	W	W	W	W	D	D	W	
Units	km	N	m	m	N			m/s	m/s	m/s	m/s	g/(km k	m		kEUR	
Target values	6000	7000	500	500	60000	1	1	200	70	70	140	0.50	3	3	80000	
Functional char. priorities	23.14	17.14	12.86	12.86	6.00	18.00	0.00	0.00	12.86	12.86	0.00	16.71	24.43	24.43	24.43	
Normalized system char. priorities	1.69	1.25	0.94	0.94	0.44	1.31	0.00	0.00	0.94	0.94	0.00	1.22	1.78	1.78	1.78	

Fig. 4. The house of quality matrix showing the relationship between the customer/stakeholders requirements and the functional characteristics. Hard requirements (demands) are greyed.

In this way the weighting for the functional characteristics can be obtained, that can be used in the formulation of the objective function. The model to describe the relationship between the design parameters and the functional characteristics are for this case based on simple models for aircraft conceptual design as can be found e.g. in Torenbeek [8] and Raymer [6]. More advanced models can of course be used, e.g. simulation based methods as in Krus [5].

First the aircraft is optimized using the target values and weights from the House of quality. This gives the objective function summation in Fig. 5. The optimized aircraft has a low fidelity geometric definition use full to obtain the major characteristics such as size and general plan-form. It is shown in Fig. 5.

										objective function	
										22.80	
System characteristics	Target value	Adjusted target value	Actual value	Direction	Unfullfillment	System characteristics priorities	Demand (1) or wish (0)	Exponent		Sub objectives	Objective sensitivities
Range	6000.00	6000.00	5060.40	1.00	1.19	1.37	1.00	2.37		3.34	1.37
Payload	7000.00	7000.00	7473.00	1.00	0.94	1.02	1.00	2.02		0.88	1.02
Liftoff distance	500.00	500.00	495.47	-1.00	0.99	0.76	1.00	1.76		0.98	0.76
Landing distance	500.00	500.00	495.84	-1.00	0.99	0.76	1.00	1.76		0.99	0.76
Takeoff weight	60000.00	60000.00	66727.09	-1.00	1.11	0.36	0.00	1.36		1.15	0.36
Required weight quotient	1.00	1.00	1.28	-1.00	1.28	1.07	1.00	2.07		4.42	1.07
Buffeting speed quotient	1.00	1.00	0.72	-1.00	0.72	1.00	1.00	2.00		0.52	1.00
Optimum cruise speed	200.00	200.00	98.13	1.00	2.04	1.37	0.00	2.37		5.41	1.37
Landing speed	70.00	70.00	39.59	-1.00	0.57	0.76	0.00	1.76		0.37	0.76
Liftoff speed	70.00	70.00	43.79	-1.00	0.63	0.76	0.00	1.76		0.44	0.76
Stall speed cruise	140.00	140.00	48.18	-1.00	0.34	0.76	0.00	1.76		0.15	0.76
Consumtion	0.50	0.50	0.79	-1.00	1.59	0.99	0.00	1.99		2.51	0.99
Rotation	3.00	3.00	3.83	1.00	0.78	1.45	1.00	2.45		0.55	1.45
Stability	3.00	3.00	3.07	1.00	0.98	1.45	1.00	2.45		0.95	1.45
Cost	80000.00	80000.00	36277.56	-1.00	0.45	1.45	0.00	2.45		0.14	1.45

Fig. 5. The set up of the objective function. It is a weighted function of the functional characteristics.

A sensitivity analysis gives us the design matrix. This is obtained through numerical differentiation by varying one parameter at a time to obtain the local sensitivities. Here the influence of different parts of the design on the functional characteristics can be studied.

The power of local sensitivities should not be underestimated. First, they can be presented in a form that is easy to grasp, since a single matrix captures the relationships between design parameters and functional characteristics in a two dimensional way. It is also computationally cheap, and that alone means that there is little reason not to do it.

There is the argument that local analysis does not show combination effects of several parameters changed at the same time, but for smooth functions the interactions between parameters are low, for small variations. It can also argued that in other fields, such as the field of automatic control, linearised system models are very often used, since there is a lot of theory is based on this assumption, although almost all real systems are non-linear, and the assumption is, at best, only valid for small signals. Nevertheless, it has been a very successful field, and it can be argued that linearised models should be the basis for analysis since it can so easily be analysed, while more complete non-linear models can be used when appropriate.

In Fig. 6. the normalized sensitivity analysis is shown. The areas with strong influence are colour coded. In the next step the correlation matrix can be derived. This is essentially the roof of the house of quality, since it always symmetric and only half has to be presented. Here the interactions between different functional characteristics can be studied. This is obtained just using the sensitivity matrix. The result is shown in Fig. 7. Here the correlations between certain functional characteristics are evident. E.g. the landing speed, take off speed and stall speed in cruise are all highly correlated. On the other hand a high optimum cruise speed has a negative correlation with those. We can also see that range have a negative correlation to everything except energy consumption.

				B	Cr	taper	tc	lambda	xw	Bht	We	T	Wf	Vcruise	halt
Functional Characteristics	Units	Target value	Actual value	19.57	3.93	0.06	0.05	0.00	6.59	9.59	24354.17	14645.23	30641.49	101.73	2413.97
Range	km	6000.00	5082.60	0.12	-0.34	-0.04	-0.14	0.00	0.00	-0.37	-0.31	-0.16	1.00	-0.09	0.13
Payload	N	7000.00	7473.00	0.00	0.00	0.00	0.00	0.00	0.00	0.00	0.00	0.00	0.00	0.00	0.00
Liftoff distance	m	500.00	499.44	-1.16	-0.95	-0.14	0.00	0.00	0.00	0.00	0.77	-0.72	0.98	0.00	0.00
Landing distance	m	500.00	499.81	-0.57	-1.24	-0.17	0.00	0.00	0.00	0.00	0.37	0.19	0.46	0.00	0.00
Takeoff weight	N	60000.00	66727.09	0.00	0.00	0.00	0.00	0.00	0.00	0.00	0.37	0.19	0.46	0.00	0.00
Required weight quotient		1.00	1.28	0.04	-0.02	0.00	-0.03	0.00	0.00	0.00	-0.56	0.06	0.38	0.00	0.00
Buffeting speed quotient		1.00	0.72	-0.50	-0.51	-0.07	0.00	0.00	0.00	0.00	0.18	0.09	0.23	-1.00	0.12
Optimum cruise speed	m/s	200.00	98.52	-0.61	-0.11	-0.03	-0.04	0.00	0.00	-0.12	0.18	0.09	0.23	0.00	0.12
Landing speed	m/s	70.00	39.75	-0.50	-0.51	-0.07	0.00	0.00	0.00	0.00	0.18	0.09	0.23	0.00	0.00
Liftoff speed	m/s	70.00	43.96	-0.50	-0.51	-0.07	0.00	0.00	0.00	0.00	0.18	0.09	0.23	0.00	0.00
Stall speed cruise	m/s	140.00	48.37	-0.50	-0.51	-0.07	0.00	0.00	0.00	0.00	0.18	0.09	0.23	0.00	0.12
Consumtion	g/(km kg)	0.50	0.79	-0.12	0.33	0.04	0.13	0.00	0.00	0.37	0.31	0.16	0.00	0.09	-0.13
Rotation	m	3.00	3.83	-1.13	-1.16	-0.17	0.00	0.00	-4.71	2.23	0.00	0.00	0.00	0.00	0.00
Stability		3.00	3.07	-0.94	-0.96	-0.14	0.00	0.00	2.11	1.87	0.00	0.00	0.00	0.00	0.00
Cost	kEUR	80000.00	36277.56	0.00	0.00	0.00	0.00	0.00	0.00	0.00	0.69	0.35	0.00	0.00	0.00

Fig. 6. The normalized sensitivity matrix for an aircraft showing the relationship between system parameters and functional characteristics.

Functional Characteristics		Range	Payload	Liftoff distance	Landing distance	Takeoff weight	Required weight quotient	Buffeting speed quotient	Optimum cruise speed	Landing speed	Liftoff speed	Stall speed cruise	Consumtion	Rotation	Stability	Cost
		5082.60	7473.00	499.44	499.81	66727.09	1.28	0.72	98.52	39.75	43.96	48.37	0.79	3.83	3.07	36277.56
Range	5082.60		0.00	-0.41	-0.37	-0.43	-0.69	-0.25	0.22	-0.29	-0.29	-0.31	0.55	-0.09	-0.13	0.29
Payload	7473.00	0.00		0.00	0.00	0.00	0.00	0.00	0.00	0.00	0.00	0.00	0.00	0.00	0.00	0.00
Liftoff distance	499.44	-0.41	0.00		0.78	0.47	-0.09	0.62	-0.75	0.84	0.84	0.83	-0.04	-0.21	-0.31	0.17
Landing distance	499.81	-0.37	0.00	0.78		0.41	-0.02	0.58	-0.63	0.95	0.95	0.94	-0.21	-0.26	-0.37	0.27
Takeoff weight	66727.09	-0.43	0.00	0.47	0.41		-0.05	0.24	-0.43	0.39	0.39	0.39	0.35	0.00	0.00	0.66
Required weight quotient	1.28	-0.69	0.00	-0.09	-0.02	-0.05		-0.02	0.06	-0.04	-0.04	-0.04	-0.40	0.01	0.01	-0.70
Buffeting speed quotient	0.72	-0.25	0.00	0.62	0.58	0.24	-0.02		-0.52	0.61	0.61	0.62	-0.16	-0.17	-0.24	0.16
Optimum cruise speed	98.52	0.22	0.00	-0.75	-0.63	-0.43	0.06	-0.52		-0.82	-0.82	-0.84	-0.09	0.14	0.20	-0.29
Landing speed	39.75	-0.29	0.00	0.84	0.95	0.39	-0.04	0.61	-0.82		1.00	0.99	-0.08	-0.27	-0.40	0.26
Liftoff speed	43.96	-0.29	0.00	0.84	0.95	0.39	-0.04	0.61	-0.82	1.00		0.99	-0.08	-0.27	-0.40	0.26
Stall speed cruise	48.37	-0.31	0.00	0.83	0.94	0.39	-0.04	0.62	-0.84	0.99	0.99		-0.11	-0.27	-0.39	0.26
Consumtion	0.79	0.55	0.00	-0.04	-0.21	0.35	-0.40	-0.18	-0.09	-0.08	-0.08	-0.11		-0.16	-0.24	0.53
Rotation	3.83	-0.09	0.00	-0.21	-0.26	0.00	0.01	-0.17	0.14	-0.27	-0.27	-0.27	-0.16		-0.21	0.00
Stability	3.07	-0.13	0.00	-0.31	-0.37	0.00	0.01	-0.24	0.20	-0.40	-0.40	-0.39	-0.24	-0.21		0.00
Cost	36277.56	0.29	0.00	0.17	0.27	0.66	-0.70	0.16	-0.29	0.26	0.26	0.26	0.53	0.00	0.00	

Fig. 7. The adjusted correlation matrix for an aircraft showing the correlation within the functional characteristics. Note that this matrix is always symmetric.

5 Conclusions

In this paper a range of methods for design analytic has been shown. The example illustrates how an objective function for optimization can be defined using the House of quality. The sensitivity analysis can be used to produce a map of the dependencies in the design. Functional correlation, derived from the sensitivity analysis, give a map of interaction between functional characteristics that is useful e.g. for requirement negotiation. The methods are computationally efficient and can visualize interactions also in a complex design.

Acknowledgments. The project has partly been funded by the Swedish-Brazilian Research and Innovation Centre, CISB.

References

1. Haskins, C.: Systems Engineering Handbook. Incose, p. 185 (2006)
2. Hauser, J.R., Clausing, D.: The house of quality. Harward Bus. Rev. (1988)
3. Martin, H.: Electric flight - potential and limitations. In: Energy Efficient Technologies and Concepts of Operation, STO (2012)
4. Krus, P.: Computational tools for aircraft system. In: 26th International Congress of the Aeronautical Sciences, pp. 1–9 (2008)
5. Krus, P., Ölvander, J.: Simulation based optimisation for aircraft systems. SAE Trans. J. Aerosp. **2004**, 445–453 (2003)
6. Raymer, D.P.: Aircraft design: a conceptual approach. AIAA Education Series, Delft (1992)
7. Suh, N.P.: The Principles of Design. Oxford University Press, New York (1990)
8. Torenbeek, E.: Synthesis of Subsonic Airplane Design. Kluwer Academic Publishers, Delft (1982)
9. Uolevi, A.: Comparison of low- and high-fidelity approach in model based design in the case. In: ICED13: 19th International Conference on Engineering Design, pp. 1–10 (2013)
10. Varma, A., Morbidelli, M., Wu, H.: Parametric Sensitivity in Chemical Systems. Cambridge University Press, Cambridge (1999)

Analysis of the Application of a Product Development Method Called Meta Project

Aline Fernandes(✉) and Franciane Falcão

Universidade Federal do Amazonas - UFAM, Manaus 69067-005, Brazil
alineaferns@gmail.com

Abstract. This work aims to analyze the ways that Dijon de Moraes' meta project method has been used for the development of innovative products. In order to do so, a systematic review was performed using scientific articles databases, which were analyzed the steps followed, techniques applied to the method and the purposes of each work. Two databases were consulted, LivreFGV and Blucher Proceedings. First, an expanded search was performed, using the project's subject keywords and obtaining 227 results. After the first selection process, containing a preliminary reading of each work and searching for relevant content, this number was restricted to 37 articles. The next step of selection was a full reading of the articles, which selected only 11 studies, used the basis for this research. Succeeding analysis and study of them, some differences in application of meta project method were identified and claimed attention, especially in techniques selected to accomplish some steps, since depending on the final product generated and viability of execution, chosen techniques have an enormous influence on the result, having a significant impact, positive or negative, on the resulting level of innovation. Therefore, with this study, it is possible to identify which technique is more appropriate for each application and thus improve the results obtained.

Keywords: Metaproject method · Product development · Design

1 Introduction

Before the emergence of the meta project method, a conventional methodology was used, characterized by being predictable, objective and linear, aiming to analyze the project itself and not the context that it was inserted, since the scenario was static and did not have as much influence on product design. Nevertheless, globalization and technological advance changed the context in which design was framed, becoming involved and dynamic. Consumers who previously did not have varieties of options to choose from started to have a diversity of products and materials within their reach and it also could come from other countries, as a result of a free circulation of raw materials on the market. After that, the conventional methodology was no longer appropriate for the situation, and it was necessary to create a new one that would support reality.

Therefore, the metaprojectual methodology was created. However, Dijon de Moraes was not the only author who defined this concept, several other authors did. Among them

L. Pereira et al. (Eds.): IDEAS 2018, SIST 198, pp. 219–227, 2021.
https://doi.org/10.1007/978-3-030-55374-6_22

is Deserti (2007), who states "The meta project is constituted from the research phase to the concept design, the conceptualization of the project, and it is pointed out as a process capable of promoting innovation, not being necessarily tied to the executive process of the project, but rather to the process of idealization through research and development".

Franzato (2011), another author who studied the meta project method, states that the method has a transdiciplinary character and does not stick to a specific phase or stage, but may pass through the whole process, where the research and the project will merge. For him, the meta project allows the development of innovative projects.

Also, Celashi (2007) defined a meta project as "(…) the idealization and organization of research and project process, also called the "project of the project," consisting of phases, times and financial and human resources, aiming to innovation."

Like the other authors, Dijon de Moraes also studied the concept of a meta project and from it created a new methodological procedure, for which he gave the same name. According to de Moraes (2010), "The meta project goes beyond the project because it transcends the design act. This is a critical and preliminary reflection on the project itself from an assumption scenario …". He states this by considering some elements such as productive, technological, market, material, environmental, sociocultural and aesthetic-formal factors and using as a base some analysis done through a data survey.

Therefore, the purpose of the method is to show a scenario that exists or will exist based on a general analysis, in which the previous evaluation related to the development of the products under study and the circumstances to which they are involved is demonstrated.

In order to better understand the scenario in which the product is found, the meta project is based on six topics: market factors, product/design system, social and environmental sustainability, sociocultural influences, productive technology and materials, and typology and ergonomics.

Since the meta project method does not follow a linear sequence, these topics do not need to be in this same order, being analyzed according to the author's interest. Intending to understand each step better, they are described below:

- Market factors: scenario, vision, and concept (demanding prior knowledge of the interaction between these three items), identity (integration between product, production, sales, and communication; definition of a company profile) and mission (describing the company's mission, producing and commercializing the product).
- Product/design system: design is analyzed through the interaction between design, communication, market, and service.
- Social and environmental sustainability: sustainable environmental alternatives to apply in the development of products.
- Sociocultural influences: analysis of sociocultural facts occurred by the time the product is being developed or studied.
- Productive technology and materials: besides analyzing functional and aesthetic characteristics, it also considers the emotion and feeling wanted for the product to cause in the client.
- Typological and ergonomic: it analyzes and improves the relation of man/space/product and helps to define the formal typology of the product.

2 Methods and Materials

Search delineation

For the qualitative and quantitative survey of the ways of applying the metaproject method for the development of different design products, a systematic literature review method proposed by Brereton et al. (2007) was used. The systematic analysis consists of identifying and evaluating, both qualitatively and quantitatively, the research related to the project. To perform the review, it was necessary to follow the steps below:

- Research question: The research problem is: What steps and techniques are being adopted for the development of different design products with the metaproject method? From this point, it was done extensive research on the subject until able to move to the second stage.
- Database: Websites of 10 national design magazines from A1 to B3 (Infodesign, Ergodesign HCI, Design & Tecnologia, Produção, Estudos em Design, Applied Ergonomics, Design Issues, Gestão Produção, Revista DAPesquisa e Revista Projetica) were consulted. In these, preliminary searches were made using the following keywords: product development, meta project, design, product, and methodology.

Once the articles were collected, those that presented the development of a product using the metaproject method or theoretically described it were selected for prior reading. Subsequently, the indexers of these journals were tabulated, and a new search was applied through them using a structured formula of expressions. Indexers' databases that listed articles on the research's theme were chosen for systematic literature review.

They are: Blucher Proceedings and LivreFGV.

The following table shows the number of articles obtained through the formula and the final number of articles that contain similar content to this study, being used as the basis for this work (Table 1).

Table 1. Number of articles found on databases

Databases	Number of articles found	Number of articles after preliminary reading	Number of selected articles
Blucher Proceedings	68	20	5
Livre FGV	155	17	6

Source: elaborated by authors.

- Search strategy: After reading the selected articles, the information about what is meta project method and its purposes was extracted, allowing the selection of used keywords: methodology, method, meta project, innovation, new design model, design process, project, product, design, model, sustainability, ecodesign, sustainable design, materials, ergonomics, human factors and market. These terms were grouped according to the thematic areas they correlate (Table 2).

Table 2. Expressions for search strategy

Project methodology	Product
• metodologia OR methodology	• produto* OR product*
• método* OR method*	• design
• metaprojeto OR meta project	• modelo* OR model*
• inovação OR innovation	• sustentabilidade OR sustainability
• "metodologia projetual" OR "projectmethodology"	• ecodesign OR ecodesign
• metadesign OR metadesign	• "design sustentável" OR "sustainable design"
• "novo modelo projetual" OR "new projectualmodel"	• materiais OR materials
• "processo projetual" OR "design process"	• ergonomia OR ergonomics
• projeto OR project	• "fatores humanos" OR "humanfactor"
	• mercado OR market

Source: elaborated by authors.

From this table, the following search formula was defined: (metaprojeto OR meta project) AND (produto OR product OR design).

- Inclusion and exclusion criteria: Articles in Portuguese or English, works that presented the development and evaluation of design products and those who had keywords similar or equal to this study were included. Articles that did not describe the methodological techniques used and that were not available for download were excluded.

Data analysis

- Criteria for articles selection by the quality of study: In this analysis, the following aspects were considered: most cited articles, countries with the highest number of articles and knowledge area with the most significant number of articles. Thus, after quality evaluation, works that had questionable results or were incomplete were excluded.
- Data organization resources: To organize selected articles, tables in Microsoft Office Excel and Microsoft Office Word were created. After that, it was possible to collect, structure, and analyze data, facilitating the development of the research.
- Results synthesis: Data synthesis was obtained from the content analysis, and the following points were considered:

  - Publication's identification – Author, title, year, study area, where it was published and journal.
  - Type of work – What type of research was used in the article, being them: theoretical or practical.

- Method steps – Name, description, and purpose of each stage, indicating techniques.
- Applied techniques – Identification of stages of the method applied in the reported project, description of fulfilled objectives of each step and respectively applied techniques, such as: case study, experimental study, questionnaire, interview, focus group, use test, functional and typological analysis, among others.

3 Results

As previously mentioned, in order to better understand the meta project method and the best way to apply it, all articles were analyzed methodically. In order to better clarify the results, three tables are presented, comparing the articles in different fields.

Table 3. Identification table.

Author/Year	Where it was produced	Title	Study area	Where it was published
Gil et al. (2014)	Sao Paulo	Considerations about design of interactive products in the contemporary world Crewee application creation and development process	Application design	Sao Paulo
de Moraes (2010)	Minas Gerais	Metaproject as project model	Theoretical design	Minas Gerais
Reyes (2012)	Rio Grande do Sul	Projecting by the exteriority of the project	Theoretical design	Rio Grande do Sul
Celi (2012)	Busto Arsizio - Italy	Design, metadesign and the importance of vision	Theoretical design	Milan - Italy
Nicolò (1993)	Rome - Italy	Metaproject analysis: multiagent virtual project networks for strategic decisions in preplanning	Theoretical design	Rome Italy
Tamekuni (2014)	Rio Grande do Sul	Metaproject -design in search of innovation through reflection	Theoretical design	Rio Grande do Sul
Machado et al. (2014)	Rio Grande do Sul	Fashion design and Strategic design: analysis of the displacement of the moulage technique for the metaprojectual stage	Fashion design Strategic design	Rio Grande do Sul
Market et al. (2016)	Mato Grosso do Sul	Understanding Digital Design: the designer in digital project processes	Theoretical design	Buenos Aires - Argentina
Calado et al. (2016)	Paraiba	Design and reverse logistics in the company of planned movies	Product design	Minas Gerais
Barros et al. (2016)	Minas Gerais	Mini coffee grinder: A project of food design connected to coffee consumption with an emphasis on slow movement and the user experience of the product	Product design	Minas Gerais
Dias et al. (2016)	Minas Gerais	Creation of a tactile map through assistive technology: more accessibility to the visually impaired using 3D printing	Product design	Minas Gerais

Source: elaborated by authors.

The first one, Identification Table, contains data of the 14 selected articles: author, year, where it was produced, title, study area, and where it was published (Table 3).

Based on this table, it is seen that Minas Gerais and Rio Grande do Sul are the Brazilian states that have produced and published more articles.

Besides, it is also worth noticing that most of the works are theoretical, that is, explain or describe the method, being only three related to product design, fashion and strategic design, and application design.

In the table below it is possible to observe more clearly what each work is about, not only if it is theoretical or practical, but also the results of the project (Table 4).

Table 4. Articles results.

Author/Year	Article type	Described results	Results characteristics
Gil et al. (2014)	Practical	Crewee application creation and development process	A group creation and support for projectual decision platform, aimed at design studios, advertising agencies, artists, architects,and other professionals involved in the design culture
de Morais (2010)	Theoretical	Explanation of the meta project method	Comparison between the metaproject method and the conventional method, with emphasis on how the metaproject approach is made
Reyes (2012)	Theoretical	Description of the meta project method	Use and analysis of the meta project method as a study base to present a proposal of a method with a broader project view
Celi (2012)	Theoretical	Description of the meta project method	Analysis of the importance of a metacognitive approach and the central role of forecasting activities
Nicolò (1993)	Theoretical	Description of the meta project method	Description of the meta project analysis considering scenarios and activities
Tamekuni (2014)	Theoretical	Description of the meta project method	A literature review of metaproject's concept approaching several conceptions to show that the method not only accompanies the project before, but also during and after
Machado et al. (2014)	Practical	The effects of the displacement of the moulage technique to the meta projectual stage in the construction of the design concepts	Application of the moulage technique during the meta projectual stage
Market et al. (2016)	Theoretical	Presents aspects of how a designer's figure acts and influences the digital design process and, therefore, its result	Presentation of how the designer acts and influences the digital design process and its outcome
Calado et al. (2016)	Practical	Develops a study table following the requirements of sustainability and Ecodesign	Development of a study table in accordance with the requirements of sustainability and Ecodesign
Barros et al. (2016)	Practical	Development of a manual grinder	Development of a manual coffee grinder
Dias et al. (2016)	Practical	A practical tactile map for the visually impaired, easy to replicate in multiple locations and at a low cost	Elaboration of a practical tactile map for the visually impaired, easy to replicate in multiple locations and at a low cost

Source: elaborated by authors.

Table 5. Level of innovation table.

Author/Year	Steps	Pointed techniques
Gil et al. (2014)	1 Strategic level – contextual and open research, which also defines the concept of the project	Through experience, research and analysis the problem was identified between designers and clients
	2 Tactical level – systematization of data or "how to"	Review of other similar apps
	3 Technical and operational level – generation of alternatives, finalization, and evaluation	Generation of alternatives, prototype, and experimentation
de Morais (2010)	1 Marketing phase 2 Product/design system 3 Environmental sustainability 4 Socio-cultural, typological-formal and ergonomic influences 5 Productive technology 6 Materials used	Not applicable
Reyes (2012)	1 The reality of the company/customer 2 Market 3 Technology available 4 User context 5 Innovation trends	Not applicable
Celi (2012)	1 Marketing phase 2 Environmental sustainability 3 User Context 4 roductive Technology 5 Materials used	Not applicable
Nicolò (1993)	1 Study of company/customer 2 Market 3 Technology 4 User experience 5 Innovation	Not applicable
Tamekuni (2014)	1 Previous research 2 Emphasis on the production of meaning 3 Relationship between user and product 4 Consequences 5 Construction of scenarios	Not applicable
Machado et al. (2014)	1 Information search - think and question about the aspects of the design problem to direct to the generation of conceptual design proposals	Printed material on the problem of the exercise
	2 Generation of alternatives - application of the technique used	Moulage technique
	3 Representation – the result of the application	Videos, photos, interviews, notes, and documents
Market et al. (2016)	Method description only	Not applicable
Calado et al. (2016)	1 Case study	Visit and questionnaire
	2 Sustainable materials research	Through books, articles, and internet
	3 Choose and search for a specific mobile	Analysis of the methodology
	4 Development of alternatives	Requirements and parameters
	5 Choice of alternative	Which uses less material

(*continued*)

Table 5. (*continued*)

Author/Year	Steps	Pointed techniques
Barros et al. (2016)	1 Analysis of audience scenarios	Imaging Panels
	2 Analysis of the sociocultural context	Imaging panels and internet data
	3 Benchmarking	Analysis of similar and imaging panels
	4 Generation of alternatives	Not applicable
	5 Select the final alternative	Not applicable
Dias et al. (2016)	1 Search for materials/scenarios	Through books, articles, and internet
	2 Manufacturing process and materials	3D printing
	3 Benchmarking	Analysis of similar projects
	4 User's choice	Interview
	5 Production of the product	3D printing and alternative materials

Source: elaborated by authors.

Analyzing only articles with practical results, a new table can be elaborated, and it shows the stage of the meta project method used in each article, so as the techniques used in each step to analyze and compare each other to know the innovation. Leveling results based on the scale made by Davila et al. (2008). Authors classify the level of innovation in three different ways: incremental (promotion of existing products and processes); semi-radical (when there is a significant change); and radical (creation of products and processes completely innovative, which lead to changes in the sector in which they fit). Accordingly, the table is presented below (Table 5).

4 Conclusion

Based on the analysis of the articles and tables, it is possible to realize that the meta project method, although already appeared a few years ago, is still considered a new methodology and for that reason does not have many articles declaring to use it yet, and those that apply it, most describe the meta project method, so that the reader can understand its steps and basis to apply it in a study.

It is also noticed that the Brazilian Southeast region had the highest amount of works on meta project, in both production and publication numbers, being followed by the South, Northeast and Center-West regions. The North region did not contain any work with this theme, which shows the reduced use of this methodology in the region.

It is noteworthy that the applicability of this methodology is better in the product design area than the graphic area, so steps of the method can develop more and better during work. Due to the small use of the method in a practical way, it is not possible to state the best technique for each stage of the methodology, however it is conceivable to notice that in some practical projects, in the marketing phase, visual techniques are used, with printed materials and imaging panels, and also in the same phase, the use of interview for data collection was remarkable.

References

Deserti, A.: Intorno al progetto: concretizzare I'innovazione. Carocci Editore, Roma (2007)

Franzato, C.: O processo de inovação dirigida pelo design. Um modelo teórico. Redige, Rio Grande do Sul (2011)

Celaschi, F.: Dentro al progetto: appunti di merceologia contemporanea. Carocci Editore, Roma (2007)

De Moraes, D.: Metaprojeto: o design do design. Eduardo Blücher, São Paulo (2010)

Brereton, P.: Lessons from applying the systematic literature review process within the software engineering domain. J. Syst. Softw. **80**, 571–583 (2007)

Gil, A.G.S., Lourenço, A.J.M.L., Hildebrand, H.R.: Considerações acerca do conceito de projeto de produtos interativos na contemporaneidade. Processo de criação e desenvolvimento do aplicativo Crewee (2014)

De Morais, D.: Metaprojeto como modelo projetual. Strateg. Des. Res. J. **3**(2), 62–68 (2010)

Reyes, P.: Projetando pela exterioridade do projeto. Strateg. Des. Res. J. **5**(2), 91–97 (2012)

Celi, M.: Design, metadesign and the importance of vision. Strateg. Des. Res. J. **5**(2), 84–90 (2012)

Nicolò, E.: Metaproject analysis: multiagent virtual project networks for strategic decisions in preplanning. Fondazione Ugo Bordoni (1993)

Tamekuni, K.I.: Metaprojeto – o design em busca da inovação por meio da reflexão. In: Congresso Brasileiro de Pesquisa e Desenvolvimento em Design (2014)

Sousa, C.E.M., Silva, A.L.S.V., Cavalcante, R.N.: Metadisciplina: Questões contemporâneas do Design na criação de metodologias de ensino. In: Congresso Brasileiro de Pesquisa e Desenvolvimento em Design (2016)

Barata, T.Q.F., Matos, I.M.: Design e produção de protótipo de mobiliário público a partir da análise de demanda dos usuários. In: Congresso Brasileiro de Pesquisa e Desenvolvimento em Design (2016)

Machado, L., Scaletsky, C.: Design de Moda e Design Estratégico: análise do deslocamento da técnica de moulage para a etapa metaprojetual. In: Congresso Brasileiro de Pesquisa e Desenvolvimento em Design (2014)

Makert, R., Alves, G.: Entendendo o Design Digital: o designer nos processos digitais de projeto. In: XX Congress of the Iberoamerican Society of Digital Graphics (2016)

Calado, I.A., Bezerra, R.P., Clementino, I.O., França, M.L.: Design e Logística reversa na empresa de móveis planejados AIAM. In: Congresso Pesquisa e Desenvolvimento em Design (2016)

Barros, C.A., Santos, I.M.: Mini coffee grinder: um projeto de food design ligado ao consumo do café com ênfase no slow movement e na experiência do usuário diante do produto. In: Congresso Brasileiro de Pesquisa e Desenvolvimento em Design (2016)

Dias, G.S., Santos, I.M.: Criação de um mapa tátil através da tecnologia assistiva: mais acessibilidade aos deficientes visuais com a utilização da impressão 3D. In: Congresso brasileiro de pesquisa e desenvolvimento em design (2016)

Check for updates

Systematic Approach to Develop Physical Models as Creative Stimuli in Conceptual Design Phase

Matheus Gomes(✉), André Ogliari, Rodrigo Bastos Fernandes, and Karuliny Marques

Federal University of Santa Catarina, Trindade. C. P. 476, Florianópolis, Brazil
matheusgalvaogomes@gmail.com

Abstract. The conceptual design is a phase in which designers are demanded to ideate creative solutions, capable of meeting the products' design specifications. To facilitate the ideation process, analogies based on the similarities between known systems, or external information brought to the design context, are frequently used. This information consists of concepts that are presented as stimuli for idea generation. In this regard, physical models present characteristics that allow them to be effective in creativity stimulation. However, to build relevant physical models, it is necessary to adapt information from the potential source of analogy to adequately represent it as a source of inspiration. Thus, the present work proposes a systematic approach to developing context-based physical models for creativity stimulation. The systematic was evaluated in two different design contexts by a product development experts panel. The result is that the proposed approach was considered adequate to guide the development of relevant stimuli, effectively enhancing the ideation process during the conceptual phases of product design.

Keywords: Analogies · Problem solving · Product development · Design by analogy

1 Introduction

The conceptual design is a creative phase in which the solutions that best meet the design specifications are sought [1]. Due to barriers in the creative process, designers make use of creative methods to facilitate solutions' search and ideation [2].

Analogical reasoning is the basis of many of these methods, being responsible for the identification of similarities between systems, using them as sources of inspiration. In an attempt to facilitate this reasoning, additional information can be presented to designers, as creative stimuli to analogy making and, thus, contributing to the generation of an innovative solution [3].

In this context, physical models are effective stimuli to creativity, as they may facilitate the development of analogies due to their physical characteristics. This premise is based on their ability to enable concepts' exploration and refinement [4–6], as well as

L. Pereira et al. (Eds.): IDEAS 2018, SIST 198, pp. 228–238, 2021.
https://doi.org/10.1007/978-3-030-55374-6_23

to facilitate communication [7], learning [8] and also to mitigate functional fixedness [9]. Also, due to the sensorial stimuli [10] added to the ideation process [11], physical models contribute to functional thinking [12].

However, unlike for other types of stimuli, as texts [3, 13–16], images [3, 13–15, 17] or some proposals that use objects [13, 18], the development of physical models needs an approach that guides the collection of information and its adaptation to an arrangement of parts. It is essential to guarantee the model functionality without lost information during the referred development process, to enhance the stimulation of designers' creativity.

The present research has the objective of proposing a systematic approach to guide the development of physical models that serve as creativity stimuli. With this, designers can benefit from the physical characteristics of models to understand, adapt, and use them during the ideation process to develop innovative solutions.

2 Related Studies

2.1 Analogical Reasoning in Product Design

Several innovations are a result of analogies, such as Velcro®, aircraft wings, and robot manipulators [2]. The identification of functional similarities, characteristics, and other information between the problem and existing solution principles in other systems consist of the analogical reasoning [19]. By the ability of this thinking process to mitigating functional fixedness, creative solutions may be easily achieved [20].

In order to facilitate analogical reasoning, there are proposals to stimulate creativity, which are based on the addition of information in the creative process. Information in the form of texts and images is proposed in: Fernandes et al. [3], who present physical and morphological attributes of biological systems. Viswanathan et al. [13], who aside text and image also propose the use of a nonfunctional physical model as stimuli; Chan et al. [14] and Fu et al. [15] also use texts and images varying, the analogical distance and commonness of the stimulus. The research of Keshwani et al. [16] proposes the use of only texts, varying the analog domain and the detail level of the description. Chai et al. [17] evaluate the inspiration only on images, categorized by analogical distance. Song et al. [18] use physical examples, classified as sources of analogy or noise stimuli, through commercial products.

The modalities of stimuli representations discussed are simplistic and may hinder the identification of valuable information available in the source of analogy. With this respect, a systematic approach that guides the development of physical models may improve the generation of more complete stimuli, in terms of information to the creative process. These stimuli can facilitate the identification of similarities between the source of analogy and the design problem by the physical characteristics and functionality of the model.

2.2 Development of Physical Models

A model is a representation of a system to be evaluated and may exist in various levels of detail and forms [1, 2, 21, 22]. Considering physical models, the scope of this research, their development should be fast and straightforward in the early phases of design [23]. Simplified models enable effective communication, facilitating work corrections, and feedback [24]. For this, it is necessary to prioritize the simplicity of the information extracted and abstracted from the source of analogy, avoiding the detailing of characteristics that are not relevant [25]. Another way to simplify the development of models is to work with systems, or subsystems, alone, focusing only on the relevant parts to the design problem [7].

Mockups may be considered as simplified models, because they represent the characteristics of a system at a high level of abstraction, emulating the functional behavior [7]. Also, the model development approach should resemble to "Do It Yourself" practices, whereby it is possible to implement technologies that no require great skill or knowledge.

The materials that compose the models must be attractive to the designers, encouraging interaction [12]. Also, they need to be able to represent different constructive forms, stimulating the diversity of ideas, which could be otherwise restricted if the material does not allow the construction of varied geometries [8, 23].

3 Systematic Approach to Develop Physical Models as a Creative Stimuli

The objective of the proposed systematic approach is to guide the transfer of the concepts from the source of analogy to a physical model. The systematics' application occurs in the conceptual design phase, after the definition of the product functional structure, according to the activities of the PRODIP model [2] (see Fig. 1).

Activity 1 - Find sources of analogy, consists of tasks to define the objective of the ideation process and the necessary characteristics of the source of analogy. A search for systems that present the defined characteristics must be conducted for, in the end, prioritize the sources found. The main inputs of this activity are the function, which demands innovative solutions, and the design requirements. The main outputs are the prioritized sources of analogy to the problem.

The objective of the ideation process, in the context of the systematic, is the type of the solution needed, in other words, if it is a radical or incremental innovation. From this objective, the characteristics of the sources of analogy are defined. For example, if a radical innovation is needed, a distant domain (e.g., such as biology to engineering) provides a more significant potential for stimulating creative solutions. Thus, through the example, the necessary characteristics of the source would be biological systems that perform an analog function to that of interest. To this task, it is suggested the use of books, manuals, articles, functional bases, and publications as support tools for potential analog system search.

To prioritize the systems found, it is proposed the use of a comparison matrix [26], which results in a classification of the best source. This prioritization is essential to sort the systems that present the highest potential to stimulate creativity, in case it is not possible to develop models for and from all the sources.

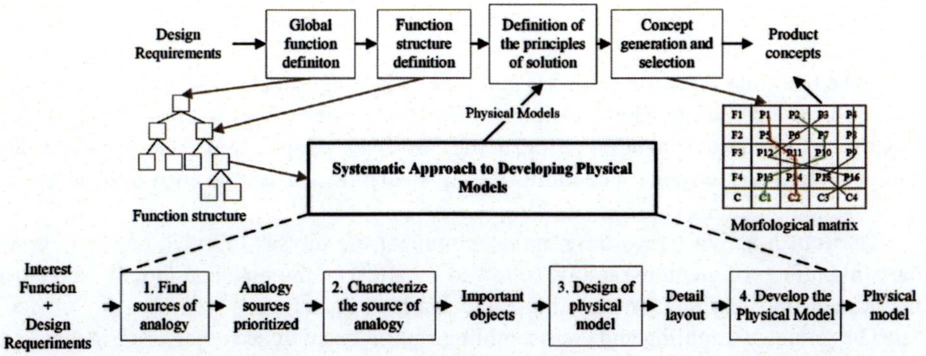

Fig. 1. Overview of the application of the systematic approach to developing physical models in the conceptual design phase. Source: Adapted from [2]

The activity 2 - Characterize the source of analogy, comprises the description of how the system meets the characteristics needed for stimulating better solutions to the design problem and the identification, in the description, of what is important to represent in the physical model.

The description must detail the analog systems' particularities, with mechanisms and functional principles. For the identification and extraction of essential characteristics from the source of analogy, it is proposed the use of the FEARB method, based on Qian [27]. This method suggests the analysis of a system according to its' Function, Elements, Attributes, Relations, and Behaviors. The definitions are as follows:

- Function: It is the primary function of the system, which relates it to the design problem.
- Elements: They are the parts that compose the system.
- Attributes: They are the characteristics of the elements, related to the functioning of the system, or their appearance.
- Relationships: They are how the elements relate to each other (e.g., positioning, actions and reactions), generating behaviors and the function.
- Behavior: They are actions that occur within the system, which result in the function.

Through this method, it is possible to map which information must be represented in the physical model to guarantee adequate functional representation of the analog system.

The third activity, which regards the design of physical model, comprise three tasks: sketch the fundamental elements of the source, define the arrangement of the elements, and detail the concept of the physical model. The sketches comprise the conception of the physical models' parts according to the characteristics identified with the FEARB model. To facilitate the representation not all identified characteristics need to be represented in the model, but rather those of greater importance in performing the function of the analog system. The assignments of values perform this evaluation according to the importance of each object to its relevance to the performance of the primary function. The objects with higher scores should be present in the model. The preliminary layout is achieved when the sketches are grouped, arranged in their shape, and when the size and material are defined.

If the resulting structure becomes complex, it is suggested the use of CAD software to support the detailing. The advantage in the use of such a tool is that the production processes, like additive manufacturing, of the parts may be facilitated. Also, if the analogy source system is extensive for handling, the physical model should be built-in scale, based on dimensional analysis [28].

The fourth activity, regarding the development the physical model, consists of the parts manufacturing and assembly, followed by a test of the physical model. The parts manufacturing should prioritize simple and available materials.

The model assembling and disassembling should be analyzed using criteria that take into account the feasibility, speed, and practicality. This analysis is necessary because the models must be freely used during the ideation sessions, allowing designers to interact with them as a whole, or only with some parts, demanding that the alternation of configurations be quick and practical.

Finally, a test of the physical model must be conducted in order to verify if it performs the function source of analogy, and remains useful during and after the usage.

4 Evaluation of the Systematic Approach to Developing of Physical Models

This section presents the application of the systematic approach, allowing the evaluation of the activities adequacy and the contribution of the guidelines and suggested tools to facilitate the development of relevant and useful creativity stimuli.

4.1 Materials and Methods

An experts' panel was formed to evaluate the systematic approach. Six mechanical engineers, with experience in product development, composed the panel. The evaluation method was the action research [29], in which two groups of work, randomly and equally distributed, each working on a different day, develop the proposed activities, following the guidelines of a facilitator, to develop a physical model to a given problem. The first problem seeks to develop a radical innovation for a function, and the second has the objective of applying available technologies of similar products to solve the problem, as an incremental innovation approach.

Table 1 shows the activities for the evaluation process, indicating the instruments and guidelines for each of them. Table 2 presents the criteria of evaluation, based on [30], and the questions applied through a form. The possible answers for each question were 1 - No; 2 - Little; 3 - Partial; 4 – In many respects and 5 – Completely. To a satisfactory result, it is considered that the average score for each question must be equal or greater than 4.

Table 1. Structure of activities to evaluate the systematic approach.

Activity	Time (min)	Instruments and orientations
Problem presentation	5	**Day 1:** Solutions to the function: Remove the peel, for an orange peeler
		Day 2: Solutions to the function: Move the user, from wheelchair to bed, and vice versa
Definition of the objective	5/per day	**Day 1:** Incremental innovation: Take advantage of similar technologies used in conventional systems
		Day 2: Radical innovation: To develop an innovative solution to the interest function
Definition of the characteristics of the analogy sources	5/per day	**Day 1:** Sources of analogy: Technical systems that perform an analog function
		Day 2: Sources of analogy: Biological systems that perform an analog function
Application of the systematic approach to develop physical models	65	**Sources of analog systems:** Day 1: Product manuals, functional basis, and specialized sites; Day 2: Biology books, articles, and specialized sites **Characteristics identification:** FEARB method and guidelines for its application; **Model analysis and test:** Guidelines to define the tests and possible corrective actions
Questionnaire application	10	Systematic evaluation: Form with questions

Table 2. Evaluation criteria and form questions.

Criteria	Question
Scope	Q1. Is the systematic approach to develop physical models aligned with the conceptual design phase?
	Q2. Is the development of activities feasible during the conceptual design, considering the number of resources required for the development of the models?
	Q3. Is it worth investing resources in developing physical models, knowing their potential to stimulate better solution?

(*continued*)

Table 2. (*continued*)

Criteria	Question
Depth	Q4. Is the level of detail of the proposed activities adequate for developing the physical models?
Generality	Q5. Is it possible to use the systematic approach to develop the physical models for any design objective?
Competence	Q6. Does the systematic approach allow the development of models based on systems of different domains?
Clarity	Q7. Is the understanding of the systematic approach easy concerning its activities?
	Q8. Is the understanding of the systematic approach easy concerning its tools?
Consistency	Q9. Are the inputs and outputs of the system consistent with the proposed activities?
	Q10. Are the tools that support the activities coherent, and they facilitate development?
Completeness	Q11. Does the systematic approach contain all the information necessary to develop physical models in the conceptual design phase?

4.2 Results and Discussion

The models developed by the participants, according to the proposed systematic, are presented in Fig. 2. The model in Fig. 2 a) consists of a saw positioned in a base, with the capacity of vertical movement. The Styrofoam ball represents the object to be cut, which has free movement, i.e., able to advance on the saw, representing the cutting dynamics of the system. This model represents the functionality of a Band Saw to stimulate ideas for the function Remove the peel (problem 1 – day 1).

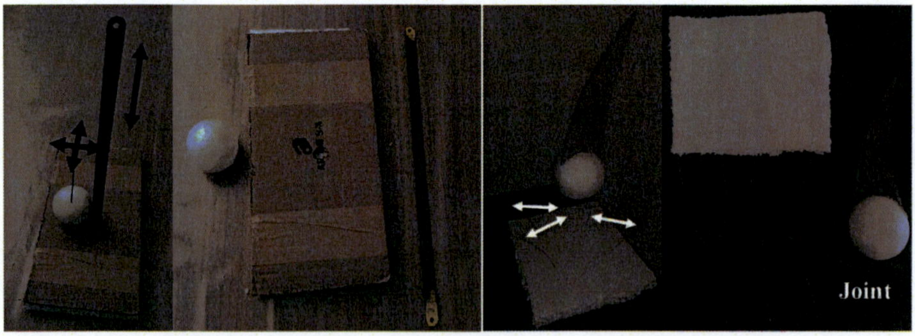

a) b)

Fig. 2. a) The physical model of a Band Saw and b) the physical model of the Spine of the European Hedgehog [31].

The model in Fig. 2 b) represents the movements in different directions of the Spine of European Hedgehog [31], which resulted from the search for an analog system, to stimulate ideas for the function Move user (problem 2 – day 2). It consists of cardboard that represents the spine, fixed on a ball of Styrofoam, analog to the spherical bulb, identified at the morphological structure of the animal. The wires play the role of muscles (two wires with a joint at the ends) that move the spine, connecting both the spherical bulb and a Styrofoam plate, which represents the animal's skin. The displacement of the wire junctions promotes the movement function. Depending on which wire is moved, the system moves in a specific direction. The various possible combinations of the four wires movement achieve displacement in several directions.

To evaluate the systematic application, after the physical models' development, the participants answered the questionnaire presented in Table 2. Table 3 presents the average score for each question.

Table 3. Experts' evaluation of the systematic application.

Criteria	Average score	Criteria	Average score
Scope	**Q1**: 4,83; **Q2**: 3,83; **Q3**: 3,67	Clarity	**Q7**: 3,83; **Q8**: 4,33
Depth	**Q4**: 4,33	Consistency	**Q9**: 4,33; **Q10**: 4,00
Generality	**Q5**: 3,33	Completeness	**Q11**: 4,00
Competence	**Q6**: 4,00		

Regarding the scope criterion, the activities proposed were considered feasible in the conceptual design, that is, all necessary information to develop the physical models exists, or can be adapted in this phase. The low scores for this criterion are resultant from the experts' panel doubt about the validity to invest resources in developing the models. This doubt can be justified by the own novelty of the proposed systematic, which has no previous application results to allow comparison.

The scores for the criteria depth, competence, consistency, and completeness were satisfactory, confirming the adequacy of tasks, tools, inputs, and outputs. Also, these results indicate that the proposed guidelines fully support the development of physical models.

The generality criterion demonstrates a doubt about the application of the systematic approach for any design objective. This doubt can be explained by the experts' difficulty in evaluating the systematic in other design contexts than those proposed. In contrast, the proposed design objectives allowed the development of two functional physical models, demonstrating that the activities are generic enough to cover these different design contexts. The criterion competence supports this assertion, because analogy sources of several domains enable different types of stimuli, facilitating the application of the systematic approach for different design objectives.

The criterion clarity evidenced that the suggested tools are easy to understand and apply during the activities. However, some activities posed some difficulty for the experts, mainly the definition of a criterion requested for the comparison and prioritization of the sources of analogy. Therefore, it is necessary to present some prioritization criteria during the explanation of the systematics, for different types of design, making this task more natural to be accomplished.

Despite the difficulties during the systematic applications, the functionality of the models developed was satisfactory in the evaluations carried out in activity 4, since they performed the function for which they were developed. This result demonstrates that the mentioned difficulties did not prevent the general objective of the systematic approach from being achieved.

Also, as a result, it was identified that in order to improve the application of the systematic, it is necessary to present the benefits of working with physical models during the ideation process. With this, the use of the proposed systematic will be perceived as an investment, which improves the design solutions. Moreover, to facilitate the understanding and execution of the systematic tasks, examples of the development of models to different design problems must be presented. With this, it is assumed that some decisions about how to accomplish the task will be made easier.

5 Final Considerations

The systematic approach to develop physical models, applied and evaluated for two different design contexts, resulted in two distinct and functional models, evidencing their ability to facilitate the adaptation of the concepts from the source of analogy into physical stimuli. Also, experts evaluated it positively, indicating its adequacy regarding the depth, competence, consistency, and completeness of guidelines and tools.

By the ability to perform the analog systems' functions, the physical models resultant from the proposed systematic approach can contribute to the development of solutions in the conceptual phase, facilitating the analogical reasoning.

Therefore, considering the experts' panel evaluation and the application results, it can be stated that the proposed systematic approach can to improve the development of stimuli that can effectively enhance creativity during ideation sessions. As future research and improvement, the systematic activities shall be more detailed, and more applications must be conducted in order to make available more examples of physical models development.

References

1. Baxter, M.: Product Design: Practical Methods for the Systematic Development of New Products. Chapman & Hall, São Paulo (1995)
2. Back, N., Ogliari, A., Dias, A., Silva, J.C.: Integrated Product Design: Planning, Design and Modeling. Manole, Barueri (2008)
3. Fernandes, R.B., Ogliari, A.: Enhancing creativity through biological stimuli during new products ideation. Int. J. Innov. Educ. Res. **6**(10), 332–350 (2018)
4. Viswanathan, V., Linsey, J.: Understanding physical models in design cognition: a triangulation of qualitative and laboratory studies. In: 41 ASEE/IEEE Frontiers in Education Conference, Rapid City (2011)
5. Viswanathan, V.K., Linsey, J.S.: Physical models and design thinking: a study of functionality, novelty and variety of ideas. J. Mech. Des. **134**(9), 091004 (2012)

6. Acuna, A., Sosa, R.: The complementary role of representations in design creativity: sketches and models. In: Taura, T., Nagai, Y. (eds.) Design Creativity, pp. 265–270. Springer, London (2011)
7. Camburn, B., Viswanathan, V., Linsey, J., Anderson, D., Jensen, D., Crawford, R., et al.: Design prototyping methods: state of the art in strategies, techniques, and guidelines. Des. Sci. **3**, e13 (2017)
8. Lemons, G., Carberry, A., Swan, C., Jarvin, L., Rogers, C.: The benefits of model building in teaching engineering design. Des. Stud. **31**, 288–309 (2010)
9. Viswanathan, V., Atilola, O., Esposito, N., Linsey, J.: A study on the role of physical models in the mitigation of design fixation. J. Eng. Des. **25**, 25–43 (2014)
10. Goldschmidt, G., Smolkov, M.: Variances in the impact of visual stimuli on design problem solving performance. Des. Stud. **27**, 549–569 (2006)
11. Serrat, O.: Knowledge Solutions: Tools, Methods, and Approaches to Drive Organizational Performance. Springer, Singapore (2017)
12. Hess, T., Summers, J.D.: Protocol analysis: studying physical manipulatives during conceptual design. In: Proceedings of the ASME 2014 International Design Engineering Technical Conferences and Computers and Information in Engineering Conference (IDETC/CIE). ASME, Buffalo (2014)
13. Viswanathan, V., Linsey, J.: Physical examples in engineering idea generation: an experimental investigation. In: ICDC 2012 - 2nd International Conference on Design Creativity, Proceedings. Glasgow, pp. 23–32, Glasgow (2012)
14. Chan, J., Fu, K., Schunn, C., Cagan, J., Wood, K., Kotovsky, K.: On the benefits and pitfalls of analogies for innovative design: ideation performance based on analogical distance, commonness, and modality of examples. J. Mech. Des. **133**, 081004 (2011)
15. Fu, K., Chan, J., Cagan, J., Kotovsky, K., Schunn, C., Wood, K.: The meaning of "near" and "far": the impact of structuring design databases and the effect of distance of analogy on design output. J. Mech. Des. **135**, 021007 (2013)
16. Keshwani, S., Chakrabarti, A.: Influence of analogical domains and comprehensiveness in explanation of analogy on the novelty of designs. Res. Eng. Des. **28**(3), 381–410 (2016)
17. Chai, C., Cen, F., Ruan, W., Yang, C., Li, H.: Behavioral analysis of analogical reasoning in design: differences among designers with different expertise levels. Des. Stud. **36**, 3–30 (2015)
18. Song, H.I., Lopez, R., Fu, K., Linsey, J.: Characterizing the effects of multiple analogs and extraneous information for novice designers in design-by-analogy. J. Mech. Des. **140**, 031101 (2018)
19. Moreno, D.P., Yang, M.C., Hernández, A.A., Linsey, J.S., Wood, K.L.: A step beyond to overcome design fixation: a design-by-analogy approach. In: Gero, J.S., Hanna, S. (eds.) Design Computing and Cognition 2014, pp. 607–624 (2015)
20. Linsey, J.S., Tseng, I., Fu, K., Cagan, J., Wood, K.L., Schunn, C.: A study of design fixation, its mitigation and perception in engineering design faculty. J. Mech. Des. **132**, 041003 (2010)
21. Hess, T., Summers, J.D.: Case study: evidence of prototyping roles in conceptual design. In: Proceedings of the International Conference on Engineering Design, ICED13, Seul (2013)
22. Isa, S.S., Liem, A.: Classifying physical models and prototypes in the design process: a study on the economical and usability impact of adopting models and prototypes in the design process. In: 13th International Design Conference - Design 2014, pp. 2071–2082 (2014)
23. Ramduny-Ellis, D., Hare, J., Dix, A., Gill, S.: Exploring physicality in the design process. In: Proceedings of the Design Research Society Conference, Sheffield (2008)
24. Gerber, E., Carroll, M.: The psychological experience of prototyping. Des. Stud. **33**, 64–84 (2012)
25. Jang, J., Schunn, C.D.: Physical design tools support and hinder innovative engineering design. J. Mech. Des. **134**, 041001 (2012)

26. Pahl, G., Beitz, W., Feldhusen, J., Grote, K.H.: Engineering Design: A Systematic Approach, 3rd edn. Springer, London (2007)
27. Qian, L.: Engineering design synthesis. In: Chakrabarti, A. (ed.) Engineering Design Synthesis: Understanding, Approaches and Tools, pp. 245–270. Springer, London (2002)
28. Szirtes, T., Rozsa, P.: Applied Dimensional Analysis and Modeling, 2nd edn. Elsevier Science & Technology Books (2007)
29. Gil, A.: How to Design Research Projects, 4th edn. Atlas, São Paulo (2002)
30. Vernadat, F.B.: Enterprise Modeling and Integration: Principles and Applications. Chapman & Hall, London (1996)
31. Healtey, J.J.: Hedgehogs. In: Mitchell, M.A., Tully Jr., T.N. (eds.) Manual of Exotic Pet Practice, pp. 433–455. Elsevier, Missouri (2009)

Development of an Automatic Machine for Sensor Manufacturing by the GOP Technique

Edmilton Stein[1] (✉), Carlos Valdiero[1], Matias Alles Hubert[1], Roberta Goergen[1], Marianna Souza[1], Luiz Antônio Rasia[1], and Antonio Carlos Valdiero[2]

[1] Regional University of Northwestern Rio Grande do Sul, Panambi, RS 98700-000, Brazil
e_stein@outlook.com

[2] Horizontina Faculty, Horizontina, RS 98920-000, Brazil

Abstract. As the industry advances toward the Advanced Manufacturing process emerges the need for more autonomous machines in order to get more agile and efficient production. Consequently, it creates the need for the development of cheap, easily manufactured and environmentally friendly sensors. Graphite on Paper (GOP) method for producing sensors is a strong candidate as an alternative to the currently available sensors since they meet all the requirements mentioned above. This technique consists of creating graphite layers onto a paper substrate in order to develop a piezoresistive sensor which shows to be suitable as temperature sensors as well as strain gauges with a wide application range, such as in force and temperature control, robotics, agricultural, biomedical engineering, automation, and control systems. Manual pencil-drawing is the simplest way to exfoliate the carbon on the paper; however, this method generates uneven graphite layers on the substrate leading to unexpected results caused mainly by the high variation of the applied force during the manual exfoliation of the carbon films. Aiming to obtain more uniform graphite layers, a machine was developed to fabricate the sensors using a commercial mechanical pencil to apply the graphite on the substrate. The machine mainframe is built in steel in order to achieve robustness. It has three axes moved by two stepper motors and one direct current motor, the latter used together with a load cell and a microcontroller Atmega328 to control the applied force during the material exfoliation.

Keywords: GOP · Piezoresistive · Automatic machine · Sensor

1 Introduction

The "Industry 4.0" concept, first conceived in Germany now spreads throughout the world. According to [1], aiming to supply the need for more individualized products as a consequence of the modern market, emerges a new industrial revolution known as the "4th Industrial Revolution", thus changing the industrial model from the large-scale production to the "batch size one" model. For this, it is necessary to introduce a set of concepts in the manufacturing system, where "Smart Factory" plays an essential role

L. Pereira et al. (Eds.): IDEAS 2018, SIST 198, pp. 239–247, 2021.
https://doi.org/10.1007/978-3-030-55374-6_24

in this new industrial era, this term represents an increase on the level of automation, thus creating a need for more affordable and ecologically friendly sensors and actuators. Intending to supply this need arises the proposal of a piezoresistive sensor, made mainly of paper and graphite.

The sensor concerned in this paper is the strain gauge type piezoresistive sensor, obtained by the GOP technique. Strain Gauges are sensors typically used to measure the strain of a component or structure under a load. The working principle of these sensors lies on the assumption that the sensor experiences all the strain caused by a load in a structure, so that resulting in a proportional change in its electrical resistance which can be easily measured [2]. To know how much strain a body is experiencing during its use is of great importance since through it one can evaluate new or modified designs of a machine component or structure to ensure the most reliable and efficient product as possible [2]. Furthermore, the use of strain gauges contributes to the enhancement and evaluation of the mathematical model used [2].

However the piezoresistive sensor obtained from the GOP (Graphite on Paper) technique itself has been proved to be flexible, low cost, lightweight, biocompatible and recyclable [3, 4] present in their paper that the sensors show some deviations that are believed to be related to the deposition method due to the manual deposition of the graphite on substrate and/or the materials nature. Therefore, aiming to minimize these deviations in the results and to help to investigate the real cause of them as well as to attempt to make a serial production possible a machine was developed to deposit the graphite in more uniform and controlled layers.

The cost of the sensors cannot be directly assessed, taking into consideration that this manufacturing technique is relatively recent or even little exploited so far. I addition, there is no other machine in the market for cost comparison due to the control system used to control the graphite deposition by mechanical exfoliation on polymeric substrates. However, as the sensors do not contain any conventional material, such as Silicon, but only organic everyday materials such as commercial A4 paper and graphite cylinder, its cost can be considered lower than the traditional sensors.

As for the methodology an analysis was carried out to identify the project needs, which are to be able to obtain a uniform deposition of graphite on the substrate through a small variation on the pressure applied as well as being able to control the position of the graphite deposition. To meet these needs the machine model developed is similar to a traditional milling machine where a tool for graphite deposition is used instead of the milling cutter, it has three axes whose control occurs by a Computer Numerical Control (CNC). Aiming to control the pressure applied during the deposition of graphite a load cell was used together with a microcontroller board Arduino Uno, which utilizes the microcontroller ATmega328P, the human-machine interface was developed using the software Java to provide a consumer-friendly interface. The machine was designed using the Software Dassault SolidWorks and its prototype developed at the University research group called SIMMER (Project in Mechanical Systems, Mechatronics, and Robotics).

Aiming to verify the machine repeatability, a set of fifty sensing elements was manufactured and, their electrical resistance was measured.

2 The Piezoresistive Sensor

According to [5], sensors are devices whose purpose is to convert energy from one domain to another. The GOP (Graphite on Paper) sensor's working principle is the piezoresistivity effect, $\frac{\Delta\rho_{ij}}{\rho}$, which according to [5, 6] is characterized by a reversible change in the material resistivity so that causing a change in the electrical resistance of a semiconductor material under mechanical stress.

The piezoresistivity is determined by Eq. 1, where π_{ijkl} is the piezoresistive coefficient tensor, which is a result of the material composition that can be changed by specific doping techniques and the T_{kl} is the mechanical stress undergone by the sensor [5]. Since the electrical resistance depends on the resistivity and the physical dimensions of the conductor as shown by the Ohm's second law (Eq. 2), it is crucial to have control over all these parameters when manufacturing a piezoresistive sensor.

$$\frac{\Delta\rho_{ij}}{\rho} = \pi_{ijkl} T_{kl} \tag{1}$$

$$R = \frac{\rho L}{A} \tag{2}$$

Usually, strain gauges comprise a cantilever beam made of any flexible material together with a material whose electrical resistance changes when subjected to deformation, so that they can measure the deformation and the force applied to the cantilever. The sensor studied in this paper is pictured in Fig. 1 where the cantilever material is cellulose paper A4 with thickness 88 μm, the graphite used is a commercial graphite cylinder whose chief properties are presented in Table 1, and copper terminals bonded by silver ink contact node.

Table 1. Graphite cylinder hardness HB properties.

Property	Value	Unit	Source
Electrical resistance	2.24	Ω	Multimeter
Electrical resistivity	9.119×10^{-5}	Ω × m	$\rho = \frac{R \times A}{L}$
Length	6.050×10^{-3}	m	Caliper
Diameter	5.600×10^{-4}	m	Micrometer
Cross-sectional area	2.463×10^{-7}	m^2	$A = \pi r^2$
Carbon	68	%	Kaneko (2015)
Clay	26	%	Kaneko (2015)
Wax	5	%	Kaneko (2015)

Source: Elaborated by authors.

The manufacturing process of the graphite sensor consists of mechanical exfoliation of a commercial graphite cylinder on the paper substrate, where the graphite layers are deposited observing that the exfoliation always occurs in the same direction. The greater the number of graphite layers depositions, the smaller is the sensor's electrical resistance. Therefore, when the sensor achieves the desired electrical resistance, it is ready for the next step of the production, where the copper terminals are attached to the sensing element by a silver ink.

Although this kind of sensor is composed of ordinary materials, it shows to have good efficiency, since it has a relatively big change on its electrical resistance over the mechanical strain (Gauge Factor) as shown in the paper [4], therefore showing to be an alternative to semiconductor strain gauges.

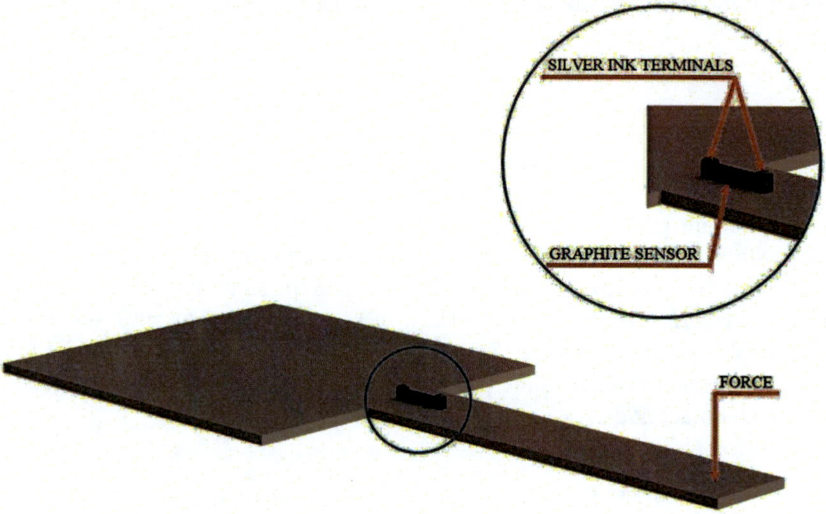

Fig. 1. Schematic of the cellulose paper cantilever with the sensor on it. Source: Elaborated by authors.

3 Machine Design

The machine development followed the project needs, where it is supposed to deposit the graphite in uniform layers along the "X" and "Y" axis (the plane parallel to the base). The deposition tool uses a hardness HB commercial 0.5 mm diameter graphite cylinder to make the manufacturing as affordable as possible. The deposition of graphite, along with other factors, is believed to depend on the exfoliation pressure while depositing the graphite on the paper, so that a pressure control system is required. All the hardware was designed using the software CAD Dassault SolidWorks.

As deformations can directly influence the machine precision, all its frame was designed in steel, thus offering the necessary robustness to support all the other elements that compose the machine. The frame comprises two main parts, the base, and the X-axis holders.

The work area composes a table made of printed ABS (Acrylonitrile butadiene styrene) covered by a flat glass aiming to have a smooth and even surface while having lightweight so that allowing the use of small power stepper motors to move it. The table slides along the Y-axis through a steel linear shaft guidance system, and it uses a threaded rod coupled to a stepper motor to move along the axis.

Stepper motors control the X-axis and Y-axis to obtain precisely the position of the table along the Y-axis and the deposition tool along the X-axis. The deposition tool slides along the Z-axis through linear shafts just like the other two axes, although its movement is provided by a DC motor, since it does not need a precise position control but a pressure control, and this is carried out by a load cell measuring the load applied by the tool. The final design, together with its prototype, is pictured in Fig. 2.

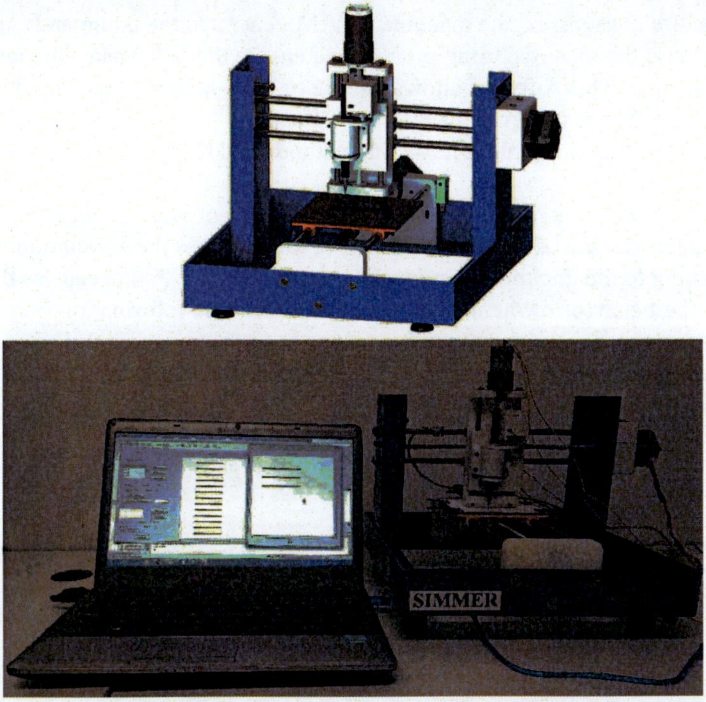

Fig. 2. Machine final design and its prototype. Source: Elaborated by authors.

4 Software CAD/CAM and Firmware

CAD/CAM software (Computer-Aided Design/Computer-Aided Manufacturing software) was developed to control the machine. This software was used to assist in the design and manufacture of piezoresistive sensors using the GOP technique. It allows one to design, automate, control, and supervise the manufacturing process of the sensors.

4.1 CAD

One can determine the material and dimensions for each part of the sensor at the software's design window, which allows one to design the piezoresistive element, its contact PADs, and encapsulation. Once one sets the desired material for the element, it is necessary to define at least three of the fundamental parameters, which are: width, length, thickness, and electrical resistance. If all parameters are set, it will regulate the resistance by modifying the length of the sensor.

During the verification of the sensor parameters, the CAD modifies them until it can be manufactured using the specified material. For example, if the sensor material uses a 0.5 mm diameter tool, the software will only allow lengths greater than 0.5 mm and multiple widths of 0.5 mm.

4.2 CAM

To manufacture the sensors, the machine's CAM generates the commands and enables one to supervise the manufacturing process and ensure that all sensors are according to their specification. This software allows the use of solid materials that are deposited by mechanical exfoliation or liquid materials by injection.

Its manufacturing principle follows the method used by the GOP method described by [8].

In the graphite mechanical exfoliation process, the software adjusts the pressure of the tool contact with the table and the number of times it does the exfoliation in the same place according to the thickness previously specified. In Fig. 3, one can see the window that shows the behavior of the machine. This window has information concerning the time interval elapsed from the beginning of the manufacturing, and the percentage of the manufacturing process that has been carried out at the time.

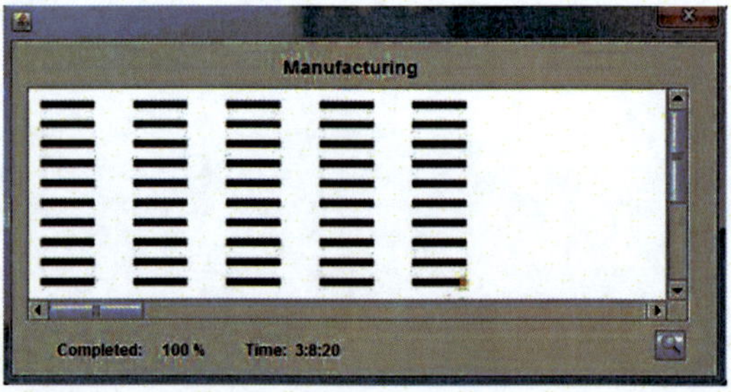

Fig. 3. Software window for production monitoring and control. Source: Elaborated by authors.

4.3 Firmware

The Firmware is responsible for controlling the machine components such as motors and sensors. It contains all the configuration of the machine components. The machine is controlled manually or automatically by the CAM through the communication protocol and sends feedback data on the position of the axes, the value of the sensor pressure and idle capacity.

Its communication with the machine occurs via serial communication on a USB port. The machine communication diagram is pictured in Fig. 4.

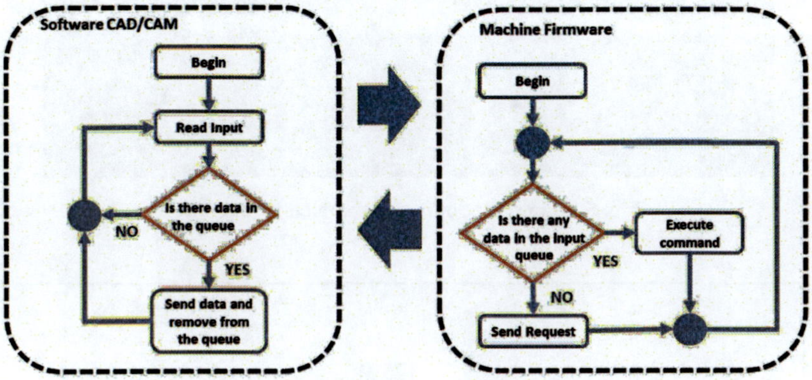

Fig. 4. Machine communication diagram. Source: Elaborated by authors.

5 Tests and Samples

A set of fifty samples was manufactured to evaluate the machine's behavior where the designed equipment makes the sensor's body by applying a desired force through the graphite on the paper in as many layers as necessary to obtain the desired resistance. These samples use a standard A4 paper whose thickness and dimensions are respectively 88 μm and 110×110 mm, the properties of the graphite cylinder used in this experiment are shown in Table 1.

Based on [6] the sensor's body size of these samples is 5 mm in its length and 0,5 mm in its width. When compared with the sensors in [6], the sensors manufactured show a decreasing in its width in order to approximate the width of the sensor to the graphite cylinder diameter, so that avoiding potential errors during the deposition. The force applied during the mechanical exfoliation of the graphite was empirically chosen to be 4.905 ± 0.029 N in a total of five layers and the space left between each sample was 3 mm in the X-axis and 2 mm in the Y-axis.

After the graphite deposition, the next step was to perform a manual adjust aiming to obtain batches of sensors with approximately electrical resistance. This manual adjustment consists of a pressure applied by a metallic tool on the sensor's body, thus decreasing the electrical resistance until achieving the desired value. The samples are pictured in Fig. 5, and their results are shown in Fig. 6, and their statistical data are pictured in Table 2.

Fig. 5. Samples. Source: Elaborated by authors.

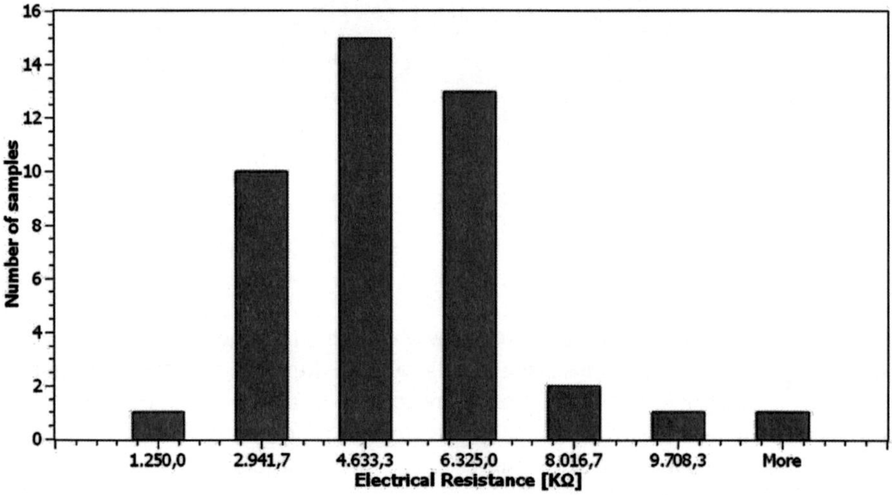

Fig. 6. Histogram of the resistor's electrical resistance. Source: Elaborated by authors.

Table 2. Statistical data of the samples.

Statistic data	Value	Unit
Maximum electrical resistance	11400	KΩ
Minimum electrical resistance	1250	KΩ
Average electrical resistance	4370,62	KΩ
Standard deviation	1956,64	KΩ

Source: Elaborated by authors.

6 Final Considerations

A machine was designed using the software Dassault SolidWorks, and its prototype built following the identified needs. This equipment can deposit graphite layers controlling the applied force with a precision of ± 0.029 N through all the sensor's length. Aiming to automate the manufacturing process a Software was developed in Java programming language to control and monitoring the process.

As the samples have shown, this machine represents an important step toward graphite sensors serial production, although, one can conclude that only force control during the graphite deposition is not enough to obtain sensors with the same characteristics since one cannot ensure paper's and graphite cylinder's homogenous composition.

Acknowledgments. The authors would like to express their gratitude to the CNPq (National Council for Scientific and Technological Development), FAPERGS (Foundation for Research Support of the State of Rio Grande do Sul) and UNIJUÍ, for the resources to purchase equipment for the construction of prototypes for master's and doctoral research and scholarships for Scientific and Technological Initiation.

References

1. Lasi, H., Fettke, P., Kemper, H.-G., Feld, T., Hoffmann, M.: Industry 4.0. Bus. Inform. Syst. Eng. **6**(4), 239–242 (2014)
2. Rasia, L.A., Mansano, R.D., Damiani, L.R., Viana, C.E.: Piezoresistive response of ITO films deposited at room temperature by magnetron sputtering. J. Mater. Sci. **45**, 4224–4422 (2010). https://doi.org/10.1007/s10853-010-4517-1
3. Kanaparthi, S., Badhulika, S.: Low cost, flexible and biodegradable touch sensor fabricated by solvent-free processing of graphite on cellulose paper. Sens. Actuators B Chem. **242**, 857–864 (2017)
4. Gabbi, R., da Silva, G.G.M., Scarton, L., Hammes, G., Rasia, L.A., Valdiero, A.C.: Modelagem Matemática de Elementos Sensores Usando Grafite. Proc. Ser. Braz. Soc. Comput. Appl. Math. **5**(1), 1–7 (2017)
5. Pedrali, P.C., Rasia, L.A., Valdiero, A.C., Fraga, M.A.: Graphite piezoresistive sensors in polymeric substrates. Int. J. Adv. Eng. Res. Sci. **5**(10), 105–109 (2018)
6. Berkenbrock, R., Rasia, L.A., Pedrali, P.C., Valdiero, A.C.: Piezoresistive graphite sensors encapsulated with Epoxy Resin Bisphenol A (BPA). Am. J. Eng. Res. **7**(4), 148–154 (2018)
7. Kaneko, S., Rachi, T., Yasui, M., Shimizu, Y., Tanaka, S., Kato, C., Endo, T.: Graphen growth: 10B lead pencil, print paper, and femtosecond laser. In: 2015 International Conference on Microwave and Photonics, ICMAP, pp. 1–2. IEEE (2015)
8. Ren, T.L., Tian, H., Xie, D., Yang, Y.: Flexible graphite-on-paper piezoresistive sensors. Sensors **12**(5), 6685–6694 (2012)

Innovation in Education with Inclusion, New Technologies, and Civil Society Engagement

Educating City: A Media for Social Innovation

Yohani Dominik dos Santos Figueiredo(✉), Márcia Aparecida Prim, and Gertrudes Aparecida Dandolini

Federal University of Santa Catarina, Florianópolis, Santa Catarina, Brazil
yohani.dominik@posgrad.ufsc.br

Abstract. Cities are composed of a diversity of elements that interact with each other in an autonomous or induced way, and can be characterized as a means. Cities are given this denomination because they are able to generate knowledge from the daily life and by the interaction of individuals. From this exchange of experience and concern for the improvement in the quality of life of the people, the term "Educating City" appears. This concept is based on the idea that educational cities are fertile ground for implementing Social Innovations, since socially innovative initiatives with new ideas (products, services, and models) satisfy social needs simultaneously. In this sense, the objective of this research is to analyze how an Educating City, acting as a media, acts in favor of the promotion of Social Innovation. For this, the bibliographical review was carried out, with a qualitative approach. It was observed that the members of this movement follow the guidelines of the Charter of Educating Cities, which defines that the city will only be an educator when recognizing, exercising and developing its traditional functions, as well as being actively concerned with the formation and development of all its inhabitants, starting with children and young people. Is concluded that the theme is closely related to the principles of Social Innovation, especially when addressing the Sustainable Development Goals (SDOs). It is also worth noting that within these two constructs, the realization of projects aims to promote the empowerment and sustainability of communities.

Keywords: Educating city · Social innovation · Media

1 Introduction

Public policies are immersed in the process of social change and transformation, in which they are inserted in the daily life of contemporary societies. In this panorama, cities play a fundamental role as entities and living organisms, allowing the exercise of citizenship [1].

Cities are centers of public policy innovation and play an increasingly important role in the sustainable development of communities [2]. While there are situations requiring national responses, many of the development challenges are most effectively addressed at the local level. It is at this level that the maximum democratic participation of citizens can be ensured and the opportunity to adapt the solutions to the needs of the population [2].

L. Pereira et al. (Eds.): IDEAS 2018, SIST 198, pp. 251–260, 2021.
https://doi.org/10.1007/978-3-030-55374-6_25

Cities can be the representation of the desire of its inhabitants since it heats the daily life and promotes the interactive exchanges, being recognized as a form of communicative experience. However, plans and utopias, imagination, and reality are mixed, where it is not always possible to find the balance between the purposes of the cities and their real needs [3].

To Messina and Valdés-Cotera [4], one way of seeking this balance is to think cities intelligently and with an inclusive and educational purpose. Educating City is a concept that presents as one of its principles of economic development in line with sustainability and social justice.

The idea of the educating city aims to improve the quality of life of citizens in society, through the solution of social problems in a systemic way [4]. The educating city is seen as a network waiting to be explored in use in terms of formal, informal, and learning opportunities. Several of its programs promote the interaction and conviviality of its inhabitants, such as youth participation programs, environmental education, values education, and the practice of democratic citizenship.

In this context, it is essential to know what conditions are necessary for a city to be considered educational, as well as concerns about the social and political processes that threaten it [4], so that public policies can make use of this knowledge, and seek for alternatives to the development of educative solutions.

Strengthening Messina and Valdés-Cotera [4], the United Nations Human Settlements Program, suggest that a city should be socially inclusive and equitable [2] and take up the challenge of reducing poverty and inequality by ensuring the rights fundamental to its inhabitants and respecting the environment. It starts from the idea that educating cities can be a fertile ground for implementing socially innovative initiatives that seek to simultaneously satisfy social needs and the creation of new social relations or collaborations. Some initiatives focus on meeting these demands, such as the United Nations Organization, through its Sustainable Development Objectives.

In this movement of cities as media and as an IS promoter, Ferrara [3] affirms that cities are not simple constructions but living organisms and representations of complex dimensions where images and sensations are mixed, which may conceal or reveal their dynamicity.

In this context, this research aims to analyze how the educating city, acting as a media, acts in favor of Social Innovation, answering the following question: How does the educating city, acting as a media, act in favor of Social Innovation?

The work is structured in six sections: 1) introduction; 2) theoretical reference; 3) methodological procedures; 4) results, 5) final considerations, and 6) bibliographic references.

2 Theoretical Reference

In order to draw considerations on the constructs in question, the theoretical framework aims to carry out a brief contextualization, and conceptualization will be presented from the perspective of educating cities as media and social innovation.

2.1 Educating City as Media

Currently, one hears incessantly the word *media.* It has been used with varying meanings, causing little clarity as to the understanding of its meaning. Many expressions used in the media illustrate the diversity of meanings, for example: "the media determines tastes and customs", "media time", "media people", "media influence on education", etc. [5]. These examples demonstrate the diversity of uses of the term media which has been adopted to represent not only the media, such as structure and technology, but also the process and the communication system, such as messages, transmitters, communicational strategies, managers [5].

Therefore, in order to understand the educating city as a media, it is first necessary to reveal about the concept of media and, second, the concept of educating city. Media is the physical part (carrier or channel) of information, in charge of storing or transporting signals, in its original form and organization, and is also understood as elements that store information [6].

The concept of Perassi and Rodrigues [6] corroborates with that presented by Briggs and Burke [7], where they affirm that media is every form or medium of expressions, such as the means of communication, the forms or vehicles used for this expression and communication, and can be a form, a system and/or a communication process.

Media can be categorized as passive and active. In the passive approach, it plays the role of support and channel of information, under the direct action of human beings. In the active approach, it performs functions of support, storage, and distribution of information, independently of direct human action, with partial or integral autonomy [6].

Ferrara [3] affirms that cities can represent a media, since they represent the customs, culture, and desires of its inhabitants, besides being a means of exchanges and interaction. They play an active role in media, since they play a role in supporting, storing, and distributing information, independently of the direct action of man.

To Ivo [5], constructed urban spaces constitute means of communication capable of both building and registering messages, regardless of their size, becoming a tool for communication, image, and expression of institutions, groups and/or of the individual himself. For this author, the urban area of cities and their different appropriations comprise one of the strategic dimensions of manifestation or expression of the various social groups, as well as of cultural minorities and individuals. It happens to the extent that the urban area of the city houses, registers, and is a source of representation of identity and relationship between the various players and the city [5].

To Ferrara [3], cities celebrate the daily life of their inhabitants by promoting interactivity and thus should be recognized as a communicative experience.

In this sense, Ivo [5] adds that the urban area becomes the center of these activities, not only as a built space, but also as a space of life, housing all forms of manifestations of cultural, political, ideological, religious groups, etc.

In view of the presented concepts, it is concluded that the city is recognized as a media, because it can be composed of isolated elements or a diversity of them, interacting among themselves autonomously or induced by the action of human agents, acting as a system, with interactions and changes in the long term [8].

In this context, the city also exercises the role of educator, when its inhabitants are treated inclusively. To Aeita and Zuin [1] the educating city is intended for education to break through the walls of schools, and for exchanges to take place enrichingly. Both the education for the city and the city for education aim to respect the knowledge generated through the daily life of the subjects, as well as respect for their language and cultural identity [9], so the city is a media/mediator, and can be an educating city.

Morigi [10] corroborates the concept of Aeita and Zuin [1], stating that it is a movement that centralizes the municipalities in search of better education for children and young people, in order to promote more social and cultural programs and services, mainly educational, thus offering a better quality of life for the inhabitants of urban centers.

The Educating City movement began in the 1990s, during the First International Congress of Educating Cities. To this movement, the concept of educating cities are cities engaged in projects to improve the lives of its citizens, integrating the offerings of social and cultural activities, in order to enhance the educational capacity, formal and informal, of urban centers. This movement resulted in the Charter of Educating Cities. This letter contains 20 guidelines for promoting lifelong learning, aimed at improving the quality of life of citizens in society acts [1]. Table 1 describes these guidelines.

These guidelines show that the movement is a network waiting to be tapped and put to use in terms of formal and informal learning opportunities such as youth participation programs, environmental education, values education, democratic citizenship practice, etc. [4]. The guidelines of the educating city are aligned with the SDGs proposed by Agenda 2030, since both are aimed at transformation for the benefit of the social, each one with its particularities.

Zitkoski [13] expects the educating city to achieve a happy human existence in harmony with nature. Taking this into consideration, he listed some characteristics of the movement that acts for the benefit of a free citizenship, being: 1) Educating City as a broad vision of education; 2) Involvement of the population in the life of the city, and 3) Need for innovation to implement the Educating City. These characteristics reinforce even more the association of the educating city with the objectives of sustainable development.

Morigi [10] points out that the city will only be an educator when recognizing, exercising and developing its traditional functions, as well as actively worrying about the formation and development of all its inhabitants, beginning with the children and young people.

Simões [13] reaffirms that one of the objectives of the educating city is that local governments seek to develop projects for the social formation and citizenship of citizens, taking special care of young people.

For example, Tehran, the capital of Iran, as a mega-city, in order to be a developed city and possess knowledgeable and capable citizens, has launched actions to improve and teach them the awareness necessary to achieve their goals. In Iran, citizenship education considers administrative, social, and cultural levels [14].

Therefore, the movement offers openness to new initiatives focused on meeting the demands of society. These social initiatives enable the realization of projects that foster the empowerment and sustainability of the community. For a better understanding of the involvement of social actions, the next chapter presents the concept of social innovation.

Table 1. Guidelines of the charter of educating cities

Guidelines	Description
1	All the inhabitants of a city have the right to enjoy, in conditions of freedom and equality, the means and opportunities of formation, personal development, and entertainment that the city offers
2	The municipalities should effectively promote their competencies in education
3	The city should look for training opportunities globally
4	Those responsible for the municipal policy of a city should have updated information on the situation and needs of its inhabitants
5	The city should know and encourage innovation, and the development of the formative action that takes place in its structured teaching centers
6	The city should evaluate the impacts of cultural, recreational, informational, and other offers, which children and young people receive without any mediation
7	The city should promote projects for the personal, social, moral, and cultural development of the entire population
8	The city should provide parents with training that will enable them to help their children grow and use the city in a spirit of mutual respect
9	The city should offer its inhabitants the possibility of occupying a place in society; advise them on their personal and professional orientation and enable them to participate in social activities
10	The city should recognize the mechanisms of exclusion and marginalization that affect them and their forms, and develop appropriate compensatory interventions
11	Interventions aimed at addressing inequalities can take many forms, but they must start from a global vision that satisfies both their interests and the set of rights common to all
12	The city should stimulate associativism, to train young people to make decisions
13	The city should teach its residents to inform themselves, to train in the information
14	The city should make available to the children a place for specialized information, orientation and help, and, if necessary, a counselor
15	The city must find, preserve and present its identity
16	The growth and transformation of the city must be in sync with the symbols and references of the city's past
17	The city must guarantee the quality of life of its residents and maintain a healthy environment
18	The municipality should encourage freedom of expression, cultural diversity and dialogue in equal conditions
19	All residents have the right to participate in the formation of educational programs and values and practices of democratic citizenship
20	A city that educates will not isolate the generations

Source: Educators, 2014 [12].

2.2 Social Innovation

Given the concept and guidelines of the Educating City, it is possible to identify the different possibilities of innovative actions in the urban space, aimed at solving social problems. Zitkoski [12] highlighted the need to innovate socially as one of the characteristics necessary to reach a City considered Educator.

Moulaert [15] presents social innovation as a tool that enables urban development, centered on human needs, and based on innovation in the neighborhood perspective and community governance.

Mulgan et al. [16] also approximates the constructs innovation and social issues, because they define SI as new ideas, worked with the aim to reach social issues. They further add that innovative activities and services are motivated by social needs, and are disseminated through community-based organizations.

Bacon et al. [17] confirm the concepts of Moulaert [15] and Mulgan [16], but complement it by saying that the term social innovation is used for new ideas, and they may be: a product; a service; or a model, provided that they are developed to meet unmet social needs.

SI also concerns the conceptual change of process or product, organizational change, and changes in funding, and may have new relationships with stakeholders and territories [18]. Still according to the OECD [19], SI identifies and delivers new answers for social problems, with the aim of improving the quality of life of individuals and communities. It identifies and implements new means of insertion in the market and promotes new jobs and new ways of participation, with the aim of improving the individual's position in the workforce.

It can be seen in the literature that each author presents a particularity in his/her thought. However, it stands out the existence and a crucial point in the concepts, which is the attendance of social needs through new ideas.

According to Bignetti [19], SI is shown as a way to reach feasible alternatives for the future of human society. The author points to an increase in initiatives to support destitute groups, the creation of non-profit organizations, non-market based business projects, the search for local competitive capacity, and sustainable forms of growth. However, the expected results, given the magnitude of the problems, are still derisory.

This context provides an opening for a greater involvement of the players in solving social problems, hoping to achieve better results in solving these issues. One of the prominent examples of this concern with social issues is the initiative of the United Nations.

In 2015, the UN brought together world leaders seeking to devise a plan of action to eradicate poverty, protect the planet, and ensure that people achieve peace and prosperity [20]. For this, the Agenda 2030 for sustainable development was developed.

This agenda is composed of 17 goals (global goals), consisting of a list of tasks, containing 169 deployments, aimed at all the people, all over the world, that must be fulfilled by 2030 [20]. The document affirms that it is necessary to take bold and transformative measures to put the world on a sustainable path. With this, socially innovative initiatives are promoted. Table 2 describes the Sustainable Development Goals - SDGs.

Table 2. Guidelines of the charter of educating cities

SDG	Descrição
1	Poverty Eradication
2	Zero Hunger and Sustainable Agriculture
3	Welfare Health
4	Quality education
5	Gender equality
6	Clean Water and Sanitation
7	Accessible and Clean Energy
8	Decent Work and Economic Growth
9	Industry, Innovation, and Infrastructure
10	Reduced Inequalities
11	Sustainable Cities and Communities
12	Responsible Consumption and Production
13	Action Against Global Climate Change
14	Life below Water
15	Life on Land
16	Peace and Justice Strong Institutions
17	Partnerships and Means of Implementation

Source: Agenda, 2030.

3 Methodological Procedures

According to Gil [21], a research can be classified as to its nature, objectives, and the means of searching the data. This research is of qualitative nature, since it seeks a better understanding of the facts investigated from contents found in a literature review [22]. As far as the objective is exploratory, it aims to gather information and describe how the investigated phenomena occur. Besides, it seeks to provide an overview of the context, from an approximative approach [21]. As for the media, this is a bibliographical review of the literature.

After defining the constructs to be searched, a search in the database Scopus and Web of Science using the descriptors "educating city" and "social innovation", without temporal delimitation, was carried out. No document was found. Then the research was carried out with the elements separated and then analysis was performed on the proposed theme.

After this search with the separated terms in Scopus and WoS, seventeen and fourteen articles were discovered, respectively, with the construct "educating city", without temporal delimitation. Duplicate articles were removed, resulting in seventeen articles. From this first refinement, nine articles that best aligned with the proposal were found, and inserted four more articles from directed search, for example, snowball.

The social innovation construct was conducted in studies that best aligned with the topic and were already part of IGTI[1] Research Group studies. Therefore, the selection of articles was carried out from a directed search, with four articles being analyzed. Thus, the corpus adopted for analysis was seventeen articles.

4 Results and Discussions

The research began with a discussion about the city and its role as a media, and only after the certification that it behaves as a mediator, the discussion began for the city as a moderator of education, conceptualizing Educating Cities. Subsequently, the concept of social innovation was discussed in the same context.

To Messina and Valdés-Cotera [4], from public policies, it is possible to list some characteristics of social and cultural significance, using the concept of educating cities for the benefit of citizens' development. For example, they: 1) can help in cases of school violence, vandalism in the educational environment during the weekends, dropping out of school, difficulty in reading, environmental pollution, among others; 2) are site-based, addressing specific socio-economic development needs; 3) can provide stimulation of educational memory and the generation of feelings of social belonging; 4) can help inclusion of citizens in active participation; 5) are promoted by local authorities, with the support of universities or civil society organizations, as well as other stakeholders; 6) adapt to different population groups, addressing specific groups that become marginalized or are at risk; 7) aim to stimulate the independence of the inhabitants of the city, especially of the young; 8) promote the creation of groups for the exchange of relationships as well as the production of goods through networking; 9) create different reception centers for societies in which there are broken communities and family ties; 10) use volunteer work, especially young volunteers; 11) use an equal gender approach, seeking equal opportunities and eliminating different types of discrimination; and 12) seek for the creation of new paths and the recovery of paths previously used, making them more accessible.

With these characteristics, it is evident that the urban space is considered as one of the main areas in which social activities are experienced, and should be seen as an appropriate environment for education to take place. It should be noted that cities allow for an expansion of the possibilities found in the public sphere, insofar as they foster interpersonal relations with each other and, at the same time, stimulate personal reflexivity [22]. With this in mind, one can on can relate the characteristics of educating cities by Messina and Valdés-Cotera [4], with the concepts presented on SI, as they aim at alternative solutions to social problems and can be achieved through socially innovative initiatives.

To Messina and Valdés-Cotera [4], programs from the Educating Cities movement have had right development conditions, especially the juxtaposition of formal and informal educational opportunities, resulting in a kind of mixed learning that has been recognized, since the 1990s, as a great potential of the cities, together with the growing value

[1] IGTI - Center for Studies in Intelligence, Management and Technologies for Innovation.

of the principle of lifelong education and knowledge society. This condition makes a direct connection with SDG 4, which addresses the issue of quality education for all.

Another highlight of the dynamics of educational cities is the fact that one of the development conditions is the rise of local government and the more significant presence of civil society organizations to complement the roles played by the city authorities. These concepts are essential for the development of socially innovative initiatives, as players from various sectors come together for a common purpose. These constructs correlate with SDG 8, 10, 11 and 17 as they foster diverse forms of partnerships for sustainable development and reduction of social inequalities.

The goals of SDG 11 are to ensure that cities and settlements are inclusive, safe, resilient and sustainable, and are of particular importance to the Educating Cities movement, as they address a number of tasks that are easily achievable through social innovations such as: 1) increasing inclusive and sustainable urbanization, and capacities for planning and managing participatory, integrated and sustainable human settlements in all countries; and (2) providing universal access to safe, inclusive, affordable and green public spaces, particularly for women and children, the elderly and people with disabilities. In carrying out the tasks proposed by objective 11, points are also reached under goal 4, which aims to ensure that education is inclusive, equitable and of quality, promoting lifelong learning opportunities for all.

This is of particular importance for this research, as it relates to the above mentioned topics of increase of inclusive and sustainable urbanization, and the capacities for the planning and management of participatory, integrated and sustainable human settlements, in all countries by 2030, in order to provide universal access to safe, inclusive, affordable and green public spaces, particularly for women and children, the elderly and people with disabilities [19].

In this context, cities seen as a form of media, which integrate the concept of educating cities and encompass the daily lives of their inhabitants, are consequently act to foster social innovation activities, as they provide countless opportunities for various types of socially innovative initiatives.

5 Final Considerations

This work sought, through a bibliographical review, to present how the selected literature relates to Educating Cities, as a media that acts for the benefit of Social Innovation.

Firstly, the concept of the media was understood, so that it was possible to represent the city as an educator and as a mediator of knowledge. After presenting this concept, its relationship with Social Innovation and the SDGs was discussed.

It is observed that the concept of SI has a strict relation with the guidelines of the charter of the educating cities and with the SDGs, since they are looking for alternative solutions to the social problems, which are demanded by the cities.

This research allows continuity of study on the theme, taking into account that the Educating Cities movement can bring numerous improvements to the education and quality of life of the community, providing sustainable development and opportunities to reach the goals proposed by the UN through the fulfillment of the 17 SDGs.

Acknowledgments. The present work was carried out with the support of the Brazilian Federal Coordination for the Improvement of Higher Education Personnel (CAPES).

References

1. Aieta, V., Zuin, A.A.L.: Princípios Norteadores da Cidade Educadora. Revista de Direito da Cidade, n. 4. (2014)
2. Unicef, Comité Português; LDA Logframe Consultoria e Formação.: Guia Para A Construção De Cidades Amigas Das CriançaS. Lisboa: Edição, Propriedade e Reprodução (2016)
3. Ferrara, Lucrécia D'Alessio: Cidade: meio, mídia e mediação. MATRIZes **1**(2), 39–53 (2008)
4. Messina, G., Valdés-Cotera, R.: Educating cities in Latin America. Int. Rev. Educ. **59**(4), 425–441 (2013). https://doi.org/10.1007/s11159-013-9369-x
5. Ivo, A.B.L.: Cidade-mídia e arte de rua. Caderno CRH, vol. 20, no. 49 (2007)
6. Perassi, R., Rodrigues, T.M.: Conhecimento, mídia e semiótica na área de Mídia do Conhecimento. Mídias do conhecimento. In: Vanzin, T., Dandolini, G.A.: Mídias do Conhecimento. Florianópolis: Pandion, vol. 1, pp. 47–72 (2011)
7. Briggs, A., Burke, P.: Uma história social da mídia, 377 p. Jorge Zahar. Rio de Janeiro (2002)
8. Perassi, R.: Princípios Teóricos Básicos da Área de Mídia do Conhecimento. Florianópolis: Egc (2017)
9. Freire, P.: Pedagogia do Oprimido. Rio de Janeiro: Paz e Terra, p. 25 (1975)
10. Morigi, W.: CIDADES EDUCADORAS: Possibilidades de novas políticas públicas para reinventar a democracia.2014. 154 f. Tese (Doutorado) - Curso de Educação, Universidade Federal do Rio Grande do Sul, Porto Alegre (2014)
11. Educadoras, A.I.: Carta das Cidades Educadoras. In: VIII Congreso de Ciudades Educadoras (2014)
12. ZitkoskI, J.J.: Cidade educadora e emancipação social: o desafio da educação fundamentada numa razão dialógica. Educação Unisinos **9**(2), 179–186 (2005)
13. Simões, J.M.S., Fortunas, C.J.C.G.: Cidades em redes e redes de cidades: O movimento das cidades educadoras. Coimbra (2010)
14. Vaziri, M., Jahani, S.: Required citizenship skills for students in the earth-quaked city of Bam. Educ. Innov. **17**(5), 163–182 (2006)
15. Moulaert, F., et al.: Introduction: social innovation and governance in European cities: urban development between path dependency and radical innovation (2007)
16. Mulgan, G., et al.: Social innovation: what it is, why it matters and how it can be accelerated (2007)
17. Bacon, C.M., et al.: Are sustainable coffee certifications enough to secure farmer livelihoods? the millenium development goals and Nicaragua's Fair Trade cooperatives. Globalizations **5**(2), 259–274 (2008)
18. Ocde, C.: CAF: Perspectivas económicas de América Latina 2016: Hacia una nueva integración con China, (2015)
19. Bignetti, L. P.: As inovações sociais: uma incursão por ideias, tendências e focos de pesquisa. Ciências Sociais Unisinos, vol. 47, no.1, pp. 3–14 (2011)
20. André, M.: Políticas e programas de apoio aos professores iniciantes no Brasil (2012)
21. Gil, A.C.: Métodos e técnicas de pesquisa social. Atlas, São Paulo (1999)
22. Oliveira, S.L.: Tratado de Metodologia Científica: projetos de pesquisas, TGI, TCC, monografias, dissertações e teses, pp. 1–26. Pioneira Thomson Learning, São Paulo (2002)
23. Amini, M., Mahdavinejad, M., Bemanian, M.R., Varzaneh, E.H.: Developing a new paradigm for performance of educating city theory in advanced technology mega-cities, case: Tehran. Iran. J. Archit. Urbanism **38**(2), 130–141 (2014)

Virtual Reality as an Educational Tool for Elementary School

Marcos Figueiredo(✉), Rovilson Mafalda, and Andrea Kamensky

Federal University of ABC, Santo André, SP, Brazil
marcos.figueiredo@ufabc.edu.br

Abstract. The new information and communication technologies (ICTs), through virtual and augmented realities, present themselves as innovative and useful digital tools in many different areas. The educational scenario, in the current context of a constant and ever-changing world, requires a pedagogical approach in constant transformation and innovation within the digital culture. This research will present an overview on the stage of VR current apps and educational relation with them and propose the beginning of adoption of this technology on classes as part of a work that seeks to create a social innovation experiment with public school teachers for the newly created "Media-Education" discipline in order to disseminate virtual reality and 360° camera knowledge and applications, as well as to understand how schools and educators are up to date and are positioning themselves in front of the new technologies. It is mandatory to be updated on the new apps and technologies and create new approaches where virtual reality can be a useful tool.

Keywords: Virtual reality · Education and technology · Digital culture

1 Introduction

Virtual Reality currently is shown as a new tool for several fields. From videogames to medicine, this technology is growing dramatically and is presenting as an innovative and exciting Information and Communication Technology (ICT) due to its immersion ability into a new environment. Rubio-Tamayo et al. (2017) comment that virtual reality and other immersive ICTs have a significant potential to transform the real world and how people interact with it. This transformation can be reached with the creation of virtual worlds or by the immersion in filmed or photos taken with a 360 camera.

Since the beginning of ages, humans communicate by the use of images, and these can depict fictions or realities. The observation opens the ways to the imaginary, and this serves to humankind as a great exercise of future speculation. Such experiences and forecasts that arise from those can become technological inventions that materialize and enchant or concern and terrify. Once not materialized, it keeps in the imaginary producing new sketches, waiting for new possibilities of becoming tangible. Duarte and Cirino (2017) mention that this virtual world can drive the user to feel in that world temporarily proposed. As a media capable of expressing these speculations or showing the reality with movies in an immersive way, virtual reality is a powerful motor to boost people's curiosity and imaginary due to its features.

L. Pereira et al. (Eds.): IDEAS 2018, SIST 198, pp. 261–267, 2021.
https://doi.org/10.1007/978-3-030-55374-6_26

As described by Grau (2003), the images universe around us never changed so fast as in the last years, as well we were never exposed to so many worlds with different images and that changes constantly and fast. It is essential to look for a sense of all these media and information, especially in education.

The ability of storytelling, excite the spectator, maintaining his attention and transform his life in some way has passed through many generations, and gained several ways according to each culture and the devices that could enhance the experience. Theater, movies, videos, and other media have grown dramatically since Internet adoption and are changing every day our way to learn, watch, and hear stories. One of the main ways of media consumption is journalism. De la Peña (2010) states that immersive journalism "is the production of news in a way people can reach first-person experiences of the events or situations described", moreover Bohrer (2016) argues that the user as an avatar, immersive VR experience the story from a first-person perspective in several forms: as oneself, an explorer obtaining first-hand access to a virtual version of the location where the story is occurring or through the perspective of a character depicted in a news story.

While VR can remove the geographic boundaries, Bricken (1991) adds that it can also eliminate boundaries of physics and the limits imposed by the physical world. In VR, it is possible to fly, teleport, change shapes, be anything, or make anything. In VR, one can execute actions that are either impossible or too dangerous to perform in the real world.

The use of this immersive approach in the educational field as an innovative tool is in the early beginning, and there are some trends to be exposed and explored for applications at schools. In this paper, it will show some of these experiences, and applications that are currently being tested or offered, challenges, and opportunities. There are platforms as UNVR, United Nations Virtual Reality- (http://unvr.sdgactioncampaign.org/vr-films/), an UN video productions since 2015 with movies where the viewer can face experiences as refugee camps and Ebola outbreaks in Africa, conflict zones as Syria and Afghanistan or the Amazonian tribe of Guajajaras effort to protect the Amazonian forest.

2 VR History and Educational VR Platforms and Experiences

2.1 Early VR Experiences and Devices

VR is not exactly a brand new field, as it remains from flight simulators built for US Air Force for pilot training during the Second World War, as Pimentel and Teixeira (1995) mentioned. The film industry became interested in 1950 and the film-maker Morton Heilig created the Sensorama device with three 35 mm cameras, and it was possible to experience the stereo sound, smell, wind, with the use of fans close to the user's head and a vibrating seat. This interface tried to reach the human senses in a motorcycle ride. The intent was to have the spectator in a multisensory trip. However, the device did not have a good acceptance due to its difficulties to operate and high cost (Fig. 1).

Many other devices were developed after Sensorama, and the researcher Ivan Sutherland, in 1968, developed the head-mounted display Sword of Damocles, bringing the first device that could render the images according to the user's movements.

Fig. 1. Sensorama Source: Wikimedia Commons

As the head-mounted displays were improving, video games and manufacturing industries appropriated from these devices while computers became smaller and powerful on graphics processing and screen quality. There some options currently in the market, since Gear VR, an HMD device used together with a high-end mobile from Samsung to Oculus Rift and HTC Vive, that requires a computer with a good graphics processor. Recently, the new device Oculus Go is trying to reach more users due to its price, as the first ones are more expensive and are adopted on a slow scale (Fig. 2).

Fig. 2. Oculus Rift Source: Wikimedia Commons

2.2 Immersion and VR as Educational Tools

2.2.1 Immersion Current Tools

Immersion is a sensation that many painters and photographers tried to bring for the viewers. It is noticeable in paintings in churches as Sistine Chapel the artist intention to immerse the spectator. Uricchio (2011) was precise when cited Robert Barker's 1787 patent for a 360-degree painting of 'nature at a glance' (*Nature à Coup d'Oeil*) emphasized the construction of a 'proper point of view' as a means of making the viewer 'experience as if really on the spot'.

Currently the immersion is possible due to new devices and is proposed with the use of the goggles, as previously shown Oculus Rift, or a CAVE system, a type of immersion provided in a room with projectors, launched in 1991. DeFanti et al. (2011) define as a walk-in virtual reality environment typically consisting of 4–6 3 m-by-3 m sides of a room made of rear-projected screens (Fig. 3).

Fig. 3. CAVE VR system. Source: Wikimedia Commons

2.2.2 VR Educational Platforms

Google Expeditions

Google Expeditions is an App used for educational purposes. The Expedition program is an app used by teachers and students to go on virtual field trips around the world. The students use a smartphone paired with cardboard, head-mounted display, and look at 360 panoramas. The students' head-mounted displays, synced to a teacher tablet that controls the scene and environment the students does or sees on their smartphone. Currently, there are more than 900 panoramas. Some options are in the Table 1 below.

Table 1. Google expeditions panorama examples.

Extraterrestrial skies	1 - Mercury 2 - Venus 3 - Mars 4 - Ceres 5 - Jupiter 6- Saturn 7 - Uranus 8 - Neptune
Wastewater treatment plant	1 - First Stop: Sewer 2 - Inflow 3 - Screening Process 4 - Sludge Conversion 5 - Updating Results 6 - Aerated Filters

(continued)

Table 1. (*continued*)

World War II	1 - Pearl Harbor 2 - The Home Front 3 - The Tuskegee Airmen 4 - Hangar 2 5 - The War Room 6 - D-Day Landing in Normandy 7 - The War in the Pacific 8 - The End of the War

Source: Google Expeditions (2018)

Nearpod VR

Nearpod VR is a VR platform for teachers formed of educational apps and various platforms and works by releasing pre-made lessons for teachers to use. Nearpod has a "virtual field trip" model within their lessons. With slideshows explaining facts about the subject at hand, quiz questions, short answer questions, digital drawings, websites, YouTube videos, and 360 images of the studied location or subject. The teacher controls the slideshow presentation linked across smartphones, tablets, laptops, and classroom computers. 360 images in high definition provide a sense of significant landmarks and locations from the studied subject.

The main principle behind this feature is that 360 breaks and "tours" enhance student engagement in the lesson and give them a sense of immersion. While the VR feature can only be used by students using a smartphone in an HMD, any student can view the 360 panoramas on their device of choice. On a tablet, a student can hold up their device and look through the scenarios, and on laptop computers can view the scene through directional input from mouse or keyboard. The content itself is identical, regardless the platform. It is possible for the teacher to create classes using pre-made classes and adding the issues he would like to.

360 Video Creation

There are several videos filmed in 360 cameras on websites like Youtube and Facebook, where the users can directly upload the videos. With the 360 cameras, one can produce and share videos easily and create content. Many traditional news companies are investing in producing content in VR, like Al Jazeera, The New York Times, VEJA and UOL, where the viewer can watch through the companies apps. Furthermore, other media producers as Discovery Channel or entities as United Nations, disclose content as Clouds Over Sidra (produced by the United Nations) which shows the daily life of Syrian refugee camps or From Waste to Taste (produced by Al Jazeera), that presents the fighting against food waste and teaching future generators about the importance of food waste management.

RoundMe

RoundMe is an app where the teacher or student can create virtual tours with the use of 360 bubbles pictures and share with others. It is possible to add legends and point interesting issues in defined places in the picture, and to zoom the picture and include facts and tips about some unique features. Initially designed to real estate market, the app started to be used for education due to its possibility to create the panorama tours in a friendly way, and as a plus, the creator can add sound files that can increase as the viewer turns his head onto some direction where supposedly is the origin of the sound.

There are other platforms and apps, as CoSpaces and EnTiTi or advanced VR software to create and edit VR worlds and spaces as Unity and Unreal engines, but the proposal is to bring the new tools that can be used in a friendly way with no need to learn programming languages.

3 Discussion and Conclusion

3.1 Discussion

Virtual reality as an educational tool and is on its early adoption. The possibilities of use grow every day, and its use in education is an important field to evaluate. Several scholars have noted the importance of recognizing various aspects of the teaching-learning relationship to promote effective practice, and virtual reality can produce this empathy.

Immerse people from different cultures, and backgrounds towards new worlds and new learning skills break barriers and opens a novel and exciting way to bring the user to a full immersion feeling. On the other hand, use the technology as the primary support for a pedagogical course is too risky and cannot achieve the expected results due to lack of knowledge, weak training, instructions, or others. This paper aims to show some of the current apps and technologies related to VR and propose to create some approaches depending on the requested objective. Furthermore, the hardware and devices are another barrier to the adoption, due to its cost and should be considered during the process as well.

3.2 Conclusion

The technology speed is breakneck. New apps, cameras and software to create virtual worlds are improving every day. It is mandatory that the user keeps in mind the research goal and available tools to develop the research. The budget for the project and methodology should be carefully studied and applied. Depending on the studied group and the participant group skills, different practices and devices would be defined where it can be created a complex matrix according to the research goal.

As in technology, some fields appropriate themselves very fast, like video games and high technology companies as aeronautical or mobile devices, others take more time to shift and dive into these devices, as educational and politic areas. Considering the innovation adoption curve above, virtual reality educators are in the beginning and inside the first or second column, but certainly before the chasm. However, big companies are betting that VR will be a leading technology adopted in several areas in the near future (Fig. 4).

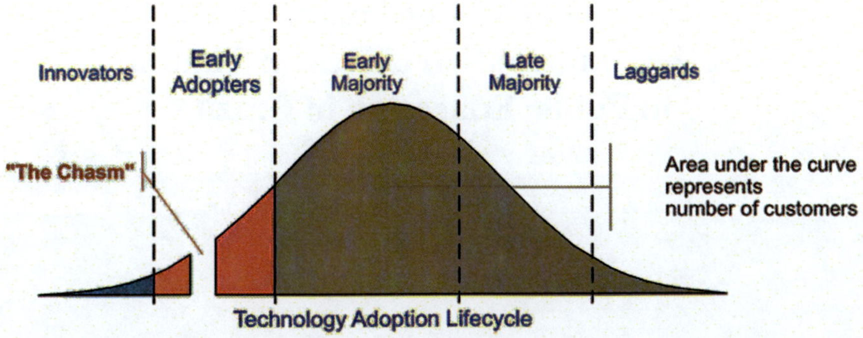

Fig. 4. Technology Adoption Lifecycle. Source: Wikimedia Commons

Considering the apps, cameras and devices available in the market, educators should start to use more accessible and cheaper solutions, like RoundMe and content creation with the use of 360 cameras, and observe the findings and audience engagement with hardware and software available, as well the how they relate the school classes and its outcomes.

References

Rubio-Tamayo, J.L., Gertrudix Barrio, M., García García, F.: Immersive environments and virtual reality: systematic review and advances in communication, interaction and simulation. Multimodal Technol. Interact. **1**, 21 (2017)

Duarte, E., Cirino, N.: A imagem além do tempo: a construção do imaginário do futuro nos produtos midiáticos. Intexto, Porto Alegre, UFRGS, no. 40, pp. 165–179, September/December 2017

Grau, O.: Virtual Art From Illusion to Immersion, 1 edn., p. 430. The MIT Press, London (2003)

De la Peña, N., Weil, P., Llobera, J., Giannopoulos, E., Pomés, A., Spanlang, B., Slater, M.: Immersive journalism: immersive virtual reality for the first-person experience of news. Presence Teleoperators Virtual Environ. **19**(4), 291 (2010)

Bohrer, C.: Experimenting with Virtual Worlds as an Emerging Journalism Platform. The JJIE Virtual World Journalism Project (2016)

Bricken, M.: Virtual reality learning environments: potentials and challenges. SIGGRAPH Comput. Graph. **25**(3), 178–184 (1991)

Pimentel, K., Teixeira, K.: Virtual Reality – Through the New Looking Glass, McGraw-Hill, New York (1995)

Uricchio, W.A.: 'proper point of view': the panorama and some of its early media iterations. Early Popular Vis. Cult. **9**(3), 225–238 (2011)

DeFanti, T., Acevedo, D., Ainsworth, R., et al.: The future of the CAVE. Open Eng. **1**(1), 16–37 (2011)

Google Expeditions. [Expeditions] List of available expeditions (2017). https://docs.google.com/spreadsheets/d/1uwWvAzAiQDueKXkxvqF6rS84oae2AU7eD8bhxzJ9SdY/edit#gid=0

Lévy, P.: O Que é o Virtual? Rio de Janeiro. Editora 34 (2011)

Conklin, H.G.: Modeling compassion in critical, justice-oriented teacher education. Harvard Educ. Rev. **78**(4), 652–674 (2008)

The "Mount Your School" Case and the Contribution of Design Thinking in Public Education in Brazil

Leandro Vasconcelos(✉) and Andres Amoedo

São Caetano do Sul, São Paulo, Brazil

Abstract. New approaches to learning are giving a new look to the search for answers to the challenges of today's world. With the evolution of open source technologies and the "design thinking" approach, we can perceive a behavioral change, in which aspects such as: Collaboration, empathy, and prototyping are increasingly present in schools. Students have an essential role in the process of solving their problems, and design is becoming an essential tool in this process. This article presents the contribution of design thinking to the Brazilian public education and the case study "Mount your School," showing the entire process in practice and the results of this operation.

Keywords: Design Thinking · Education · Design · Open source · Open design · Social innovation

1 Introduction

Ever since lecturing became the most used method in educational practices both in primary and higher education [1], one of the main problems with lecture-based learning is that it is passive [2]. In addition to not engaging students in the process, it also does not address different learning styles. How could classroom practices provide more engaging and proactive activities?

When including students in the process of searching for answers for existing problems, we utilized empathy to recover a sense of belonging and inclusion, since the entire process of discovery and solution-seeking is done collaboratively. Through Design Thinking, we attempted understanding the perspective of students in two public schools of the city of São Paulo and generate quick solutions to the problems pointed out.

The main intention of this study is to highlight this new viewpoint from the perspective of students, and to propose ways of approaching them through human-centered design, using elements such as: Empathy, collaboration, and prototyping. The goal of this paper is to reflect on the experiences of the Project Mount your School ("Set Up Your School") based on the student-centered concept of an active learning environment [3].

L. Pereira et al. (Eds.): IDEAS 2018, SIST 198, pp. 268–279, 2021.
https://doi.org/10.1007/978-3-030-55374-6_27

2 Using *Design Thinking* to Innovate in the Learning Experience

For Serrat [4], it is the change from the existing conditions to the ideal ones. For Brown [5], the migration of the designer to solve social and behavioral problems has been occurring for the last couple of years, since it has been understood that "the evolution of design to design thinking in the history of the development of the creation of products based on the analysis of the relationship between the people and the product and, finally, the relationship between people.

Design Thinking does not provide the correct answers, and it changes the focus of the questions. Even though, frequently, people cannot clarify what their necessities maybe, their behaviors can provide valuable clues about the range of essentials that haven't been satisfied. Henry Ford understood this when he observed that, when asking clients about what they wanted, they often answered: "a faster horse" [6].

Brown [5] states that Design Thinking only exists when utilizing the skills that the designers have developed across the years, such as aligning the necessities of men with the available technology in the organization, using one's intuition; recognizing patterns; forming ideas with emotional and functional meaning, and the act of expressing oneself differently, more than only with words and symbols. Brown stresses the ability to work in an interdisciplinary and not only multidisciplinary method, in which "each defends their technical specialty." In a multidisciplinary approach, "a property of collective ideas" exists, and "everyone is responsible for them [5]," that is why it is common to find designers working with psychologists, ethnographers, engineers, etc., in an integrated environment where everyone thinks about the problem and tries to reach the best solution.

Design Thinking means being more empathetic, collaborative [7], because having these characteristics means understanding and reaching people in all phases of the production of a product, since those who will use the product, being involved in the creation, not only will be satisfied with the result, but will also encourage others to adopt this improved method of design and development.

2.1 Empathy

Empathy is the attempt to see the world through others 'perspectives, understanding it through others' experiences and emotions. For Brown [5], the development of empathy is perhaps what distinguishes academic thinking and *design thinking*.

2.2 Collaboration

Collaboration is one of the key elements of Design Thinking and is characterized by the development of products "with" the clients and not "for" the clients.

People, instead of allowing themselves to be labeled as 'consumers,' 'clients' or users, can now see themselves as active participants in the creative process.

2.3 Experimentation

Experimentation refers to the act of analyzing possible results and collecting feedback. Experimenting allows you to avoid and fix significant mistakes. According to Alt [7], it is less frustrating to make a mistake early and learn from it quickly. Sometimes, it is necessary to try many times before, finally, you hit the target, because every wrong attempt puts the team one step further. It is difficult for a project to start with all the barriers and difficulties mapped. Because of this, it is crucial to experiment and reflect on profitable ways to consummate a plan.

According to Vianna [8], "The prototype is the tangibilization of an idea, the transition from the abstract to the to represent reality - even if simplified - and provide validations."

2.4 Interdisciplinarity

Always growing and allowing itself to be influenced by other areas of knowledge, the *Design* has been in a permanent process of rebuilding and evolving, expanding its limits, according to the requirements of the current era. [...] Design is a remarkably flexible discipline, susceptible to radically different interpretations in theory as well as in practice [9].

As Fazenda [10] states, 'interdisciplinarity emerges more from the encounter between individuals than between disciplines.'

2.5 Convergent and Divergent Thinking

In the design process, the activity was divided in steps of which the designer utilizes different thoughts, according to Jones [11]. It is necessary to understand the convergent and divergent thoughts presented in Fig. 1 to comprehend how to think and work with *Design Thinking.*

In the 60's and 70's, the American psychologist Joy Paul Guilford [12] introduces the distinction between convergent thinking and divergent thinking, and both originated from creative thought [13]. According to Guilford [12, 14], convergent thinking seeks a specific or conventional answer; happens when there is a problem, "when there is a standard method for solving it, and a solution can be guaranteed within a finite number of steps. It implies a single correct solution" [12, 14].

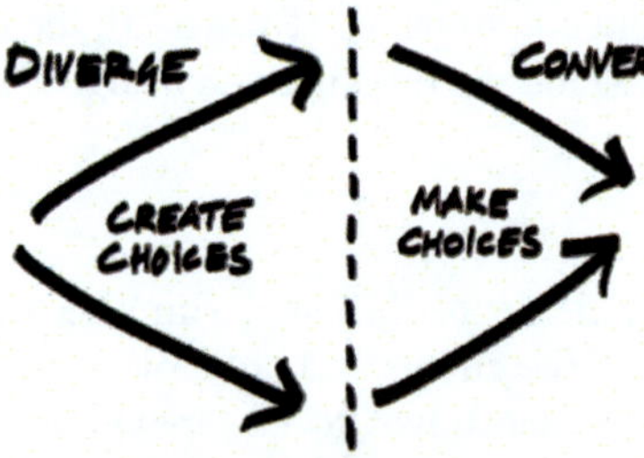

Fig. 1. Tim Brown's diagram. Source: Brown [5].

Meanwhile, divergent thinking seeks several paths to find answers; "It tends to be necessary when the problem hasn't still been fully discovered and there isn't yet a method for fixing it. It can provide more than one solution" [12, 14].

In today's environment, Brown [5] brings divergent and convergent thinking and uses it in his work, understanding that it is one of the primary steps of the design/*Design Thinking* project. The author believes these kinds of thoughts to be essential for dealing with the complexity of today's world, characterized by the intense flow of endless information (the more complex a system, the more information flows through).

For Brown [5], if the convergent problem resolution phase is the bridge to arrive at solutions, the goal of the divergent thought is multiplying options to create more. According to Brown [5], to have a good idea, you need to have many.

3 Research Design

The Project "Mount your School" was born from the necessity to bring the student as an essential agent in the concept of a co-creative school environment, together with various students, through the approach of "Design Thinking," from problematization to prototyping.

3.1 Sample Selection

It was through a public school professor from São Paulo, named Renato Lopes, that we had the opportunity to start the application of the project "Mount your School".

The project began in 2016 and has already happened twice, involving more than 700 children and professors from public schools. This year's edition has engaged 690 students from João Ramalho Public School, São Bernardo do Campo - São Paulo, and had 4 phases.

3.2 Data Collection

In the first phase, the students were asked to answer the question: "What would you do to make your school a happier place?" By expressing their concerns and how to solve them, 25 students from various classes and ages were chosen.

The second phase was a workshop with the selected students to define their daily problems and come up with creative, collaborative, and empathetic solutions.

The third phase happened inside the multidisciplinary team of "Mount you School" with the goal of selecting ideas from the workshop groups and turning them into concrete plans that could be reproduced by a CNC Machine. The fourth and last phase was another workshop of setting-up and customization with dynamics that reinforces teamwork and the importance of expression through customization.

3.3 Data Analysis

All ideas were read and separated in groups of similar interest. This way, we measured what were the main notes and in which group they belonged. The selection of the student addressed not only their vision for identifying the problem but also their initiative to suggest solutions for these issues.

3.4 Validation

The prototyping phase is a phase where all of the inputs, drawings, models and texts presented by the students in the workshop are collected and serve as reference for the start of the project, which will be entirely developed to use technology for digital manufacturing, which allows us to share the project open- source with the most significant number of people, as well as allowing us to modify and/or improve it quickly. After validating a prototype and the corresponding digital file, we can replicate it in scale anywhere in the world if a CNC Cutting Machine is available.

4 The "Mount Your School" as a Student-Centered Active Learning Experience

For people to "learn how to learn", it is necessary to stimulate creativity and the interest in participating in the construction of one's education. The simple transmission of information does not facilitate intellectual formation. First, it is essential to "learn to feel" and then to "learn to think" [15].

The focus of the educational system at the knowledge society is directed at cultivating students that are skilled to produce knowledge and associated products through transdisciplinary research. The key competence of workers in the knowledge society is the skill to create useful experience and not just experience that is governed by academic interests [16, 17].

Design Thinking is a model for thinking. It means having the possibility of believing that it is possible to make a difference by developing an intentional process to arrive at something new, with creative solutions, to create a positive impact on society. It strengthens creativity as it transforms challenges into opportunities. *Design Thinking* is the confidence that new things are possible and anyone can make them happen. It also reminds us that, in education, optimism is always welcome [18].

Based on these premises, the "Mount your School" case will be presented: A Brazilian project that takes place in public schools in Brazil, where the way of thinking uses *design thinking* as a basis and has been creating new possibilities for making a difference in public education, impacting societies and presenting new solutions to contemporary problems.

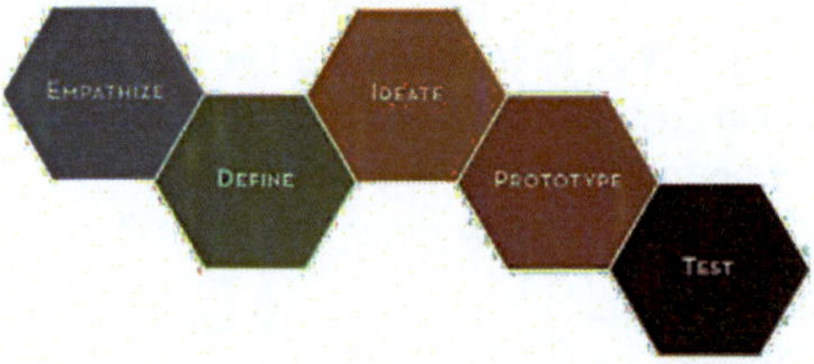

Fig. 2. The process of design thinking proposed by Stanford University

The Project "Mount your School" was born from the necessity to bring the student as an essential agent in the concept of a co-creative school environment, together with various students, through the approach of "Design Thinking," from problematization to prototyping.

The project started in 2016 and occurred twice, involving more than 700 children and teachers of public schools. The process is developed by steps that will be best exemplified by the graph above (Fig. 2) and images captured by our team.

4.1 "What Would the Ideal School Desk Be Like for You?"

In 2016, the project was carried out to the Public School Ordânia Jahone Crespo, located in the city of Santo André, São Paulo. 18 Students were involved in the project. We sought to introduce the universe of architecture and design so that these students could start an activity where they would answer the following question: "What would the ideal school desk be like to you?"

The discovery is the part where students comprehend the challenge, prepare the research, and gather inspiration. At this stage, a workshop was held at the school itself to define the problems and find solutions for them.

(a) (b) (c)

Fig. 3. Process of generating ideas in the first phase of the workshop. **(a)** Students design school desk options that they believe are ideal, **(b)** Drawing of school furniture made by one of the students, **(d)** Drawing from a student: A chair with a backrest in the shape of wings. Source: Authors

The project had, as its main goal, exercising the capacity of collective creation, seeking to rescue that sense of belonging while involving the school furniture.

In the interpretation phase, preferences and wishes are expressed through drawings or texts, documented and shared, insights are defined, discoveries deciphered, the meaning for the research is found, and the opportunity is structured in a primarily visual style, to transform insights into actions. At this stage, we have contact with a multitude of ideas and desires about what the ideal school desk would be like. As an example, we can see illustrations of a chair with a backrest in the shape of "wings," a table with a "cup holder" and a multitude of other details of this project phase (Fig. 3).

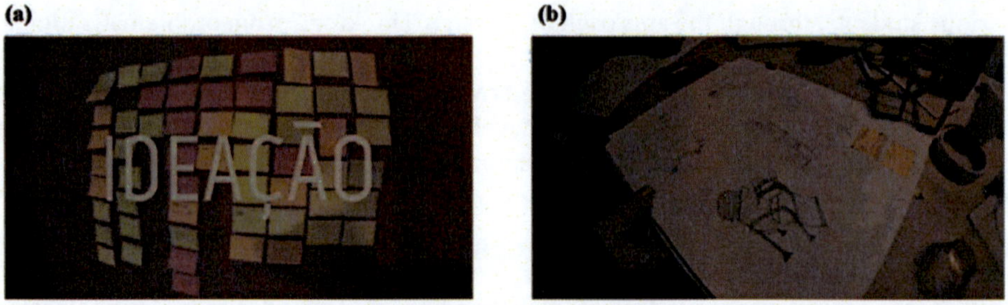

Fig. 4. Ideation process. (a) The ideas collected in the workshop are being organized in groups of common interest, (b) Student drawings being analyzed. Source: Authors

In the Ideation phase (Fig. 4), the ideals are generated, and the more promising ideas are selected, arranged in groups and refined to continue towards the next steps.

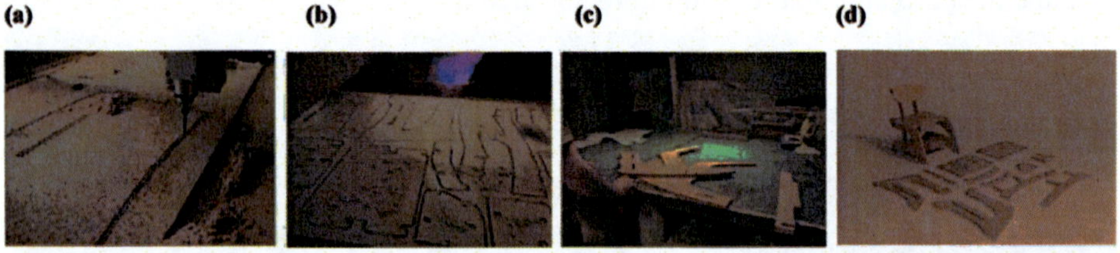

Fig. 5. Process of prototyping: (a) CNC Cutting Machine cutting the samples, (b) The MDF board image after the cutting from the CNC Machine, (c) assembling of all the pieces the prototype, (d) a view of the furniture before assembly and after the assembly Source: Authors

The experimentation phase proposes the creation of prototypes (Fig. 5). In this project phase, digital manufacturing tools and technology are used, making the project open to the public so others can have access to, replicate, and modify the design according to their needs. All of the information and knowledge produced are shared through the internet for free.

In the last phase, all children are involved in assembling parts of the desk, which fits without the need for screws or nuts. Every process is collaborative, and the teams included also participate in the customization and give their own identity to their creation (Fig. 6).

The sequence to the project is the evolution phase, where other people are involved, and the follow-up process is on the school's behalf. This phase involves teachers, new students, and everything that has been generated can be introduced to a new context with new viewpoints and new uses for the project. Every process is documented and available online. The video of this project can be accessed at https://youtu.be/B-kG1u_M2KU.

Participated in this project the students: Caique Eduardo, Christopher de Jesus, Eduardo Ferreira, Felipe Raphael, Gabriel dos Santos, Ingridi Lorrany, João Vitor, Luiz Diego, Luiz Gustavo, Marcos Cruz, Mayse Lorrane, Paola Denise, Renan Aleksandro, Samuel Candido, Victória Cristina, Vinicius da Silva, Pamela Duarte, Cassio Francisco. Teachers: Roberto Xavier, Selma de Paula, Renato Lopes.

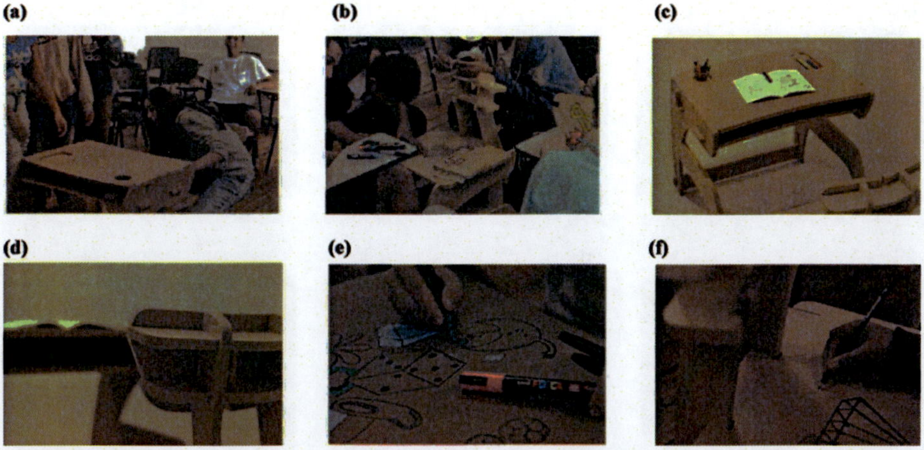

Fig. 6. Assembling and customization processes made by the students in the final phase: **(a)** student checking the joints of the table **(b)** chair customization, created by students, **(c)** image of the furniture assembled, **(d)** detail of the new seat, **(e)** personalization made by the students, **(f)** a student personalizing the school seat. Source: Authors

4.2 "How Would You Make Your School a Happier Place?"

In 2018, the question "How would you make your school be happier?" started the "Mount your School" project in 2018. The main goal of that year was to give children the opportunity to express their ideas at the João Ramalho public school, located in São Bernardo do Campo.

Fig. 7. The first phase of the process. Organizing the material with presented ideas from the 690 students of the Public School of João Ramalho. Source: Authors

In the discovery phase, a challenge was held, where 690 children from 10 to 15 answered the question: "How would you make your school a happier place?" (Fig. 7).

In the interpretation phase, 25 students were chosen to write their ideas. These students were invited to be a part of a workshop at their school. They expressed themselves,

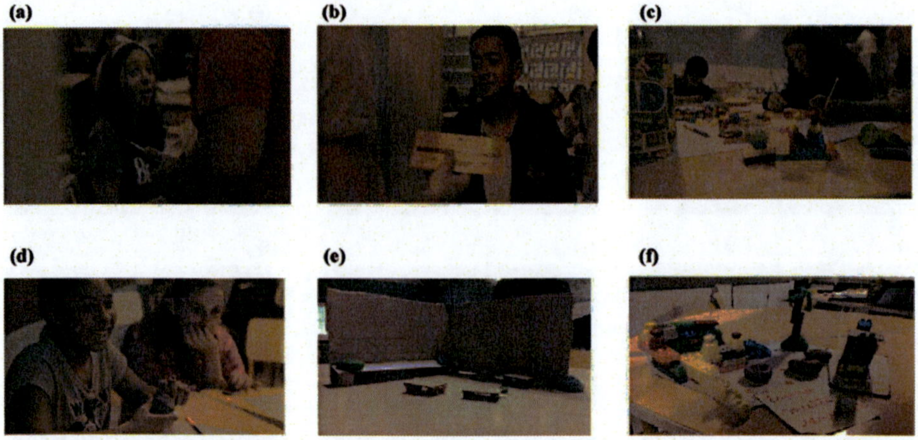

Fig. 8. Invitation and workshop with the 25 students (a & b) two selected students being invited to attend the workshop (c) during the workshop, children using LEGOs to present their ideas, d) students separated into groups on the day of the workshop, (e) the model as a result of one of the ideas presented by one of the groups, (f) a prototype of a collective use space developed by one of the groups. Source. Authors

defining daily problems at school and presenting alternative paths and possible solutions through prototypes, drawings, and a lot of creativity (Fig. 8).

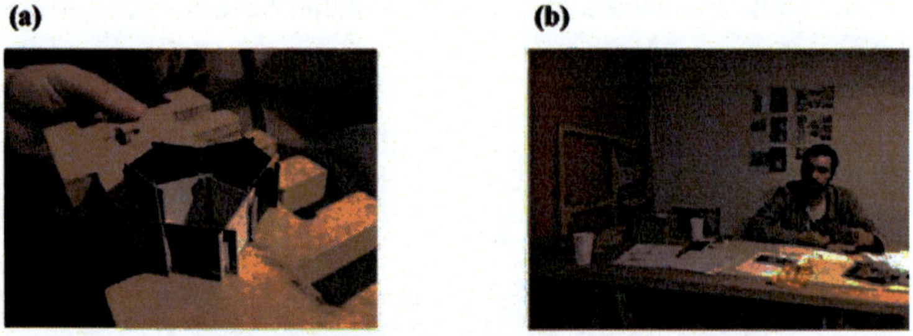

Fig. 9. In the ideation phase, all ideas are crucial to the development of the project on a real scale. (a) The first studies developed by the architecture team and designers of the Mount team, (b) Designer Marcelo Santana on the Mount team's office. Source: Authors

The ideation phase is the third big stage of the project and was held internally among the members of the "Mount your School" project. In this phase, all of the content that resulted from the workshop is analyzed and further used to convert into a tangible product (Fig. 9).

In the experimentation, real-scale prototypes were created using digital manufacturing, and the production was handled with CNC Machines (Fig. 10). The files are available for free to be downloaded on the "Mount your School" website. With this, people can download the designs anytime and anywhere, and reproduce and/or modify them.

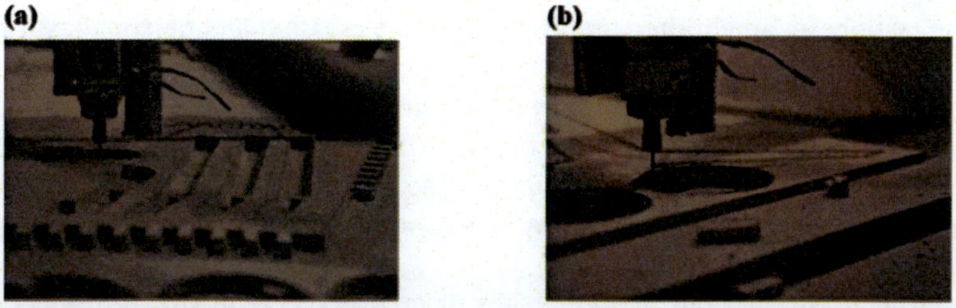

Fig. 10. In the experimentation phase, the first parts begin to get manufactured (a) details of the machining process using a CNC Cutting Machine to give the product its form (b) First real-scale prototype being fabricated. Source: Authors

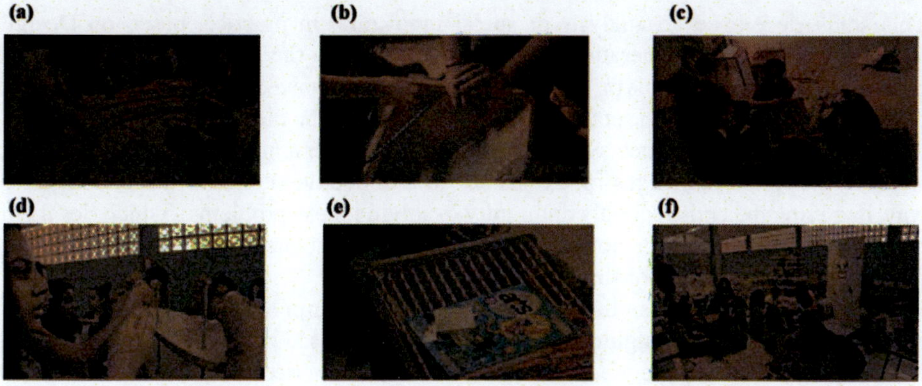

Fig. 11. Final stage, with the assembling and personalization involving all of the students (a) All of the pieces arrive disassembled and the process begins (b) The process of assembling the seats, (c) A group of students works collectively in assembling the furniture (d) A group of students is joining a shelf to accommodate books (e) Finished seat (f) During the process of creating the space. Source: Authors

Fig. 12. Students using the space after finishing the entire process. Source: Authors

In the last phase, all children are involved in the creation and personalization of the school space (Fig. 11). The project resulted in the construction of a space that can be used by everyone, made with *open-source* furniture, produced by laser-cutting machines and CNC Cutting Machines. The collection of different furniture and items resulted in a playful, creative space (Fig. 12) that amplifies the access to literature and the possibility of interaction between students of that school. The video of this project can be accessed at https://youtu.be/BHQvj9TInww.

The project "Mount your School" takes place every year in Brazil to this day, and can be seen in detail on the site www.mountsuaescola.com.br [19].

5 Discussions and Conclusion

This study developed an analysis of social innovation in practice, based on Design Thinking as a tool for generating innovative solutions in the learning environment of public schools in Brazil. This process made evidence and showed us that Design Thinking has much to contribute to the educational context, and can be fundamental in the changes that education needs so much. School managers can, for example, use Design Thinking to solve the main challenges of the school in a collective, creative and participative way, involving the entire school community in solving these problems - which ties them together, gives everyone the notion of belonging and stimulates proactivity and zeal for the environment and relationships.

The connection between the two cases showed us common results, where collaboration and a sense of belonging were present at all stages. The project showed us new forms of organization generated by the co-creation of experiential space, reproduced to be shared on an open digital platform and thus be manufactured anywhere in the world at a low cost.

In both cases, the objective was not to solve problems, but to facilitate the resolution of the issue through collective participation. In this sense Design Thinking transforms the student into a participative participant in the school context, developing his ability to work as a team and put himself in the other's place, essential skills for life in society. The experience is very rich for school-age students, since the vast majority of people only come in contact with similar approaches in the university or the job market.

Therefore, we believe that it is very important that managers and teachers take these tools, seeking new ways of solving problems, and organizing thought. Thus, they will be able to transform the school environment, regardless of the context in which they are inserted, into a space that prioritizes, awakens and stimulates empathy, creativity, and collective work. We are sure that the results achieved will be very positive.

As suggestions to deepen future studies regarding the use of design as a strategic tool in public education, we suggest a closer approximation of universities in practical actions, enabling and involving students and teachers outside the academic environment, bringing to practice the real experience of how to solve problems in a collaboratively. The project "Mount your School" can be viewed at www.mountsuaescola.com.br [19] and still happens in Brazilian public schools.

Ethics Declaration

Conflict of Interest. The authors declare that they have no conflict of interest.

Compliance with Standards Involving Humans as Subjects. All procedures performed in studies involving human participants were in accordance with the ethical standards of the institutional and/or national research committee and with the 1964 Helsinki Declaration and its later amendments or comparable ethical standards. Informed consent was obtained from all individual participants involved in the study.

References

1. Freeman, S., Eddy, S.L., McDonough, M., Smith, M.K., Okoroafor, N., Jordt, H., Wenderoth, M.P.: Active learning increases student performance in science, engineering, and mathematics. Proc. Natl. Acad. Sci. **111**(23), 8410–8415 (2014)
2. Klahr, D., Nigam, M.: The equivalence of learning paths in early science instruction: effects of direct instruction and discovery learning. Psychol. Sci. **15**(10), 661–667 (2004)
3. Oliver-Hoyo, M.T., Allen, D.: Attitudinal effects of a student-centered active learning environment. J. Chem. Educ. **82**(6), 944 (2005)
4. Serrat, O.: Design Thinking, p. 78. Knowledge Solutions (2010)
5. Brown, T.: Design thinking: uma metodologia poderosa para decretar o fim das velhas ideias. Elsevier, Rio de Janeiro (2010)
6. Brown, T., Wyatt, J.: Design thinking for Social Innovation. Stanford Social Innovation Review. Leland Stanford Jr. University, California (2010)
7. Alt, L.: Empatia, Colaboração e Experimentação (2011). Accessed 24 June 2014
8. Vianna, M., et al.: Design thinking: inovação em negócios. MJV Press, Rio de Janeiro (2012). 162p.
9. Couto, R.M.S., et al.: Gustavo Amarante Bomfim: uma coletânea. Rio Book's, Rio de Janeiro (2014)
10. Fazenda, I.C.A.: Práticas Interdisciplinares nas escolas, 6th edn. Cortez, São Paulo (1999)
11. Jones, C.: Métodos de diseño. Gustavo Gilli, Barcelona (1978)
12. Guilford, J.P.: Creative Talents: Their Nature, Uses and Development. Bearly Limited, Buffalo (1986)
13. Tschimmel, K.: O Pensamento Criativo em Design: Reflexões acerca da formação do designer. In: Use(r) Design. Congresso Use(r) Design, Lisboa (2003)
14. Bosch, M.T.: Adaptad from Educadigital (2013 thinking como abordagem para gerar inovação - Uma reflexão (2012). https://www.academia.edu/6055173/O_Design_Thinking_como_abordagem_para_gerar_Inova%C3%A7%C3%A3o_-_Uma_Reflex%C3%A3o. Accessed 12 May 2014
15. Toro, M.E.B.: La formacion de la competencia investigativa en los estudiantes del Instituto Tecnologico Metropolitano [archivo de computador] (2002)
16. Beretter, C.: Education and Mind in the Knowledge Age. Lawrence Erlbaum, Mahwah (2002)
17. Valmara, J., Hoffman, D.: Knowledge society discourse and higher education. High. Educ. **56**, 265–285 (2008)
18. Educadigital, Instituto: Design thinking para Educadores. Versão em Português: Instituto Educadigital (2013). Accessed 25 June 2014
19. Site www.mountsuaescola.com.br

Faculdade Zumbi dos Palmares Case Study of Racial Integration in the Advertising Area

Lina Maria Moreira Garai da Silva[1](✉), Alexandre Acácio de Andrade[2], and Patrícia Gonzaga Cesar[2]

[1] Universidade Metodista de São Paulo, R. Alfeu Taváres, 112 - Rudge Ramos, São Bernardo do Campo, São Paulo, Brazil
linammoreira@gmail.com

[2] Universidade Federal do ABC, Rua São Paulo, s/n, Bairro Jardim Antares, São Bernardo do Campo, São Paulo, Brazil
aacacio@ufabc.edu.br, patriciagonzagacesar@gmail.com

Abstract. The decade of 2000 was marked, in Brazil, by changes of the government policies of expansion of the higher education; the accessibility of the low-income population and concomitantly, the discussions of racial quotas and the increase of self-declared blacks in the country. In this scenario, the first Brazilian educational institution to implement racial quotas was created, allocating 50% of its vacancies to students who declare themselves to be black: Zumbi dos Palmares College. Its first courses were implemented to enable the inclusion of blacks in strategic spaces of society, one of them being communication. This essay describes the process of structural changes implemented during the creation and consolidation of the course and how concepts of innovative ambience were decisive in the development strategies of this educational product to guarantee differential and competitive advantage for trainees to compete in a highly competitive sector.

Keywords: Innovation in advertising education · Innovative · Black in advertising

1 Innovate in Education

In a globalized economy, companies compete with each other and seek in innovation the alternatives to develop competitive advantages, overcome uncertainties and expand opportunities. "Innovations allow us to increase the level of employment and income generation, as well as access to the globalized world" [1].

The purpose of this essay is to describe how strategic decisions in the learning proposal, close relationship with students, and teachers are in keeping with the concepts of innovative ambience, using the innovative and innovative experience of the implementation of the advertising and advertising course at Zumbi dos Palmares College. Concepts that make microenterprises more competitive, sustainable and also feasible to promote innovative higher education.

Considering the principles of the Charter of the United Nations, the Universal Declaration of Human Rights, the International Covenant on Economic, Social and Cultural

L. Pereira et al. (Eds.): IDEAS 2018, SIST 198, pp. 280–289, 2021.
https://doi.org/10.1007/978-3-030-55374-6_28

Rights and the International Covenant on Civil and Political Rights, and in particular Article 26 of "Everyone has the right to education" and that "higher education should be equally accessible to all on the basis of merit" and endorsing the basic principles of the Convention on Discrimination in Education (1960), which, through of Article 4: commits Member States to "making higher education equally accessible to everyone according to their individual capacity", UNESCO convened the World Conference on Higher Education in 1988. From this event came the World Declaration on Higher Education in the 21st Century: Vision and Action of UNESCO (1998), considered one of the most important references for the emergence of new proposals and changes universities for an education of the future. The statement addresses the mission of the university in the highly qualified and citizen-oriented training of individuals, as well as providing continuous learning, dissemination of research and collaborative proximity to the world of work to identify and prevent the needs of society.

From this period onwards, it was highlighted the registrations of innovative higher education projects and the Education of the Future. In Brazil, the first concepts of innovation in education are also emerging, such as [2, 3], whose proposals show that there is a convergence between the authors' ideas about how innovation encompasses process actors, reorganizing resources, content, learning methods, policies and even society. Authors such as [4–8] also agree that educational innovation takes place in a conducive environment.

In studies of innovation there is a line of study that is dedicated to understanding the environment and the relationship of the actors of the internal and external environment of the company, as being of innovative ambience. This environment influences and is influenced by the development strategies of the innovation culture, strategies, techniques and processes in the organization [9]. The environment also considers the national innovation system and the relations that determine the capacity of learning and innovation of a country [10, 11]. For the neo-schumpeterian line of study, innovation systems can be understood as a basis or institutional arrangement that seeks to consolidate an innovation-friendly environment at the national level, involving an institutional base composed of government, universities, research and technological centers, incubators [12].

The University is identified with one of the links of the innovation environment of a country, because it presents itself as one of the suppliers of knowledge for the development of researches and interactions [13] (The companies are motivated to seek the universities due to the need to count on the intellectual capital for the development of new projects, to minimize courses of equipment implementation, laboratories and improvement of their productive processes [14]. But will all the Brazilian Universities present themselves able to serve as a support of competitiveness for this scenario?

2 Brief History of Brazilian Higher Education from the 1990s

The decade of the 90 was marked by the decision to reduce expenses with the public universities of the Minister of Education, Paulo Renato de Souza. These measures created favorable conditions for the expansion of private higher education [15]. Another factor that favored the warming of this sector was the National Education Plan (PNE) for the decade 2011–2020 which presents goals such as: universalization and expansion of access and attendance at all levels of education. The use of the national high school examination (ENEM) as a criterion for access to higher education, the provision of financing for higher education, the improvement of the installed capacity and the human resources of public institutions of higher education; expansion of vacancies through the expansion and internalization of the federal higher education network, the Federal Network of Vocational, Scientific and Technological Education and the Open University System of Brazil; the provision of free public higher education as a priority for the training of teachers for basic education; programs to encourage student and teacher mobility in undergraduate and postgraduate courses, nationally and internationally; increase of the qualification required for action in higher education by expanding the performance of masters and doctors in higher education institutions.

This governmental positioning of incentive to education caught the attention of groups of foreign investors. According to the KPMG (2011) report from 2008, a scenario of mergers and acquisitions began in the education sector, when the sector ranked third in the ranking of mergers and acquisitions, behind Information Technology and Food, Beverages and Tobacco [16].

Private equity funds already had holdings in important educational groups, such as American Capital Group, a shareholder in the Kroton Educacional group, and the Cartesian Group, which acquired part of the Northeastern group Maurício de Nassau and Laureate Education, one of the largest private educational institutions in the world, has acquired control of Anhembi Morumbi College. Another group from GP Investimentos bought 20% of Estácio de Sá; the Fund Pátria, which participated in the capital of Anhanguera, and UBC Pactual, with 38% of the Faculty of the Northeast (FANOR), in Fortaleza [17].

Up to this point we wanted to see an economy of scale for the education sector, because it is a business in which the biggest cost is the personnel sheet and in this one did not apply the gain in scale. But the insertion of the new information technologies in the context of the teaching/learning relationship made possible the integration of projects with the unification of curricular matrices, paving the way for the gain in scale in the educational sector.

Moehlecke and Catani (2006) [18] cautioned against the risks of disorderly growth in private sector education. Tachizawa and Andrade (2002) [19] emphasized that higher education institutions needed to adapt their models of market management and the construction of attributes that would guarantee a competitive advantage, especially those that survived the acquisitions. Goulart, Macedo and Zurquim (2014) evaluate that the understanding of the strategic decisions of each institution started with the goal of attracting students to the different courses, seeking to offer a differential linked to quality or cost, and also to invest in educational marketing. The authors also comment that in the process of expansion of higher education it was not anticipated that the increase of educational institutions would provoke this competition and that not all those who had access to university education would seek it.

With the rise of Luiz Inácio Lula da Silva to the Presidency of the Republic in 2002, there was a strong influence of the social movements, we can see the increase of the public policies directed to specific groups of the population in which age, income/gender, or race/ethnicity besides heavy capital contributions via BNDES in educational groups with highlight to the Kroton group.

And in this scenario of great changes in the educational sector that appears the Zumbi University of Palmares. A private, community institution, inaugurated on November 20, 2003, Black Consciousness Day. Located in the city of São Paulo, the Institution of Higher Education (IES) is the only socio-educational project belonging to a Non-Governmental Organization of Latin America.

With accessible monthly fees, the institution strategically planned its first courses aiming to associate aspects of ethnic identity and the training of qualified professionals to the strategic niches. The first ones planned to start the institution's activities were: Administration aimed at a generation of Afro-entrepreneurs; Right to guarantee the rights of the black population, Pedagogues who taught African history and culture in the classroom and Advertising, to have blacks generating black business communication and increase their presence in advertising production.

2.1 Innovative Features of the Zumbi Advertising Course

As already mentioned, the environment in which the company is inserted influences the capacity for innovation. To study them, some proposals indicate understanding the company's innovative behavior (strategy and other environmental variables, such as government incentives for innovation) and the innovative activities generated for the company; and the object, which privileges the understanding of the specificities of the innovation produced (OECD 2005). But the proposal of this essay brings the descriptive of the dimensions that impact the ambience of intra-organizational relations - in this case, the relationship university government companies.

For reporting this experience we chose to use the dimensions will innovate present in the Oslo Manual "provide guidelines for the collection and interpretation of data on internationally comparable way to innovation" (OECD 2005, p. 12). As it is a scenario and a educational one, the innovation focused on the dimensions foreseen in the Manual of processes and organizational management. In these dimensions, the following aspects of innovation are pointed out in the case:

2.1.1 Innovation, Entrepreneurial Culture and Identity

The creation of the Zumbi dos Palmares College alone is already an example of the study of interaction with society. It is born during the period of changes in government policies for the expansion of higher education in the country; the accessibility of the low-income population and concomitantly, the discussions of racial quotas and the increase of self-declared blacks in the country. It was the first Brazilian institution to implement racial quotas, allocating 50% of its places to students who declare themselves black. Since her first college entrance exam in 2003, she records on average about 90% of self-declared black students, with most students in the socioeconomic class "C/D".

Santos (2012) has made one of the few records on the history of Zumbi's implantation, and it is necessary to highlight aspects about the support of the municipal government and agreements with companies according to the concept of the innovation trinomics of Audy et al. (2002, p. 43). From its inauguration in 2003 until 2009, the Faculty changed its address three times. With the support of the City Hall of São Paulo, the institution installed itself in the old Olympic sports building of the Tietê Regattas Club, which had its reintegration of possession with the end of the loan of 100 years.

Santos (2012, p. 43) also reports that the institution maintained a series of agreements with other educational institutions (public and private) and Brazilian companies as well as the international black movement. The contacts with these institutions generated cultural and educational projects of the NGO Afrobrás and that influenced in the relation with the project of the Faculty generating initial actions of research and inclusion of the students, it became still one of the entities to first promote the black in the organizations.

But it is in the history of the implementation of the Advertising and Propaganda course that it is possible to identify the concepts of entrepreneurial culture and a process of co-creation of the innovation and identity environment. Etzkowitz (2003) further states that the institution, by combining its resources and research potential for the economic and social development of society, allows an entrepreneurial culture.

The Advertising and Publicity course began its activities in 2010 with the perspective of disseminating the entrepreneurial and Afro-Brazilian culture to form entrepreneurial advertisers.

2.1.2 Elaboration of the Pedagogical Project, Socioeconomic Challenges

The biggest challenge identified when designing the Pedagogical Project of the Course was how the student of Zumbi, Afro-Brazilian Class C/D could have a competitive day in an area where the formation of advertising is costly?

The history of the emergence and development of the educational area of Advertising in Brazil has always been limited to an elite. A striking feature of the Advertising area is the glamorization culture of the profession. Carrascoza (2011) exemplifies this glamour in the autobiographical accounts of big advertisers and highlights in his narrative a conception of the high standard of life that advertising provides. For Oliveira (2015, p. 5), the American Advertising model did not influence only the technical approach, but also the culture of the professional environment.

(…) These foreignisms gave a further charm to the profession by making it clear how much advertising was cosmopolitan and linked to a world much wider than common sense … Oliveira (2015, p. 5).

In search of a new meaning to elaborate an educational project aiming at the competitiveness of the egress strategy was to opt in teaching theories and techniques of Advertising from the knowledge of the student's cultural context. An example of this strategy in the classroom was to propose an advertising creation exercise for the cosmetics sector using the mythology of the religions of African matrices. This strategy allowed the Pedagogical Project to create its first innovative action in relation to other courses in the market.

When the Advertising and Propaganda course of Zumbi started its activities in 2010, there was already an intention not to repeat the models of curricular matrices that already exist in the market. The initial matrix of the Pedagogical Project of the Advertising and Propaganda course was elaborated by a nucleus of collaborators, in which the majority remained in its implementation or at least, carried out collaborative activities.

In the elaboration of the Pedagogical Project of the Course there was no intervention of the maintainer, which generated a confidence in the team to innovate freely, respecting the Institutional Policies and its goals. Like any company that intends to implement the culture of innovation, projects when being taxed from the top down or from outside to inside the organization do not bring the expected results.

"The feeling of" belonging "to the project is a basic requirement to be experienced and worked from the beginning. The interaction between teachers and teachers, teachers and students, teachers and management are presented as the interrelationships that can support the construction of an innovative project" (MASSETO 2012, p. 209).

The interaction between "teachers and teachers" pointed out by the author was still essential in the group's decision to determine learning based on real projects focused on the care of small and medium businesses in the community. From the first semester of the course, the projects were evaluated by a bank formed by external guests, for the student's exercise and argumentation of students' ideas. The strategy was to bring students closer to a market niche little explored by the big market agencies and which it points out becomes profitable in the following years.

2.1.3 Lack of Resources Stimulates Creativity

The change in the Pedagogical Project of the Zumbi Advertising and Propaganda course began in 2013. They resulted from some internal factors and external influences. The first occurred as the team of teachers interacted with each other and with the students. It was clear that it would not be possible to develop some interdisciplinary projects without access to certain specific knowledge, such as those related to technology and communication.

On the institutional side, the course was not financially viable, it was not possible to count on all audiovisual production studios, enough equipment to develop advertising production. As for the students, they could not afford to have access to events for students in the area; most of them did not have computers or cameras.

In writing about innovative environments, Porter (1989) dealt with the question from the idea of a construct formed by a set of conditions (limits, obstacles, possibilities, stimuli) of innovation in a given social and cultural formation. Thus, barriers are elements that can foster innovation when associated with the creative abilities of the company or a country to create new products or services. Such limitations of the advertising course infrastructure did not allow another option to the teachers' team and coordination other than to think about how to redesign the course proposals.

One of the strategies used was the search of vacancies in events by some faculty and partner companies of the Faculty. Activities such as this, as well as exchanges and agreements involving companies, governments, universities, incubators and research centers, and the increase of networks has become a recurring theme in research on innovation.

But a new obstacle has been identified: a refusal of most students to participate for emotional reasons. The students described being "afraid" of attending sophisticated spaces. Meeting with students from other renowned institutions of the advertising course, the fact of attending an institution little known by the advertising market or even the name Zumbi became mockery among the people were arguments also presented by the students. The constitution of the subjectivity of individuals, either implicitly or explicitly determined and reach standards in the areas of education, work and quality of life "beyond the socially widespread expectations, in the case of black individuals, are also situations that can (and usually do) re-signify the way these individuals perceive themselves in society "(apud Sansone 2007; Gomes 1995; Parente 2005).

2.2 Communicate to Co Create Restructuring

Aiming to broaden the vision of the communicational process, the course's team of teachers was inspired by the discussion about the future of advertising and the possibilities of mapping the journey of consumer experience. To connect all these challenges, contents, workshops, the dialogues between teachers, students and coordination have become even more frequent. The actions created by the team to sustain the changes were:

1. The preparation of the project manuals became the responsibility of the group of teachers working in the semester and no longer discussed only among those who belong to the Structuring Teaching Core.
2. Communication between students and teachers for the implementation of interdisciplinary projects was carried out in groups closed by Facebook. The contents of the disciplines, delivery date of each phase of the project and references that could collaborate for the development of the project were disclosed in the virtual community.
3. Students also began to evaluate their learning as of their classmates, teachers and the participation of external guests. Learning the feedback has become part of the assessment of the learning process of both training and attitudes. Information that is not obtained in evaluation processes proposed by the evaluation instruments suggested by the MEC.
4. Teachers and coordination carried out studies on Digital Marketing and implemented the methodological approach of Design Thinking in the elaboration of the projects and reorganization of the new curricular matrix.

From these changes in the learning process of the course, the third group of graduates then had three modalities of Conclusion: the proposal of elaborating an advertising campaign was maintained, but the alternatives of the elaboration of the own business model were also established, including the communication of the same and also the development of a scientific article, accompanied by a video article. The proposal of the video article arose with the contact with the first scientific magazine with this pattern of diffusion of scientific communication.

The general coordination of the three modalities was the responsibility of the Course Conclusion Work discipline, It is worth here that the orientation of the Course Conclusion Work was no longer centered on a teacher, but all the teachers of the semester and the course, from the knowledge of the subject to the conclusion of the course was an article or monograph.

It was observed that the offer of three modalities of conclusion of course stimulated the interest in new knowledge of the students. From the group of 25 students from the third class, three scientific articles with video were produced in 2016. One of them was published in 2018, in a book by the Brazilian Institute of Data Analysis that focuses on the participation of Candomblé practitioners in social networks.

Three technology-based business projects have been developed:

- Logistics reserve of fantasies for samba schools and a collaborative platform of contents on African history for pre-school teachers.
- Insertion of student in the selection process in Google.
- Gastronomic business initiated in the classroom had its improvement in the Work.

3 Financial Viability

Another challenging moment in the history of the course was the announcement of the closure of its activities at the end of 2013. The low number of student's enrolled (total of 60 in three semesters) made it unfeasible financially. The news of the closing of the course promoted the emptying of a second semester class and a few students entry in the first semester. Practically the course declined in the average of incoming between 2013 and 2014 due to the lack of visibility in the capture of students.

They also generated dissatisfaction and insecurity among the students, teachers who understood that they were not welcome to the institution. There was always a doubt whether during the January vacation or the July recess, the course would have its activities closed.

4 Results - Former Students in the Market

During the period from 2014 to 2017 the course was never close to the Advertising agencies with the aim of promoting the insertion of students as trainees. The first students occupied spaces were in the area of communication of banking companies, already partners of the Faculty Zumbi dos Palmares and, most prominently, companies related to digital media. It is worth mentioning that this access to the digital media sector was possible because in this period the proximity of two companies from the area with the course stood out in offering training courses parallel to college. One was Google, from the group of volunteer employees entitled Afrogooglers, the other was the Institute for Analysis and Data Processing (IBPAD).

The qualification in topics such as performance and monitoring of social networks were indispensable to ensure the competitiveness of students in the first chances that arose for the students of the course of Advertising in the communication market. The area of digital media was more receptive with the diversity proposal. As the brands charged the presence of blacks in the teams of their agencies, the course of Publicity of the Zumbi of the Palmares began to be looked for.

5 Conclusions

This article aimed to relate aspects of the process dimensions and organizational management of the concept of innovative ambience in design, implementation and management actions in the development of the Advertising and Propaganda course of Zumbi dos Palmares College. It is a one-off report from the period 2010 to 2017, when the author of this essay was at the forefront of the cycle of implementation and recognition of the course.

Allowing the pedagogical and management changes to be managed without interference from the maintainer is an indication of having favored the promotion of the team's perception of belonging in the development of the project. The students' perception of not having the right to be in the advertising space as events promoted for students associated with the closing of the course activities, the lack of recognition of the course by the publicity medium, the name of the Faculty suggest jokes, were obstacles to be considered during the course changes. Also added here is the announcement of the end of the course activities in the institution.

These factors should be a reflection of future educational projects aimed at the black community, because they indicate that it is not enough to have an ethnoracial proposal to generate engagement or even its adoption.

However, it is important to point out that the identification of market trends and the strategy of exploiting the socio-cultural knowledge of the students has re-signified the practice of teaching the course. Expanding the classroom with the presence of professionals and companies from the market and research for the exchange of specific knowledge were situations that allowed to enable the improvement of students, promote the dissemination of the course and generate networking for indication of vacancies. It is noticeable to note that there is a favorable arrangement for an innovation environment, but one that does not comprehend the absence of obstacles and challenges.

The interaction between the team of teachers, student-teachers, university-business and collaborative management of coordination, market trends studies are also requirements for innovation. In this sense, it is suggested that future studies can explore what are the elements of greater balance for educational projects with a focus on the black population.

References

1. GLOBAL ENTREPRENEUSHIP MONITOR - GEM. Empreendedorismo no Brasil (2010). http://www.sebrae.com.br. Accessed Jan 2019
2. Masetto, M.: Innovation in higher education. Interface - Comunic. Saúde Educ. **8**(14) (2004)
3. Carbonell, J.: A aventura de inovar: a mudança na escola. Artmed, Porto Alegre (2002)
4. Hernandez, F.: Aprendendo com as inovações nas escolas. Artes M.dicas Sul, Porto Alegre (2000)
5. Imbernón, F.: Amplitude e profundidade do olhar: a educação ontem, hoje e amanhã. In: Imbernón, F. (org.) A educação no século XXI. Artmed, Porto Alegre (2000)
6. Thurler, M.G.: Inovar no interior da escola. Artmed, Porto Alegre (2000)
7. Camargo, C.R. (org.) Experiências inovadoras de educação profissional
8. Cebrian, M. (coord.) Enseñanza virtual para la Innovación universitária. Narcea de Ediciones, Madrid (2003)
9. Bessant, J., Tidd, J., Pavitt, K.: Gestão da inovação. Bookman, Porto Alegre (2008)
10. Lundvall, B.A.: National Innovation Systems of Innovation. Pinter, London (1992)
11. Freeman, C., Soete, L.: The Economics of Industrial Innovation. The MIT Press, Cambridge (2000)

12. De Oliveira, R.S.: Proposta de avaliação das incubadoras de empresas de base tecnológica como mecanismo indutores da inovação tecnológica. 231f. Dissertação (Mestrado em Administração). Universidade Federal de Viçosa - UFV, Viçosa (2010)
13. Freitas, R.C., dos Santos Ferreira, G.M., Moreira, L.C.P.: Inovação, TIC e docência: práticas e concepções de professores em uma IES privada. Revista Internacional de Educação Superior **4**(1), 25–51 (2018)
14. Rosa, E.O.R., Hemais, C.A.: A dinâmica do relacionamento Universidade-Empresa na visão de seus atores: um estudo de caso. In: ENCONTRO NACIONAL DA ANPAD, Anais... Brasília, vol. 24 (2005). CD-ROM
15. Monteiro, C., Braga, R.: O mercado da educação superior particular no Brasil, 12 edn, pp. 13–23. Revista Aprender Virtual, São Paulo (2003)
16. KPMG CORPORATE FINANCE. Pesquisa de fusões e aquisições: espelho das transações realizadas no Brasil. http://www.kpmg.com/BR/PT/Estudos_Analises//fusoes_aquisicoes_4o_trim_2008.pdf. Accessed 24 Oct 2018
17. De Oliveira, R.P.: A transformação da educação em mercadoria no Brasil. Educação e Sociedade Campinas **30**(108), 739–760 (2009). https://www.cedes.unicamp.br/publicacoes/edicao/82. Accessed 20 Jan 2019
18. Moehlecke, S., Catani, A.M.: Reforma e Expansão do Acesso ao Ensino Superior: balanço e proposições. In: Oliveira, J.F., et al. (eds.) Política para as Instituições Federais de Ensino Superior (IFES). Políticas de acesso e expansão da educação superior: concepções e desafios. INEP, Brasília (2006). 103
19. Tachizawa, T., Andrade, R.O.: Gestão de instituições de ensino. Editora Fundação Getúlio Vargas, São Paulo (2002)

Check for updates

Engrena ITA: Alliance for Innovation Prospection on Gear Technologies

Ronnie Rodrigo Rego(✉)

Aeronautics Institute of Technology, São José dos Campos, SP 12228-900, Brazil
ronnie@ita.br

Abstract. Over a hundred million gears are annually produced in Brazil. They are applied to automobiles, wind turbines, aircraft and sugar and alcohol energy systems, among others. Despite the variety of applications, these segments share similar demands for technological solutions within the same economic challenges to investing in innovation. This context engendered the launch of "*Engrena ITA*", in 2017, as an alliance among organizations with the aim of enhancing research prospections about gears and power transmission systems. Sharing benefits and responsibilities among its members, *Engrena ITA* assumes the integrative role established by the Open Innovation concept. The operational essence is the systematic interaction among members, who receive the right of defining guidelines of the gear-dedicated research group of ITA. The fundamental interaction occurs in workshops twice a year. Deep technical discussions precede the definition of a priority list of research topics to be prospected. The top priority is taken to further discussions on the scope content, and its generated project proposal is submitted either to government calls or to a pool of companies. The first year of operation, executed as a pilot plan, joined 24 companies of the segment, from tooling developers to OEM's. In one year, 15 projects were prospected, involving 29 organizations worldwide. Projects were prospected on a partnership basis, involving Brazilian and foreign research institutes. The systematic integration which drove new ideas creation is now the basis for the initiative's continuity and for multiplying the model to further technologies.

Keywords: Entrepreneurship · Open innovation · Cooperation model · Gears

1 Introduction: The Opportune Scenario

It is estimated that over a hundred million gears are annually produced in Brazil [1]. They are applied to industry segments such as automotive, aeronautical, wind energy, and sugar and alcohol energy. This market plurality, evidenced by a wide range of sizes, geometries and materials, reveals the remarkable capacity of the Brazilian gear industry.

Despite the diversity of applications, these segments share similar challenges for innovation and demand for technological solutions. They claim for more productive teeth cutting processes, improved strength against fatigue failures, less expensive materials, and lubricants for reduced power losses, among others [1, 2].

L. Pereira et al. (Eds.): IDEAS 2018, SIST 198, pp. 290–297, 2021.
https://doi.org/10.1007/978-3-030-55374-6_29

The similarities are not limited to the technical scenario. A common challenge became even more prominent with the onset of the economic crisis in Brazil in 2015 [3]. It struck the industry to one of its deepest recessions throughout history. This particular segment faced the crisis utilizing a conventional approach to reducing expenses and canceling investments. Not different was the attitude of the government. The annual budget addressed for the Ministry of Science, Technology, Innovation and Communication of Brazil was reduced by more than 50% [4, 5]. Restricted with the funding possibilities, efforts towards innovation and research were, on a large scale, interrupted.

The crisis, however, highlighted that the conventional cooperation models for innovation [6, 7] and research demanded changes and new alternatives. Bilateral projects between Academy and Industry pursue the advantage of a straightaway beginning, without needing the approval process from *ad hoc* reviewers. On the other hand, they are limited to the available investment of one company. The decision of a joint application for a call of government funding agencies may leverage the available investment, but the approval time may not be synchronized with the innovation timing demand. Furthermore, the described economic scenario did not improve the approval chances.

The picture gets even more critical when considering that innovation processes are majorly conducted as an "on-demand approach" in large companies. Task forces are organized to promote innovation as a technical challenge appears, an abrupt change of the market share is imminent, or a financial action is mandatory. Once the process is overcome, the dedicated staff turn to take part in the daily attributes of the company [8]. This issue is the basis for the Incubation phase of the DNA innovation concept proposed by O'Connor et al. [9]. It is assumed that discoveries need to be isolated from a company's daily demands, so they can gain enough potential.

The lack of an innovation management model inside the industry was characterized by [7] also for Brazilian companies. They concluded that the isolation needed from the internal attack of production priorities is not enabled with constituting a formal innovation department. Therefore, despite the need for a systematic model for innovation, it needs as well an external agency to lead or to support it [10].

The conclusions, coupled with the observed considerable potential to be obtained with the integration of the gear industry, were the basis for conceiving the initiative "Engrena ITA". Launched in July-2017, the initiative is an alliance among organizations to enhance research prospections and knowledge dissemination about gears and power transmission systems. The initiative is managed by the gear-dedicated research staff of ITA, Brazil.

Sharing benefits and responsibilities among its members, *Engrena ITA* assumes the integrative role, established by the "Open Innovation" concept. According to Chesbrough [10], knowledge is progressively more widely distributed, which forces companies not any longer to rely solely on their own research. The conception describes that, instead of creating the best ideas, market leaders will be the ones who make the best use of internal and external ideas. Systematically circulating ideas is the role addressed to universities in this conception, integrating key partners for enabling innovation. *Engrena ITA* represents the integration between the Academy, the Brazilian gear industry, governmental agencies, other research institutions, and their partners, especially the international ones (Fig. 1).

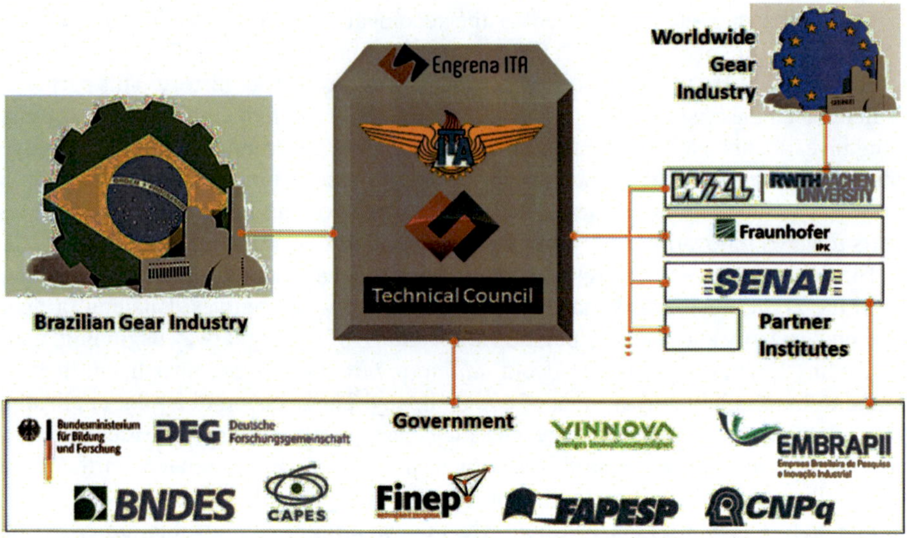

Fig. 1. The integration role conducted by *Engrena ITA*, following the concept of Open Innovation. Elaborated by author.

The integration, primary purpose of this initiative, drives the creation of new ideas and the circulation in collaborative action. Different companies identify potential partners onto a technical challenge to be researched either with shared expenses or with enhanced scope. The establishment of *Engrena ITA* inside an academic environment, the reason for a continuous connection to the government, allows the identification of investment models that enable audacious projects, proportional to the power of this segment in Brazil. The systematic approach of technological demands in an open model results in affordable projects to its members. The feasibility comes either because they identify the possibility of joint participation towards a common goal, or by including research partner institutions with complementary competencies or even once the deep discussions come up with a more suitable model for funding an idea. The organic consequence is the transformation of new ideas into real innovation projects.

2 Modus Operandi

The key concept of *Engrena ITA* is to systematize the actions for research prospection. Through an adhesion contract, the partners commit themselves to be regularly in contact with the organization staff and to be present in the promoted events. It replaces the conventional on-demand approach, in which representatives of the Academy and industry are brought in contact by the advent of production problems and technical challenges. Additionally, the individual communication to each company is enriched by a global discussion with representatives of several organizations.

From these events, the "*Reunião de Trabalho*" (RT), is the most important interaction moment among members, consisting of the operational essence of the initiative. Organized every semester at ITA, the RT's provide to the members a top attractive benefit:

the possibility of defining research directions for the gear-dedicated research staff of ITA. Along an RT, the members are encouraged to expose their technical challenges and current demands. The natural consequence of joining the entire gear production chain is the wide range of topics initially considered to be necessary. The main goal of an RT is, therefore, of defining a priority list of topics. After stimulating directed discussions, a voting session is organized and each company, independent on its size, history or origin, has the right for two votes.

The technical discussions of each RT are promoted in two different modes, one at each event. The first one is represented by topics that are brought by the research staff of ITA or by researchers of partner institutions. These topics are characterized by ideas already academically considered as feasible and already with a relatively matured planning. On the other hand, they do not necessarily represent the demands of the industrial segment, which opens the opportunity for the ideation sessions. In this second mode, the initiative organization inserts the member inside creative environments in which collective activities engender the creation of topics representing the real demands of the gear industry. The prioritized topics, independent of the mode they were raised, become now a directive to the research staff of ITA.

At this point, the topic is still on a high level, needing to be further explored in order to become a project scope. Audio conferences, organized within a month after the RT, join all the potentially interested members. They have the aim of collecting detailed information to simplify the convergence of expectations and project scope generation. The topic is discussed in the perspective of state-of-the-art, application, accepted simplifications, and risks to the execution.

The outcome of the audio conferences is a map of demands that supports the definition of potential partners and a business model. Once they are confirmed, the researchers of ITA or the partner institution lead the project proposal preparation. Depending on the scope content, the proposal is submitted either to government calls or to a pool of companies. The entire procedure is illustrated in Fig. 2.

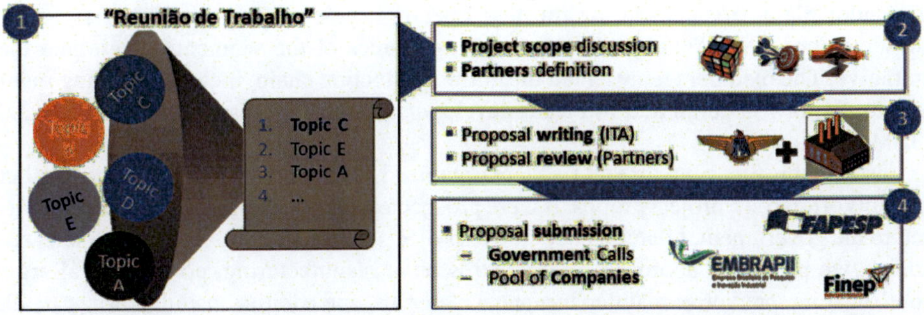

Fig. 2. The prospection mechanism organized by Engrena ITA. Elaborated by author.

The involvement with several companies brings the first impression of an ordinary consortium model. However, the described mechanism shows a differential of *Engrena ITA* that must be highlighted. Its main objective is to systematize the prospection of projects. Rather than conducting the research project, the members are joined to define, as a voice of the entire segment, what needs to be researched from that point on. Therefore, the initiative is not based on one unique research objective with a fixed time schedule and defined deliverables. It is a continuous mechanism of creating project proposals, enhancing the chances of the gear industry to be granted with resources that will be converted into innovation and thus global competitive leadership.

From an administrative perspective, the focus in prospection turns *Engrena ITA* into a dynamic model. Without approaching the execution phase of projects, activities are conducted without the need for discussing intellectual properties and costs negotiation. These become activities of the moment in which a project contract is defined, which is already out of the scope of *Engrena ITA* and responsibility of the research team defined to execute the project.

To stimulate the adhesion of a large number of companies, further benefits are offered to members. Non-confidential state-of-the-art material is shared, technical seminars are organized, and technological workshops allow the demonstration of experimental and virtual solutions. At the same time, these activities characterize an environment of continuous networking among members and expose the students of the research staff for the potential recruitment of qualified human resources.

The analysis of the depicted model recalls similar cooperation structures in the international scenario, such as the "*Advanced Manufacturing Centre*" from the *University of Sheffield*, the "*MIT Media Lab*" and the "*Gear Research Circle*", organized by the *WZL-RWTH Aachen University*. These models could not be straightforward reproduced in Brazil. There is a mindset challenge in a society in which the cooperation between the Academy and industry is not only unusual but in several instances seen as a threat or a waste of time. One of the particular actions taken was to launch the initiative with a short-time and costless phase, called pilot plan.

3 Partial Results and Outlook

The pilot plan, executed during the first operation year of Engrena ITA, has ended in September 2018. Within this phase, 24 companies of the segment have joined the initiative. The members represented the entire production chain, including tooling manufacturers, raw material and lubricant developers, gear and transmission factories and OEM's.

During this first year, three RT's were organized. Out of them, and due to the parallel benefits offered, 15 project proposals were prospected. They generated proposals submitted to the government, bilateral cooperative projects, and consortia. Among the selected topics, the proposals approach residual stresses in manufacturing, powder metallurgy, tooth coating, predictive maintenance, new materials, and additive manufacturing. In all, the projects involved 29 corporations, being 23 from Brazil and six foreign ones (Fig. 3). At the moment of this publication, three proposals were already approved to start, three were declined and the rest in under discussion.

Considering the integration role, prospections of *Engrena ITA* were also conducted in partnership with other research centers. *SENAI Institute of Innovation for Laser*, from Joinville (SC), *SENAI Institute of Innovation for Surface Engineering* and *SENAI*

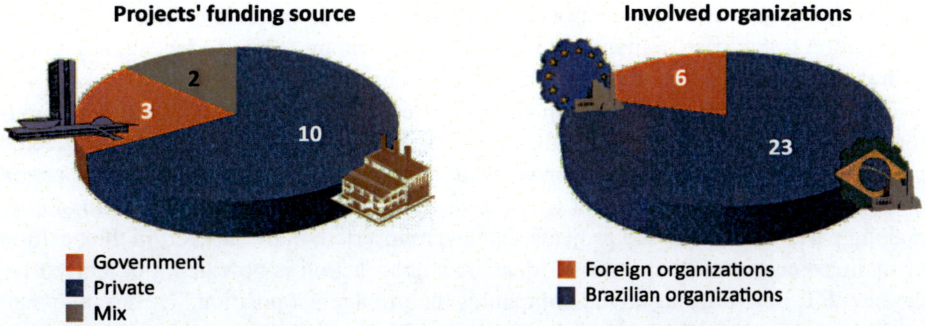

Fig. 3. Summary of the prospected projects along the first year of operation of Engrena ITA. Elaborated by author.

Institute of Innovation for Metallurgy, both from Belo Horizonte (MG) represented the national partners. Internationally, the prospection was performed with the German institutes *WZL-RWTH Aachen University* and *IPK Fraunhofer Institute*, from Berlin. The same plurality was reproduced to the governmental agencies to which proposals were applied, In Brazil, projects were prepared to *FAPESP*, *CAPES*, and *EMBRAPII*, while a portion was submitted to *DFG* and *BMBF*, from Germany.

The prospected projects cover technological challenges on manufacturing, design, and testing of gears. Among them, the majority is approaching state-of-the-art topics, which reveals the mentioned involvement with companies and funding agencies from abroad. Examples of the explored thematic are the residual stress interaction in-between each manufacturing step, powder metallurgy with hybrid material for reducing noise in the e-mobility context, advanced teeth coating for increased torque efficiency, and additive manufacturing for repairing of special gears or toothed-tools.

The exposed results exceeded the optimistic expectations, especially considering that the current gear-dedicated research staff of ITA is composed of only ten collaborators directly involved, among them four Ph.D. students, four M.Sc. students, and two undergraduate students. Nonetheless, since they are granted with real challenges that will derive their academic works, and considering the complexity of the prospected projects, these numbers are much increased. Several collaborators from other research groups in ITA and from the research partner institutes are benefitted from this initiative.

However, the conclusion from the prospection metrics cannot still reproduce the qualitative results. The feedback statements from the members majorly brought enthusiastic content to be part of the initiative, stimulating the model to be applied to further technologies rather than gears and even out of engineering. This feedback was not only individually provided, but spontaneously discussed between several members, highlighting the alliance character that the initiative claims for. Certainly, the statements also contributed with constructive criticism to support the creation of the second phase of *Engrena ITA*.

Planned since the pilot plan, "Phase 2" was structured to turn *Engrena ITA* into an economically sustainable model, with the members contributing to financial support. The annual fee was defined based on the investments of the first year and, above all, shared and agreed with the members of the pilot plan. Immediately after its start, the expectation for at least 16 members was overcome, and the initiative has already joined 19 members, among participants of the pilot plan and new partners.

The actions for a sustainable innovation system start already in Phase 2. A straightforward analysis of the direct collaborators involved against the amount of prospected projects leads to a potential excessive workload in the short-term. It derived the action on strengthening the involvement with the research partner institutions. Instead of only making part of the discussed projects, they were inserted, in the last RT, in the position of main executors of the proposed ideas. A similar action is now being organized for the next RT, now offering to the companies the protagonist position. The arrangement is planned to intersperse the lead position with ITA, maximizing the initiative efforts to engender further new project proposals.

With the well-succeeded initial phase of Engrena ITA, it is foreseen the possibility of start producing research results in addition to the projects prospection. If Phase 2 was designed for three years, this outlook moved the organization to works with the strategy of "Phase 3". The central differential of the third phase is to complement prospection activities with general and short-time projects execution. Although being of execution character, they would reinforce prospection. The planned concept is to perform feasibility studies enriched by tests and quick investigations, which will create a robust database to feed the next project proposals and enhance the chances of granting. The structure of Phase 3 has recently gathered the attention of a public funding agency, from which the systematic innovation model was of particular interest. The proposal of a co-funding cooperation model was already approved by the current members of the initiative, and it is in a step of model structuring with the agency. If approved, the contributions of more concrete and immediate results for the companies will be the basis for a sustainable system in the long-term.

4 Conclusion

An industry segment historically characterized by similar technological challenges for widely different applications, suddenly involved with a deep economic crisis, engendered the launch of *Engrena ITA*. The alliance for innovation prospection on gear technologies brought together representatives of the entire production chain, encompassing companies with a supplier-client relationship, but also competitors.

The 15 prospected projects along one year, involving 29 organizations from Brazil and abroad, represent the success of a model to overcome funding limitations for research and innovation. Efforts joined within a concept of open innovation show that technological leadership is an outcome from which organization can assume a more relevant criterion even than the availability of resources.

The initiative has now started a second phase, while plans for a third phase are already drawn. Nonetheless, it is expected that *Engrena ITA* inspires further research groups and companies to form similar initiatives for several other technological matters demanded by society.

Ethics Declaration

Conflict of Interest

The authors declare that they have no conflict of interest.

Compliance with Standards Involving Humans as Subjects

This article does not contain any studies involving human participants performed by the author.

References

1. Schulhouser, R.F.: Drive technology – gears. Capital Mind - Sector Report (2016)
2. Goldstein, M.: State of the gear industry. J. Gear Technol. **35**(1), 2018 (2018)
3. Oliveira, G., Coronato, M.: Como o Brasil entrou, sozinho, na pior crise da história. Revista Época (2016). https://epoca.globo.com/ideias/noticia/2016/04/como-o-brasil-entrou-sozinho-na-pior-crise-da-historia.html. Accessed 24 Feb 2019
4. Mugnatto, S.: Cortes no orçamento de Ciência e Tecnologia podem inviabilizar pesquisas. Câmara dos Deputados (2017). http://www2.camara.leg.br/camaranoticias/noticias/ciencia-e-tecnologia/546380-cortes-no-orcamento-de-ciencia-e-tecnologia-podem-inviabilizar-pesquisas.html. Accessed 24 Feb 2019
5. Escobar, H.: Orçamento de ciência e tecnologia pode encolher ainda mais em 2018. Estadão – Portal do Estado de São Paulo (2017). https://ciencia.estadao.com.br/blogs/herton-escobar/orcamento-de-ciencia-e-tecnologia-pode-encolher-ainda-mais-em-2018/. Accessed 24 Feb 2019
6. The United Kingdom, Department for Business, Energy & Industrial Strategy – DTI: Innovation report: competing in the global economy: the innovation challenge (2003)
7. Bagno, R.B., Salerno, M.S., Dias, A.V.C.: Innovation as a new organizational function: evidence and characterization from large industrial companies in Brazil. Production **27** (2017)
8. Kelley, D.: Adaptation and organizational connectedness in corporate radical innovation programs. J. Prod. Innov. Manag. **26**(5), 487–501 (2009)
9. O'Connor, G.C., Leifer, R., Paulson, A.S., Peters, L.S.: Grabbing Lightning: Building a Capability for Breakthrough Innovation. Jossey-Bass, San Francisco (2008)
10. Chesbrough, H.W.: Open Innovation: The New Imperative for Creating and Profiting from Technology. Harvard Business Press, Boston (2003)

How to Map Employees' Competencies for More Innovative Higher Education Institutions? The Case of a Brazilian Interdisciplinary University

Mauricio Wojslaw(✉), Luciana Pereira, and Júlio Francisco Blumetti Facó

Universidade Federal do ABC – UFABC, Santo André, Brazil
mauricio.wojslaw@ufabc.edu.br

Abstract. A large portion of an innovative organization relies on its human resources. In other words, organizational success depends on the value added by its collaborators [1]. Considering Higher Education Institutions (HEI) is not different. In order to achieve its three contemporary missions-teaching, research, and social and market-oriented and knowledge transfer activities-HEIs need to increase emphasis on the development of back office employees management competencies. Therefore, this study aimed to identify the management competencies considered necessary for HEI to function effectively within the higher education context. In order to understand this phenomenon, we have mapped the competencies of a Brazilian federal higher education institution. As a result, we have proposed a framework that can map HEI core competencies.

Keywords: Organizational knowledge · Competencies management · Public higher education institution · Interdisciplinary

1 Introduction

There is no doubt that a significant parcel of organizations' success comes from their structures and systems, whose essential components are the individuals. In this sense, the interaction between subsystems, people, and organizations enhances or bounds the dynamic capabilities in organizations and, consequently, of its entrepreneur's potential [2].

By observing traditional theories in innovation, considering this phenomenon as favorable to the context of organizations, we have that individuals with entrepreneurial profiles always seek to overcome corporative challenges through organizing different and untested combinations of resources. In this sense, these actors are continually modifying and developing new paradigms considering, making use of technological advancement and adjusting organizational structures in order to improve their production management, thus bringing their organizations to prominent levels [3]. However, for these individuals to achieve these essential abilities focused on assertive changes, which are expected to be in line with the overcoming demand from the competitive environment we regularly experience, it is not enough to have only basic technical knowledge, but specialized

L. Pereira et al. (Eds.): IDEAS 2018, SIST 198, pp. 298–313, 2021.
https://doi.org/10.1007/978-3-030-55374-6_30

skills, such as academic and managerial or technical and behavioral competencies [4], which can lead the organization to an enhanced dynamic capacity in order to overcome the challenges faced in the competitive environment we live in [1].

In common sense, research has suggested that an innovative model in universities, which seek for effectiveness in their missions, should consider the entrepreneurial capacity of managers [5] considering competencies favorable to risk-taking [6], to the support of innovative ideas, and the use of opportunities [7], and also competencies could reinforce leadership and motivation, communication, strategic vision, and planning capacity, as well as decisive assertiveness [2]. Therefore, as pointed by Spendlove [8], the attention to identifying or developing leadership skills among universities needs a more proactive approach, which points to a demand for more profound studies on how to do it.

Based on this gap, this study aims to answer the following question: how to identifying managerial competencies that are essential to enhance the ability of an HEI to fulfill its three contemporary missions - teaching, research and social and market activities and knowledge transfer? To answer this, and taking the literature of entrepreneurial universities as a guide, this article will present a study realized in a Brazilian HEI through which managerial competencies, considered fundamental for the institution to reach its objectives, were mapped. As a result of this study, we also suggest a model that can collaborate with these institutions when mapping and ranking their essential managerial competencies.

2 Modeling an Entrepreneur University Through Managing Competences

Argote and Ingram [9] consider that the transfer of knowledge and technology in an entrepreneurial society only occurs when the experiences of some of its actors influence the behavior and activities of others.

In this perspective, universities emerge as one of the main actors that influence changes. According to Guerrero *et al.* [10], the interest in entrepreneurial universities is being discussed by several authors, however, the sense is also expressed through the capacity of these organizations to serve as both, producers and disseminators of knowledge, thus becoming highly influential in a country's innovative capacity (US Patent and Trade Department - USPTO, Rothwell 2013).

Charles E. Eesley and William F. Miller [4], in the book " Impact: Stanford University's Economic Impact via Innovation and Entrepreneurship " cite Stanford University as an example of a transforming university: *Stanford University has a seductive history of entrepreneurship and technological innovation. For over a century, the university has incubated ideas, educated entrepreneurs, and fostered innovative technologies that have been instrumental to the rise and steady regeneration of Silicon Valley and, at the same time, have contributed to the global economy [4].*

The authors complement the statement by citing examples of some of the world's most highly regarded innovative businesses founded, built, or run by Stanford undergraduates, such as Google, Nike, Cisco, HP, Charles Schwab, Yahoo!, Gap, Netflix, and Tesla. However this context turns up into a question that seeks to explain this success - How does Stanford shape this entrepreneurial success? Besides the fact Stanford is grounded in a very favorable environment to its mission, once it is surrounded by a community engaged in a high technology industry, Eesley and Willian [4], simply and considering the mantra that all research universities perform multidisciplinary research, suggest an answer to this question considering differential features of Stanford. For them, the key for its success resides in its focus on fostering the development of skills which pin the entrepreneurial culture, and this point lies most in the ability of your employees to create an environment that stimulates innovation at all times - both in the classroom and beyond - *Stanford's strategy is to attract the best people to do cutting-edge work and provide an environment that encourages innovation and support the free flow of information [4].*

According to the authors, this model changes the culture in research since it allows the Centers to seek and involve an also multidisciplinary faculty, engaged with academic and managerial competencies or behavioral techniques necessary to bring ideas at an early stage, even if they are considered risky, while providing them with greater access to resources.

In another approach, Seeber *et al.* [11] carried on a survey among academic leaders and principals of twenty-six universities in eight European countries seeking, from their managerial visions, the organizational factors which would enable the university as "a full organization". The result brought three key dimensions - identity, hierarchy, and rationality - directly related to the following competencies:

The identity refers to the symbolic and cognitive profile that differs the organization from others in the same field. It presupposes the ability of reflection and managed, through which an autonomous and protected structure is strengthened from external influences, mainly considering keys decisions [11]; *The hierarchy* comes as a key point in structuring the organization considering coordinating actions, tasks, and responsibilities [12, 13]. It leads to the development of management and strategic skills linked to the profile of the organization [11]; and *The rationality*, as a manner of characterizing organizations as "structures of ends and means", whose collective goals are made possible by the adoption of formal and rational means. In this sense, rationality is adjusted according to the leadership capacity, considering the evaluation of results and control of employee's behavior through the use of power [11].

Guerrero *et al.* [10], emphasize the recognition of several authors on the importance of human capital regarding leadership and autonomy in the process of creating an entrepreneurial university.

Boyett [14], in her work *The public sector entrepreneur - a definition*, presents a research carried out among 106 universities deans in England and Wales. In the research, these leaders identified and ranked, in order of importance, the top 10 attributes of an educational entrepreneur (Table 1). The objective was to reflect the expected competencies of managers who supposedly enhance the potential of these institutions and, consequently, their successful results.

Following this line of reasoning, and regarding the question of this study, it is clear that it is crucial for the entrepreneurial potential of universities that their leaders are fully dedicated to their duties and actively trained in managerial skills, highlighting the ability to create and lead multifunctional teams composed of many types of human capital.

Table 1. The top ten competencies of university managers

Position	Competence
1	Systematic view
2	Ability to allocate resources for quality services
3	Ability to delegate
4	Organization
5	Management of individual and team conflicts
6	Ability to think long term
7	Responsibility and leadership
8	Ability to motivate
9	Ability to select teams
10	Ability to develop teams

Source: Adapted by the author (2017) from Boyett (1993).

2.1 The Dimension of Management in HEIs

Guimarães [15] addresses the dimension of management based on the decision-making capacity of its managers. In this context, and considering an approach from the corporate world, the author points that decision making is guided according to the manager's ability to solve situations, and this results to the success or not of the institution's strategies and missions.

In the case of the IFES, this decision-making capacity is influenced by a tendency towards omission, unconcern and risk aversion [15], probably due to the absence of charging, excessive delegation or even unpreparedness of its leaders.

Regarding this last point, Guimarães [15] presents arguments that suggest that the best managers for HEIs would be those with both academic and administrative knowledge, being this last accessible through institutional practices, what could be obtained by procedural and or gradual manner, translating the importance of learning in the operational context in form and content. In this sense Guimarães [15] rescues the "creative destruction" brought by Schumpeter in 1942 that is based on the integration between the original manager skills and the process of incorporating new knowledge and competencies, making it able to lead the unit in a less insecure way through innovative and unplugged management.

Fact that in the context of the HEIs this is a desired reality, since that not only culturally but above all legally, as of Decree 1.916/96, which regulates the process of choosing the leaders of federal HEIs, only teachers who are members of the Higher Magisterial Career, occupying the positions of Full Professor or Associate Professor 4, or PhDs, are eligible Rectors. That is, the Brazilian universities must have as top leaders, only professionals connected to the Higher Education segment, and since they are actively connected to research and held at the highest academic level. Point that these leaders specialties are often unrelated to administrative competences, but are expected to be acquired through the practice, as coordination of courses, management position, chairs in committee, etc.

In the field of innovation this factor becomes essential, since the federal public universities, as particularly academic structures, must be prepared to act in the role of organizations focused on the generation and diffusion of knowledge, so they must have a leadership that follows the deep and dense changes induced by technological, cultural and social advances [16], brought by the several fields of research and development.

Retracing this tendency of association between academic and administrative competences in the “top management”, Table 2 displays the doctoral area of the Deans of Federal Brazilian HEIs (2018). It is clear the interdisciplinarity between the academic performance of the Deans and the exercise of administrative management. The sample considered the five regions and one university per State of the Federation.

As can be seen from Table 2, none of the pointed universities has as its top leader a specialist in the area of public administration or management, which would undoubtedly provide fundamental skills to enhance the exercise of its duties. Regarding that, only two universities in the northern region, specifically the Federal University of Amapá and the Federal University of Rondônia, have their leaders in the areas of Social Sciences and Human Sciences respectively, what brings them closer to a formation more consistent with typical attributions of managers. Still, in the northern region, other universities have their rectors acting academically in economics and experimental psychology, areas that address competencies favorable to the exercise of managerial functions, although in indirect relations.

In the South region, it is possible to observe specialists in the area of law and education, areas also favorable to the various technical skills required for the management exercise, but also this formation does not fully provide important managerial competencies.

In the other regions there is a clear separation between the academic formations/specialties of the managers and the field of the administration, suggesting that, in general and due to this scenario, there is a need for continuous training of the managers, since the demands by specific competences to perform their positions follow the evolution of work systems and organizational strategies [17].

Regarding these new mechanisms, the university management starts to look for the strategic approach, but no longer the bureaucratic one [18]. The decision-making process begins to observe data gathered from university relations not only with its students, but with society as a whole, its technical-administrative staff and teachers [19].

In this context, it is noted that the profile of management of HEIs has followed a trend towards the need to overcome traditional models through new concepts and innovations on hierarchical relations, design of tasks, information flow, rules and work rules, team dynamic, power relations and interaction with society, which should be guided by the interdisciplinarity, which in turn demands compatibility of skills and management and academic potential to factors that dictate their specificities and differences, so that these latter ones must always be focused on a fundamental goal of HEI, which is to generate benefits and values to the social ecosystem which surrounds, legitimizes and maintains them [20].

Table 2. Doctoral area of the Deans of Federal Brazilian HEIs (2018)

Region	State	IFES	Education Doctoral degree/Specialty
Midwest	Goiás	UFG	Genetics and plant breeding
	Federal District	Unb	Geology
	Mato Grosso	UFMT	Food and nutrition
	Mato Grosso do Sul	UFMS	Software engineering
Northeast	Alagoas	UFAL	Social service
	Bahia	UFBA	Philosophy
	Ceará	UFC	Medicine (nephrology)
	Maranhão	UFMA	Geography - spatial information analysis
	Paraíba	UFPB	Natural and synthetic bioactive
	Pernambuco	UFPE	Transportation
	Piauí	UFPI	Organic chemistry
	Rio Grande do Norte	UFRN	Education
	Sergipe	UFS	Pharmacology
North	Acre	UFAC	Nursing
	Amapá	UNIFAP	Social sciences
	Amazonas	UFAM	economy
	Pará	UFPA	Experimental psychology
	Rondônia	JOIN	Humanities and social sciences
	Roraima	UFRR	Agronomy
	Tocantins	UFT	Geography
Southeast	Espírito Santo	UFES	Physics
	Minas Gerais	UFMG	Letters
	Rio de Janeiro	UFRJ	Education
	São Paulo	UNIFESP	Pharmacology
South	Paraná	UFPR	Right
	Rio Grande do Sul	FURGS	Education
	Santa Catarina	UFSC	Right

Source: Prepared by the Author from data andifes.org.br (2017)

3 Research Design

As brought by the introduction of this work, to answer the question laying on the challenge to identify core competencies which are able to bring up management to a effectiveness, this study considered the Federal University of ABC as the research environment regarding its effort to construct a Competency Map focused in gathering managerial competencies according to identity, hierarchy, and rationality of the organization itself [11], fundamental elements to the attainment of its objectives.

The structuring of competences in this HEI can be summarized into two stages:

I) **Competency Mapping** - this project consisted of various sub-steps and resulted in indicators which would compose 4 (four) management dimensions to be used in the elaboration of a framework aimed at identifying core competencies and also at the development of a training plan based on the real gaps of competencies. Its focus was detailing tasks and attributions of the various players invested in each role that composes the management of the organization, besides raising the respective essential competencies to its achievement. For this mapping, a standardized survey containing conceptual elements was elaborated and applied to 355 administrative functions of the entire university, including Coordination and Direction posts - CDs, Gratified Functions - FGs as well as support functions. In this same stage, a survey was conducted along with the leaders of the organization regarding the Institutional Behavioral Competencies.

II) **Institutionalization of the Referential Framework of Management Positions** - this project consisted of 3 stages: the definition of core competencies, the creation of a competency leveling criteria, and finally its application. This stage resulted in the University Matrix of Competencies for Functions and Positions adjusted to the superior hierarchical and also the support levels.

During the sub-steps of the mapping, specifically those used to consolidate the description of job positions and functions, 4 management dimensions used to build up the *framework* were pointed out and weighted by top management as being priorities and adherent to the organization's context: the decision-making capacity, leadership, complexity of duties and strategic profile of functions.

Decision Level - According to Stoner and Freeman [21], the decision-making process involves the identification of a specific problem and the choice of action to solve it. Mañas [22] supports decision-making as a conscious way to guide a course of action, among the available alternatives, in order to reach the desired result. Mintzberg (2010), as cited by Faraco *et al.* [23], states that decision-making under conditions of uncertainty is one of the most important skills for a manager. LaValle *et al.* [24] argue that organizations which take decisions based on rigorous analyzes demonstrate twice as much performance and are better oriented towards both future strategies and day-to-day operations. Thus, the Decision-Making Level of the position represents the power and decision-making capacity granted to a position/function so that its agent fulfills its responsibilities and attributions in an assertive manner and lines with the organization's strategies, being then one of the dimensions considered of high impact on management.

Leadership - Although it has been the focus of many studies, there is no universal conclusion about the concept of leadership as well as the behavior and style of the leader [25–27]. For this dimension, with the objective of standardizing concepts in order to maintain the same line of thought and criteria for the study, the concept of "leadership" was used as the ability of an agent A to interfere in the choice and opinion of an agent B.

In other words, it can be considered as a process of influence and confidence of a person on the way an individual or group of individuals conduct their efforts to establish and achieve goals [28]. In this context, taking into account the need for symmetry regarding the alignment between leaders and subordinates, and also considering that the leadership exercise demonstrates the potential of influence that each hierarchical level exerts on the organization as it controls the operational, tactical, or strategical administrative levels, this dimension was also pointed out as having a very high impact for structuring the management of organization.

Complexity of Assignments - The complexity of assignments is directly related to the level of value-added and the management of competency-based development [29], thus representing a fundamental element for organizational management. Lawrence and Lorsch [30] dealt with complexity through what they called differentiation, that is, differences in the work styles and structure of an organization that directly influence the degree of governance over its activities. Hall [31] states that complexity is represented by three elements: the horizontal differentiation or specialization, the vertical ones, and the spatial dispersion. The first one consists in the subdivision of a complex task into small tasks - in which case the management can assess its available competencies and model actions according to division or agglutination of tasks. The second represents the number of hierarchical levels in the organization that divide top management from operational support levels - this variable then represents the size and complexity of management. Finally, the third element or spatial dispersion refers, according to the author, to the number of spatially separated places in which members of the organization are employed, thus constituting a fundamental element for organizational architecture.

Strategic Profile of Attributions - According to Hox e Candea when quoting Anthony [32], "strategic assignments" focus on tactical and strategic levels, and demand differentiated performance once they extrapolate the operational activities common to mere routine. In other words, the strategic profile of a position/function is defined while its main attributions focuses on improving management processes and/or the overall performance of the area.

The Three Levels of Attributions, According to Antony (1965): **Operational level** - means the effective and efficient use of existing facilities and all resources to perform operations. Operational control uses pre-established decision-making procedures and rules, with generally very stable procedures, and their actions usually result in an immediate response; **tactical level** – relates to the generic acquisition of resources and the respective tactics for the acquisition, localization of projects and new products. They are usually related to administrative control. At this level, information about the planned operation (norms, expectations, assumptions), its variations and respective analysis of possibilities along decision in the course are necessary; **strategic level** - involves the definition of objectives, policies and general criteria for planning the course of the organization. Activities at this level do not have a uniform cycle period - some strategic plans are made within annual or pre-established planning.

4 The Case Study of the Federal University of ABC

This work welcomes the case of the Federal University of ABC (UFABC) considering its profile of a multidisciplinary organization emerged from "the new public administration" and that "started from scratch.

These features brought the university a differentiated and innovative proposal as well as infinite possibilities and freedom to construct a new model of thinking public management. However, this scenario also brought the institution the challenge of proposing solutions that foster a managerial ecosystem to maximize the delivery of quality services.

Based in this context the university sought alternatives for its management structuring under the understanding that it would be necessary a detailed study and planning which would lead the administration in the creation of an organizational model linked to the necessary competencies for the university to fulfill its institutional role.

In this sense, in 2010, using benchmarking with the other federal HEIs, the administration proposed a model to reference the commissioned positions for the university. The model was presented for the administration of UFABC together with organizational chart templates to base areas for self-structuring. However, after a first exercise, the results showed that it was very challenging for the administration to create a basic structural format suitable to its needs, or even to match the number of positions and functions available to the university. The main reason detected for this difficulty was that the benchmarking methodology brought a standard structural universe, but far from the specificities necessary for the model of this university.

For this reason, the HR area proposed to the university's administration a work aimed ar improving the organizational structuring based on managerial competencies. The model would focus on the four dimensions of management. This work, which was carried out by all employees involved in university administration, resulted in a Matrix of Competencies and Positions.

The matrix was elaborated after mapping organizational competencies, as well as the individual technical and behavioral competencies demanded by all the 355 administrative functions existing in the university, that were then evaluated under the scope of the four dimensions of management identified as priority for the development of successful institutional governance: decision-making level, leadership, complexity of assignments, and strategic profile of attributions.

The result of this process resulted in a competence-based framework to be suggested to the public administration, especially federal HEIs, which suggests a mechanism for modernization, innovation and optimization of human and administrative resources, as well as a methodology for the strategic composition to be used in the organizational architecture of the university.

4.1 The Mapping

A Working Group on Competency Mapping was set up to make the mapping process under a more participative and legitimate scope while most activities of the process would be carried out by the own actors of the management structure.

In order to collect the data, a qualitative questionnaire was applied by the group to all the collaborators of the administrative structure. For didactic purposes this questionnaire was called "Functional Descriptive", and so applied to 576 administrative workers, including all 185 commissioned positions from the university board of administration. The sample represented 40% of the university's total servers and 75% of the management functions (25% unmapped refers to academic management functions). Anonymization techniques have been applied during data analysis process.

The questionnaire contained standardized elements that maintained the homogeneity of concepts and was elaborated considering the need to survey the constitutive requirements of the management architecture. Its focus was the detailing tasks and attributions of the actors considering the four dimensions of the management used to elaborate this model, which involve the respective levels of complexity and strategic profile, their decision prerogatives, and leadership levels. All the functional descriptions were validated in face-to-face interviews with the responsible leaders.

4.2 Leveling Job Positions

Each position in an organizational structure has a hierarchical level corresponding to the influence its inherent competences exert on the organization. In Federal HEIs, some positions such as Dean and Deputy Dean, Pro-rectors and Directors of Academic Centers are called elective or approved, so the purpose of this methodology is to suggest an appropriate profile for the institutionalized posts, that is, those in hierarchical sequence that was created according to the necessity and specificity of the organization. So, the elective or approved positions were mapped exclusively with the objective of validating the methodology and reference the post of responsibility for the university's strategic positions.

In order to subsidize the positioning of all working posts of the management role, the functional descriptions mapped were first validated by its upper leader. In such phase, objective and precise information about each position has been thoroughly evaluated and considered in determining the level of influence that each position imposes upon organization management, as well as the necessary and available conditions and competencies for them to achieve their goals. This level of influence was nominated Function Positioning Quotient (FPQ) for methodological purposes, and it was instituted observing four leveling indexes (LI) synthesized from the central management dimensions considering the viewpoint of the university: 1 - LI by Decisive Level; 2 - LI by Leadership Exercise; 3 - LI per Complexity of function; and 4 - LI per Strategic Profile of Assignments. It should be noted that the stage of organization and leveling of functions was carried out exclusively by the HR area in order to maintain the standardization of the works, the convention of concepts and the preservation of impartiality concerning the results.

4.3 Weighting Competences

LI By Decisive Level: During the validation process, the following decision levels of each position in management role were agreed by the managers [33]: *Prescriptive* – Absence of autonomy to decide duties. It is controlled directly by the superior; *Routine* – Decisions correspond to routine activities. Receives superior supervision; *Limited* - Allowed to decide according to norms and procedures of the institution. Receive periodic supervision from superior; *Normative* - Makes decisions that define norms and procedures of work. It is limited by the functional policies brought by superior directions; *Functional* - Makes decisions that determine functional policies for the area. It is limited by the General Policy of the Institution, but it can also participate in its definition; *Generic* - Makes decisions defining the institution's ultimate goals and strategies. It formulates general policy and is limited by the decisions of the High Councils; *Free* – Owns wide open decisions, limited only by organization priorities, norms, and public interest. The LI by Decisive Level were converted into standardized indexes from 0 to 5, according to Table 3.

Table 3. Assignment of value to IN by decisive level.

Decision level	IN by decisional level
Prescriptive	0
Routine	0
Limited	1
Normative	2
Functional	3
Generic	4
Free	5

Source: Author (2018)

LI By Leadership Exercise: It demonstrates the potential influence of each managerial position on the organization as it controls the operational, tactical, or strategic administrative levels. Therefore, the LI by Leadership Exercise does not refer to the number of servants led by each function, but rather to the number of hierarchical levels over each position. At this stage, the denominations for job positions were not considered due to their "de-ordination", and the numbers from 1 to 4 levels were adopted considering the university's organizational macro structure bias, i.e., Dean, Pro-Rector and Director of Center, Coordinator, Head of Division, head of Section. Table 4 demonstrates the allocation of lanes considering the LI of Leadership Exercise.

Table 4. Allocation of values to the LI of leadership exercise.

Number of levels led	IN leadership exercise
1	1
2	2
3	3
4	4

Source: Author (2018)

LI per Complexity of Function: During the validation process, this indicator was extracted from the analysis of the Impact and Difficulty levels related to the performance of each job position in the university. For analytical equity, the following definitions have been agreed [33]:

Impact of Function – it can be low, medium or high, and represents the influence that "errors and hits" in the performance of functional assignments entails over the organization and community. At this stage, the evaluators were oriented always to consider a direct chain of events, that is, actions and consequences directly linked to those responsible and the intended objectives. In this case, in the execution of the assignments, if the impact of the actions is limited to the sector or the image of the server, it is considered LOW. That is the typical case of job positions that necessarily submit all their actions to the higher appreciation, having as the only evaluation body the action restricted to the area itself; If the impact of actions is limited to the organization internally, it is considered MEDIUM. In this case, the results of the actions extrapolate the area itself and reach other units of the organization, but with no external repercussions. In the chain of events, this is a typical, but not exclusive, level of impact of leadership; If the impact of actions extrapolates the institution's gates, directly evidencing its image, it is considered HIGH. Generally linked to actions related to the product that the organization intends to provide to the community.

Difficulty of Function [33] – it can be low, medium or high, and represents the availability of the needed elements so the server can execute its assignments. In this case, in the execution of its assignments: if the resources or procedures are fully available so the server can easily make use of it, the difficulty level is considered LOW. That is the case for activities involving mechanical or automatic procedures, called routine activities; If the resources exist, however, the procedures are not available to the server, demanding some application of its competencies, in full or part, the level of difficulty is considered MEDIUM. This level of difficulty is typical of activities that require a process of analysis and/or elaboration by the agent; If resources and procedures are not available, requiring the server to use its competencies to the extreme, including the need for breaking and/or deploying new paradigms in the institution, the level of difficulty is considered HIGH. It is a typical decision-making position on high impact matters. Indexes from 1 to 5

Table 5. Evaluation of complexity level.

Impact	Difficulty	Complexity
Low	Low	1
Low	Medium	2
Medium	Medium	3
Medium	High	4
High	High	5

Source: Author (2018)

(from lowest to highest) were agreed to represent the ratio "impact X complexity" of each activity, considering the combination as Table 5:

LI per Strategic Profile of Assignment: Considering the common characteristics of assignments. *Operational level* - complexity level reduced to 2 or less; *Tactical level* - intermediate level of complexity, or equal to 3; *Strategic level* - high complexity level up to 4 or greater. In fact, the levels of impact and/or difficulty of the strategic assignments increase their LI by Complexity to 3, 4 or 5 gradually demonstrating the elevation of the strategic tendency of the job position. On the other hand, assignments with low impact, and/or difficulty levels with IN per Complexity reduced to 1 or 2 are considered activities of operational level or mere routine, it means, without strategic tendency, bringing the LI to zero or null. The values of the LI indicators by Strategic Profile of Assignments, demonstrating the strategic trend of the job position in the management role, are arranged in Table 6:

Table 6. Assignment of values to IN by strategic profile.

Mode of the level of complexity of post/function assignments	IN by strategic profile of the assignments
1	0
2	0
3	1
4	2
5	3

Source: Author (2018)

4.4 Map of Institutional Competences

During the process of the Functional Descriptive Validation, each job position in the management role was weighted within the Leveling Indexes considering its attributions and technical and behavioral management competences.

This process constituted the Map of Institutional Competences; within this map each validated position was grouped in its respective management position considering the situational organizational structure; from this grouping it was observed the mode on the FPQ of each existing hierarchical level; Finally the set of FPQ modes determined the level of responsibility of the respective position at each hierarchical level in the management structure. From this sequence it was possible to identify an institutional-grade for each managerial positions already practiced in the structure, from where a Matrix of Competencies and Positions was elaborated, constituting the institutional panorama of all the positions evolved in the management of the university, arranged in each level of complexity standardized in the organization. This procedure was also worth to validate the method, once a parallel study made it possible to compare the results brought by the model with the management positions already established by best practices.

5 Results

During the mapping process, it was notable that most areas in the university were playing volatile organizational structures pointed to situational needs.

Considering a general view, there was no such structural organization or working plan which could favor the connection between the academic and administrative performance. As one consequence of this, solutions tended to be drawn based on moment resources, once gaps in identifying and working out appropriate technical, behavioral and managerial skills prevailed, and this is the sort of problem which leads to the question of this paper.

In order to suggest a tool to solve these points, the Institutional Map of Competencies and the UFABC Matrix of Competencies and Positions were created, and during the process to establish them, several other welcomed results were raised, as well as the reorganization and optimization of job positions, the adjustments in the hierarchical structure, delegations and delimitation of competencies, redistribution of activities and duties between areas, the identification of gaps for the development of management training plans, and the principally the improvement of the connection between academic and administrative missions.

Concerning the people of the organization, after mapping the functions and associating them to their profile, it was possible to stipulate their levels of competence to perform their functions as well as other eventual needs of the organization, besides subsidizing development plans based on the gaps of competencies found.

6 Conclusion

The question of this study settles on finding a way to identifying managerial competencies in order to push up the university potential to effectively reach its contemporary mission.

Regarding this and as a first point, the process of mapping suggested that a successful university management demands different skills from other contexts, which involves additional competencies, especially regarding to the ability to link academic and administrative interests, bringing up the policy- making as a very special skill to its management.

The Institutional Map of Competencies and the *Matrix of Competencies and Positions* here suggested can, apparently, help the administration to refer actions for its functional arrangement in a systematic way considering both, academic and administrative bodies, and also bring the organization to a dynamic and continuous manner to reorganize its needs for specific skills, showing then to be efficient tools to identifying essential managerial competencies to empower university to accomplish its mission.

At this point, this paper mainly suggests that the competency mapping considering customized leveling indexes can be an efficient tool in modeling the management skills and the organizational structure to benefit common academic and administrative interest, empowering university effort and consequently to being a great ally of public policies.

However, the process has shown that the success of this framework is directly connected to the support of senior management bodies. In this sense, this study concluded that in addition to any methodology aimed at empowering management structuring, it is imperative that complementary studies focus on the influence of political power on the organizational structure of HEIs, since these are positions called "in trust" which are commonly nominated considering competences as background, what adds a debate on whether HEIs' leaders should be academics, politicians or professional business administrators.

Ethics Declaration

Conflict of Interest. The authors declare that they have no conflict of interest.

Compliance of with Standards Involving Humans as Subjects. All procedures performed in studies involving human participants were in accordance with the ethical standards of the institutional and/or national research committee and with the 1964 Helsinki Declaration and its later amendments or comparable ethical standards. Informed consent was obtained from all individual participants involved in the study.

References

1. Lytras, M.D., de Pablos, P.O.: Competencies and human resource management: implications for organizational competitive advantage. J. Knowl. Manag. **12**(6), 48–53 (2008)
2. Ferigotti, C., Fernandes, B.: Management competencies and capacity for innovation: the case of electrolux do Brasil S/A. J. Manag. Innov. **11**(5) (2014)
3. Schumpeter, J.A.: The Theory of Economic Development: An Inquiry into Profits, Capital, Credit, Interest, and the Business Cycle. Routledge, New York (2017)
4. Eesley, C.E., Miller, W.F.: Impact: Stanford University's economic impact via innovation and entrepreneurship. Found. Trends® Entrepreneurship **14**(2), 130–278 (2018)
5. Mazzucato, M.: The Entrepreneurial State: Unmasking the Public Sector vs. Private Sector. Portfolio Penguin, São Paulo (2014)
6. Guterres, A.: Taza Koom Conference About Public Service, Bishkek, Quirguistão (2017). http://nacoesunidas.org/secretario-geral-da-onu-destaca-importancia-de-servicos-publicos-para-desenvolvimento-sustentavel/
7. Amdam, R.: An integrated planning, learning and innovation system in the decentralized public sector: a Norwegian perspective. Public Sect. Innov. J. **19**(3), 1–16 (2014)
8. Spendlove, M.: Competencies for effective leadership in higher education. Int. J. Educ. Manag. **21**(5), 407–417 (2007)
9. Argote and Ingram (2000)
10. Gerrero, M., Urbano, D., Cunnigham, J., Organ, D.: Entrepreneurial universities in two European regions: a case study comparison. J. Technol. Transf. **39**, 415–434 (2014)
11. Seeber, M., et al.: European universities as complete organizations? Understanding identity, hierarchy and rationality in public organizations. Public Manag. Rev. **17**(10), 1444–1474 (2014)
12. Thompson, V.A.: Bureaucracy and innovation. Adm. Sci. Q. **10**(1), 1–20 (1965). Special Issue on Professionals in Organizations
13. Mintzberg, H.: Creating effective organizations: structures in five configurations. Translation: Cyrus Bernardes. Atlas, São Paulo (1995)
14. Boyett, I.: Int. J. Public Sect. Manag. **9**(2), 36–51 (1996)
15. Guimarães, J.C.: Management in IFES: the reasons that determine the doctor as a manager – the approach under the perspective of individual skills. Rio Grande do Norte **5**(2) (2013)

16. Vieira, Vieira: Organizational structure and performance management in Brazilian federal universities. RAP – J. Public Adm. (2003)
17. Zarifian, P.: The Competency Model: Historical Trajectory, Current Challenges and Proposals, 2nd edn. SENAC São Paulo (2011)
18. Júnior, F., et al.: Restructuring the Brazilian public university model to meet the new managerial challenges. In: XV International Conference on University Management – CIGU, Challenges of University Management in the 21st Century, Mar del Plata, Argentina, December 2015
19. Borges, M.C.: Functions of trust "*stricto sensu*" and functions of trust "in commission": an approach constitutionally adequate. TCEMG Magazine Minas Gerais **82**(1), 45–54 (2012)
20. Amarante, J.M., Crubellate, J.M., Júnior, V.M.: Strategies in universities: a comparative analysis from the institutional perspective. University Management in Latin America – GUAL, Florianópolis (2017)
21. Jamil, G.L.: Aspects of the managerial environment and its impacts on the use of competitive intelligence systems for decision-making processes. Perspect. Inf. Sci. **6**(2) (2001)
22. Mañas, A.V.: Administration of Information Systems: How To Optimize the Company Through Information Systems, 3rd edn. Érica, São Paulo (2002)
23. Faraco, M.M., Lavarda, R.A.B., Glebcke, F.L.: Net of rational and intuitive aspects in the process of strategic decision making: multiple case study. In: XX SEMEAD Seminars in Administration (2017)
24. LaValle, S., Lesser, E., Shockley, R., Hopkins, M.S., Kruschwitz, N.: Big data, analytics and the path from insights to value. MIT Sloan Manag. Rev. **52**, 21–32 (2011)
25. Burns, J.M.: Leadership. Harper & Row Publisher, New York (1978)
26. Bowditch, J., Buono, A.: Elements of Organizational Behavior. Pioneira Thomson, São Paulo (2002)
27. Robbins, S.P.: Administration. Saraiva, São Paulo (2005)
28. Fairhurst, G.T.: The Power of Framing: Creating the Language of Leadership. Jossey-Bass, San Francisco (2011)
29. Dutra, J.: Management of people in contemporary public organization. Corporate School – Fiocruz, Rio de Janeiro (2016)
30. Lawrence, P.R., Lorsch, J.W.: Business and the Environment: Administrative Differentiation and Integration. Editora Vozes, Petrópolis (1973)
31. Hall, R.H.: Organizations: Structure and Processes, 3rd edn. Prentice Hall, Rio de Janeiro (1984)
32. Anthony, R.N.: Planning and Control Systems: A Framework for Analysis. Harvard University Press, Cambridge (1965). Mass, apud in Hox & Candea (1984)
33. Leme, R.: Rediscovering the Matrix Nine in Box, 1st edn. Qualitymark, Rio de Janeiro (2013)

A Case Study on Hackathon # Develops and a Hackathons Contribution to Innovation

Deise Carolina de Souza Silva(✉), Clarissa Stefani Teixeira, and Jatyr Ranzolin Junior

Federal University of Santa Catarina – UFSC, Florianópolis, SC, Brazil
deisekarolinass@gmail.com

Abstract. In recent years there has been a significant increase in hackathons in the Brazilian public sector, especially between the period from 2012 to 2016, where more than 51 hackathons have been identified. The present study sought to analyze the potential of these events for innovation within the public sector and carried out a research with the managers of the case Hackathon #DesenvolveSC, which was the first hackathon of the Government of Santa Catarina - Brazil. Thus, five government officials were interviewed in order to identify their perceptions regarding the practices of hackathon and changes occurred in the public organization. Santa Catarina began to perceive the need to promote events that provoked the interaction of the state with the citizen and the triple helix, and that through collaborative environments, exchange of experiences, sharing of resources and creation of ideas, encourages solutions to be to fill gaps in the public sector. This process of interaction and collaboration where there is an exchange of information between the internal environment that is the public sector and the external environment is where open innovation occurs. Hackathon is not just a marathon of technological artifact competition, but it is also an open innovation tool perceived by the servers. In addition, the main results indicate the interaction potential that events such as hackathon provide in the ecosystem, not only in connection and support with other state folders, but also with universities and private companies that accompany the marathon in its execution.

Keywords: Hackathon · Hackathon in the public sector · Open innovation and hackathon

1 Introduction

In today's information-rich environment, companies can no longer rely entirely on their own ideas to advance their business, nor can they restrict their innovations to a single path to the market. The traditional model of innovation - which has been largely internally focused, closed off from outside ideas and technologies - is becoming obsolete. Emerging in its place is a new paradigm, "open innovation", which strategically harnesses internal and external sources of ideas and leads them to the marketplace in multiple ways [1].

Citizen sourcing strategies, such as hackathons, have been increasing in the Brazilian public sector. Hackathons have a clear focus on inviting citizens to participate in problem

L. Pereira et al. (Eds.): IDEAS 2018, SIST 198, pp. 314–321, 2021.
https://doi.org/10.1007/978-3-030-55374-6_31

solving and building solutions that can benefit society. Hackathons are a strategy for open innovation. However, its format still presents weaknesses in relation to the costs involved in its organization and the sustainability of the solutions created [2].

In Brazil, hackathons events within the public sector have been growing in recent years. States have sought to find solutions to the problems surrounding them using hackathons as a tool. In the period from 2012 to 2016 51 hackathons were identified in the Brazilian public sector, most of which were carried out at municipal level, orchestrated through prefectures[3]. However, some initiatives at the state level were located and, in this context, the present article sought to analyze the potential of actions in stimulating innovation through a collaborative environment between government, citizen, university and company under the focus of the Hackathon case #DesenvolveSC.

2 Hackathon and Its Characteristics

The hackathon concept came about when independent developers created a cryptographic development event held in Calgary on June 4, 1999. A small number of developers came together to avoid legal issues arising from the United States cryptographic software export regulations. The word hackathon is combined with the words hack and marathon, where the hack is used in the exploratory sense and investigation of programming (not as a reference to cybercrime) and "marathon" and means a long run to create something useful in a single event [4].

The initiatives present similar characteristics as to the format of the event, opening lectures are held and the teams that will participate in the marathon are formed. Teams are evaluated at the end of the event and rewarded with financial and/or non-financial rewards (travel, "mini-internships", courses, mentoring and incubation programs, etc.). However, in Brazil these values are still differentiated in each case[2].

One of the reasons for the popularity of hackathons is its symbiotic or collaborative nature where everyone involved benefits. These benefits are associated not only with the participants, but also with the organizers and sponsors. Many hackathons are geared toward college students and the event provides an opportunity to learn aspects of software development that are difficult to teach as part of their regular classes. Thus, in some hackathons, students have free access to software and hardware considered to be expensive or for products that have not yet been released [5].

2.1 Hackathons in the Open Innovation Context

Open innovation according to Almiral et al. [6] will succeed when the needs of the entire ecosystem and supporters are organized in order to promote competition and collaboration. In the meantime, Ferreira [2] states that the promotion of hackathons is used in the context of open innovation, analyzing the impact perceived by the end customer of the product or service. For the same authors, practices tend to be positive and the procedure helps to remove barriers and accelerate ideas.

The relaxed organizational structure of hackathons events encourages participants to innovate and create an environment that can sustain innovation (that is, it can manage the failure needed for innovation to emerge). In addition, the relative ease of hosting such events allowed many events to be realized, therefore, for a variety of experience and skills to be applied within the reach of different cultures. Likewise, the ease with which they can be maintained and the diversity of events that have emerged make them challenging. It is estimated that there is an average hackathon per week in London, with participants from different agencies and departments [4].

In the public sector, according to [3], governments seek support from citizens to provide assistance in solving problems. Such a practice is called by some authors of citizen sourcing, which functions as a resource to collaborate and find answers to the difficulties encountered in the public sector in an efficient manner. From this, one of the initiatives to contemplate this scenario of cooperation between government and citizen has been occurring in the public spheres through hackathons.

Contests that involve greater citizen involvement in solving a problem, such as hackathon, may be important to enable a greater level of open innovation in the public sector. In this way, it can be said that the public agent plays an important role in promoting open innovation initiatives, and it is necessary to discuss their perceptions about how citizens' initiatives can promote open innovation in the public sector. However, the hackathon, as a citizens' initiative for open innovation, is still a little studied by the academy considering the current interest of the government in the initiative, mainly in Brazil [2].

3 Methodology

The methodology applied to conceptualize hackathon was carried out through a bibliographic review [7] with the keyword "hackathon" in international databases and was also used Google Scholar and Portal Capes to search the scientific information about hackathons performed in national territory. A bibliographic survey on hackathon in the Brazilian public sector and the relationship of hackathon with open innovation was also carried out.

In order to carry out the study of the case study, the two editions Hackathon #DesenvolveSC of the Government of the State of Santa Catarina - Brazil, carried out through the Center for Information Technology and Automation of the State of Santa Catarina (CIASC), which is a company founded more than 40 years ago. In the year 2016, CIASC launched the first edition of Hackathon #DesenvolveSC, with 120 participants and in 2017 launched the second edition of Hackathon #DesenvolveSC with the involvement of more than 150 participants. The events cited present a methodology from the Federal University of Santa Catarina that conducts open innovation events for governing bodies in the State of Santa Catarina [8]. For the data collection, a questionnaire was applied with the internal committee of the CIASC, responsible for the promotion of the Hackathon #DesenvolveSC. The questions that led to the pursuit of knowledge about employee perceptions were: 1) Do you consider Hackathon #DesenvolveSC to be an open innovation method? 2) Describe the benefits of Hackathon #DesenvolveSC for the CIASC, and if there has been any significant change to the company; 3) Describe what in your view did not bring a positive result and what needs to be reviewed or improved in the Hackathon method #DesenvolveSC. Thus, five questionnaires were applied with the strategic leaders who participated in the constitution of the two editions of Hackathon

#DesenvolveSC. The questionnaire was sent via email and prepared using the Google Forms tool. The answers were analyzed qualitatively and interpretatively and used for the elaboration of this article.

4 Hackathon #DesenvolveSC: The First Hackathon of the Government of the State of Santa Catarina

Hackathon #DesenvolveSC was the first hackathon carried out in the state of Santa Catarina, through the Information and Automation Center of the State of Santa Catarina (CIASC), a public company responsible for the execution of the company's policies, management and technology services, information and electronic governance of the organs and entities of the State Public Administration, direct and indirect [8].

According to information contained in the official site of the event, official government source, the first Hackathon #DesenvolveSC, was held in September 2016, with more than 41 (forty one) participants, 10 (ten) multidisciplinary teams formed, with the participation of 23 mentors (business, developer and usability), five speakers, one of them international. The competition aimed to strengthen the relationship between the public sector and the citizen through projects in the areas of economic development, sustainability and innovation. The three best works received the award in the amount of R $ 5 thousand each. The hackathon was carried out by the CIASC in partnership with the Federal University of Santa Catarina (UFSC), with support from the State Secretariat for Sustainable Economic Development (SDS) and the Santa Catarina Foundation for Research and Innovation (FAPESC) [9].

The second edition of Hackathon #DesevolveSC took place in August 2017 and counted on 150 participants, among students, mentors and speakers and aimed to promote the development of technological solutions mobile, web and internet of things (IoT), with a focus on public safety. The 2nd Hackathon #DesenvolveSC was held in partnership with the State Department of Public Security (SSP) and Federal University of Santa Catarina (UFSC) [8]. Information on 2016 and 2017 events can be seen in Table 1.

According to information contained in the official website of the event Hackathon #DesenvolveSC, it was possible to observe that both editions sought to promote an environment of innovation and collaboration between government, university and professional market practices, stimulating the creation of technological solutions for the use of the state and of the citizen. In both events, there was connection of the triple propeller where members of government, universities and companies could act as mentors of the solutions proposed during the event [10].

The results obtained in the questionnaire applied with CIASC employees, directly involved in the organization of the event, are shown in this Table 2:

In the analysis extracted from the form, 95% of the respondents answered that they consider hackathon an open innovation method. Authors [6] consider that hackathon practices are carried out in government settings, such as in Amsterdam, Barcelona, Boston, Helsinki, New York and Philadelphia. These practices are understood as being of open innovation since they present the real problems faced by public institutions. In the case of the state of Santa Catarina, two institutions were involved in opening up the problems faced at the governmental level, being the State Secretariat for Sustainable Economic Development and the State Secretariat of Public Security.

Table 1. Hackathons numbers made in the years 2016 and 2017.

Year	2016	2017
Entity receiving the event	Center of Informatics and Automation of the State of Santa Catarina	Center of Informatics and Automation of the State of Santa Catarina
Problem solver	Secretariat of State for Sustainable Economic Development	Secretary of State for Public Security
Problems	Environment Economic development Innovation in citizen's relationship with government	Public security
Number of participants	41	47
Number of teams	10	10
Number of mentors	23	35
Number of speakers	05	07

Source: Prepared by the authors.

Table 2. Results of the applied questionnaire

1) Do you consider Hackathon #DesenvolveSC method of open innovation?	2) Describe the benefits of Hackathon #DesenvolveSC for the CIASC, and if there have been any significant changes to the company	3) Describe what in your view did not bring a positive result and what needs to be reviewed or improved in the Hackathon method #DesenvolveSC
YES = 95% NOT - 5%	Respondents answered that the benefit was not the creation of a new product, because there was no sustainability to continue the development of the prototypes. For the employees the reports are for the gain associated with the interaction of the triple propeller and the awakening to a new culture of innovation within the company	Respondents answered that the public company was unable to absorb the ideas generated in the event and turn into business and that the profile of the participants of the event was not an audience with an entrepreneurial vision, did not envisage a business opportunity and therefore had no interest in developing the solution and post-event method must be forecasted so that hackathon is not only an encouragement of ideas, but the transformation of good ideas into solutions

Source: Prepared by the authors.

[2] also considers that one of the goals of the hackathon marathon is the very achievement of innovation in organizations. In the case of the practices of Santa Catarina it is possible to identify that the greatest changes were not in the ambit of the Secretariat that opened the problems, but in the receiving entity. This is mainly due to the interactions carried out before and during the event with the team of mentors and the team that holds the methodology at the university.

However, other benefits are still found. A study such as [2] considers that hackathon practices were associated with: i) an increase in the number of new businesses in the region; ii) greater appropriation and openness of data of the public organization; and iii) changes in the perception of the organization's employee innovation culture. In terms of the hackathon analysis of Santa Catarina, employees of the CIASC (organization hosting the event) perceive that the action at the state level did not bring a resolution of the problems, due to the difficulty of continuing and developing post-event solutions. In this context, [2] already indicates the difficulties encountered in terms of sustainability in the development and execution of solutions. According to the same author, the government has difficulties absorbing the initiatives developed in hackathons, mainly related to the necessary contracting procedures after the development of the solutions, emphasized by the interviewees as "bureaucracy".

Both editions of the hackathon were carried out in partnerships with secretaries of state who are clients of the CIASC and had as objective to validate the technological solutions with their clients, during the marathon and to sign post-event partnerships for the conclusion of the solutions. Due to all the impediments and barriers, both political, management and bureaucratic, the solutions were not absorbed and made feasible. However, in the case of the 2016 event, they were made available to society [9].

However, employees indicate that hackathon has initiated a process of innovation within the company itself. [11] consider that the most valued results are associated with the community spirit generated from the event, but that lasts after the event. The same authors consider as benefits the educational proposals with the actors involved in the practice of the activity. These issues are true given the employee engagement and the second event - which took place in 2017.

For public employees, the benefit was not the creation of a new product, because there was no sustainability to continue the development of the prototypes. For employees the reports are for the gain associated with the interaction of the triple helix and the awakening to a new culture within the company. Authors such as [11] indicate that marathons provide those involved the perception of new forms of intervention and professional activities, directed to the construction of creative answers to socially relevant problems. Moreover, in the study by [2] the main purpose of the Brazilian public sector in promoting hackathons from the perspective of public officials is to generate innovations, promote social participation, greater transparency, environmental strengthening and innovation network, actors' approach to innovation (private sector, public sector and educational institutions, research and development) and acceleration of internal organizational changes.

One of the problems pointed out by the servers is associated with the non-incorporation of the solutions by the company. Moreover, according to perceptions the solutions were not turned into business. Studies already report the difficulties of governing bodies in stopping the solutions in open innovation processes [2]. According to the same author, the perennial or sustainability of the solutions developed by the participants is a critical aspect highlighted by the interviewees. However, in the event of the Santa Catarina event, what is observed is that by the proposal of the announcement of the event, the solutions corresponding to the version fully implemented in the winning proposal should be licensed to the CIASC, and should not be used for commercial purposes.

Another point considered for the non-continuity of the proposed solutions was the profile of the participants of the event who were not an audience with an entrepreneurial vision, did not envisage a business opportunity and therefore had no interest in developing the solution. The study by [2] indicated that some organizations try to reduce these risks and weaknesses of the format of hackathons by working with partners, mainly in incubation stages of the winning solutions of the contest. However, it should be noted that the main public of the event was a university which may also interfere with future visions.

5 Final Considerations

Hackathon, as identified in the research, is a method of open innovation [2], as it is possible to bring together different actors, such as academia, business, university and participants from different places, with different cultures and abilities, who share experiences, ideas, materials and deliver a value-added product for the state and the citizen.

The CIASC held two editions of Hackathon #DesenvolveSC, between the period of 2016 and 2017, being the pioneer in the Government of the State of Santa Catarina. The results were positive regarding the promotion of innovation within the company and the process of acculturation of the servers, who began to prioritize innovation and were instigated to create tools to support open innovation.

In addition, the interaction with other organs (secretaries) that together proposed the event was observed. The results of the two hackathons were beneficial to the process of innovation within the company and interaction with the players of the innovation ecosystem. Not only did the connection between the government, considering different portfolios occurred, but also with actors such as universities and private companies who participated as mentors.

One of the relevant points is associated with the post-event context that needs to be defined in order to achieve the expected results, that is, the delivery and development of a value-added technological solution to the state and the citizen. However, new studies are being conducted in order to identify that the format of events such as hackathon can somehow meet this demand or whether other forms of innovation are necessary.

The hackathon for the Brazilian public sector still needs to be improved in relation to sustainability in the development of ideas, a great challenge today proposed for public administration, due to bureaucracy and management in hiring procedures and partnership models with private entities.

Ethics declaration

Conflict of interests. The authors declare no conflict of interest.

Compliance with norms involving human beings as subjects. All procedures performed in studies involving human participants were in accordance with the ethical standards of the institutional and/or national research committee and with the 1964 Helsinki Declaration and its subsequent amendments or comparable ethical standards. Informed consent was obtained from all individual participants involved in the study for the application of the questionnaire and the article does not have a description of the names, it is anonymous.

References

1. Chesbrough, H.W.: Open Innovation: The New Imperative for Creating and Profiting from Technology. Harvard Business School Press, Boston (2003)
2. Ferreira, G.D.D.: O papel dos Hackathons promovidos no setor público brasileiro: um estudo na perspectiva de inovação aberta, citizen-sourcing e motivação dos participantes (2017)
3. Moraes, M.P.G.: Hackathons: um estudo das iniciativas promovidas pelo setor público brasileiro (2017). 38 f., il. Trabalho de conclusão de curso (Bacharelado em Administração) - Universidade de Brasília, Brasília (2017)
4. Briscoe, G., Mulligan, C.: Digital Innovation: The Hackathon Phenomenon. Creativeworks London **6**, 1–13 (2014)
5. Lara, M., Lockwood, K.: Hackathons as Community-Based Learning: a Case Study Tech Trends. TechTrends **60**, 486–495 (2016)
6. Almirall, E., Lee, M., Majchrzak, A.: Open innovation requires integrated competition-community ecosystems: lessons learned from civic open innovation. Bus. Horizons **57**(3), 391–400 (2014)
7. Godoy, A.S.: Introdução à pesquisa qualitativa e suas possibilidades. Revista de administração de empresas **35**(2), 57–63 (1995)
8. Centro de Informática e Automação do Estado de Santa Catarina - CIASC (2016), Governo do Estado promove Hackathon #DesenvolveSC. Recuperado de: http://www.ciasc.sc.gov.br/334-governo-do-estado-promove-hackathon-desenvolvesc
9. Centro de Informática e Automação do Estado de Santa Catarina - CIASC (2017), Governo do Estado promove Hackathon #DesenvolveSC. Recuperado de: http://www.ciasc.sc.gov.br/388-confira-como-foi-a-2-edicao-do-hackathon-desenvolvesc
10. Hackathon #DesenvolveSC.homepage. disponível em: http://www.ciasc.sc.gov.br/hackathon/. Acesso em 20 mar 2018
11. Guizardi, F.L., Santos, K.F.D., Lemos, A.S.P., Severo, F.M.D.: Maratonas hackers no Brasil com desafios no campo da Saúde. Interface-Comunicação, Saúde, Educação **22**, 447–460 (2018)

Editorial Production of Scientific Journals: The Influence of Technological Development on Scientific Certification, Editorial Management, and Post-publication

Tito Lívio do Nascimento Fernandes(✉) and Franciane da Silva Falcão

Department of Design and Graphic Expression, Federal University of Amazonas, Av. General Rodrigo Octávio, n. 3000 CEP: 69.077-000, Manaus, AM, Brazil
titofern@gmail.com

Abstract. The scientific journal is the primary means of communication of science, acting as a qualitative filter, guided by specific rules to scientific communication, validating and establishing the degree of content originality. Its communicational efficiency is essential for the diffusion of scientific knowledge, acting as an indicator of the evolution of the different areas of science, being part of the process of scientific and technological development of a region. This paper aims to address the technological, normative, and theoretical issues that today involves the process of editorial management of scientific journals. Through the literature and chronology of technological development that is applied in this field, it is possible to map process transitions. We can highlight the leading technologies and the main events that guide the way of scientific journals in this scenario, such as the inclusion of manuscript management systems, the use of object identifiers, profile and serial publication, bibliometric indexes, editorial design tools, among others. As a result, it is possible to identify the impact on the process flow of editorial production which led nowadays to include in the editorial process the stages of scientific certification, editorial management, and post-publication that make up the editorial production of scientific journals in the contemporary communicational ecosystems. The submitted proposals require a management team with multidisciplinary background training and editorial production, indicating the need to professionalize this area. This information contributes to the production of guidelines for efficient production management of scientific journals.

Keywords: Editorial process · Editorial technologies · Editorial design · Scientific journal

1 Introduction

1.1 Scientific Communication and Scientific Journals

The scientific journal is the primary means of communication of science, acting as a qualitative filter, guided by rules of scientific communication, validating and establishing

L. Pereira et al. (Eds.): IDEAS 2018, SIST 198, pp. 322–331, 2021.
https://doi.org/10.1007/978-3-030-55374-6_32

the degree of originality of the content. Its communicative efficiency is essential for the diffusion of scientific knowledge, acting as an indicator of the evolution of a broad range of areas of science, as part of the process of scientific and technological development of a region [1, 2].

Periodicals play an essential role as a means of communication between research institutions and society, and authorial content, from the moment of its conception and elaboration, tends to be aligned with institutional and governmental policies. We also observe the hierarchies between institutions and government, academics and institutions, organized mainly by economic and social capital, which is also used to create a positional power relation within the fields. However, the scientific publication is directly linked to economic and social capital, and the fields of higher education and academic publication or scientific publication present relations of dependence and interdependencies, and cannot be treated in a separate way [3, 4].

The beginning of the scientific communication took place mainly in an oral form and through a meeting of scientific associations. Scientific dissemination via letters was inefficient, circulating only among acquaintances. In the middle of the seventeenth century, the first scientific journals appeared, the first two being the *Journal des Sçavans* and the *Philosophical Transactions* of 1665. They were motivated by the difficulty of communication on the part of the editors and printers of the time, the presumption that new knowledge would only emerge in the face of scientific debate [5–7].

The gradual signing of the scientific journal, to the detriment of oral communication, personal correspondence and books have had remarkable implications for science, forming itself as the primary means of scientific communication. This format has remained unchanged for centuries, adding in this process, in 1731, the *Peer Review*, where the first peer-reviewed journal, the *Medical Essays, and Observations* of the *Royal Society of Edinburgh*, was published [5, 7–9].

In the twentieth century, with the acceleration of technological development, several innovations emerged that influenced new approaches, bringing speed and quality to the editorial process of a scientific journal. From the second half of the twentieth century, serial publications with the configuration of scientific periodicals have grown exponentially, requiring the need for bibliographic control.

During the development of computer science, especially in the 1980s, several initiatives emerged to try to computerize the whole editorial process, but only with the advent of the Internet, especially in the 90's, with the use of computer network for electronic distribution of scientific articles, which effectively brought about significant changes in the editorial process.

The change from print to online directly interfered in practices reviews for the production of scientific journals. Despite this type of circular publication in specific social space, the scientific community, which maintains its configuration based on the structure and function established in the early seventeenth century [5], the emergence of online journals became the present-day standard. In the late 1990s circulation in electronic format increased the citations by 336% relative to the same printed content, says Lawrence [12], making an inevitable path to visibility.

Even though there are some printed journals, the online scientific journal is a hegemonic standard. In a printed publication, the existence of its online version is mandatory.

This happened by transforming it into the only way to ratify its visibility and notoriety in the competitive environment that has become scientific communication among peers.

In a study presented by Clarivates Analytics, using data from one of the primary databases, Web of Science - WoS, presents a growth rate of new journals of 3.5% in the world. That means, given the number of journals available, the appearance of approximately one thousand journals per year, and this is a trend that will continue to increase [13].

In the international context, considering the proportion of submissions between 2012 and 2017 presented in Fig. 1, Brazil is in the sixth position, with the growth of 136%. A parity of article submissions between the US and China is observed in 2017 [13].

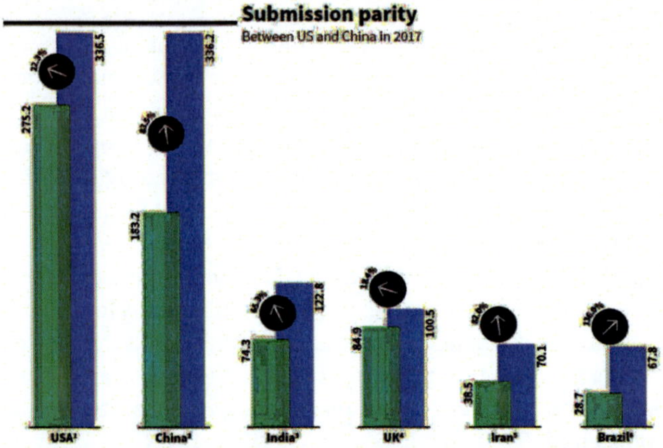

Fig. 1. proportion of manuscript submissions between 2012 and 2017. Source: Dudziak [11].

This paper aims to address the technological, normative and theoretical issues that today involves the process of editorial management of scientific journals in a way that contributes to the qualification of Brazilian scientific journals in the context of the internationalization of academic production.

2 Materials and Methods

The methodology used was based on bibliographical researches related to the subject, and the collection of information from provider and indexing institutions, establishing the theoretical foundations and categories for analysis.

This survey was also used to describe a chronology of the emergence of key technologies and significant events in the editorial production of scientific journals. The categories defined were: bibliometric indices and evaluative systems; repositories and database indexes; financing agencies; identifiers; search systems; Copyright.

After this survey, a representative sample of Brazilian scientific journals with the international scope was defined, where we adopted the sampling by judgment and defined by all Brazilian journals that have an impact factor. The journals that make up the sample were identified by the publication of the Journal Citation Reports - JCR in 2017. Through this survey, it was possible to cross-reference information and serve as a subsidy for analysis, contributing to the understanding of the contemporary technological scenario.

3 Results and Discussion

3.1 Editorial Production and New Technologies

The editorial production of scientific journals has specific stages and characteristics of this type of publication, since it has its characteristics that differentiate it from other literary genres. One of the editorial structuring approaches of this genre is the editorial flow presented by Gruszynski *et al.* [3], composed of editorial planning, editorial flow, and circulation (Table 1).

Table 1. The flow of the editorial production of scientific journals. Source: Gruszynski *et al.* [3]

Editorial planning			
Editorial management			
	Editorial Board Composition		
	Definition of ad hoc evaluators		
Infrastructure			
	Physical space		
	Equipment and technological resources		
Specialized technical services			
	Financial resources for hiring		
	Partnerships, promotion or sponsorship		
Editorial policy			
	Journal title and subtitle		
	Knowledge area covered		
	Editorial project		
Editing criteria (arising from the editorial policy)			
	Guidelines for authors		
	Minimum number of texts per volume		
	Organization/editing of the contents in order		
	Schedule/ deadlines per stage		
Editorial flow			
	Text editing		
	Peer evaluation		
	By accepting, spelling and grammatical review		
	Through the acceptance, technical standardization		
		Elements that must appear on the site	
			Journal identification data
			Institution responsible
			Management policy and editorial data
			Circulation data
			Summary of issue
			Datasheet

(*continued*)

Table 1. (*continued*)

		Elements to be included in the article	
			authors information
			Hierarchical levels of text
			Metadata for article identification
Layout Editing			
	Information architecture and site interface		
		Consistency between pages	
		Types of resources used by articles	
		Format of issues and articles	
		Search system	
		Contextual tools	
		Accessibility	
	Publishing the issue/articles		
	Proofreading		
	Online availability/publication		
Circulation			
Regularity			
	Frequency		
	Continuity		
Distribution - Ways of access			
	Free (OAI-PHM protocol)		
	Opened my own site		
	Direct mail		
	Signature		
Available in another format			
	Presence in full text databases / References / Citations		
	Access statistics and impact factor		

The editorial flow presented in Table 1 was simplified by Rodrigues *et al.* [4] as shown in Fig. 2. According to Rodrigues *et al.* [4], the flow remained constant in its larger composition, regardless of printed or online support, pointing out some modifications in the execution of some tasks according to the techniques adopted.

The phases of Editorial Planning, Editorial Flow and Circulation presented by Gruszynski *et al.* [3], does not contemplate a strategic vision for decision making, since several technologies and events have emerged and impacted the editorial production of scientific journals (Fig. 3). This event has increased competitiveness, directly influencing the criteria adopted in these three phases of the editorial flow.

Figure 3 shows the ScholarOne and Open Journal Systems (OJS) as the central manuscript management systems. It stands out ScholarOne as the most diffused manuscript management system among journals that perform better in the Journal Citation Reports (JCR) ranking. On the other hand, the OJS (SEER or Submission/SciELO) is the most used in the periodical portals of the Institutes of Higher Education, motivated mainly by free use. Of the 129 Brazilian scientific journals with Impact factor in 2016, 58% use ScholarOne and 9.3% use OJS.

The use of object identifiers (DOI) and profile (ORCID), presented in Fig. 3, have become mandatory, providing usability and persistence, quality essentials for network

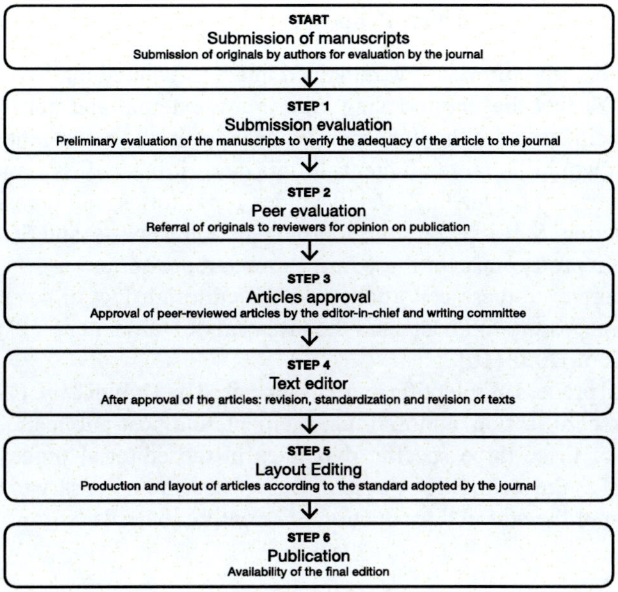

Fig. 2. Editorial simplified flow of scientific journals. Source: based on Rodrigues *et al.* [4] e Gruszynski *et al.* [3].

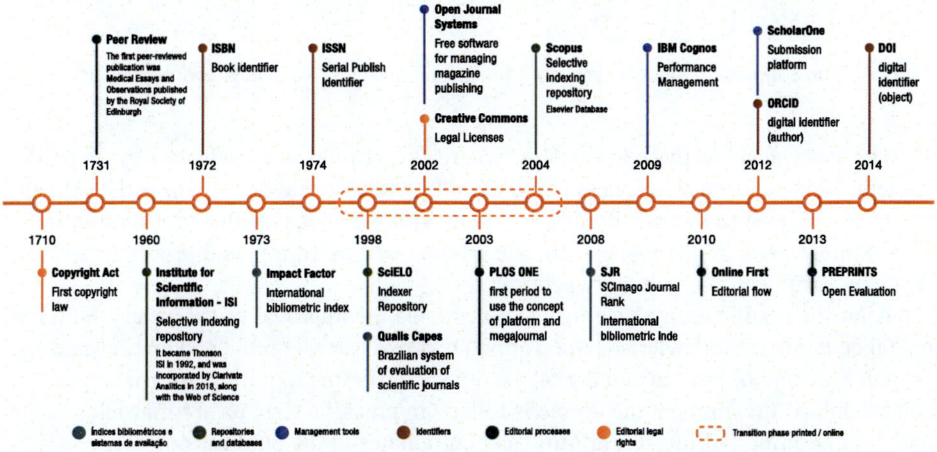

Fig. 3. Timeline presenting the chronology of the emergence of the leading technologies and events that influenced and influenced the editorial production of scientific journals. Source: Elaborated by authors.

environment [14, 15]. All Brazilian journals that have Impact Factor presented use DOI, and only 2 use ORCID, a fact that will change due to the need to adapt to indexing criteria [16].

3.2 Editorial Production, a New Proposal

The assertions that the editorial flow remained intact as pointed out by Rodrigues *et al.* [4] disregards the fact that the indexing criteria are external and not internal. Therefore, an ecosystem structure that incorporates changes in the environment of scientific communication would be more efficient. In this case, it has a direct relationship with the indexing criteria of the leading selective index repositories, and repositories focused on index production, SciELO Citation, Journal Citation Reports, and SCImago Journal Rank as the main protagonists of these international repositories.

How to incorporate these conditions into the editorial flow of scientific journals? Previously it is necessary to understand the proposal of Dubini *et al.* [17] together with the propositions of Diniz [16].

The editorial process of scientific journals proposed by Dubini *et al.* [17] is composed of three phases: certification, editorial management, and post-publication (Fig. 4). The understanding of these three specific dynamics in the editorial process of scientific journals, as well as the technologies used in each phase, are necessary to understand the current scenario of the editorial production of scientific journals.

Fig. 4. The editorial production process of scientific journals. Source: Elaborated by authors.

The first phase is the certification, where the articles are evaluated by scientific content. This peer review process is managed by the editorial board where the primary purpose is to ensure the quality of the article with the focus on the contribution in its area. Editors perform the work, associate editors, ad hoc editors and proofreaders, and in the most of time is a voluntary activity.

After the certification phase, which constitutes the approval of the article, editorial management is started, where are performed text revisions, graphic and text formatting, graphical design of printed and digital versions, conversion to multiplatform languages, and others. At the moment the article is found "in press," a term used for articles in the process of editing/publication/printing and culminates in the publication.

This phase is multidisciplinary and involves professionals in some areas such as design, language translation, information technology, and computer science. The professionalization of this phase induces a higher quality of presentation, adaptation to the requirements of more selective indexers and as a consequence contributes to the internationalization process [18].

Finally, we have the post-publication, which is guided by the strategy of dissemination of the journal, seeking to increase the visibility of the articles. The monitoring of indicators is essential for this activity.

Editorial management plays a significant role in all editorial production. It is not by chance that it is at the center of the process, it influences both certification and post-publication, it has the role of coordinating the set of activities that are interrelated [18].

Considering the two proposals and inserting external forces into the journal, we have the journal within an editorial process that incorporates the concepts of Dubini *et al.* [17], the flow proposed by Gruszynski *et al.* [3], and the propositions of Diniz [16]. This dynamic is shown in Fig. 4.

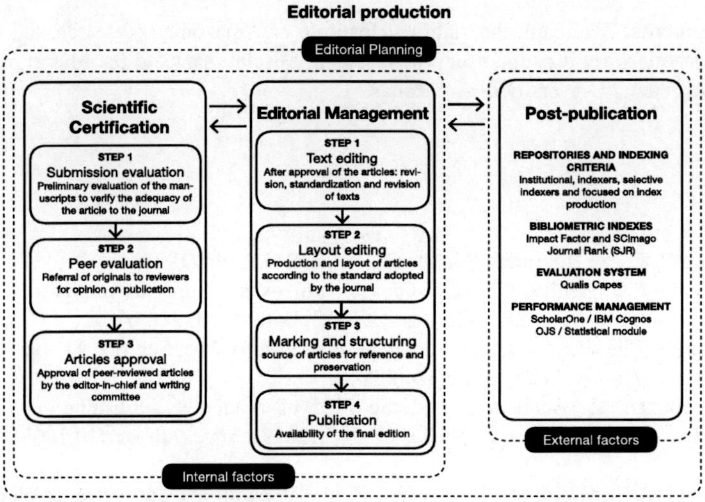

Fig. 5. The editorial production process of scientific journals considering the external factors. Source: Elaborated by authors.

Within the context of the internationalization of scientific journals, the embody and appraise of post-publication includes factors necessary for strategic decision-making. It brings to the editorial production of scientific journals the contemporary competitive environment and divides the internal stages to scientific certification and editorial management. This restructuring process impacts the actions of editorial production positively, because it brings light to the paths to be followed within this literary genre (Fig. 5).

4 Final Considerations

At the certification stage, it is essential to have a multi-institutional scientific body that is composed of prominent representatives from the areas determined in the journal's scope and who are committed to meet editorial deadlines.

Understanding the technological scenario that interferes in the phases of certification, editorial management, and post-publication, and also understanding that this scenario points the way of internationalization, it is clear that Brazilian journals have to follow this path, otherwise the risk of losing national relevance.

The constant monitoring and dissemination of the qualitative and quantitative indicators extracted from all three phases of editorial production must be used to improve its efficiency and transparency, being fundamental for the recognition of the journal in the scientific community and editorial management has this role. This action, incorporated into the editorial production is related to the competitive environment of the periodicals and contributes to the decision-making process that allows the authors to keep searching.

The journal should be submitted to the most significant number of possible repositories, if it is in the ecosystem of scientific communities. This allows entering a large number of academic networks, thus improving their visibility.

Acknowledgments. We thank the National Institute of Amazonian Research and the Federal University of Amazonas for supporting the research. Special thanks to the Master's Program in Design of the Federal University of Amazonas.

References

1. Bourdieu, P.: Pierre Bourdieu: Sociologia. Ática, São Paulo (1983)
2. Gruszynski, A.C., Golin, C.: Periódicos científicos nos suportes impresso e eletrônico: apontamentos para um estudo-piloto na UFRGS. Eptic **8**, 15 (2006)
3. Bourdieu, P.: The social space and the genesis of groups. Soc. Sci. Inf. **24**, 195–220 (1985). https://doi.org/10.1177/053901885024002001
4. Padmalochanan, P.: Academics and the field of academic publishing: challenges and approaches. Publishing Res. Q. **35**(1), 87–107 (2019). https://doi.org/10.1007/s12109-018-09628-2
5. Gruszynski, A.C., Golin, C., Castedo, R.: Produção editorial e comunicação científica: uma proposta para edição de revistas científicas. E-Compós **11**, 1–17 (2008)
6. Rodrigues, R.S., Quartiero, E., Neubert, P.: Periódicos científicos brasileiros indexados na web of science e scopus: Estrutura editorial e elementos básicos (2015)
7. Meadows, A.J.: A comunicação científica. Briquet de Lemos, Brasilia (1999)
8. de Frazão, S.O.: A contribuição das coleções de periódicos científicos do Portal de Periódicos da CAPES para a produção científica na área de Farmácia. Universidade Federal do Estado do Rio de Janeiro (2017)
9. Donato, H.: Revisão por Pares: o Coração das Revistas Científicas. Gaz Médica **4**, 15–20 (2017)
10. Stumpf, I.R.C.: Passado e futuro das revistas científicas. Ciência da Informação **25**, 1–6 (1996). https://doi.org/10.18225/ci.inf..v25i3.637
11. Lancaster, F.W.: The evolution of electronic publishing. Libr Trends Urbana **43**, 518–527 (1995)
12. Lawrence, S.: Free online availability substantially increases a paper's impact. Nature **411**, 6837 (2001)
13. Dudziak, E.A.: Da submissão à decisão - como está a aceitação dos artigos de autores brasileiros? - SIBiUSP - Sistema Integrado de Bibliotecas da Universidade de São Paulo. (2018). http://www.sibi.usp.br/noticias/conectando-os-pontos-da-submissao-a-decisao-como-esta-a-aceitacao-dos-artigos-de-autores-brasileiros/. Accessed 3 July 2018
14. SciELo: Critérios, política e procedimentos para a admissão e a permanência de periódicos científicos na Coleção SciELO Brasil. São Paulo (2017)

15. Scielo: Critérios SciELO| Nova versão a partir de 2018. In: SciELO Bras (2017). https://mailchi.mp/scielo/critrios-scielo-nova-verso-a-partir-de-2018?e=9e4d51a2b7. Accessed 26 Jul 2018
16. do Fernandes, T.L.N., da Falcão, F.S.: Technologies applied to the editorial production of scientific journals and a case study of the jornal Acta Amazonica (in press). Brazilian J. Inf. Des. **16**, 111–128 (2019)
17. Dubini, P., di Trani, F., Micheli, M.R.: PEER Economics Report. Art, Science and Knowledge (ASK) (2011)
18. Diniz, E.H.: Periódicos brasileiros da área de administracão no contexto de internacionalização da produção científica. Rev Adm Empres **57**, 357–364 (2017). https://doi.org/10.1590/s0034-759020170406

Design of Sustainable Business, Products, and Services for Rural, Forestry, and Agriculture Innovation

Knowledge Sharing Structure of Agricultural Products: Case of *KokuzoYuzu* (Citrus)

Shintarou Mori[1](✉), Akane Matsumae[2], and Yukari Nagai[1]

[1] Japan Advanced Institute of Science and Technology, 1-1 Asahidai, Nomi, Ishikawa 923-1292, Japan
s-mori@jaist.ac.jp

[2] Kyushu University, 4-9-1 Shiobaru, Minamiku, Fukuoka 815-8540, Japan

Abstract. Agricultural products in rural areas are developed mainly for urban consumers. However, consumers often find themselves forced to make purchase decisions based on only price information and appearance of products, lacking the necessary contextual knowledge about products. Furthermore, in Japan, it is becoming increasingly difficult for farmers and primary processors of the food crop to generate sufficient profits, which puts the agricultural production system at risk.

Although many studies examine agricultural products' marketing [1], few focus on its knowledge sharing structure. The authors took an empirical approach to address this problem, focusing on the knowledge sharing structure among stakeholders (farmers, primary processors of the food crop, distributors, and consumers), under the assumption that an intrinsic response to consumer needs would result in reasonable profit for each stakeholder. In the current knowledge sharing structure, only sellers respond directly to needs of consumers, who focus less on products' intrinsic value.

Therefore, the authors implemented the value co-creation process among stakeholders to share contextual knowledge of *KokuzoYuzu* (Japanese citrus) by making a promotional video and selling it together. This increased its sales unit price, diversified the production area's impression, and increased interaction between consumers and farmers.

This case study empirically indicates that value co-creation has changed the knowledge sharing structure among stakeholders, which increases the shared contextual knowledge about products among stakeholders while improving consumers' experience and behaviors. These findings will contribute to the development of sustainable agricultural systems.

Keywords: Knowledge sharing structure · Contextual knowledge of products · Value co-creation

1 Introduction

1.1 Background

In the annual report on food, agriculture and rural area in Japan for FY 2018 [2], Japanese government focuses on "hilly, mountainous and rural areas", which occupy roughly 70%

L. Pereira et al. (Eds.): IDEAS 2018, SIST 198, pp. 335–345, 2021.
https://doi.org/10.1007/978-3-030-55374-6_33

of Japan's total land mass, as having the potential to utilize local resources as a treasure to increase profitability" [3]. These rural areas often produce agricultural products that are high in value, but they do not generate sufficient income as the crop production is small.

Agricultural products from rural areas (APRs) are produced primarily for consumers in urban areas [4]. In such cases, consumers must be made aware of the value-addition provided by distinctive regional resources. However, analysis of the actual purchase process shows that consumers often lack the necessary knowledge and context to judge value and base their purchases on price information (cheapness) or partial knowledge, typified by product appearance. Therefore, APRs' true value is not shared properly with those who determine prices; this makes it difficult for the products' farmers and primary processors of the food crop (PPF) to generate a profit commensurate with their value, thereby putting the long-term viability of APRs in doubt.

1.2 This Study's Viewpoints

In a typical price-determination mechanism for agricultural products, wholesalers and retailers who are closest to consumers often wield a large degree of influence in the distribution process; APRs basically follow the same distribution mechanism. The current price-determination mechanism and knowledge-sharing process are summarized below (Fig. 1).

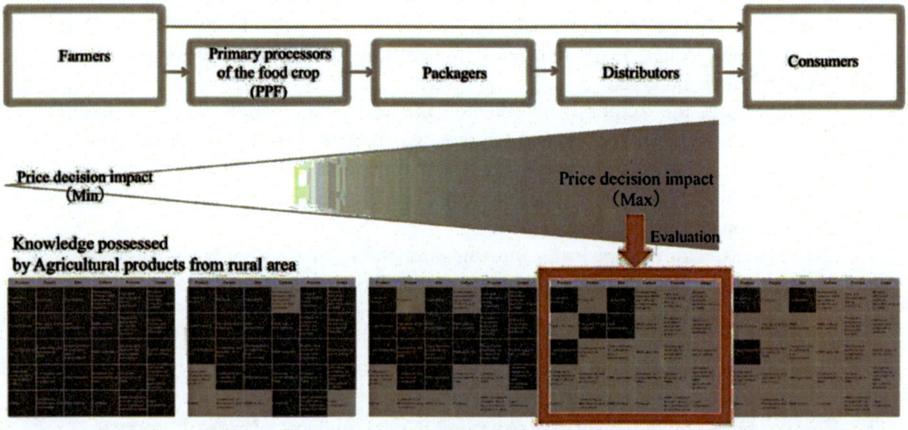

Fig. 1. Typical distribution process of the current marketing system for Japanese agricultural products (The dark-shaded area indicates shared contextual knowledge of products.)

In this study, the authors propose and focus on the following viewpoints: the *knowledge sharing structure* (KSS) and the *contextual knowledge of products* (CKP), under the assumption that a fundamental relationship with consumers, that is, sufficient level of knowledge sharing about APRs, will produce reasonable benefits for each stakeholder.

Knowledge Sharing Structure (KSS). KSS is the structure how the knowledge is shared among stakeholders (farmers; PPFs; distributors, such as wholesalers and retailers; and consumers). The KSS is either merely a state of being passive by having distributors share knowledge with the farmer or a state in which each side can actively share knowledge with one another. The authors consider that KSS is a state wherein all parties involved can objectively grasp the directionality of shared knowledge, flow of communication, and relationships, and consider which state these are actually in.

Contextual Knowledge of Product (CKP). CKP referred to here is the context and knowledge of regional farm products in which a consumer may find value, such as function, stories during development, manufacturing district's background, and farmer's particularities.

Distribution Process. Generally, Japanese wholesalers and retailers (buyers to farmers) close to consumers in the distribution process determine value-added price. However, each stakeholder sorts and selects the CKP of APRs during the distribution process. As a result, only a portion of the CKP is transmitted by wholesalers and retailers, who have a strong degree of influence on price determination. Naturally, consumers or retail buyers tend to set lower prices, and farmers tend to receive less price than the value they offer with their products.

As Fig. 1 shows, there is a gap between the CKPs that consumers receive from APRs and their own CKP. One reason for this gap is that APRs constantly cater to the distribution mechanism of mass-produced agricultural products [5]. For example, referring to *KokuzoYuzu*, the specialty of Nomi (Ishikawa prefecture) covered in this study, the interviews revealed that yuzu fruit with blemishes is generally sold at a discount for fruit-juice processing; however, there are cases wherein unblemished, good-quality fruit is shipped at a discount to meet the quota set by the production process of the juice-processing company. Naturally, the CKP is not delivered to consumers at all since the good-quality product is sold as discounted fruit for juice processing, and that too at a lower price.

In this case, there is neither knowledge sharing of the CKP of APRs with consumers nor an appropriate valuation. In particular, lack of knowledge sharing of the CKP means that consumers cannot assess the value-addition; specifically, this refers to the large amount of time and effort spent on APRs' detailed, valuable information about the production area and production process. As a result, the APRs' incidental value is not reflected in the price.

Therefore, the authors propose the concepts of CKP and KSS to visualize the correlation between price-determination and quality of relationship among stakeholders.

2 Related Studies

There have been active discussions on knowledge sharing in the context of cooperative work. In recent years, researchers have stressed the importance of repeating questions and discussions and, therefore, deepening mutual comprehension to share knowledge among a team of engineers [6]. Cash et al. [6] argue that development of a shared understanding within a team is supported by effective communication and consider this idea to be applicable across different fields. Particularly, various discussions on inter-organizational relationships pertinent to this study's focus have been conducted, such as studies on the significance of knowledge sharing for inter-organizational cooperation [7–9].

Further, researchers have long studied marketing efforts for agricultural products; there have been studies on agricultural products from the perspective of product characteristics and sales promotion [5]. However, despite the abundance of studies on marketing methodology, there has been lack of research on knowledge sharing.

Matsumae et al. focused on the relationship between the design process and formation of a co-creation entity. Their study discusses the following: mechanism of co-creation between organizations; how the co-creation process leads to mutual subjectivity; and the co-creation entity, which becomes the foundation that drives continuous co-creation [8]. However, they deem the value generated through co-creation (a focused aspect of this study) as being outside the co-creation mechanism and do not discuss it from the viewpoint of KSS.

Therefore, in this research, the authors raise the research problem of whether knowledge sharing improves shared CKP of APRs and the relationships and reasonableness of profit distribution among stakeholders.

3 Methodology

In this study, participant observations and interviews were conducted among stakeholders of *KokuzoYuzu*, and the obtained data were analyzed from the marketing and knowledge science viewpoints. The authors propose the concepts of CKP and KSS to analyze the case.

The case study was conducted for *KokuzoYuzu*, an APR from the Kokuzo-producing district of Nomi in Ishikawa prefecture; the district has been designated for promotion by the Mountain Villages Development Act. In FY 2016, the Kokuzo district invited a national coordinator of agricultural food processing as an external consultant to identify agricultural products that urban dwellers might find appealing. *KokuzoYuzu* is considered the most appealing because it possesses inherent value as a brand that represents hilly and mountainous area. Thus, *KokuzoYuzu* is selected as the APR of this district. This study's hypothesis is verified by assessing the current state of KSS and implementing strategies that alter the KSS.

Regarding the method of selecting the distributers to be interviewed, the choice was made based on whether the person in charge of sales is actually a part of the company selling to the consumer of agricultural products face-to-face, for example, a person in charge of sales such as a general retailer or trade show personnel.

4 Case Study: *KokuzoYuzu* Project

4.1 Case Outline

The authors conducted interviews with and observed stakeholders participating in the *KokuzoYuzu* production process: farmers, local citizen groups, and local government officials. However, understanding all the CKP and KSS with the parties involved would require an enormous amount of work, so this research deduced the relationship between the farmer and the distributor who claims to have a strong influence on pricing.

The survey and observation were conducted in three types of KSS (Fig. 2).

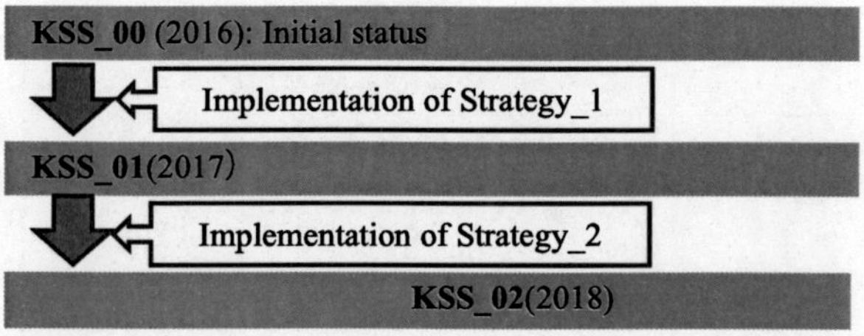

Fig. 2. Shift of the knowledge sharing structure (KSS)

4.2 CKP of the Case

To set the CKP inherent in the *KokuzoYuzu*, the authors visited the production area during the KSS_00 period with an external consultant who works as a national coordinator of agricultural food processing, as stated above.

The authors interviewed and observed the external advisor for three days to understand the perspective using which he regularly evaluates rural areas throughout Japan; the CKP was determined by focusing on the following six items: product, people, area, culture, process, and usage. With these items, the authors conducted interviews with key personnel associated with the production of *KokuzoYuzu*: farmers (yuzu farmers), local citizen groups (Nomi Yuzu YuYu Club), and local government (City of Nomi) officials. Factors relevant to each item are listed as the CKP of *KokuzoYuzu* (Table 1).

4.3 Strategies to Change the KSS

Strategy_1. PPFs, distributors, and consumers visited *KokuzoYuzu* production and processing area with invited experts on food processing and use.

Date and Time: November 14, 2017, 10 am–4 pm
Location: Kokuzo district
Details: Mini tour of yuzu farm (farm tour and farm experience), lunch with farmers, viewing of yuzu fruit-juice-processing facility, and meeting for exchange of ideas
Target: Stakeholders working in Tokyo who may not be familiar with *KokuzoYuzu* (PPFs, distributors, and consumers)

Strategy_2. PPFs, distributors, and consumers visited the *KokuzoYuzu* production and processing area and *co-created* its promotional video.

Period: March 2017 – December 2017

Table 1. Contextual knowledge of product for the *KokuzoYuzu*

Product	People	Site	Culture	Process	Usage
Production method	Farmers	Traceability	Relationship between residents and the rural products	Care and consideration when producing	Utilization methods/features unique to the rural area
Taste/texture	Farmers' thoughts	Environment of the region	Culture and customs of the rural area	Farmers' way of enjoyment during production stages	Delicious recipes and cooking methods
Types/features	Communities of the related areas	Growing environment of the production site	Daily life in the rural area	Devices and methods for sustainable production	Stylish ways to eat or use
Featured components of the products	Community centering farmers	Features of the related areas	Initiatives for sustainability in the rural area	Children growing up in the rural area	Examples of the consumers' usage
Safety	Community centering consumers	History of the related areas	Traditional festivals	Situation brought about by sustainability of the production area	Legal compliance

Source: Elaborated by authors

Details: Joint video production by experts and tour participants with respect to content, composition, video title, caption, and narration
Target: Stakeholders (PPFs, distributors, and consumers) who participated in Strategy_1 as well as farmers

5 Results and Discussions

5.1 Shift of CKP Between KSS_00, KSS_01, and KSS_02

Shared CKP in KSS_00. KSS_00 is initial KSS, wherein knowledge was primarily conveyed in a one-way direction from farmers to distributors prior to 2016 (Fig. 3). The shaded areas in Table 2 represent the shared CKP of *KokuzoYuzu*.

Fig. 3. Initial knowledge sharing structure (KSS_00)

Table 2. Shared contextual knowledge of product of *KokuzoYuzu* during KSS_00

Product	People	Site	Culture	Process	Usage
Production method	Farmers	Traceability	Relationship between residents and the rural products	Care and consideration when producing	Utilization methods features unique to the rural area
Taste/texture	Farmers' thoughts	Environment of the region	Culture and customs of the rural area	Farmers' way of enjoyment during production stages	Delicious recipes and cooking methods
Types/features	Communities of the related areas	Growing environment of the production site	Daily life in the rural area	Devices and methods for sustainable production	Stylish ways to eat or use
Featured components of the products	Community centering farmers	Features of the related areas	Initiatives for sustainability in the rural area	Children growing up in the rural area	Examples of the consumers' usage
Safety	Community centering consumers	History of the related areas	Traditional festivals	Situation brought about by sustainability of the production area	Legal compliance

■indicates shared content

Source: Elaborated by authors

Shared CKP in KSS_01. Through Strategy_1, KSS was changed into KSS_01, wherein farmers and distributers exchange knowledge mutually (Fig. 4). The shared CKP is increased, as shown in the shaded areas in Table 3.

Fig. 4. Knowledge sharing structure after strategy_1 (KSS_01)

Shared CKP in KSS_02. Through Strategy_2, KSS_01 was changed into KSS_02, wherein farmers and distributors started to jointly come up with ways to appeal to consumers (Fig. 5). The shared CKP is increased, as shown in the shaded areas in Table 4.

5.2 Relationships Between Stakeholders and Factors Determining Price

The authors interviewed "Katakana" (that sells miscellaneous food items at Jiyugaoka, Meguro-ku, Tokyo), and observed stakeholders in each KSS. Interviews and observations were conducted in terms of the following: changes in the relationship between farmers and distributors in each KSS period; whether the distributor selected the product to sell; sale price; and factors considered for price determination.

Table 3. Shared contextual knowledge of product of *KokuzoYuzu* during KSS_01

Product	People	Site	Culture	Process	Usage
Production method	Farmers	Traceability	Relationship between residents and the rural products	Care and consideration when producing	Utilization methods/features unique to the rural area
Taste/texture	Farmers' thoughts	Environment of the region	Culture and customs of the rural area	Farmers' way of enjoyment during production stages	Delicious recipes and cooking methods
Types/features	Communities of the related areas	Growing environment of the production site	Daily life in the rural area	Devices and methods for sustainable production	Stylish ways to eat or use
Featured components of the products	Community centering farmers	Features of the related areas	Initiatives for sustainability in the rural area	Children growing up in the rural area	Examples of the consumers' usage
Safety	Community centering consumers	History of the related areas	Traditional festivals	Situation brought about by sustainability of the production area	Legal compliance

■ indicates shared content
Source: Elaborated by authors

Fig. 5. Knowledge sharing structure after strategy_2 (KSS_02)

Table 4. Shared contextual knowledge of product of *KokuzoYuzu* during KSS_02

Product	People	Site	Culture	Process	Usage
Production method	Farmers	Traceability	Relationship between residents and the rural products	Care and consideration when producing	Utilization methods/features unique to the rural area
Taste/texture	Farmers' thoughts	Environment of the region	Culture and customs of the rural area	Farmers' way of enjoyment during production stages	Delicious recipes and cooking methods
Types/features	Communities of the related areas	Growing environment of the production site	Daily life in the rural area	Devices and methods for sustainable production	Stylish ways to eat or use
Featured components of the products	Community centering farmers	Features of the related areas	Initiatives for sustainability in the rural area	Children growing up in the rural area	Examples of the consumers' usage
Safety	Community centering consumers	History of the related areas	Traditional festivals	Situation brought about by sustainability of the production area	Legal compliance

■ indicates shared content
Source: Elaborated by authors

Relationships and Price Determination in KSS_00. Initially, farmers had unilaterally conveyed a minimal number of production area resources to stakeholders, as they had tied up with wholesalers and retailers to ensure shipments that met the distribution flow.

As a result, as shown as Table 2, when retailers used the judgment criteria (4 out of the 30 CKPs presented in Table 2) to compare the Kokuzo products with those from other regions, distributors did not select the Kokuzo products as they could not be differentiated.

Relationships and Price Determination in KSS_01. By using Strategy_1 to invite wholesalers and retailers to the production area, more production area resources were shared; more importantly, a mutual interaction was created between distributors and farmers as new production area resources (CKPs) farmers had not deemed important were shared with wholesalers and retailers upon their request.

In addition to the production area resources communicated by the farmer, retailers requested those that might be appreciated by consumers (20 out of the 30 CKPs presented in Table 3); these were shared through Strategy_1. As a result, distributors selected the product because it was different from that of other regions. The requested production area resources included multiple points with which Katakana customers could identify. Normally, Katakana customers are unable to obtain this information; thus, the buyer determined the expected increase in customer demand and set the retail price at 2.8 times the local price.

Relationships and Price Determination in KSS_02. Strategy_2 encouraged the sharing of various characteristics and needs of customers (consumers) with wholesalers/retailers. Farmers and wholesalers/retailers both presented which production area resources should be emphasized. As a result, farmers became more aware of what information might be useful to customers and changed their methods of showcasing production area resources, such as photographing usual tasks or trying new things that might appeal to consumers (e.g., releasing honey bees during honey collection to experience pesticide-free farming).

Meanwhile, wholesalers and retailers anticipated greater consumer demand owing to the increase in content and promotional methods of production area resources that they could use to appeal to consumers' diversifying tastes and needs. As a result, the price set by distributors increased. Changes were also seen in the proactive marketing of production area resources to consumers (e.g., promoting the production area by using customized paper bags). A telling change in the quality of the relationship between farmers and wholesalers/retailers was that they began to jointly come up with ways to appeal to consumers.

Strategy_1 in the previous year affected the sharing of production area resources by farmers and those recommended by retailers as being beneficial for consumers. Moreover, Strategy_2 led to knowledge sharing about production area resources that farmers and retailers jointly determined would appeal to consumers (28 out of the 30 CKPs presented in Table 4).

During production of the promotional video, farmers and retailers debated which production area resources would resonate with Katakana customers and provide an experience they would not have in their daily city life. Sales staff at the retailing company Katakana determined that the video is likely to yield further customer demand and set a retail price that was three times the local price.

6 Discussion

The survey results showed an increase in the number of items (as shown in Fig. 6) for which knowledge was shared with wholesalers or retailers who are influential in price determination.

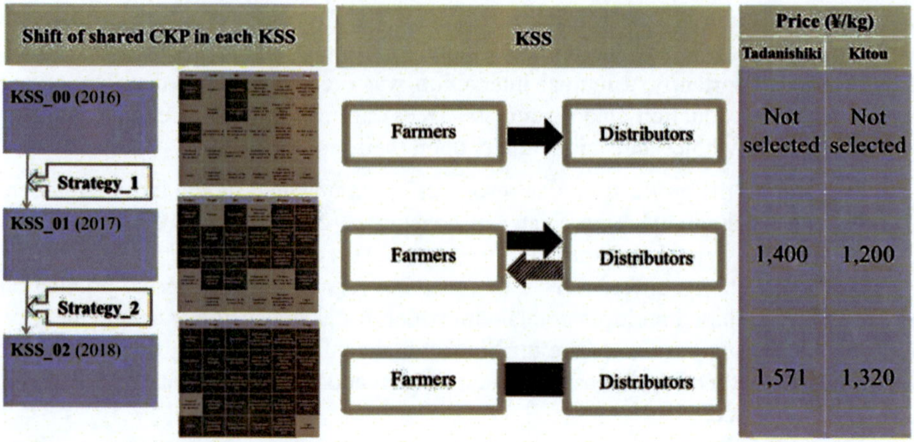

Fig. 6. Summary of case study focusing on contextual knowledge of product (CKP) and knowledge sharing structure (KSS)

In this situation, authors conducted an interview in each KSS period with the selected shop, Katakana as a sample distributor to study the impact of knowledge sharing on price. Changes in the quality of the relationship among stakeholders were observed; two-way knowledge exchanges, knowledge co-creation furthermore, enabled the distributor to supply consumers with many of the CKPs they required. As a result, the distributor selected *KokuzoYuzu* as an item to be sold in their stores, and, further, its retail price increased.

This suggests that each stakeholder can gain reasonable profits from a fundamental relationship based on sharing knowledge of intrinsic CKP; further, farmers and wholesalers/retailers (distributors) can jointly identify ways to appeal the CKP to consumers. However, even after sharing CKP, it is not clear if the information is valuable to consumers; there is also a possibility that really important CKPs may be ignored. On the other hand, this implementation provides limited clarification on the relationship between farmers and distributors (wholesalers or retailers).

7 Conclusion

The authors proposed the concepts of KSS and CKP to analyze the relationships among stakeholders. This case study suggests that the strategies used to alter the KSS among stakeholders could improve the CKP shared with consumers via distributors; this would ultimately lead farmers to receive reasonable compensation for their agricultural products.

The authors believe these findings will contribute to the development of a sustainable and reasonable agricultural system.

References

1. Hellin, J., Lundy, M., Meijer, M.: Farmer organization and market access. Leisa Mag. **23**(1), 26–27 (2007)

2. Japan's Ministry of Agriculture, Forestry and Fisheries, "About the type of agricultural area". http://www.maff.go.jp/j/tokei/chiiki_ruikei/setsumei.html. Accessed 21 Dec 2018
3. Japan's Ministry of Agriculture, Forestry and Fisheries, "Summary of the annual report on food, agriculture and rural areas in Japan", Tokyo (2018)
4. Sakurai, S.: A study of farmers' market from the viewpoint of urban-rural cooperation. J. Rural Plann. Assoc. **20**(3), 203–208 (2001)
5. Katsura, E.: The properties of agricultural products and sales promotion. Res. Agric. For. Issues **62**, 8–14 (1981)
6. Cash, P., Dekoninck, E.A., Ahmed-Kristensen, S.: Supporting the development of shared understanding in distributed design teams. J. Eng. Des. **28**(3), 147–170 (2017)
7. Dyer, J.H., Singh, H.: The relational view: cooperative strategy and sources of Interorganizational competitive advantage. Acad. Manage. Rev. **23**(4), 660–679 (2009)
8. Matsumae, A., Nagai, Y.: The function of co-creation in dynamic mechanism of intersubjectivity formation among individuals.In: Proceedings of the DESIGN 2018 15th International Design Conference, pp. 1925–1936 (2018)
9. Wenger, E.C., Snyder, M.W.: Communities of practice: the organizational frontier. Harvard Business Review. vol. Jan-Feb (2000)

Blockchain and Conservation: Why Does It Matter

Applications in Payments for Ecosystem Services and Bolsa Floresta Program

Luiz Villares(✉)

Fundação Amazonas Sustentável, Manaus, AM, Brazil
luiz.villares@fas-amazonas.org

Abstract. Blockchain has created an innovative framework for safe, efficient and transparent chain of transactions, with the impressive rise of more than 2.100 cryptocurrencies traded anytime, worldwide. These transactions are registered on the web, with no central server, or proprietors. Blockchain also offers exchange and data solutions for businesses and society with tokenized currency, immutable data registry, smart contracts and funding sources.

The Amazon region and forest conservation have a positive benefit from Payment for Ecosystem Services (PES) models that reward landowners and users for keeping the standing forest, thus providing services for the climate and biodiversity. A Blockchain proposition addresses key difficulties for PES models, such as lack of trust, high transaction costs and distance among parties.

In the Brazilian Amazon, Bolsa Floresta Program (Standing Forest Program) benefits nearly 9.500 families living in traditional riverine communities, rewarding them for their role as the tropical forest keepers. A Blockchain proposition for Bolsa Floresta indicates positive impacts for forest conservation. In project finance, it allows the alternative funding with cryptocoins, observed possible regulatory issues. In project activities, Blockchain provides platforms for a safe exchange of information, goods, and possibly, monetary values, among families and associations, thus improving the standing forest community model. The technical configurations for these proposals include Application Program Interface solutions with online and offline mobility among members. Overall, an analysis of key issues linked to a Blockchain proposition indicates a favourable deployment for pilot projects in the Amazon region, and likely, to conservation projects with forests and communities, worldwide.

Keywords: Blockchain · Standing forest · Conservation · Payment for ecosystem services · Amazon · Communities

1 Introduction

The Amazon region has 40% of the world's rainforest, 1/6 of the world's fresh water, providing environmental services in carbon storage climate balance, rain cycle maintenance and biodiversity, directly benefiting the South regions of Brazil [1, 2].

L. Pereira et al. (Eds.): IDEAS 2018, SIST 198, pp. 346–355, 2021.
https://doi.org/10.1007/978-3-030-55374-6_34

Strategies to avoid deforestation include law enforcement against illegal logging and conservation initiatives to place more value on the standing forest than cut. Effective law enforcement depends on government funding, structure and anti-corruption practices. Conservation initiatives can employ Payments for Ecosystem Services (PES) models in projects to reward forest inhabitants and improve their livelihoods. These initiatives normally face implementation challenges because of the nonbanking situation of beneficiaries, difficulties in the exchange of goods and project reporting.

The recent rise of cryptocurrencies and non-currency features based on Blockchain has opened solutions for the exchange of values and information, fundraising and project reporting, perfectly applicable to conservation initiatives. This paper proposes Blockchain based solutions for PES models and the Bolsa Floresta Program in Brazil.

PES models face common difficulties in the dispersion of providers and buyers, verification of information and payment issues. Blockchain offer appropriate solutions to these issues. Bolsa Floresta Program improves the forest communities' livelihoods with projects, mainly, in income generation, social assistance and education. A Blockchain proposition offers solutions in project information, data reporting and values exchange. It requires a technical framework of phones connected to a central application program. The proposed Blockchain solutions will be initially implemented in areas with internet coverage, and extended to all communities assisted by Bolsa Floresta, with a positive outlook for connectivity and adoption of offline technologies.

Following, these solutions are further explored with a Blockchain overview, identification of PES and Bolsa Floresta initiatives, connectivity and implementation issues.

2 Blockchain, Cryptocurrencies and Applications

Blockchain is presented as a ledger book that records transactions with data stored across multiple computers. Intrinsically linked to cryptocurrencies, such as Bitcoin (BTC), it records transactions of values and information in an open and immutable platform called the Distributed Ledger Technology (DLT). All transactions placed in the Blockchain are gathered in blocks, linked to each other by a cryptographic code called "hash". Together, they form a chain of blocks, the Blockchain. Any transaction in the Blockchain is valid once a block is filled and closed with information or values.

Blockchain ensures all transactions to arrive at the correct address, just once, without duplication, due to a complex set of strings in any hash code, plus a proof of identity verification for each sequential block.

This encryption process may require complex mathematical calculations of high computational cost and energy demand, such as for Bitcoin transactions. Nevertheless, in this paper's propositions, the forecast is for a very low energy demand.

The blockchain began in 2008 to allow the Bitcoin to be launched. Its creator is an anonymous person (or a group) with the pseudonym of ***"Satoshi Nakamoto"***, who developed and authored the Bitcoin white paper [3], created and deployed Bitcoin's original reference implementation, and devised the first Blockchain database.

Cryptocurrencies, or just "coins", use cryptographic functions in the authorization and verification processes for a successful transaction to be validated. Cryptocurrencies have no control of a central counterparty, like a Central Bank; being non-discriminatory to public access, and secure against fraudulent spending [3]. Soon after Blockchain made the Bitcoin possible, hundreds of new cryptocurrencies have emerged using the Blockchain logic. As of July, 2020, there are more than 5.500 cryptocurrencies in the world, with a market value above U$ 270 billion. The market value of Bitcoin is U$ 168 billion, about 62% of all coins [4].

Cryptocurrencies are much more volatile and less regulated assets than fiduciary currencies (U$, €, £, R$, etc.), thus not a reliable store of value for a regulated entity. Nevertheless, the Blockchain technology under coins is a safe tool for multiple businesses and project applications. The four relevant applications explored in this paper are: **tokenization, immutable registry, smart contracts and funding sources**. These applications are already in use by corporations, banks, governments and institutions under specific Blockchain platforms, requiring low energy demand, such as ***Ethereum*** and ***Corda***. As an example, *Consensys,* a venture production company, builds decentralized applications and develops end-user tools for Blockchain ecosystems [5] and IBM already offers Blockchain solutions for businesses in general [6].

Tokenization has more than one definition. For this paper, it can be explained as the digital conversion of a fiduciary currency into a digital coin with an initial fixed value. These tokens are not tradeable for other digital or fiduciary currencies for speculation. A "tokenized" coin may be used to facilitate transactions without the over regulated conditions imposed on fiduciary currencies. Such features offer easy transaction and exchange of values in situations where banks are not available, like in the remote Amazonian communities. Tokens can be used for purchases in local markets. Since a tokenized currency facilitates the exchange of goods, it also supports "circular economy" practices, helpful for communities and riverine populations.

Blockchain offers the **immutable register** technology, allowing data to be perpetually stored in the web, with no private servers, possible by the encryption of key information in immutable blocks. Examples of safe and open storage applications are: financial data, transactions registers, government data, election results, property rights, identity systems and perpetual healthcare data. In this paper's context, attention is placed on data reporting applied to projects related to beneficiaries and donors.

Project reports can have a set of key data information stored in "Distributed Ledgers" with connections to all relevant databases, thus providing transparency and data safety to all parties involved in the project.

On top of secure records, Blockchain provides the framework for **smart contracts** representing a set of logical conditional functions among parties, enabling automatic commands and approvals to be performed without no central or third party control. These logical and automatic commands reduce bureaucracy and possible human mistakes. Giving its features, smart contracts can solve the lack of trust and hinder corruption practices among participants of any trade. In addition, it places key issues involved in multiple relationships under one platform, such as loan clauses, exchange conditions, inspections, fiscal agents, thus eliminating intermediaries, like notaries and banks, by bundling them in a single structure of automatic permits.

Blockchain technologies coupled with coins offer alternative **source of funds** for many projects. Briefly, possibilities are in Initial Coin Offerings (ICOs), grant coins and hybrid donations (combining coin grants and fiduciary currency). An ICO is the launch of new coin, for instance, a "project coin". It may be considered a valuable asset for coin investors because of their positive perception about the project, not necessarily related to an economic return. However, it represents a liability between the issuing entity and its coin investors, due to regulations over cryptocurrencies [7].

3 Payment for Ecosystem Services - A Blockchain Perspective

Payments for Ecosystem Services (PES) are payments to landowners who have agreed to take certain actions to manage their land or watersheds to provide an ecological service. As the payments provide incentives to land owners, PES may be comparable to a market-based mechanism, similar to subsidies and taxes, to encourage natural resources conservation [8]. PES may also represent streams of funding from those benefiting from the ecosystem services, either directly or indirectly. The concept of payments may also be identified as a reward to landowners and users, like the Bolsa Floresta Program, in which a conservation entity rewards forest families, without market based mechanisms, rather as a public policy to support the standing forest.

In general PES models involve groups of "buyers" (or donors) flowing funds to landowners acting as natural resource keepers, the "providers". This is a complex process because of difficulties, such as: (1) tracking real ecosystem services provided, considering their low visibility; (2) structuring and maintaining an efficient platform of information among parties; (3) identification and segregation of the different services, thus matching them with the corresponding payments; and (4) in some situations, land titles and property rights are not clear enough to secure a safe payment from buyers to the services' providers. Blockchain addresses these situations enabling a network of multiple stakeholders, not necessarily known to each other, transacting currency, assets and information with transparent and trustful actions.

These transactions can be placed in a low cost web platform, with a central party in charge of technical operations. In addition, PES may involve the "tokenization" process to facilitate some exchanges, and secure data storage, encrypted in the Blockchain, thus providing all parties with immutable and transparent information.

Another set of challenges, concerning payments, are (1) the "bankless" situation of landowners, such as riverine people in the Amazon, unable to access formal banking; (2) high transaction costs, decreasing the PES feasibility; (3) possible low traceability of payments; (4) costly validation of payments; and (5) possible unclear identification of buyers and some values of services. Blockchain allows continuous and unique identification of participants through a secure transaction platform where participants exchange values and information under clearly defined conditions. Finally, issues related to verification of services, like establishing baseline situations and storage of reports, are facilitated by Blockchain, with the use of image location devices with secure identification. The features presented can be observed, below (Figs. 1 and 2):

Further analyses of Blockchain for PES models must take into consideration the particularities of each program, such as purchase conditions, the number and diversity of parties involved, geographical dispersion, internet, and others. Session 6 proposes the use of a Blockchain evaluation matrix containing these subjects, applied to Bolsa Floresta Program. This matrix can be framed as subsidiary tool for a Blockchain utility and solutions analysis for PES models and other nature conservation projects.

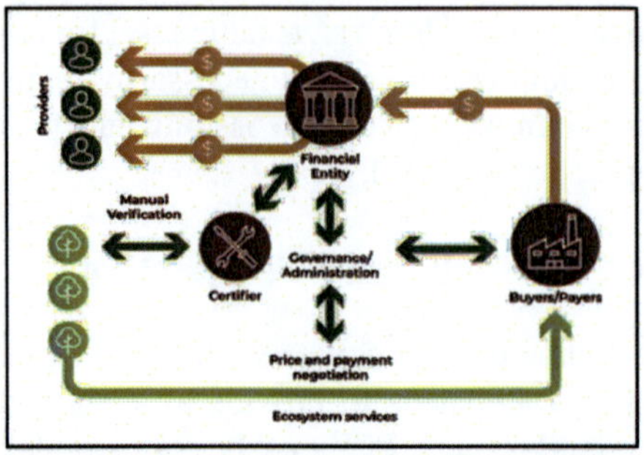

Fig. 1. Traditional PES transactions (Author)

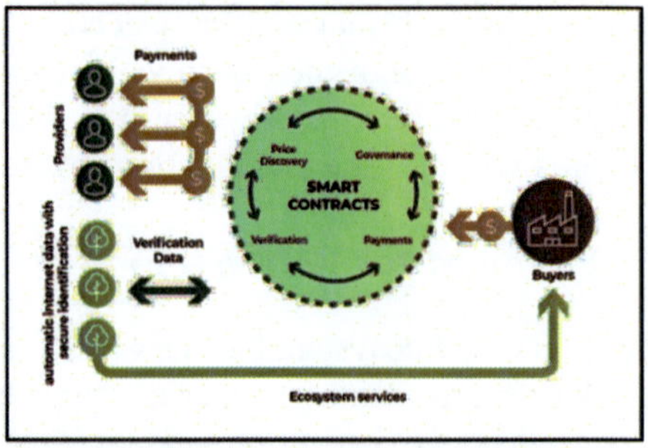

Fig. 2. PES under smart contracts (Author)

4 Forest Conservation in the Amazon – Bolsa Floresta Program

In the Amazonas State, Brazilian Amazon, the Bolsa Floresta Program (Standing Forest Program) promotes the sustainable development of 580 forest riverine communities, living in 16 Conservation Units. Managed by Fundação Amazonas Sustentável (FAS), a Brazilian nonprofit organization, the Program rewards 9,420 families for keeping the forest standing (July, 2020). The Amazonas State has an area of 1.57 million km^2, with 95% of the original forest remaining. The Conservation Units assisted by Bolsa Floresta have 110 thousand km^2, with nearly all original forest coverage (Fig. 3).

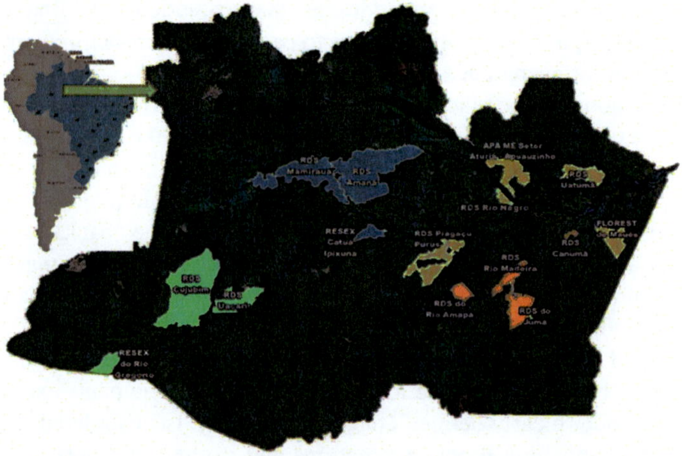

Fig. 3. Amazonas State Conservation Units assisted by Bolsa Floresta (Author)

Bolsa Floresta assists local communities with four subprograms: Family, Income, Social and Association. The Family initiative is a direct reward to resident families for their role in maintaining the standing forest. The other components provide local businesses support, community based social benefits and assistance to local associations. Activities include workshops, trainings, biannual leadership meetings and monitoring.

Membership is voluntary, but families are required to keep their children in school, and not allowed to deforest new areas for small scale agriculture.

Bolsa Floresta is among the largest PES project in the world, in number of landowners/residents receiving rewards, with U$ 49 million historically applied to the Program in 11 years. Fundraising is a high strategic issue for conservancy entities, demanding their active relationship with funds, corporations and individuals.

Bolsa Floresta Program and other conservancy projects could benefit from a **cryptocurrency funding** source, with simple donations from coin based philanthropies and complex structures like an Initial Coin Offering (ICO). The tasks and investments associated with each solution are directly linked to their funding potential. Conservancy entities should first approach coin based initiatives that raise and donate funds from coins purchased by citizens, worldwide. In this system, project entities are not liable to coin regulations, neither they have direct obligations with coin investors.

The engagement in an ICO strategy would require a project entity to develop a sophisticated financial and operating system, in partnership with outsourced financial services. In an ICO, coin revenues have to be exchanged into fiduciary currency and coin investors demand attention from the "project coin" entity. These are issues of high attention in compliance with regulations, demanding further risk evaluation for project entities aiming at funding resources with their own cryptocurrencies.

A Blockchain proposition for the Bolsa Floresta Family component considers the substitution of a fiduciary currency payment system for a tokenized currency.

Presently, the Program monthly rewards each family with U$ 13, holding a debit card. Each card has to be delivered to each household every five years. Most cardholders need to travel to distant cities with bank tellers or ATMs to get their cash. In addition, beneficiaries purchase merchandise in local retailers, using their cards as an automatic debit device under informal instalment payment plans. The travel and informal credit situations could be solved through "tokenization", where cards would be changed for tokens. Unofficial credits could be regularized with smart contracts.

A Blockchain proposition for Bolsa Floresta Income initiatives will provide more transparency in purchases, safer exchanges of goods and project data. Typical projects enroll local producers in the production chains of sustainable forest products, such as nuts, manioc flour, managed wood, fruit species and fishing. Producers receive equipment and assistance for their activities, demanding a high volume of equipment purchases and deliveries to communities. In addition, the project assists producers with entrepreneurship practices and supports the commercialization of local products. Blockchain solutions can improve the controls of equipment donations, shipments to communities, and collective transactions involving producers and buyers. A reasonable lack of trust and high paperwork demands are addressed by simple immutable registers of goods and sales. In addition, smart contracts will help the organization and efficacy of collective sales of fish, wood and other products.

A Blockchain proposition for Bolsa Floresta Association and Social initiatives will provide transparency and registry in community meetings and infrastructure registries. These components assist communities with infrastructure support, such as clean water, electricity and social centers. It also provides administrative guidance for their representative associations. The benefits are represented by clerical assistance, trainings and donations of office equipment. Blockchain provides immutable registers for equipment supplies and data storage with shared information among members, partners and stakeholders in the project. This transparency adds credibility to project reporting, contributing for a positive evaluation from donors and partners.

Field and management operations can be streamlined with Blockchain, resulting in economies of resources. Bolsa Floresta personnel provide guidance to families and associations in distant and remote areas, requiring air and river travels. Field missions to the riverine communities represent a relevant cost in logistic and staffing resources. Blockchain solutions could help the economy of resources with automatic and immutable data registers for field missions and key expenses, facilitating controls and audits, thus enhancing the Program´s credibility.

Forest conservancy programs, like Bolsa Floresta, frequently **relate to land use, ownership** and **illegal logging** issues. Land registry and titles, particularly in the Amazon, are imprecise and subject to fraud. Blockchain can assure landowners to have an immutable register for their properties. Past illegal possessions cannot be solved by this technology, but future immutable land registers will provide a positive outlook for less deforestation pressure due to secure land ownership in the Amazon.

The same solutions are available to illegal logging controls. Blockchain provides safety data storage for timber production and transportation sources, addressing data veracity and transparency. Immutable registers will help the lack of trust between locals, authorities, project teams, and stakeholders to combat deforestation with trustful tools. The implementation of such initiatives is complimentary to the Bolsa Floresta Program, recognized as part of a conservation project, in advocacy for effective public policies for a standing forest.

5 Use Analysis, Connectivity and Project Implementation

Blockchain for riverine communities can be analyzed under a framework consisted of a matrix showing the initiatives and key issues associated with such proposition (see Annex 1). As mentioned in Session 3, decentralization, dispersion and distrust among parties are positively solved by smart contracts and immutable registers. This matrix exposes the proposed Bolsa Floresta initiatives with key items like: databases; members' access and permits; hardware; software; connectivity, and ownerships. It serves as a support tool for an adequacy analysis of each proposition, with qualitative and quantitative data. Each initiative is related to: (1) the degree of decentralization among members; (2) transparency needs; (3) number of participants; (4) assets and information involved; (5) number of transactions; (6) disagreements between members; (7) intermediaries; and (8) management of permits. The collected data indicate a high adequacy of the Blockchain solutions proposed for the Bolsa Floresta initiatives.

The general system architecture for the mentioned projects will include a web interface utilizing an **Application Program Interface** (API), and mobile phones. The API will perform a set of routines, protocols, and tools with software applications under an *Ethereum* Blockchain platform, probably using the *Hyperledger* open source collaborative database. This system will be managed by a central party (project entity or a separated association), connected to mobile phones (using apps or SMS protocols) and personal computers. Families and communities will relate to each other using these devices. The API will store the key immutable information in the Blockchain. Further definitions under this project are to be developed with a technical implementation guide and continuous project guidance with communities.

Presently, about 70% of the 9.420 families assisted by Bolsa Floresta have mobile phones, but few communities have internet coverage. The Blockchain projects will begin with connected locations, and progress in stages, to the entire group, as follows:

1) Implement tokenized cards, immutable registers and first smart contracts at Rio Negro Communities, covered by internet, with 19 communities and 626 families. In this stage, immutable registers will be also implemented for associations to store key data and equipment deliveries, representing all 9.420 families, in the 16 conservation units. These registers can be recorded with offline devices (tablets or smartphones), with data later uploaded to the project web interface in urban locations, where associations keep an office or representative.
2) Next, tokenized exchanges and smart contracts will be extended to about 150 communities and 3.000 families living close to urban settlements. These locations have a positive outlook to receive telephone and internet connectivity with antennas or satellites, in three to five years. It is presently visible the efforts of governments and public policies toward connectivity in remote areas in the Amazon.
3) Finally, permanent offline Blockchain solutions are in accelerated stage to become effective, enabling Blockchain for remote communities. As an example, Cryptofuse, Inc. [9] has launched an offline/online connectivity technology, the *ByteBlock Protocol*, which enables users to start a Blockchain transaction offline, using portable proprietary devices, performing temporary block storages, to be closed when the devices are connected to the internet. This technology (and similar to come), will deliver a positive outlook for the Amazon forest communities and others in remote areas, worldwide, to be engaged in Blockchain.

6 Conclusion

The preliminary results of a Blockchain proposition for PES models and forest communities in the Amazon indicate the creation of electronic platforms for local communities to exchange values and information with markets and projects' stakeholders. These exchanges will reduce financial intermediaries and transportation needs between communities and local cities. Blockchain will enable communities and local producers to have more resources and information under a circular economic model, supported by trust, based on immutable registers and smart contracts. This is a realistic scenario to be introduced in five years, in a phased approach guided by local established project entities, using pilot projects to learn, adapt and grow new technologies to more beneficiaries under local conservancy programs. These entities have a key role to overcome expected weaknesses in bad connectivity, possible distrust from communities, partners, and lack of funding. These issues will require temporary offline solutions for internet connections, the consent of associations and communities in every project step, and secured project funding. Finally, unexpected situations will be addressed with project management practices, under the common knowledge of entities with positive project implementation records. A Blockchain proposition for PES and the Amazon region and conservancy projects, worldwide, does matter.

Annex 1 - Blockchain Use Analysis for Bolsa Floresta Program

	Bolsa Floresta	What Problem will be solved?	Degree of Decentrali-zation	What is the Transparen-cy Needed?	Participants	Assets (annual avg)	Transactions (annual avg)	Opposite interests Smart Contracts	Intermedia-ries involved	Who will have access	Who will manage permits
1.	**Bolsa Floresta - Family Program**										
1.1	Tokens instead of cards	Discontinue plastic debit cards, valid for 5 years and not necessarily controlled by benefitiaries	high	not relevant	above 8.500	R$ 5.200.00 (USD 1,3 million)	above 1.500 deliveries	not relevant	bank	tokenholders. But may be off line	FAS and benefitiaries
1.2	Tokens as credit instruments	Possibility of better use of cards as collateral for microcredit finance	high	credit and interest incurred in each card	above 8.500	R$ 5.200.00 (USD 1,3 million)	3.000 (estimated)	credit issuers and cardholders	financial services, bank	tokenholders. But may be off line	Residents' associations and FAS
1.3	Fair credit contracts with retailers	Eliminate unfair trade & credit practices from Retailers holding resident's cards	high	credit and interest incurred in each card	above 8.500	R$ 5.200.00 (USD 1,3 million)	30.000 yearly (estimated)	retailers and cardholders	retailers, bank	tokenholders. But may be off line	Residents' associations and FAS
2.	**Income Generation & Entrepreneurship Component**										
2.1	Supplies and shipments control	Lack of trust among people involved in receiving donated goods	high	key deliveries' information	580 communities	USD 600.000	600	FAS and communities	no intermediaries	community reps., FAS	project entity (FAS)
2.2	"Cantinas" transactions	Commercial coditions defaults from suppliers	medium	suppliers' information on deliveries	presently, 2; potentially, 50	presently, USD 400.000	200 (tbc)	Buyers and vendors	river freights	buyers, sellers and freighters	buyers
2.3	Collective production sales - distribution of income	assure correct payments to producers on collective sales	medium	Producers acknowledge their share on collective sales	2.000 producers (tbc)	USD 400.000 (tbc)	800 (tbc)	community Sellers and producers	river and city freights, sales points	buyers, sellers and freighters	Sellers' associations
2.4	Forest Management source & delivery control	assure correct wood extraction and processing information to avoid illegal logging	medium	Regulators, buyers, third parties know correct information	start with 8 producer areas	USD 50.000	40	producers and buyers	environmental agency and certificators	producers, buyers, certifiers	regulators
2.5	Tokenization with suppliers	decrease the cost of financial transactions	medium	match purchase volume in fiat currency with tokens	10 (for FAS case)	USD 300.000	200	buyers and vendors	donor of funds (the "tokenizer")	buyers, sellers and donors	donor
3.	**Association & Social Component**										
3.1	Supplies and shipments control	Lack of trust among people involved in receiving donated goods	high	key deliveries' information	580 communities	USD 200.000	600	FAS and communities	no intermediaries	community reps., FAS	project entity (FAS)
3.2	Association meetings' key decisions	Ensure registration of key commitments in communitarian meetings	high	show all participants, key decisons	580 communities	to be estimated, per community	not relevant	communities	possible, local governments	association leaders	FAS and leaders
4.	**Project Monitoring and Management**										
4.1	Key data safety	Assure immutable and general public storage of projects' key data	low	entity's projects and transactions data	FAS, donors and benefitiaries' reps.	USD 8 million	400 (100 indicators per quarter)	not applicable	not applicable	donors, project entity, benefitiaries	project entity (FAS)
4.2	Land Monitoring	Assure immutable and transparent information of forest coverage with geo-location sensors with cryptographic identification	high	key data link w/ image spots	580 communities. State Environment Agency and FAS	intangible	600 images	not applicable	State Environment Agency	open to public through internet platform	FAS
5.	**Project Financial Reporting**										
5.1	Key financial data in Blockchain	Assure immutable and specifica storage of projects' key financial data to donors	low	project key financial data	Project entity and donors	USD 8 million	60 (15 indicators per quarter)	not applicable	not applicable	donors and project entity	donors
5.2	Key project indicators in Blockchain - project platform	Assure immutable and specific storage of projects' key technical data to donors	low	project key financial data	Project entity and donors	USD 8 million	80 (20 indicators per quarter)	veracity	not applicable	donors and project entity	donors

References

1. Zorzetto, R.: Um rio que flui pelo ar, Revista Pesquisa Fapesp, pp. 63–65, April 2009
2. Aguiar, M.S.: Valor econômico da Floresta em Pé, II Sem. de Governança de Terras e Desenv. Econômico, 212
3. Nakamoto, S.: Bitcoin: A Peer-to-Peer Electronic Cash System (2009). https://bitcoin.org/en/bitcoin-paper
4. https://markets.bitcoin.com/
5. https://consensys.net/about/
6. https://www.ibm.com/blockchain
7. Virtual Currencies, Monetary Dialogue, European Parliament (2018)
8. https://www.iied.org/markets-payments-for-environmental-services
9. CryptoFuse, Inc.: CryptoFuse, The World's First Offline Blockchain, Whitepaper Version 1.2, May 2018

Development of a Human–Machine Interface Implemented in Smartphone for a Variable Rate Fertilizer Applicator

Ben-Hur Maciel[1(✉)], Ivan Mantovani[2], Matias Alles Hubert[1], Roberta Goergen[1], Carla Rannov[1], Luiz Antônio Rasia[1], and Antonio Carlos Valdiero[3]

[1] Regional University of Northwestern Rio Grande do Sul State, Ijuí, RS 98700-000, Brazil
begonhur@gmail.com
[2] Federal University of Santa Catarina, Florianópolis, SC 88040-900, Brazil
[3] Horizontina Faculty, Horizontina, RS 98920-000, Brazil

Abstract. This paper presents the development of a smartphone application to be a human-machine interface for a solid fertilizer dosing system with electric drive for seeder-fertilizer with dosage variable rate. Many facilities are available in applications in the commerce, communications, industry and services sectors, such as Uber, whatsApp and Airbnb. However, there are many innovation opportunities and challenges in family agriculture for sustainability and food security. We use the design methodology to develop a modular smartphone application with ergonomics and usability with the App Inventor 2 for Android operating systems and is suitable for a variable rate fertilizer doser. The dosing system consists of a feeder, driven by an electric motor of direct current powered by a high current battery, the drive is given through a power drive commanded by an Arduino microcontrolled platform of the MEGA series. And also has accessories such as Bluetooth, GPS module and Incremental encoder to give feedback in the control system. The application makes Bluetooth communication with the Arduino, which allows the user to enter the line spacing and amount of fertilizer, and display on the screen the values of the angular velocity, speed of travel, latitude and longitude of the seeder-fertilizer. It is intended to provide technological support so that workers can have a practical system of strategic decision making.

Keywords: Sustainable agriculture · Ergonomics · Usability · Design methodology · Farm automation

1 Introduction

This paper aims to present the development of a human-machine interface for a variable rate fertilizer applicator implemented in smartphone. In many business sectors, such as education, communication, tourism and transportation, mobile applications show exponential growth, but the mobile agricultural app does not show such growth despite the importance of agriculture in the economy. Agricultural applications can be developed

L. Pereira et al. (Eds.): IDEAS 2018, SIST 198, pp. 356–362, 2021.
https://doi.org/10.1007/978-3-030-55374-6_35

through better access to market information, weather and pests, better access to financing, payment methods, simpler and more efficient communication between producers, suppliers and buyers, improve access to agricultural extension services, such as good agricultural practices [1].

A survey of 57 British and French farmers showed that fifty-nine percent use field-specific applications. Research shows that application engagement varies on farms so respondents have named four applications in the field in the media, ranging from one to ten named applications. The most popular applications among respondents are weather applications, rain radars and weather alerts, soil mapping applications and weed control applications [2]. This suggests a shortage of applications that are a human-machine interface between farmers and agricultural machinery.

A common practice in the field is to ignore specificities of the field to treat it homogeneously in the act of not considering a series of site and soil specific factors to adequately treat the plant, such as the application of fertilizers and pesticides, considerable costs and a source of environmental pollution. The application of precision agriculture technologies, such as variable rate fertilizer applicator, allow a suitable treatment for each site in the field. Even in computer-controlled machines, adjust the amount of fertilizer applied to the soil as the machines move through the soil [3].

An experiment carried out in citrus fields in São Paulo, Brazil, for six years, found to be higher productivity and a reduction in the consumption of inputs when variable rate fertilization was used in comparison to the application of fertilizers through a uniform rate in areas with great soil and topographic vagueness [4].

Over the years, the emphasis of precision agriculture has changed, through variable-rate technologies, vehicle guidance systems for product quality and management. So, the implementation of precision agriculture increases the number of correct decisions per unit area of land per unit time. Thus, applying the use of specific technologies, so the decisions that can be taken by electronic sensors can also be made by humans [5]. That said the developed smartphone application has automatic dosing options, after a soil mapping, or manual, in which the operator defines the spacing between the lines and the amount of fertilizer.

Firstly, it will be described the dosing system in which the human-machine interface was developed. Secondly, the human-machine interface implemented in smartphone will be described, then your operation will be displayed.

2 The Dosing System

The dosing system is composed of a helical feeder with overflow of the Fertisystem line. That is connected, through a system of gears and to the Bosch DC electric motor. An Arduino MEGA microprocessor platform is used to control motor through PWM signal that changes its rotation. The interface between the Arduino and the Motor is performed by a power drive connected to a high current battery, depending on the motor's operating current being high and thus being higher than that supported by the Arduino's digital output port.

The system has an incremental encoder installed in the axis of the doser, which collects angular displacement data, which when applying a derivative in this displacement results in the angular velocity, the collected data are sent to the Arduino. For the application of a PID control in closed loop. To obtain geo-referencing data and speed of travel of the seeder-fertilizer, a GPS device was implemented to the system. Another accessory of the system is the Micro SD Storage Board that is used to save in the Micro SD card all the data of position, quantity of fertilizer and speeds of the process of fertilization.

To interface between the smartphone and Arduino was added a Bluetooth device HC-05, which connects only with devices with the android operating system and has the master-slave operating model. This is just the smartphone sends commands addressed to the Bluetooth device.

Figure 1 shows the diagram of the design of the dosing system, which has its operation initiated by the smartphone that communicates via Bluetooth with the Arduino. It receives data from the GPS and activates the DC motor coupled to the feeder. The process ends when the data generated by the encoder is sent to the Arduino and it uses it as feedback of the PID control and sends together the speed of movement and location for the smartphone. After finalizing the fertilization, the operator can send the data to the SD card.

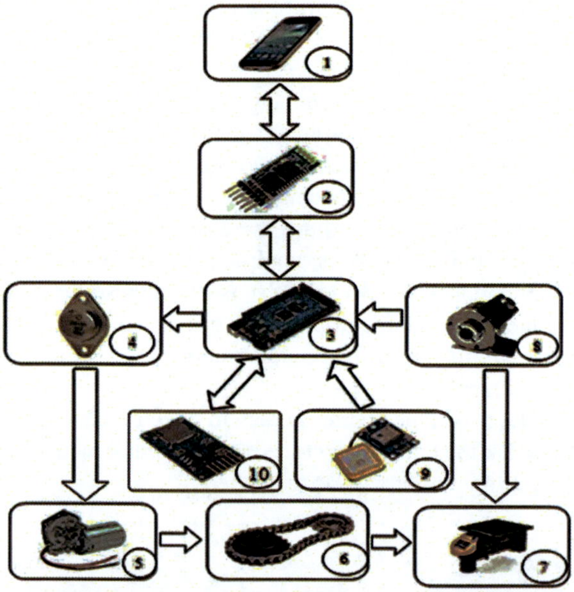

Fig. 1. Diagram of the design of the dosing system Where (1) is the smartphone, (2) a Bluetooth HC-05 module, (3) a Arduino Mega Microcontroller, (4) a power driver, (5) a DC motor, (6) a chain drive system, (7) the FertiSystem feeder, (8) an incremental encoder, (9) a GPS module for Arduino, (10) Micro SD Storage Board for Arduino. Source: Elaborated by authors

The control system shown in the Fig. 2 below consists of a system in which there is the input of amount of fertilizer, spacing between lines through the smartphone. Along with the location data that are sent from GPS to Arduino Mega Microcontroller.

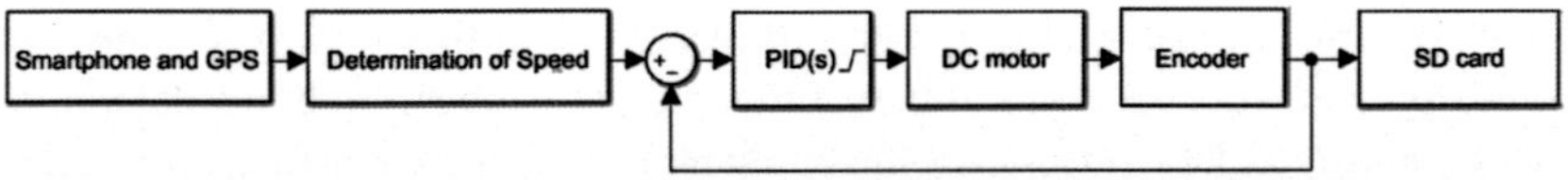

Fig. 2. Closed loop diagram of the control system. Source: Elaborated by authors

From the amount of fertilizer to be deposited, the desired angular velocity of the doser is determined. Which is activated by a control signal sent to the DC motor. The system is fed back with the angular speed relayed by the meter, which is collected through an encoder connected to its axis. In order to perform closed-loop PID control, all collected data is finally sent to the SD card.

3 Human–Machine Interface

Table 1. Application icons and descriptions.

Icon	Description
	Amount of fertilizer
	Line spacing
	Travel speed
	Angular speed
	Geographic positioning
	Latitude
	Longitude

Source: Elaborated by authors

The human–machine interface implemented in the smartphone consists of an application whose programming was developed in the software AppInventor2, developed by the Massachusetts Institute of Technology - MIT, whose one of its limitations is that it only applies to devices with android operating system.

Firstly, a series of icons were created to represent, in this way, it was possible to reduce the space of occupation of these functions, allowing for future enlargements. Also, with the use of the icons leaves the application with a greater intuition, in Table 1 is presented the icons next to their respective descriptions.

The top part of the home screen is composed of buttons of the functions connect, record, automatic, manual and mapping. Pressing the "CONNECT" button will open a list of paired devices in the device, that is, before starting the application the smartphone should pair with the Arduino, in the list one must select the device named "doser". This will display a message that the Bluetooth is connected. At this moment, the angular velocity, travel speed, and geographic positions will already be displayed in the four fields below according to the icon of each field, the travel speed of the seeder-fertilizer is also displayed at a speedometer on the right side of the screen.

After connecting, you must choose between manual and automatic mode of fertilization. When the manual, mapping, or automatic option is selected, a dark green circle to the right of the option will become light green. The manual mode is simple and intuitive. First must select the manual option to then determine the line spacing and the amount of fertilizer, the determination is made by selecting blank fields next to the respective icons, and after entering the value select the "OK" buttons.

From the amount of fertilizer to be dumped the Arduino sends the PWM signal to change the rotation of the electric motor, and consequently of the feeder, which has its angular velocity verified through the angular displacement derivative collected via incremental encoder, for PID control in closed loop. When finalizing the fertilizer process in manual mode, the angular value, fertilizer quantity, travel speed and geographic position data can be saved to the SD card. The saved data is used in the automatic mode.

The Fig. 3 below shows the manual mode interface on the smartphone with the fields of the values of the feeder angular velocity, speed of travel, latitude and longitude of the seeder-fertilizer that are updated during the fertilization process. And, the fields for the inputs of the fertilizer amount and the line spacing.

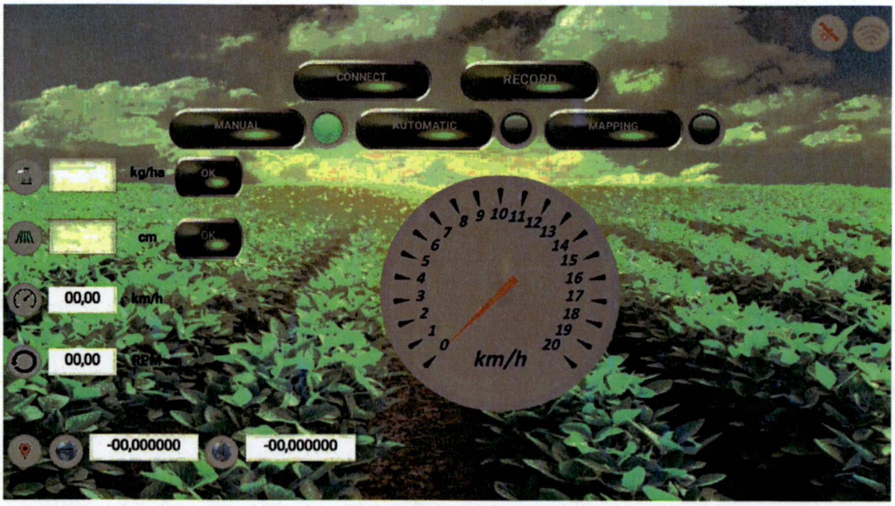

Fig. 3. Interface in manual mode on smartphone horizontal position. Source: Elaborated by authors

Before using the automatic mode, the mapping of the area must be done. To perform the mapping, must be selected the mapping option that will change the initial screen, so that the options to mark a point and restart will appear, and the options of determining the line spacing and the amount of fertilizer will disappear. The latitude and longitude geographic information fields from the GPS will remain as long as the angular velocity and speed travel information fields will disappear. The figure below shows the interface to the mapping option (Fig. 4).

Fig. 4. Interface in mapping mode on smartphone horizontal position. Source: Elaborated by authors

The mapping consists of marking the external points of the place to fertilize, so that by selecting the "MARK" button it will save the latitude and longitude data of each point. Once the area has been fertilized, the geographic positions of the points are sent to the SD card.

After the mapping is completed, the automatic fertilization process can be performed, just select the automatic option. When selecting the automatic option, the Arduino will receive data of the geographical position of the GPS module and check if they are inside the area mapped and recorded in the SD card.

After the check, the Arduino will search the amount of fertilizer that was deposited in this place during the manual fertilization process, which is recorded on the SD card. Having the amount of fertilizer to be dumped, the Arduino will perform the same process manually, sending a PWM signal that would change the angular velocity of the feeder, and performs closed-loop PID control with the angular velocity of the feeder.

4 Conclusion

In this paper, we present the human-machine interface for a variable rate fertilizer applicator implemented in smartphone. The implementation of this technology with intuitive aspects and that provides a presentation of relevant data to the farmer, has the potential to assist the work in the field in order to make it more automated and facilitating the taking of decisions.

The development of a human-machine interface for a variable rate fertilizer applicator seeks to make agriculture more sustainable and productive, to guarantee food safety, with the adequate dosage of solid fertilizers on the farm.

Acknowledgments. The authors would like to express their gratitude to CNPq (National Council for Scientific and Technological Development), and to Unijuí (Regional University of North-western Rio Grande do Sul State), and to FINEP (Studies and Projects Financing Agency) for the support to the Nucleus of Innovation in Automatic Machines and Servo Systems (NIMASS), through the public call MCTI/FINEP/CT-INFRA - PROINFRA - 02/2014 – Multiusers equipment, Ref.: 0141/16 (Electronic protocol: 124), with the approval of resources to purchase equipment for the construction of prototypes for master's and doctoral research. This work had the financial support through the project called: "Research on Mechatronics geared to the Challenges of Society" (Grant term no. 17/2551-0001014-0) no EDITAL FAPERGS 02/2017 - PqG (Gaúcho Researcher Program).

References

1. Costopoulos, C., Ntaliani, M., Karetsos, S.: Studying mobile apps for agriculture. IOSR J. Mob. Comput. Appl. **3**(6), 44–49 (2016)
2. Dehnen-Schmutz, K., Foster, G.L., Owen, L., Persello, S.: Exploring the role of smartphone technology for citizen science in agriculture. Agron. Sustain. Dev. **36**(2), 1–8 (2016)
3. Aubert, B., Schroeder, A., Grimaudo, J.: IT as enabler of sustainable farming: an empirical analysis of farmers' adoption decision of precision agriculture technology. Decis. Support Syst. **54**(1), 510–520 (2012)
4. Colaço, A.F., Molin, J.P.: Variable rate fertilization in citrus: a long term study. Precision Agric. **18**(2), 169–191 (2017)
5. McBratney, A., Whelan, B., Ancev, T., Bouma, J.: Future directions of precision agriculture. Precision Agric. **6**(1), 7–23 (2005)

Comparing Aerial Platforms for the Amazon Application Scenarios

José Reginaldo H. Carvalho[1](✉), Samuel S. Bueno[2], and Josivaldo Modesto[3]

[1] Instituto de Computação, UFAM, Av. Rodrigo Otávio, 6200, Manaus, AM 69077-000, Brazil
reginaldo@icomp.ufam.edu.br
[2] CTI Renato Archer, Rod. D. Pedro I (SP-65) km.143.6, Campinas, SP 13069-901, Brazil
[3] IDSM-Instituto de Desenvolvimento Sustentável Mamirauá, Tefé, AM, Brazil

Abstract. This paper presents a comprehensive comparison of most known aerial platforms: fixed-wings, multirotor, and airships. The goal is to evaluate each platform´s characteristics with the operational conditions of the Amazon forest. Among the most challenging aspects, one may cite the large area (almost all Europe fits in its territory), lack of access by land, a typical of twelve meters displacement annual flood cycle, hot and humid environment. These conditions impose a set of design requirements that no current aerial platform can meet. Instead of declaring a winner, the intention is to define the limits of each platform, showing their commentary aspect, and list recommendation of how and where to apply them. The analysis is restricted to small to medium equipment for civil applications. Moreover, the paper reflects about the combination of aerial vehicles with wireless sensor networks to do risk management of large areas of forest. The purpose is to help users such as environmental agencies, non-governmental organizations, and other civil organization to benefit from these new technologies, but cannot afford to acquire an unmanned aerial system form a private or military company.

Keywords: Environmental monitoring · Aerial systems · Unmanned aerial vehicles

1 Introduction

The Amazon region is rich in natural resources and well known for its ecological relevance. There are several programs worldwide aiming to protect its biome, however, the protection of the Amazon region poses challenges comparable with the size of the forest. The Amazon biome occupies an area comparable to all Western Europe, with wetlands, floodplains and dry land. There are very few roads; the most common transportation means is through the river. The anthropogenic pressure for deforestation in certain regions aggravates the scenario.

There are technological-based initiatives focused in the Amazon region. Among them, we have LBA/INPA [1]. With unquestionable value in bringing the knowledge about the deforestation process, the LBA plays an important role in the definition of long-term policies. Most of the information come from satellite images made available

L. Pereira et al. (Eds.): IDEAS 2018, SIST 198, pp. 363–373, 2021.
https://doi.org/10.1007/978-3-030-55374-6_36

at intervals of days. Despite their importance, they have little effectiveness for short-term actions, such as starting a firefight before it spreads, or sending law enforcement agents to an area where deforestation is in course. In such cases, low altitude unmanned aircraft systems (UAS) has indisputable advantages. These systems provide multimodal data with high regularity, small granularity and low.

We can divide UAS into two large classes: based on fixed and mobile aircrafts, applied either alone or combined in collaborative systems. Fixed systems consist on aerostats. They are flexible, versatile and relatively simple to deploy. The envelope's volume depends on the desired operating condition. Aerostat results in much less aggressive intervention to the environment than a conventional communications tower, with comparable reliability and resistance to atmospheric incidents. They can host several types of sensors, and provide telecommunication infrastructure [2].

Mobile systems, based on unmanned aerial vehicles (UAV) (aka drones) have shown continued market growth penetration. Initially aimed at military missions, they cover a wide and relevant range of civilian applications. Nowadays one can buy off-the-shelf fully featured UAS from the Amazon.com.

The contribution of this paper is two-fold. First, we cover the most common aerial platforms, comparing their characteristics, considering the application in the Amazon operation scenarios. Then, we will discuss case-studies of threats to the forest, and how unmanned aircraft systems could be of help.

The remaining of the paper has the following organization: In Sect. 2 the reader will find the fundamental concepts related to unmanned aircrafts systems. In Sect. 3 these systems will be further detailed, and we will compare their main characteristics against the application in the Amazon; In Sect. 4 we will list two case studies of real threats to the biome and how these aerial platform could be of help. In Sect. 5 we will cite ongoing initiatives on the subject of aerial systems for the Amazon. Finally, in Sect. 6 we conclude the paper.

2 Fundamentals About UAS

In this section we briefly discuss the most important concepts about Unmanned Aerial Systems (UAS). Usually, they are divided into two perspectives:

1. Command and Control Perspective (C&C): it is the pilot's perspective. The C&C is responsible to fly the equipment safely, and in accordance to the flight plan approved. The C&C has the authority to abort the mission in case of risks to life or to the equipment. Also, the C&C may deny changes in the ongoing flight plan;
2. Mission Payload Perspective: The payload is the entire load that is not involved in making the aircraft to fly. However, all aircrafts takes-off for a reason, and the payload concerns this reason. Cameras, spreading pesticides, of even (and sadly) ammunition, are example of payload. The payload has no authority to change the current flight plan. It has to request permission to the C&C. The principle is that mission safety has higher priority than mission completion.

The total load of an aircraft is the sum of vehicle, fuel, and the payload weights. The load of fuel plus payload it the useful load, which decreases as the mission progresses.

For aircrafts propelled by electric motors (powered by batteries or solar panels), payload and useful load are always the same and one refers only by payload. Figure 1 shows the main subsystems of a UAS, detailed as follows:

- **UAV or the vehicle itself:** A UAS has one or more vehicles, usually with the same characteristics, although heterogeneous systems are increasingly available. There are three types of airframes: fixed wings (airplanes), rotating wings (helicopters) and lighter-than-air (airships). There are hybrids projects and, the experts might review this classification in a near future.
- **Ground station:** Fixed or mobile computing system that hosts the necessary equipment and software to support the UAV. The tasks of a ground station include planning and monitoring the flight execution, real-time visualization of data, and the capability to intervene in the ongoing mission;
- **Telecommunication subsystem:** To enable communication between UAV and ground station. It has at least four links: C&C, telemetry data, the payload data-link, and a fourth for back up, in the event of a failure.
- **Data Center:** information system that will gather all data acquired and processed. This subsystem was recently incorporated into the UAS due the need of specific systems to deal with the large amount of available data.

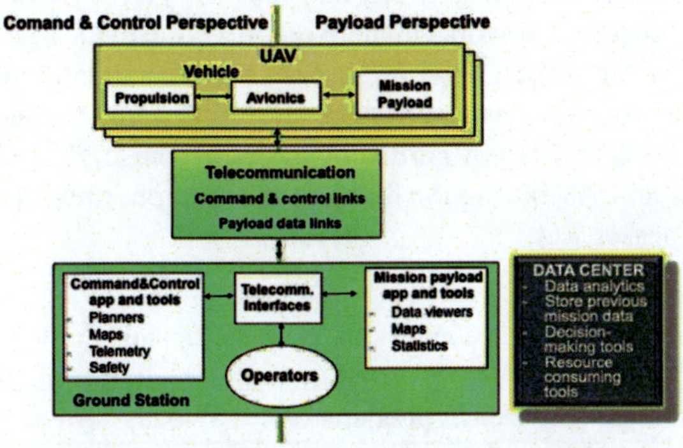

Fig. 1. Main sub-systems of a typical UAS.

Mission phases are common for any aircraft (e.g. takeoff, landing). The specifics are the necessary tasks to perform them (e.g. taxing, avoid obstacles). Mission profile is the cross combination of phases and tasks, from preparation to mission finalization. Figure 2 shows an example a mission profile matrix for an airship.

Concerning decision-making capacity, UAS can be Remotely Piloted Air Systems (RPAS), or autonomous systems. There was a third class, the semi-autonomous systems. However, the other two incorporated it. That is, every RPAS system has some autonomy and none autonomous system is completely autonomous. All these components are intensive in the use of computational resources for their reliability and robustness. As a result, their design takes into account mission critical requirements.

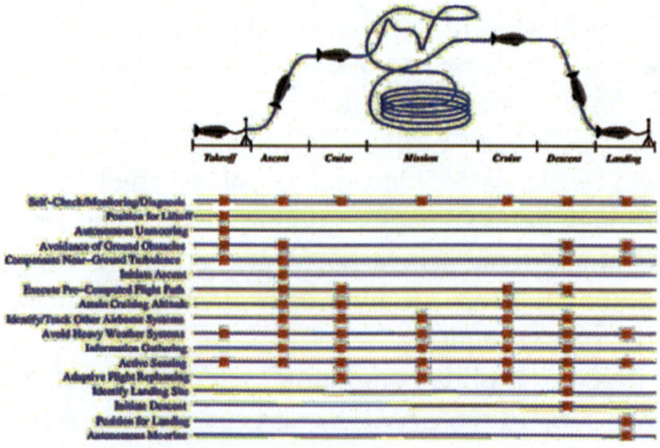

Fig. 2. Mission profile matrix for an airship.

3 UAV Platform Comparison

In this section, we compare the airframes previously mentioned. We will focus in the Amazon rainforest operational scenarios. Figure 3 shows three examples of existing UAV per airframe. The three fixed-wings and the helicopter are all military. The two quadcopters and the airships are for scientific purposes. Concerning the application in the Amazon rainforest, we believe that all frames have its value. Table 1 compares each airframe with respect to its maneuverability and capability fly, as VTOL (vertical takeoff and landing). An airship has the worse maneuverability due to its high virtual inertia (the effect of the air in the envelope), however, it can take off and land in different terrains. Multirotors is the best ones in terms of maneuverability, being the most flexible among all airframes. However, this is not sufficient to define the flight plan. Table 1 shows important aspects of an UAV during the flight, with airframes presenting their respective strengths and weaknesses.

Table 1. Comparing maneuverability and mission capability of UAV airframes.

Airframe	Example	Maneuv.	Endurance	Velocity	Hover	Coverage	VTOL
Fixed wings	Airplanes	Good	Medium/high	High	No	High	No
Rotating wings	Quadrotor	Excellent	Low	Low	Yes	Low	Yes
Lighter-than-air	Airships	Fair	High	Low	Yes	Medium	Yes

Source: Elaborated by authors

One also has different UAV with respect to the operational conditions. Table 2 shows the most used classes. For the Amazon scenarios, we understand that, with the exception of the HALE (high altitude, long endurance) vehicles, the others have their application. Note that there is not winner in all aspects. Instead, one would combine the UAVs to benefit from its strengths. Section 4 will discuss this in detail.

There are several different classifications of UAS, elaborated mostly but the defense and military organizations around the world. Although, there is no consensus, all of them

Fig. 3. Three examples of fixed-wings, rotating wings, and lighter than air UAV.

define classes in several tiers, with respect to the goal application. Maybe, one of the most recent classifications is with respect to its readiness to use [2]:

Table 2. Operational conditions.

	Altitude (m)	Coverage radius	Endurance
Hand launched	600	2 km	Low
Short-range	1,500	10 km	Low/Medium
Tactical	5,500	150 km	Medium/high
MALE	9,000	200 km	High
HALE	>9,000	>200 km	High

Source: Elaborated by authors

- Ready-to-fly (RTF)/Commercial-off-the-shelf (COTS): They include software that allows any person to use it. Examples are the DIJ Phantom end-user series;
- Bind-and-fly (BNF) – require minimum knowledge to fly the platform. Most research UAV fall in this category. The manufacturer assumes that research staff wants to have some flexibility in changing parts and/or software components;
- Almost-ready-to-fly (ARF)/Do-it-yourself (DIY) – require significant knowledge to get in the air. They are becoming common, as UAV is becoming known equipment. Examples are the numerous UAV kits one may find at Amazon;
- Bare frame – requires significant knowledge and your own parts to get it in the air. They are for those interested in specify and design their own equipment. In this case, knowledge includes specification of propellers, thrusts, speed controller, autopilot, and everything that composes an operational UAS.

We believe that, in the Amazon scenario, one has to select ARF based UAS. In our experience, both RTF and BNF will be soon or latter left in a corner, out-of-office. There is no repair office in the jungle. Scientists and research staff have to train technicians capable to, at least, assemble DIY kits. We have several examples of colleagues who had to spend more that the UAV costed to send it back for maintenance of to bring replacement parts. Moreover, to build an ARF might be fun.

In the next section, we present our configuration of UAS for selected application scenarios. Cost, flexibility, and the capacity to do the assembly and maintenance were the non-negotiable requirements. In this paper, we selected a combination of tactical airship with hand-launched airplane (adapted to a higher endurance) and hand-launched or tactical quadrotors. The support of a tethered aerostat was also of use.

4 Application Scenarios

Our proposal is to consider all platforms and to search for complementary use in dividing to conquer [3]. Any Amazon region of interest is extremely large to a single platform. For instance, the Mamirauá reserve is comparable to one third of Portugal.

One may mention the satellite monitoring. It coves almost the entire region at one passage. Of course, it has its value on monitoring the forest, but in the context of this work, we are interested on scenarios that demand data with three attributes:

- High regularity: it means short interval between samples of the same subject. High regular data has the capacity to detect changes faster. Fire, deforestation and flood are example of events with high changing rate;
- Small granularity: The granularity defines the quantity of the real world measured by a unit of data sample. For images granularity is the amount of square meters per pixel. Image resolution, combined with the distance from the subject, defines the granularity of an image;
- Low latency: Latency is the elapsed time between data acquisition and availability. In image data, for instance, as larger the resolution is as larger will be the latency, as the equipment has to manage more information to have a single shot ready for use.

Although satellite information has unquestionable importance to provide information about the Amazon region, it just cannot trigger immediate actions to cope with ongoing harmful events. The application of UAS adds value, however, no one on its good sense will to monitor any Amazon region using a specific UAS solution. Fixed-wing's mission profile is restricted, multirotors, although maneuverable, have low autonomy and airships have issues with wind and turbulence. Those platforms can cooperate, and their strengths overcome the mutual weaknesses. Combined they provide multimodal data with high regularity, covering a sufficient large area. In the next subsections we mention three scenarios where the joint application of different platforms are more beneficial that a specific one. Due to the space limitation, we will detail the first one alone: the risk management approach.

4.1 Risk Management Against Deforestation

We consider deforestation as a risk management problem. As such, it consists of a list of risks mapped in both impact and probability to occur. Every risk has events that start a sequence of mitigation actions. We propose the combination UAS platforms to trigger those actions. We define the concept of risk perimeters. They are borders correlated with deforestation risks, as follows.

Imminent Risk Perimeters: They are regions where deforestation is almost certain. Examples are forest segments bordering areas with human influence, such as urban areas, or access to roads. This perimeter demands monitoring with high regularity.

Usually perimeters of imminent risk are relatively small when compared to the forest regions. This perimeter has typical length in the order of kilometers. The aim is to detect small changes in the biome using image, sound, or other sensory information.

An interesting solution is to combine wireless sensor networks associated with rotating wing UAS. Quadrotors would overfly blind spots or verify if a detected event is indeed deforestation.

High-Risk Perimeter: Areas of relatively easy access by humans that did not fall into the imminent risk class. Thus, monitoring needs less regularity. We propose the combination of the autonomy of fixed wings vehicles (to overfly the perimeter), with the flexibility of rotating wings aircrafts (for local inspections). The course of navigable rivers and roads, the limits of farms, logging and mining companies, are examples of perimeters of high risk.

Permanent Protection Areas: Areas of low probability of deforestation, but that still demands care. In this case, the system monitors the region with less regularity. The flight autonomy of an airship is a key aspect in this scenario. It would perform regular overflights to monitor large areas per mission.

As case study, let us take the reserve Adolpho Ducke, in Manaus (Fig. 4, left). This reserve is under such deforestation pressure that its distinct green square gives the impression of an artificial artifact. Figure 4 (right) shows irregular deforestations

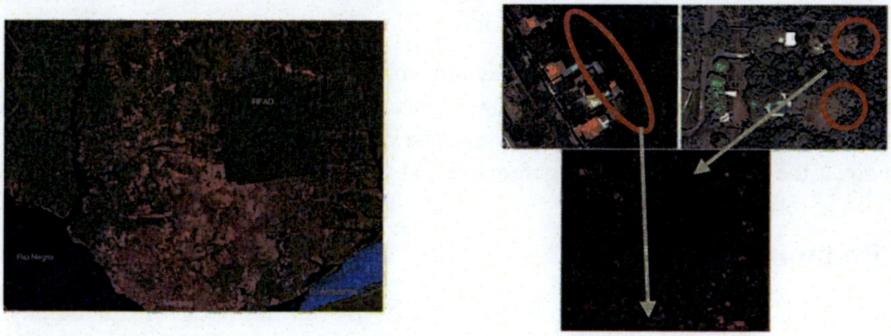

Fig. 4. The Adolpho Ducke reserve (left), and events of deforestation (right).

by residents neighboring in an urban protected reserve. Hidden by the walls of their residences, they make the forest part of their backyard.

Another example is the deforestation known as "fishbone", as in the area of influence of BR 174 (Manaus – Boa Vista) (Fig. 5). A paved road opens many possibilities for criminal deforestation, by offering easy access of men and machines.

It is important to remark that, not all deforestations are illegal. The Amazon has large infrastructure projects that will certainly produce high environmental impact.

Fig. 5. Fishbone deforestation phenomenon.

It is part of governmental responsibility to approve the project. The problem is that these projects tend to promote disorderly economic changes. Such changes will certainly extend the environmental impact far beyond the original project.

In this case the, risk management policy would be dynamic. At the project start-up, the whole perimeter is of imminent risk. As it progresses, the stakeholders reevaluate the risks, by assessing the status of the environmental impact. In the end of the project, the region receives the status of permanent protection. The proposal is to contain the deforestation to what is strictly necessary.

4.2 Other Relevant Application Scenarios

Fighting Bio Piracy: Another relevant scenario is the combat of bio-piracy along the Amazon River and tributaries. They are extremely dangerous. Combining the flexibility of a multirotor with the longer range of an airplane the law-enforcement agents would be well prepared to combat these criminals;

Early Fire Detection: It is of fundamental importance to detect a fire in the forest before it spreads. In this case, long-range cameras positioned in threes or towers, with image based smoke detection, and airplanes to verify the extension of the fire. Airship will hover the fire, giving to firefighters the real-time status of the situation.

5 Preliminary Results

In this section, we will present three low cost UAV platforms under construction by our R&D network. They are a zagi flying wing, an airship, and a tethered aerostat. We also tested three models of quadrotors.

UAS Based on a Fleet of Hand-Launched Mini UAV: This project aimed to build a mini-UAV with around 1 h autonomy. The researchers in charge of the vehicle design decided to use the zagi flying wing airframe. The final prototype had 2.2 m wingspan, powered by one rear brushless motor and guided by an autopilot from Micropilot, or a gumstix with own autopilot (Fig. 6). The payload consisted by a GoPro Hero 3+ pointed to the ground. The system can control up to four UAV per mission. The researchers proposed a bi-criteria optimization problem to minimize the number of UAV applied and the total mission time, simultaneously [4].

Fig. 6. Zagi flying wing.

UAS Based on an Airship: This project consists of a four propelled 11 m long airship (Fig. 7). The thruster configuration confers to the vehicle higher maneuverability than typical for airships. Onboard software and autopilot consider state-of-the art algorithms for both stand-alone and cooperative missions [5, 6]. Payload includes camera, Lidar, and a set of atmospheric sensors. The vehicle has a long-range data link based on Microhard solutions. The expected endurance of the final prototype is of up to 2 h. The airship inaugural flight is available at https://youtu.be/Ld8Hsmeak2U. The logos of involved institutions may be seen in the banner on the airship envelope.

Fig. 7. Airship for environmental monitoring.

A Nomadic Aerostat System: Researchers and riverbank communities in the deep jungle are often isolated by weeks. The information acquired by UAS might take days to arrive in the data center. This situation hurts both data latency and regularity. The aerostat is a multi-purpose platform (Fig. 8-left) designed to stay at up to 150 m from the ground, having its own payload (cameras), and data-radio links to the ground (IEEE 802.11n) and to a fixed tower (IEEE 802.11ac), as in Fig. 8-center.

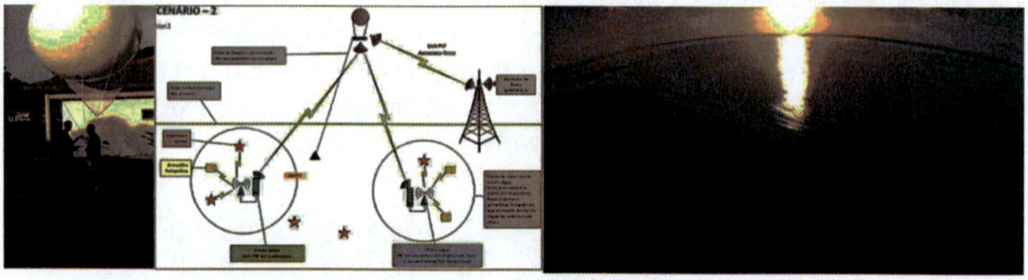

Fig. 8. Aerostat (left). Utilization scheme (center). Image from the top (right).

It also works as data relay to UAVs, extending their communication range. Being nomadic means it is easily deployable and recovered. The users might move from one site to another. A GoPro Hero 3 at 130 m took the picture of Fig. 8-left during field trial. We designed these systems with focus on cost, and simplicity. The bill of material consists of standard off-the-shelf components. We also have instructed our partners to apply quadrotors; however, instead of doing a new project we suggested the well-known F450 or F550 DIY kits.

6 Conclusion

This paper presents a sound review of the most important UAS concepts, along with a comparison of flight profile aspects of the three air frames. The operation scenarios of interest is the Amazon region, and the mission in focus are those where data regularity, granularity and latency matters. Examples are monitoring missions of high rate changing environments. The authors understand that in the deep jungle, it is crucial for the technical and scientific staff not to depend on manufacturer expertizes to operate and maintain their equipment. The paper proposes to combine different and complementary types of low-cost and easy to use UAS platforms to cover a large area of forest in a divide and conquer approach.

As a concrete contribution to the subject, the authors detailed a specific and relevant application, the risk management against deforestation. There are also two other detailed applications, however, due to space limitations, the authors just mentioned them. Furthermore, the authors are involved in the design of the different UAS platforms, all based on affordable off-the-shelf components. Future works include more field trials and the availability of the detailed project to the open community.

Acknowledgments. This work is sponsored by ARTES (FAPEAM 087/2014, 114/2014), DRONI (CNPQ 402112/13-0), and INCT (CNPQ 465755/14-3, FAPESP 2014/50851-0), Brasil.

References

1. Avissar, R., Silva Dias, P.L., Silva Dias, M.A.F., Nobre, C.A.: The large-scale biosphere-atmosphere experiment in Amazonia (LBA), J. Geophys. Res. **107**(D20), 8086
2. Dronewallah (2015-02-23). Knowledge Base: What are RTF, BNF and ARF drone kits? rcDroneArena. Accessed 3 Feb 2016
3. Carvalho, J., Moutinho, A., Azinheira, J.R.: Integrating mission planner to the flight control system of a robotic airship, XXI Congresso Brasileiro de Automática - CBA (2016)
4. Freitas, E., Carvalho J.R.H. Genetic algorithm approach for a class of multi-criteria, multi-vehicle planner of UAV. In: Evolutionary Multi-criterion Optimization. Springer (2015)
5. Azinheira, J., Moutinho, A., Carvalho, J.: Lateral control of airship with uncertain dynamics using incremental nonlinear dynamics inversion. In: 11th IFAC Symposium on Robot Control SYROCO, Salvador - BA, Brazil, August 2015
6. Dos Santos Jr, C.R.P., Carvalho, J.R.H., Souza, F.O., Savino, H.J.: Exponential consensus with decay rate estimation for heterogonous multi-agent systems. J. Intell. Robot. Syst. **95**, 543–553 (2018)

Designing Efficient Energy Policies, Systems, and Tools

The Diffusion of Solar Photovoltaics in Brazil: A Technological Innovation System Approach

Mauricio Uriona-Maldonado[1](✉), Thiago Caliari[2], Luiz H. de Souza Costa[1], and Caroline Rodrigues Vaz[1]

[1] Federal University of Santa Catarina (UFSC), Florianopolis, SC 88040-900, Brazil
m.uriona@ufsc.br

[2] Aeronautics Institute of Technology (ITA), Sao Jose dos Campos, SP 12228-900, Brazil

Abstract. One of the most significant challenges nowadays is the transition to more sustainable energy models. In this context, renewable energies stand out, among them, solar photovoltaic (PV) technology has evidenced accelerated growth in the last years. However, Brazil has a robust foreign dependence on the required components for the manufacturing of PV modules which might jeopardize the future growth of this energy source in the future. This paper aims at understanding the critical processes behind the PV supply chain and at assessing its performance using the 'functions of technological innovation systems' - fTIS approach. Through the fTIS, it is possible to diagnose the functioning of the system and the linkage between several influencing actors and networks. Therefore, the paper identifies barriers and drivers for the development of a PV domestic supply chain, whereby there are a large amount of PV installation projects, with low or no participation in the intermediate processes between metallurgical silicon and module manufacturing. The technology diffusion is recent and faces several tax obstacles to its development. However, the growth of PV systems projects hiring and the increasing of panel installers may represent an expansion in this industry participation.

Keywords: Technological innovation system · Solar PV · Supply chain

1 Introduction

Energy use has grown steadily due to the increase in consumption levels and the quality of life of the population. In recent decades, we are experiencing a revolution in energy supply, with the continuous and growing inflow of renewable sources in the energy matrix all around the world [1]. In this sense, one of the significant challenges of our time is the transition to more sustainable energy models and hence, less dependency on fossil fuels without harming economic growth [2].

Thus, renewable energy is the most promising alternative for two reasons: i) the fact that they are virtually inexhaustible sources of energy, and ii) the low or zero environmental impact produced in their generation processes [3]. Brazil has a considerable proportion of renewable energy, and hydroelectric power, about 60% of Brazil's installed

L. Pereira et al. (Eds.): IDEAS 2018, SIST 198, pp. 377–385, 2021.
https://doi.org/10.1007/978-3-030-55374-6_37

power capacity [4]. When considering this strong dependence, there is a pressing need for diversification through the use of solar, wind energy, and other sources [5, 6].

The photoelectric effect generates solar photovoltaic (PV) energy in a photovoltaic cell. This technology is relatively new, the first photovoltaic cells with efficiency greater than 6% emerged in the 1950s, and the photovoltaic industry for power generation on a large scale only gained strength with the oil crisis in the 1970s [7].

The annual growth in the number of photovoltaic facilities as well as the technological development of the energy source has been important factors for the inclusion of the technology in the Brazilian energy matrix. The country stands out for its intense level of solar radiation it receives each year due to its location about the Equator [8], being among the countries with the world's most considerable solar PV potential [9]. This energy technology is also promising for servicing locations with difficult access to the electrical transmission network.

However, Brazil has a robust foreign dependence on the required components for the manufacturing of PV modules which might jeopardize the future growth of this energy source in the future. In this sense, there are two research gaps we want to fill: i) to what extent does a domestic supply chain has been developed in Brazil? Also and: ii) which are the main drivers and barriers for the domestic PV supply chain to develop in the country fully?

Thus, this paper aims at diagnosing the maturity level of the domestic PV supply chain in Brazil. In order to do so, we use the 'functions of technological innovation systems' – fTIS approach.

2 Theoretical Framework: The TIS Approach

The main idea behind the Innovation System approach is that innovation and technology diffusion act on both individual and collective ways [10]. Bergek et al. [11] define a technological innovation system (TIS) as the set of elements (technologies, agents, institutions, and networks) who actively contribute to the development of a technological field.

The TIS approach has been widely used to study generation and diffusion of renewable energy technologies [12], as well as innovation policies [13], subsidies and support mechanism [14], the differences on diffusion in countries [15] and case studies on less developed countries [16], among others.

Using a TIS perspective to study the diffusion of solar PV in Brazil may be justified through the inherent features of the approach, which is, emphasizing the systemic dependence among agents – public and private ones – who generate and use knowledge and the regulatory and/or institutional framework in which the PV technology is embedded [17].

On the other hand, Tsoutos e Stamboulis [13] highlight that technologies' specificities of each renewable energy are essential, since specific idiosyncrasies may be considered.

Additionally, within the TIS framework, authors as Hekkert et al. [10] and Bergek et al. [11] propose a method to measure performance. The rationale is based upon evaluating the performance of a set of activities (or key-processes) that synthesize the functioning of a technological innovation system. That key-process are known as 'functions of technological innovation systems' – fTIS, since they stress the importance of diagnosing and managing system functionality, that is, "how well" the TIS works. Such functions may serve to identify the fundamental driving forces as well as barriers behind the idea of a domestic PV supply chain.

3 Method

We first surveyed the main characteristics, macro processes, actors (firms, universities, research centers, government organizations, etc.), networks (research and supply) and institutional settings (national and international legislation) that comprise and have an influence on the PV supply chain. Secondary sources were used, in order to obtain information regarding the PV supply chain, including articles, books, technical notes, and international publications related to the subject.

Then, we carried out the mapping and analysis of the current performance of the domestic PV supply chain. For this, we used the diagnostic procedure of the functions of the technological innovation system (fTIS) based on Bergek et al. [11]. The functions refer to the critical processes necessary for the well-functioning of a technological innovation system. They are: i) entrepreneurial activities, ii) development of knowledge, iii) dissemination of knowledge through the network, iv) guidance to research, v) training market vi) resource mobilization, and vii) the creation of the industry's legitimacy.

Finally, we identified the barriers and inducing mechanisms for the development of a domestic PV supply chain in Brazil.

4 Results and Discussion

4.1 The PV Supply Chain

The PV supply chain is composed of several supplier levels [3] at the upstream. Currently, 90% of the modules are made of a silicon semiconductor element. After extraction, silicon is then purified by chemical or metallurgical procedures before becoming ready for photovoltaic applications. Later, ingots and wafers are manufactured, which then compose crystalline photovoltaic cells. These cells are then assembled with electronic components, glass, plastics, and other, in order to form a PV module, ready to be installed (See Fig. 1).

At the downstream level, there are many services associated with the implementation, design, and maintenance of PV systems. According to EPE [3], the main activities and agents are:

i) Project Developer: Usually it operates power plants projects, identifies the location, conducts feasibility studies and develops projects, participate in auction and prospecting process of EPC companies and O&M;
ii) Engineering, Procurement and Construction (EPC): Responsible for the design and construction of the plant, it is the developer agent;
iii) Operation and Maintenance (O&M): involve energy production control as well as preventive maintenance, repairs, and others. Usually outsourced service by the plant owner;

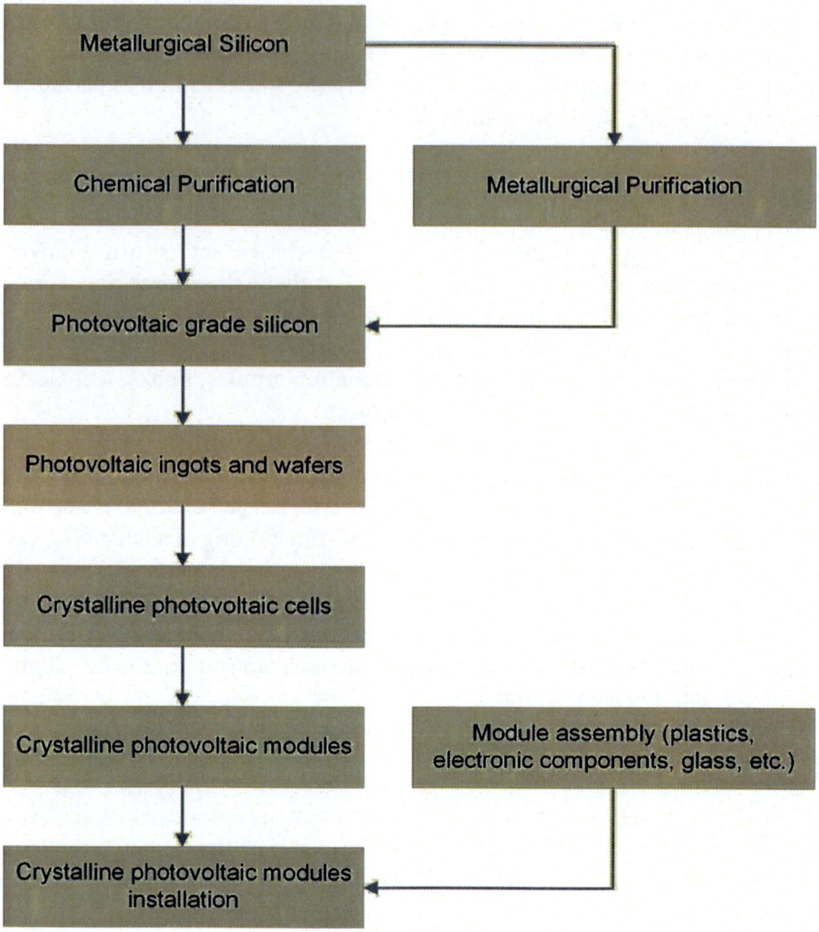

Fig. 1. The PV module manufacturing supply chain.

iv) Integrator Agent: Responsible for contacting suppliers, perform installation projects, dealing with bureaucratic and legal procedures in order to provide a complete solution to its customer, working in small systems as well as the below;
v) Installer: Companies responsible for the installation of photovoltaic panels, which may or may not be part of the companies that carried out the project;
vi) Maintenance: Hardly a photovoltaic system has some maintenance. Regarding the cleaning module, the owner is responsible for the case of faulty equipment, the company that installed the system is triggered.

SEBRAE [18] also highlights companies that provide services such as power producers, publishing and educational and research institutions, as well as financial agents and consulting.

4.2 Mapping of the Critical Processes in the PV Supply Chain

Considering the seven functions proposed by Bergek et al. [18], the analysis of the solar PV supply chain is presented below.

Function 1 – Entrepreneurial Activities: Growth insertion of module manufacturer firms in Brazil. The systems installation support industry presents firms producing inverter manufacturers, batteries, and trackers, for example. The downstream supply chain is well established. However, there is no domestic manufacturing industry on solar grade silicon, and cell manufacturing, as well as other goods for the composition of the module (glass, encapsulant, junction box and back sheet).

Function 2 – Knowledge Development: R&D activities are concentrated in universities and research institutes. For the purification process of solar grade silicon, a partnership was signed between Minas Liga and Rima with IPT-SP and BNDES, as well as a study for the purification through the chemical route by PUC-RS, CEPEL, and CETEC-MG. Richard Louis Anderson Photovoltaic Energy Laboratory for custom cells. ANEEL R&D initiatives of 2011 and 2016, focused only on the application of photovoltaic systems, being the first call for the construction of power plants and the second for educational institutions.

Function 3 – Knowledge Diffusion Through Agents and Networks: Besides the government agencies, two civil organizations – ABSOLAR and Instituto Ideal – are the primary agents. ABSOLAR is the National Association of Solar Energy and has partnerships in projects regarding challenges and opportunities in the sector, besides disclosures and industry recommendations defending and supporting the interests of its members. The Instituto Ideal provides information to facilitate the understanding of the technology for the general public, as well as the promotion of certifications for companies in the chain (Solar Seal).

Function 4 – Direction of Technological Efforts: Guidelines drawn up by the Paris Agreement (COP 21), where Brazil committed itself to reduce greenhouse gas emission levels by 37% to 2025 and 43% to 2030 as compared to 2005. It leads to the use of renewable energies, in addition to hydroelectric, to 23% in 2030. PDE 2024 and 2026 show the government's interest in decreasing dependence on hydroelectric plants and contracting 1GW of solar energy per year between 2020 and 2026. LER 2014 and 2015 considered solar energy separated from other energy sources and LEN of 2017 and 2018 with significant participation of solar energy. PROGD to stimulate distributed generation.

Function 5 – Markets: There are regulations, incentives and subsidies (Normative Resolution 482/2012 amended by RN n° 687/2015 and net metering) that support microgrid production, which resulted in more than 15,000 connections in 2017 (against 1,500 in 2015). Discounts in the TUSD and TUST, application of ICMS only in the net injection in all states, PNP BNDES, created in 2014 and made flexible in 2017 for the financing of module manufacturers and PV generator systems. Exemption of ICMS and IPI on modules and cells (CONFAZ 101/97), REIDI grants PIS/COFINS discharged for final products in the GC. Tax exemption by PADIS, but the annexes of Decree n° 6.233/207 do not grant an exemption for a series of inputs. LER 2014/2015 resulted in the contracting of 2,652.8 MW and LEN A-4 2017/2018 resulted in the contracting of 1,380.6 MW in PV power.

Function 6 - Resources: Companies in the downstream chain, as presented in the study developed by SEBRAE in Bahia, are satisfied with access to suppliers of essential goods and services. In addition, it is worth to highlight the partnership between ANEEL, BNDES and Finep, which resulted in Inova Energia, with an investment of R$ 3 billion for the promotion and selection of businesses that contemplate R&D and innovative activities for the electric sector, as the support to the development and technological domain of the solar PV production chain.

Function 7 – Formation of Legitimacy and Trust: The legitimacy of the Brazilian photovoltaic sector has grown in recent years. The increase in the number of auctions, which has caused the high import of panels and cells, has also aroused the interest of foreign companies to settle in Brazil. Programs and partnerships, such as the one promoted by ENGIE and Celesc (in Santa Catarina State), with bonuses to facilitate the purchase of 1,000 PV systems, stand out. Binational partnership (Brazil/Paraguay) such as the Green Silicon program, is established to develop a vertical industry. The are facilitating the implementation of large plants using BNDES financing, such as the largest solar park in Latin America.

It is noticed that the diffusion of the technology in Brazil is very recent, being the first significant governmental effort for a large generation instituted in 2012. Energy auctions that considered solar energy as a separate product began to appear in 2014.

The manufacturing activities showed that the country is well established with its traditional companies in the transformation of metallurgical silicon, but with no presence in the subsequent manufacturing phases. When it comes to module assembly, on the other hand, significant growth has been seen in recent years, after the positive results of the energy auctions and flexibilization of the financing mechanisms.

Regarding service delivery, there is an adequate presence throughout the country, with companies working in the design, maintenance, and installation. Finally, support industries for composite modules lack firms producing glass, encapsulant, back sheet, and junction boxes.

4.3 Barriers and Inducing Factors

Based on the information above, we present a list of the most relevant barriers for the development of a domestic PV industry in Brazil and consequent technological diffusion, in Table 1.

5 Final Remarks

This study aimed to carry out a diagnosis of the current situation of the Brazilian PV supply chain using the TIS approach, mainly the functions of TIS.

Table 1. Summary of barriers and inducing factors for the domestic PV supply chain

Barriers	Inducing factors
Delay in market making at the expense of other energies [19]	COP 21 guidelines
The tax burden on capital goods	Need to reduce dependence on hydroelectric sources
Inadequacy of PADIS	High level of solar radiation
access to credit difficulties and lack of alignment with the distributed generation ventures	Active metallurgical silicon extraction and high-quality quartz industry [9]
A significant proportion of renewable energy sources (mainly hydro)	The joint project between Brazil and Paraguay vertical integration of the PV industry, the Green Silicon

Source: Elaborated by authors

It was identified the presence of numerous companies (mainly with only a few years of existence) in the service sector scattered out around the country, as well as a consolidated market in the processing of metallurgical silicon, whereby the country has excellent quality quartz reserves with high annual silicon production. The following links transformation, processing, and building cells lack entrepreneurial activities only in the assembly segment of photovoltaic modules, which had its installed capacity increased especially after 2015, with the annual addition of new foreign companies. However, the country has no representatives in the manufacturing of essential parts in the module components such as glass, encapsulant, back sheet, and junction box. Moreover, in the process of installation of PV systems, the country has the presence of inverter manufacturers, string box, and trackers.

The extension of the concession agreements with lower costs in 2013 (Law n°12.783) discouraged investment in R&D by distributors. Research activities do not appear highlighted on the development of knowledge only some educational and research institutions have conducted studies of solar grade silicon purification with some partnerships formed with companies operating in the country.

Finally, ANEEL held R&D calls in 2011 (focused on solar energy) and in 2016 (energy efficiency and mini-generation) but were not focused on the PV supply chain but for application-ready systems.

Entities operating in the diffusion of knowledge in solar PV, such as ABSOLAR and the Ideal Institute, have shown significant activity regarding clarification of the public and political mobilization for the increased solar source in the energy mix (leading also to creating legitimacy of the PV sector in the country). There are evidence of initiatives promoted by the Institute Ideal as the Solar Seal and Guide microgenerators Photovoltaic while ABSOLAR regularly participates with its working groups, along with federal agencies in the debate to legislation and institutional changes to break barriers down for the PV industry.

The direction of technological efforts, which presents itself as a guide to the growth of technology diffusion, was diagnosed with the long-term goals set by the government at COP-21, with the Brazilian commitment to reduce their greenhouse gas emission levels greenhouse effect compared to the base year 2005. It should also be noted a significant share of solar energy in virtually every energy auction that has happened in recent years, in addition to the PDEs that present the government's interests in the coming years to increase participation renewable energy. It stands still the PROGD, which aims to stimulate the distributed generation to meet the goals set between Brazil and the United Nations.

The formation of the market has been driven by the results of new energy auctions (hiring of 1.38 GW) besides reserve (hiring of 2.65 GW) and institutional frameworks, such as RN No. 482/2012, relaxed in 2015 with NR 687, which boosted the number of GD connections (about 15,000 connections in 2017). However, there is a lack of complementarity of taxation applied, since the exemptions for imported modules outweigh the expected national production, lack of aggregation of certain goods and inputs in the PADIS. The flexibility of the BNDES in providing financing to the original provisions of 2014 shows a regression in the nationalization of the intermediate links in the chain at the same time giving more enormous scope for input modules manufacturing companies.

In addition to the funding made available to actors responsible for a centralized generation in the market making and module assembly companies, there is the development program innovation in the electricity sector in the country, with a 3 billion investment. Regarding the downstream chain, the survey by Sebrae in Bahia shows the satisfaction of businesses with access to essential goods and services, as well as ease of location.

The sector's legitimacy could be diagnosed by the interest of the insertion of new entrepreneurial activities, a significant increase in the market brought about by energy auctions, passive acceptance of the population and international partnerships to piggyback on the chain. Noteworthy are projects developed between federal institutions such as the MME, institutions responsible for the dissemination of knowledge, as ABSOLAR, and distributors and installation companies (project "Photovoltaic Bonus" between CELESC and *Engie*). It is important the discussions held by different agencies and companies in the sector to discuss the growth of supply and partnership signed by Green Silicon for the vertical chain.

The application of innovative functions led to the identification of barriers and inductors where there are the current charges which do not reflect the country's development potential, performing with many obstacles and lack of specific legislation for the sector, in contrast with the increasing share of solar energy in energy auctions, a fact that has perspective to attract more companies to the country, including interest manufacture of photovoltaic cells.

In terms of the research gaps this paper fills in, we conclude the domestic supply chain has been poorly developed. This, in fact, opens up windows of opportunity for entrepreneurs and the policy-making actors to establish policy mechanisms and incentives in order to develop domestic supply better.

On the other hand, the main barriers are also, primarily, of policymaking nature (research question #2). We argue that better incentives and policy mechanisms should also contribute to reducing the dependency on hydro sources. Also, our TIS analysis leads to recommend a revision of the current incentives and policies, which in some cases, such as PADIS seem inadequate to the current picture of solar PV in Brazil.

Finally, there is the importance of establishing the solar photovoltaic industry to the socio-economic development of the country, with its potential for job creation, as well as diversification of energy sources, reduction of hydroelectric dependence and guarantee of natural resources.

References

1. REN21. Renewables 2017 Global Status Report, Paris (2017)
2. Silva, R.M.: Energia Solar no Brasil: dos incentivos aos desafios. Núcleo de Estudos e Pesquisas/CONLEG/Senado, Brasília (2015)
3. EPE: Energia Renovável - Hidráulica, Biomassa, Eólica, Solar, Oceânica 2016
4. Mitscher, M., Rüther, R.: Economic performance and policies for grid-connected residential solar photovoltaic systems in Brazil. Energy Policy **49**, 688–694 (2012)
5. Youssef, Y.A., Guerra, J.B.S.: Renewable Energies Market needs: A Perspective from Europe and Latin America. Hamburg University of Applied Science, Germany (2011)
6. Kumar Sahu, B.: A study on global solar PV energy developments and policies with special focus on the top ten solar PV power producing countries. Renew. Sustain. Energy Rev. **43**, 621–634 (2015)
7. Luque, A., Hegedus, S.: Handbook of Photovoltaic Science and Engineering. Wiley, Hoboken (2003)
8. da Silva, G.D.P., de Souza, M.J.R.: Análise de variáveis de projeto de sistema solar fotovoltaíco utilizando o modelo sam: uma comparação entre Belém, Fortaleza e Brasília. Revista Brasileira de Energias Renováveis **5**(2), 297–312 (2016)
9. Carvalho, P., Mesquita, P., Rocio, M.: A rota metalúrgica de produção de silício grau solar: uma oportunidade para a indústria brasileira? BNDES Setorial **40**, 205–234 (2014)
10. Hekkert, M.P., et al.: Functions of innovation systems: a new approach for analysing technological change. Technol. Forecast. Soc. Change **74**(4), 413–432 (2007)
11. Bergek, A., et al.: Technological innovation systems in contexts: conceptualizing contextual structures and interactions dynamics. Environ. Innov. Soc. Transit. **16**, 51–64 (2015)
12. Jacobsson, S., Johnson, A.: The diffusion of renewable energy technology: an analytical framework and key issues for research. Energy Policy **28**(9), 625–640 (2000)
13. Tsoutsos, T.D., Stamboulis, Y.A.: The sustainable diffusion of renewable energy technologies as an example of an innovation-focused policy. Technovation **25**(7), 753–761 (2005)
14. Del Río, P., Bleda, M.: Comparing the innovation effects of support schemes for renewable electricity technologies: a function of innovation approach. Energy Policy **50**, 272–282 (2012)
15. Gosens, J., Lu, Y.: From lagging to leading? Technological innovation systems in emerging economies and the case of Chinese wind power. Energy Policy **60**, 234–250 (2013)
16. Tigabu, A., Berkhout, F., van Beukering, P.: Development aid and the diffusion of technology: improved cookstoves in Kenya and Rwanda. Energy Policy **102**, 593–601 (2017)
17. Yaqoot, M., et al.: Review of barriers to the dissemination of decentralized renewable energy systems. Renew. Sustain. Energy Rev. **58**, 477–490 (2016)
18. Bergek, A., et al.: Analyzing the functional dynamics of technological innovation systems: a scheme of analysis. Res. Policy **37**(3), 407–429 (2008)
19. SEBRAE: Especialistas em pequenos negócios, p. 53 (2017b). http://agenciasebrae.com.br/asn/Indicadores/NovoMPEIndicadores-11112016.pdf. Accessed 20 Apr 2018

Design Management for Sustainable Development: Concepts and Examples

Ana Carolina Correa de Medeiros, Claudete Barbosa Ruschival(✉), Eminy Laís Silva da Costa, and Luciana Kurack da Silva Misucochi

Universidade Federal do Amazonas, Manaus, Amazonas 69080-900, Brazil
claudete@ufam.edu.br

Abstract. The Design has proven to be a strategic tool for the management of several organizations, which have been looking for ways to distinguish themselves within an increasingly competitive market. The main objective of this paper is to validate the best practices that make environmentally responsible companies prove themselves on the market. The focus of this paper is to highlight the Design Management strategies for sustainable development of industries, demonstrating possible solutions for any organization to follow. The research uses the Visual Data Mining methodology to search within databases for Sustainability, Design Management, and Sustainable development. Additionally, it makes use of case studies (companies: Suzano, Braskem, Natura, and Fibria) to list Design Management characteristics for environmentally responsible companies to obtain competitive advantages. As results, this article lists the minimum resources to lead an environmentally responsible company to its success. Companies should improve environmental performance through sustainable use of inputs, and add environmental values and product differentiation values to goods and to services. They should also apply product lifecycle management to their products, and seek sustainable technological innovations. It is also essential to observe local and international environmental regulations and establish ethical and moral standards for all employees of the organization and provide a safe space for all employees. This Study exhibits the positive aspects of environmental responsibility and points criteria for new studies.

Keywords: Design management · Sustainable development · Case studies

1 Introduction

The Design has proven to be a strategic tool for both scientific research and business development. Organizations have the rising need to differentiate themselves in the market in order to achieve competitive advantages and the society's concern with maintaining a healthy environment that can be enjoyed by future generations is growing correspondingly. Within the model of sustainable development and environmental responsibility that permeates the discussions about human production. Several companies and organizations have tried to meet the world's expectations on the upkeep of the environment. However, what makes an organization environmentally responsible?

L. Pereira et al. (Eds.): IDEAS 2018, SIST 198, pp. 386–395, 2021.
https://doi.org/10.1007/978-3-030-55374-6_38

The World Commission on Environment and Development coined the terms environmental sustainability and sustainable development in 1987 in the WCED Annual Report: *Our Common Future*. These concepts are, in fact, general structures that integrate the policies of protection of the environment with development policies [1]. Leite and Sehnem describe that sustainable development is the realization of the basic needs of human beings, coupled with the maintenance of nature [2].

The main objective of this paper is to point the best practices within Design Management to ensure environmentally responsible enterprises can keep competitive advantages within the market. This article aims to analyze these strategies and establish minimum parameters for an environmentally responsible company to generate profits while outstanding the market.

2 Research Protocol: Visual Data Mining

The Design is interdisciplinary, and sometimes it can be challenging to find a proper scientific method to research. For that reason, this research is based on Blum, Merino, and Merino visual method for systematic review in design that uses the concepts of Data Mining [3].

On the first step, there was a database selection. The selection of "Capes Periódicos" is reasonable, since the research aims to show results related to Brazil. The second step was choosing the research period, from 2014 to 2019. The third step was the selection of a combination of descriptors that would form a pattern within the selected articles (Sustainability AND Management model AND Design Management AND Sustainable production OR Sustainable development). The results were 44189 articles that related to those descriptors. As the fourth step, the post-processing of data returned with the twelve most relevant articles. The results were separated by year for visual analysis (Tables 1, 2, 3, 4 and 5).

Table 1. Results of data mining process for the year of 2014

Authors	Keywords
Braga [4]	Ecodesign, life cycle, sustainable production
Plentz and Tocchetto [5]	Environmental management, reverse logistics, life cycle, productive processes

Source: Elaborated by authors

The last phase (fifth step) of this research protocol is the discussion of the articles. This research protocol permitted the creation of a listing of the parameters that lead an environmentally responsible company to obtain a competitive advantage, since this is the main focus of this paper. Other documents will be analyzed such, as the selected organization publications.

Table 2. Results of data mining process for the year of 2015

Authors	Keywords
Silva, Moraes, and Machado [6]	Matrix Leopold, cleaner production, ecodesign, reverse logistics
Laruccia and Garcia [7]	Ecodesign, environmental management, sustainability

Source: Elaborated by authors

Table 3. Results of data mining process for the year of 2016

Authors	Keywords
Brocco, Heemann, and Chaves [8]	Design management, sustainability, furniture development
Oliveira, Correia, and Gomez [9]	Consumer patterns; companies; sustainable symbolic value
Cuneo, et al. [10]	Design management, diagnosis, CDS model, handicraft

Source: Elaborated by authors

Table 4. Results of data mining process for the year of 2017

Authors	Keywords
Barreto and Moreira [11]	Strategic, lead, environmental responsibility
Zacchi, Merino, and Merino [12]	Design management; co-creation; sustainability
Kibrit, Machado, and Kruglianskas [13]	Innovation, sustainability, startup, single-case study, sustainable packaging, automotive industry

Source: Elaborated by authors

Table 5. Results of data mining process for the year of 2018

Authors	Keywords
Leite and Sehnem [2]	Crafts. Sustainability. Competitiveness. Management model
Tiburtino-Silva, Maciel, and Costa [14]	Sustainable design; ecodesign potential; minimal impact on the environment

Source: Elaborated by authors

3 Organization Selection: A Randomization Approach

The selection of organizations took the EXAME Sustainability guide as a reference since this list only brings to light enterprises that worry about sustainable development in Brazil (Table 6). Among the several listed companies, each company received a number and was randomly drawn like a blind selection of objects [15]. The companies that appear more than once in the table were placed only once within the draw so that the selection was indiscriminate enough. The selection within the target population of firms reduced to a sample that followed the randomization sampling plan [15].

Table 6. Companies awarded by the EXAME sustainability guide, on the last three years.

2018 [16]	2017 [17]	2016 [18]
Enel Brasil, Coruripe, Baterias Moura, Weg, Boticário, Verde Ghaia, Newinc, Siemens, Eurofarma, Grupo CataraRetas, Aegea, Itaú Unibanco, Duratex, Arcelor Mittal, Fibria, Termotécnica, Sabin, Algar Telecom, Renner, Basf, Whirlpool, EDP, Nestlé, Banco do Brasil, L'Oréal, Santos Brasil, Mapfre, Eletropaulo	Natura, Elektro, Sabin, Boticário, Coca-Cola, Basf, HP, Siemens, L'Oreal, Beraca, Aperam South America, EDP, Merck, Grupo Cataratas, EcoRodovias, Banco do Brasil, Mapfre, Duratex, Klabin, Basf, EY, Algar Telecom, Newinc, Raízen, Renner, Volvo, Nexa Resources, Weg, Whirlpool	Alcoa, Amanco, Anglo American, Bradesco, Santander, Braskem, HSBC, Fibria, EDP, CPFL, Bunge, Itaú Unibanco, Masisa, Natura, Phillips, Promon, Suzano, Unilever, Walmart, Whirlpool

Source: Elaborated by authors

There were 52 different organizations awarded on those three years. For this paper, we needed a limited number of enterprises to study, so, the team decided we should study four organizations. The random selection picked Suzano, Natura, Braskem, and Fibria among those organizations.

4 How to Make Organizations More Sustainable Through Design Management

Companies focus on making profits at fewer costs. Even if this is the main objective of any enterprise, many of them also aim to make profits and respect the environment. Each of the selected organizations can relate the suitability and making of profit in their way, for instance, being also can do the most of sustainable development.

Design management is one way of maintaining competitive advantages within the market, especially about the advantages related to sustainable development. According to Sobreira and Delgado, Design Management is "the identification and communication of processes and ways in which design can add value to the company" [19, p. 1791]. Therefore, it is necessary to observe companies whose competitive differential lies in both the environmental values and goals assumed. By doing so, we can better understand how to achieve sustainable development.

4.1 Sustainable Use of Inputs and Sustainable Industrial Processes

The industrial production causes many harmful impacts to the environment, so, the better way to mitigate those effects is to look for materials that cause a low impact on the environment [4]. The reduction in the consumption of raw materials and other resources is also a way to diminish the human impact on ecosystems [4], by the reduction of exploitation of raw materials and energy [8]. The industry should improve productive efficiency [12], reducing non-renewable energy consumption [2].

The United Nations - Sustainable Development Goals on its Goal number 12 "Responsible Production and Consumption" [20] only adds to *Our Common Future* goals that also appear on Brazilian Agenda 21 in Earth Summit, objective 16 - forest policy, deforestation control, and biodiversity corridors [21] (Table 7).

Table 7. Sustainable use of inputs by selected companies

Companies	How they apply this goal
Suzano	*FuturaGene project.* The program focuses on discovering genes for increasing productivity, improving wood processing capacity, crop protection, and the development of new biotechnology tools applicable to the forest sector [22] Suzano also considers nature as an essential capital for business, so that they commit to conserving biodiversity, mitigate the negative impacts of climate change, and sustainably use of resources. Thus, they ensure that future generations also have access to a healthy environment [22]
Braskem	*Braskem's purpose* is "to improve people's lives by creating sustainable solutions in chemistry and plastic, and [their] strategic vision is to be the world leader in sustainable chemistry" [23]
Natura	*Brazilian biodiversity is a value in Natura's products,* and for this, the company invests in researches and tests in laboratories, and in sustainable input deployment [24]
Fibria	*Fibria considers the forest its most significant capital,* having the ecological footprint as one of the fundamental strategies of the company. The focus is the conservation and maintenance of ecological corridor reserves, in order to protect the endemic species of the region and the biodiversity of the biomes of the vegetation areas [25]

Source: Elaborated by authors

4.2 Environmental Values as Product Differentiation

Environmental values are now an essential part of products. People seek more eco-friendly products, even if they pay more for them [9]. The costs do not seem to matter for a portion of the consumers that are worried about their habitat and with the next generations [11]. The sustainable society seeks environmental management to produce goods that have significant commercial value and high environment value [11]. Some of Brazilian Agenda 21 in Earth Summit objectives relate to this notion: Objective 9, development and sustainability must have ethical dimensions; Objective 21, pedagogy of sustainability - ethics and solidarity [21]. The sustainability emerges as a product differentiation aspect for products [9] (Table 8).

Table 8. Environmental values of the selected companies

Companies	How they apply this goal
Suzano	*Suzano also considers the communities as an essential capital for business.* Suzano upholds several social projects to ensure that future generations also have access to a healthy environment [22]
Braskem	*Braskem seeks the environmental preservation of areas of social interest* [23]
Natura	*To Natura, using raw materials from regional agricultural producers and maintaining contact with the scientific community are central strategies* [24]
Fibria	*Fibria maintains several social projects investing in development, respect for human beings, and the environment* [25]

Source: Elaborated by authors

4.3 Product Lifecycle Management (PLM)

The product lifecycle management (PLM) of products will make it easy to extend their lifetime, and will also enable the possibility of upcycling, recycling, and downcycling materials and products [4]. The PLM consists of a process of organization of the entire lifecycle of a product from inception, through engineering design and manufacture, to service and disposal of manufactured products as waste [4]. The product lifecycle management allows products to have higher quality imbued [10] since this management system allows products to have higher sustainable value [11]. It is also vital to apply eco-design guidelines to redesign products that have been on production lines for too long [14], as a way to reduce inputs consumption [2].

Companies can assist in conscious consumption with advertisements and stakeholder education using PLM strategies. Design management can influence conscious spending at the strategic, tactical, and operational levels [9, 11].

It is also possible to reshape consumer's habits by observation. To observe the perception of consumers and get them to consume in minor quantities through marketing tools. The contribution of design in this context can occur both in functional and symbolic factors, from product conception to ecodesign (a product that causes the least possible impact during its life cycle) with its symbolic part related to the consumer [13]. This discourse also aligns itself to the United Nations Sustainable Development Goal 12: "Responsible Production and Consumption" [20] (Table 9).

Table 9. Product lifecycle management on the selected companies

Companies	How they apply this goal
Suzano	*Suzano invested 27 million (BRL) on social and environmental projects in 2017 through the following activities:* socio-environmental educational projects and projects for the sustainable development of local communities. According to the company's discourse, Suzano's purpose is to be "strong and kind," so that sustainable growth is part of the company's primary goal [22]
Braskem	*Braskem endeavors to improve the production of the green polyethylene,* produced from ethanol, so it is possible that the PVC can be biodegradable. There is a preoccupation with PLM since the polymer manufacture [23]
Natura	*Natura has established the following values:* environmental, social, and economic. Natura also believes in the brand as an amplifier of awareness and the capacity of relationships among the value chain stakeholders [24]. Another strong point of Natura is the valorization of product experience (Experience Design)
Fibria	*Fibria's business vision is "to cultivate the planted forest as a source of life,* shared wealth generation, and people's well-being," in this way, it maintains several social projects investing in development, respect for human beings, and the environment [25]

Source: Elaborated by authors

4.4 Sustainable Innovation

It is essential to take note that the innovation process via research and development (R & D) should aim sustainable development [5]. The industry should apply cleaner production tools (C+P) as a way to prevent environmental damage; to minimize waste and polluting gas emissions, and, to maximize product output [6]. The search for new technologies

Table 10. Sustainable innovation on the selected companies

Companies	How they apply this goal
Suzano	*FuturaGene project* is a R&D program that combines the findings made internally with the pursuit and acquisition of technologies [22]
Braskem	*Braskem aims to produce PVC* (polyvinyl polychloride, a material widely used in pipes, walls, and furniture) with the lowest economic and environmental cost as possible [23]
Natura	*Natura invests in consumer experience studies in partnership with MIT* (Massachusetts Institute of Technology), a leading reference in innovation, design, science, and technology in the world [24]
Fibria	*Fibria's focus is on the conservation and maintenance of ecological corridor reserves* in order to protect the endemic species of the region and the biodiversity of the biomes of the vegetation areas [25]

Source: Elaborated by authors

must be constant and aim at more sustainable solutions to solve wicked problems [7]. A critical point of investigation is the design for reuse, which is a sustainable innovation that aims to make new products for disassembly [13]. The Sustainable Development goals also point for these actions on Goal 9, Industry innovation and infrastructure [20] (Table 10).

4.5 Ethical Standards and Observance of Regulations

Design can help companies on the creation of better working conditions, especially using Design Management technics [11]. The ethical standards begin within the organization [2], so it is imperative that all stakeholders understand what the main goals of a company are.

With sustainable goals is no different, the industry has to use natural resources without harming their possession by future generations. The production has to reduce the negative impacts caused by the productive process [10] and also aim to use local/regional inputs; to employ lighter raw materials and to manufacture lighter products [4]. Thus, to be able to commit with sustainable development, all organizations have to understand the partners in the value chain [7]. They also have to seek partners who have the same environmental values [6]. Organizations should also apply eco-design guidelines to redesign the production chain [10].

To obey the laws is also a necessary moral standard. Organizations have to recognize and to apply the national legislation. It is significant to know that legislation, and preventive action are ways to make sustainable development a reality [14]. The companies cannot only observe the regulations for lawful pressure, since regulatory milestones become more dynamic [7], day by day, enterprises have always to stay put into applying the mandatory regulations (Table 11).

Table 11. Sustainable innovation on the selected companies

Companies	How they apply this goal
Suzano	*Among Suzano's corporate policies are:* responsibility towards society; health and safety of its employees; and better practices with suppliers. Suzano's ethical principles are: corporate governance, integrity, sustainable development, equality, professional valorization, and transparency [22]
Braskem	*Braskem, in its corporate policy,* states that all decisions accord to positive and negative factors. The company also states that legal regulations are always part of every decision [23]
Natura	*Natura uses raw materials extracted by extractive communities* (local, regional). Natura also aims to add sustainable qualities to the value chain by seeking partners with the same environmental concerns [24]
Fibria	Fibria protects forests, having the ecological footprint as one of the fundamental strategies of the company. The company also invested in the following activities: monitoring and conservation of natural resources and biodiversity, and environmental certification demands [25]

Source: Elaborated by authors

5 Contributions and Final Words

Design Management can help companies in obtaining competitive advantages in their target market and, also, can be associated with environmental management so that an environmentally responsible company can sustain itself in the market without losing sight of its commitment to the planet Earth. Following Suzanos's, Braskem's, Natura's and Fibria's examples, it is possible to make the industry more sustainable and ecofriendly.

As a result of this perception, it was possible to build a table that relates theoretical and practical aspects to generate minimum parameters that allow an environmentally responsible company to generate profits while maintaining a leading position in the market. It is crucial that environmentally responsible companies aim:

a) To improve environmental performance through sustainable use of inputs;
b) To add environmental values and product differentiation values to goods and to services;
c) To apply product lifecycle management (PLM) to the manufacturing process;
d) To seek technological innovations aiming at environmental sustainability;
e) To observe local and international environmental regulations and establish ethical and moral standards for all employees of the organization and provide a safe space for all employees.

Those are the minimum parameters necessary for a company to be considered environmentally responsible. This study demonstrates the positive aspects of environmental responsibility. In this way, new studies on other companies are encouraged in order to expand this research even further. Other companies may be considered on future researches, remarkably, one of the other 52 organizations received awards on the EXAME's sustainability guide. Other parameters also may be established as milestones for environmental responsibility. There is still a vast field for research on this subject, expressly regarding other solutions of Industry 4.0 and sustainable development.

References

1. WCED. World Commission for Environment and Development: Report of the World Commission on Environment and Development: Our Common Future (1987). http://www.un-documents.net/our-common-future.pdf
2. Leite, A.A.V., Sehnem, S.: Proposition of a sustainable and competitive management model for handicrafts. Cadernos EBAPE.BR **16**(2), 264–285 (2018)
3. Blum, A., Merino, E.A.D., Merino, G.S.A.D.: Visual method for systematic review in design based on data mining concepts. DA-Pesquisa **11**(16), 124–139 (2016). https://doi.org/10.5965/1808312911162016124
4. Braga, J.: Ecodesign: case study of strategies applied to national products. Revista Portuguesa e Brasileira de Gestão **13**(2), 28–40 (2014)
5. Plentz, N.D., Tocchetto, M.L.O.: Ecodesign in the footwear industry: proposal for a changing market. Revista Eletrônica em Gestão Educação e Tecnologia Ambiental **18**(3), 1022–1036 (2014)
6. Pereira, C.M., Engler, R.C., Martins, D.M.: Design, social innovation and sustainability: the concept of creative communities in Nova Lima – MG. Janus **12**(21), 34–47 (2015)

7. Silva, A.L.E., Moraes, J.A.R., Machado, E.L.: Cleaner production proposal for ecodesign and reverse logistics practices. Engenharia Sanitária e Ambiental **20**(1), 29–37 (2015)
8. Laruccia, M., Garcia, M.: An analysis of the perception and use of ecodesign practices by companies. Braz. Bus. Rev. **12**(3), 1–15 (2015)
9. Brocco, R., Heemann, A., Chaves, L.I.: Fundamentals of design management and sustainability in furniture design. In: Blucher Design Proceedings, Congresso brasileiro de pesquisa e desenvolvimento em design 2016, Belo Horizonte, vol. 9, pp. 1–8 (2016). ISSN 2318-6968
10. Oliveira, V.M., Correia, S.E.N., Gomez, C.R.P.: Consumer culture, sustainability and business practices: how can companies contribute to promoting the symbolic value of sustainability in consumer activities? Revista de Gestão Ambiental e Sustentabilidade (J. Environ. Manag. Sustain.) **5**(1), 61–77 (2016)
11. Cuneo, M., Hinnig, R., Merino, G.S., Triska, R., Figueiredo, L.F.G., Silva, C.S., Merino-díaz, E.A.: Design management: diagnosis based on competitiveness, differentiation and sustainability in an association of artisans in Southern Brazil. Revista online de la Red Internacional de Investigación en Diseño **2**(2), 207–226 (2016)
12. Barreto, J.M., Moreira, M.S.: Sustainable management: a socio-environmental commitment in the strategy of the companies. In: Eixo Temático: I Estratégia e Internacionalização de Empresas (online) 6º Fórum Internacional Ecoinovar, Santa Maria, Brazil (2017)
13. Lima, B.L., Camargo, C.W., Barp, D.R.A.: Criteria for evaluating sustainability in fashion brands. Design e Tecnologia **7**(14), 59–68 (2017)
14. Zacchi, G.P., Merino, E.A.D., Merino, G.S.A.D.: Co-creation and design management in small rural and fishing enterprises: a sustainable approach. Mix Sustentável **3**(1), 52–63 (2017)
15. Kibrit, E., Machado, R.J., Kruglianskas, I.: Sustainable innovation management - single case study in a startup. In: Encontro Internacional sobre Gestão Empresarial e Meio Ambiente, XIX ENGEMA, Brazil (2017)
16. Tiburtino-Silva, L.A., Maciel, J.C., Costa, R.B.: Ecodesign from the perspective of local development and sustainability. Interações (Campo Grande) **19**(1), 93–102 (2018)
17. Wild, C.J., Seber, G.A.F.: Chance encounters: a first course in data analysis and inference. Translation: Cristiana Filizola Carneiro Pessoa; Djalma Galvão Carneiro Pessoa. LTC, Rio de Janeiro (2004)
18. EXAME: The companies awarded by the EXAME Sustainability Guide (2018). https://exame.abril.com.br/negocios/as-empresas-premiadas-pelo-guia-exame-sustentabilidade-2018/
19. EXAME: EXAME awards the companies that stand out in sustainability (2017). https://exame.abril.com.br/negocios/exame-premia-as-empresas-que-mais-se-destacam-em-sustentabilidade/
20. EXAME: The companies awarded by the Sustainability EXAME Guide (2016). https://exame.abril.com.br/negocios/as-empresas-premiadas-pelo-guia-exame-de-sustentabilidade/
21. Sobreira, M.A.S., Delgado, P.S.: Sustainable design and design management: interfaces and edges. In: Blucher Design Proceedings, Congresso brasileiro de pesquisa e desenvolvimento em design 2016, Belo Horizonte, Brazil, vol. 9, pp. 1790–1798 (2016). ISSN 2318-6968
22. UN: Sustainable Development Goals (N. D.). https://www.un.org/sustainabledevelopment/sustainable-development-goals/
23. BRAZIL. Brazilian 21 Agenda (1992). www.mma.gov.br/responsabilidade-socioambiental/agenda-21/agenda-21-brasileira.html
24. SUZANO: Sustainability report (2017). http://www.suzano.com.br/relatoriodesustentabilidade2017/
25. BRASKEM: Sustainability report (2018). https://www.braskem.com.br/estrategia-sustentavel
26. NATURA: Sustainability report (2017). https://www.natura.com.br/sustentabilidade
27. FIBRIA: Sustainability report (2017). http://www.fibria.com.br/sustentabilidade/ambiental/

Investigation of the Classical Control Design Methodology to Improve the Performance of the Photovoltaic System on Grid

Marenice Melo de Carvalho(✉) and Renan Landau Paiva de Medeiros(✉)

Federal University of Amazonas, Manaus, AM, Brazil
marenice.m.carvalho@gmail.com, renanlandal@ufam.edu.br

Abstract. Nowadays, several ways of electric generation have been investigated, one of the most promising is the photovoltaic system. They are renewable systems because they use the energy of the Sun as an input, and they are also an alternative to the types of sources more polluting like the combustible ones. This system produces electric energy in DC format, however, commonly several grids are AC grids, so it is necessary to transform DC to AC format, then it is necessary to use a device to convert DC/AC, i.e., namely voltage inverter, aiming to make the connection between the photovoltaic source for conventional grid. This device operates by using a set of semiconductors static key based on IGBT, which has a goal of shifting the DC voltage produced by the photovoltaic panel and transform in the sinusoidal output. Furthermore, the voltage inverter is used to connect the photovoltaic source with the grid. Then, several undesired problems appear like the loss of the setpoint and coordination between the grid and the protection system, and the instantaneous recloser and islanded operations of the inverter. Thus, in this work, it was proposed an analysis to improve the performance of the voltage inverter when applied on the photovoltaic system connected in the grid, so it was used the current mode control and Phased-Lock-Loop (PLL) to ensure the connection of the photovoltaic panels on the grid. Therefore, a classical design methodology was investigated to ensure the connection with grid and system stability. Finally, several tests are performed in the developed simulator aiming to show the possible uses of the classical control design methodologies to improve electric generation in photovoltaic systems.

Keywords: Renewable energy · DC/AC power converters · Current mode control · Phase-Locked Loop · Classical control design methodology

1 Introduction

There are many ways to generate energy, depending on the source. Among them are hydroelectric, thermal, geothermal, wind, solar, among others. These types of generation can be classified based on the renewal capacity of the material to be used as input for the generation of energy, classifying them into two groups: renewable energy sources and non-renewable energy sources. It should be noted that energy sources classified

L. Pereira et al. (Eds.): IDEAS 2018, SIST 198, pp. 396–405, 2021.
https://doi.org/10.1007/978-3-030-55374-6_39

as renewable, are considered as inexhaustible sources, for example, solar energy, wind energy, biomass, among others.

Many of the human activities generate negative impacts on the environment, and energy production does not escape from it [1]. Although renewable energy classification is essential, it is necessary to investigate and classify the sources of energy generation according to the capacity to generate pollutants, in this way one can still classify the sources in a source of clean energy or source of energy dirty. As the leading representatives of dirty energy sources, we have fossil fuels [1]. Several negative impacts on the environment are caused by the generation of electrical energy, such as flooding, water pollution, air pollution, radioactive material pollution, acid rain, damming rivers that have a direct impact on fauna and flora, besides sociocultural impacts, among others [2].

Given the mentioned negative aspects, it is necessary to search for sources of energy generation that are renewable and do not harm the environment. In this way, several studies seek besides the development of discoveries of new sources of clean energy generation [3] and be developing sustainable applications based on renewable energy sources. As mentioned earlier, an example of renewable energy is that which has the Sun as the source. Which provides daily energy through solar radiation, it is naturally used as a food source for plants, is also used to warm the planet among other forms of influences on all forms of life.

Photovoltaic systems are made up of devices that convert the energy from the Sun into electrical energy, from the photovoltaic effect. The voltage and current levels of the output depend on the irradiance and ambient temperature levels [4]. The voltage inverters are the equipment responsible for transforming the constant output of the solar panel into an alternating signal. They have the purpose of interconnecting systems and devices that operate in CA with DC generation sources. They work from a set of semiconductor devices of high power, such as IGBT, which has the function of switching repeatedly. So that the constant signal at the input of the device has periodic variation in its output, making it possible to get a sine function in the output [5, 6].

When connecting a solar generation system to a standard power grid, where the photovoltaic system acts as a second-generation form for a group of loads, this topology is categorized as a hybrid system; such a topology is used when a single source of energy is not enough to meet the required energy demand that the population needs [6]. In this work, the first source of energy as the conventional electricity grid, which is generally composed of large-scale thermoelectric and, or hydroelectric plants, besides to other types of generation, being the second source of energy generation through the use alternative renewable energy, such as the use of photovoltaic solar energy.

In this way, the inverter has a significant role in solar energy systems, since the interconnected use of the grid can cause some problems for the utility, such as loss of coordination in protection systems due to overcurrent, improper action of instant reclosures and islanding operations [7, 8].

Some unwanted problems appear on the power converter device. Such as the loss of setpoint and coordination between the network and the protection system and the instantaneous recloser and the inverted inverter operations. With this, it becomes interesting to use control methods to improve the performance of this device applied to solar panels. The types of controllers are current mode and Phase-Locked Loop (PLL) [9–11].

2 Voltage Converter Modeling

The DC-AC voltage converter has an alternating output signal with the null mean value. The signal can be generated from a constant independent source such as solar panels [6]. The conversion of the DC signal is performed from two modulating signals, m_d, and m_q, which are constant quantities that make up the three sine-wave modulation indices m_a, m_b, and m_c — this transformation made from the inverse transform of Park. The mathematical model presented in this article is based on the average model, since it is possible to describe the relationship between the output dynamics of the inverter and the modulation signals used, as presented in Eq. 1 [9], where V_{dc} is the constant voltage of the input.

$$v_{ta} = \frac{V_{dc}}{2} m_a; \;\; v_{tb} = \frac{V_{dc}}{2} m_b; \;\; v_{tc} = \frac{V_{dc}}{2} m_c \tag{1}$$

The converter is connected through an RL filter to the standard power grid. As shown in the schematic diagram of Fig. 1.

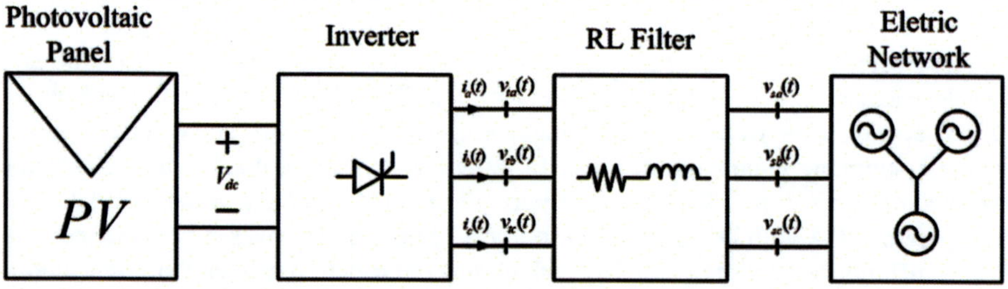

Fig. 1. Diagram of the analyzed circuit connection.

This connection can be represented from Eq. 2, where R and L are the resistance and inductance of the RL filter, respectively.

$$H(s) = \frac{1}{Ls + R} \tag{2}$$

To ensure the phase and frequency synchronism of the signal generated by the converter with the electrical grid, it is necessary to develop the Phase-Locked Loop subsystem. Its purpose is to ensure the phase and frequency adaptation of the voltage inverter with the network, using a PI-structure controller. As shown in Eq. 3. The power control transferred from the input to the output is made from the current controller, which uses the same PI control structure, as shown in Eq. 3, where K_p and K_i represent the proportional and integral gains of the controller.

$$C(s) = \frac{k_p s + k_i}{s} \tag{3}$$

3 Methodology

For the computational implementation of the simulations of the controllers applied to the DC/AC converter, the MATLAB/SIMULINK software was used, so that it is possible to perform tests and check the performance and operation of the system.

3.1 Phase-Locked Loop

O PLL is a device that tracks one signal from another. It keeps the output signal synchronized with a reference signal [10, 12]. The block diagram of the controller is presented in Fig. 1. Its goal is to keep the difference between $\rho = \omega_0 t + \theta_0$, where ρ is output from the PLL and $\omega_0 t + \theta_0$ represents the input angle.

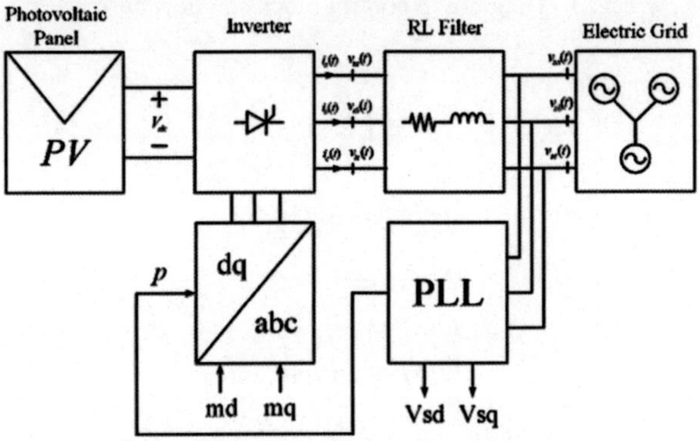

Fig. 2. PLL block diagram

Using MATLAB to find the gains of the controller, using the Geometric Place of Roots method, LGR [10, 12], considering the control structure of type PI, one can get Table 1, whose structure was presented in Eq. 3.

3.2 Current Control

A controller in the current mode controls the active and reactive powers supplied to the power grid. They protect the systems against overcurrent, and also to have the advantage of better dynamic performance and greater control accuracy [9]. The block diagram of the control system is shown in Fig. 2.

Considering that $K_d(s)$ and $K_q(s)$ are two control structures whose structure is of type PI, designed to increase the speed of the system response and grarantee the null error of permanent regime. Using the poles and zeros cancellation method, the values for K_p and K_i of the $K_d(s)$ and $K_q(s)$ controllers are obtained.

Observing the pole to be canceled, as presented in Eq. 5 (Fig. 3)

$$\Delta_d(s) = \left(s + \frac{R}{L}\right)\left(s + \frac{1}{\tau_i}\right) \tag{5}$$

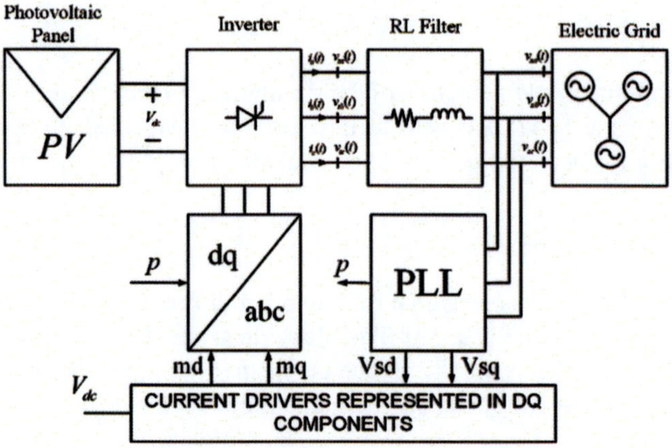

Fig. 3. Diagram in control blocks on how current

$$\Delta_d(s) = s^2 + \left(\frac{R}{L} + \frac{1}{\tau_i}\right)s + \frac{R}{L\tau_i} \tag{6}$$

The characteristic equation is presented in Eq. 7.

$$\begin{aligned} \Delta_c(s) &= 1 + CG \\ C(s) &= \tfrac{k_p s + k_i}{s} \\ G(s) &= \tfrac{1/L}{s + R/L} \end{aligned}$$

$$\Delta_c(s) = s^2 + \left(\frac{k_p}{L} + \frac{R}{L}\right)s + \frac{k_i}{L} \tag{7}$$

Comparing Eqs. 6 and 7, where the pole cancellation occurs, we get the results of the values of the controller gains in Table 1, for R = 1.63 mΩ and L = 100 μH and τ_i = 1 ms. Being R and L, the resistance and inductance of the line and τ_i are the time constant in a closed loop.

$$K_p = \frac{L}{\tau_i}; \qquad K_i = \frac{R}{\tau_i} \tag{8}$$

Table 1. Summary of the gains obtained for the PLL controller and the controller in a current mode

Parameters	PLL	Current-mode
Method	LGR	Cancellation of poles
K_p	203	0,1
K_i	19361	1.63

3.3 Description of Tests

For the PLL controller, the outputs of the voltages in the direct and quadrature axes of the model were observed, and a voltage sink was inserted in one of the phases of V_S. With this it is possible to verify the effect of this sinking in the performance of the PLL.

For the second experiment, a step variation was performed in 0.5 s and 0.7 s in the Id reference with a variation of 0.7 p.u. to 0.5 p.u., then in the time of 0.9 s a new variation of 0.25 p.u. in the reference of I_q, considering an accommodation time constant of 5 ms. A disturbance was also introduced, aiming to check the performance of the system with the controls in the current mode.

4 Results and Discussion

The model implemented in a computational environment, along with the developed PLL controller is presented in Fig. 4.

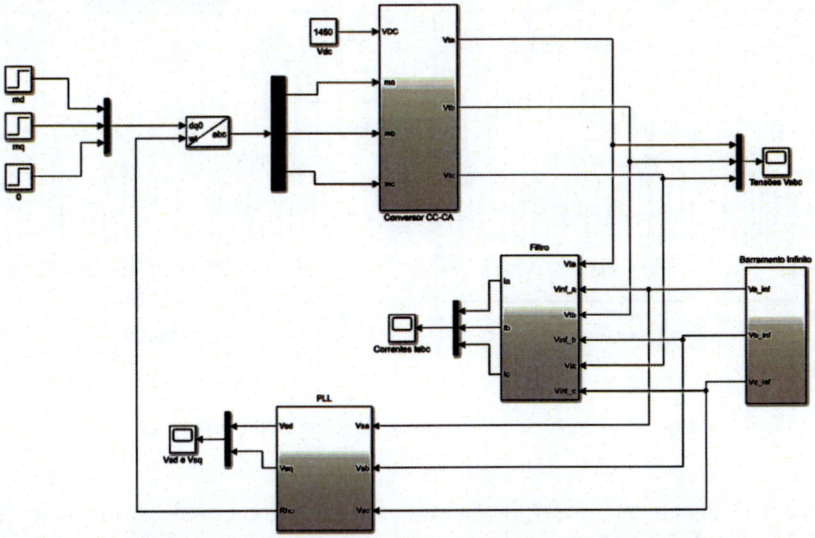

Fig. 4. PLL driver implementation in MATLAB/Simulink

Figure 5 (a) shows the output of the voltage V_S. in the direct and quadrature axes of the PLL controller. Meeting the specifications of over 5% and an accommodation time of 100 ms.

It is worth noting that when a voltage sink is included in the V_S, phase, at the time of 0.6 s, Fig. 5(b). There is an oscillation at this instant of time at voltages V_{sd} and $V_{sq,}$ and this, indicates that the disturbance inserted causes an oscillation in the system signals to try to keep it under the before specified voltage set point.

In Fig. 6, we have a comparison of the three-phase outputs of the system. After the transient period, the output of the PLL, Fig. 6(a) is equal to the three-phase voltages of the infinite bus, Fig. 6(b).

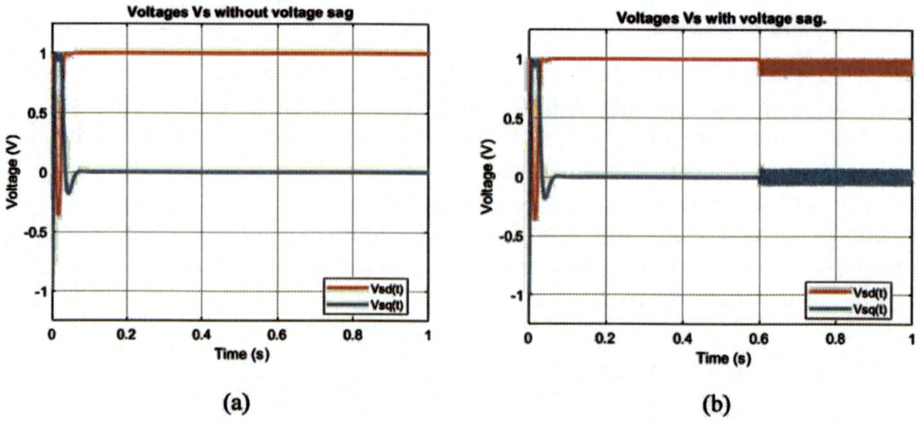

Fig. 5. V_S. voltages at pu of the PLL controller output

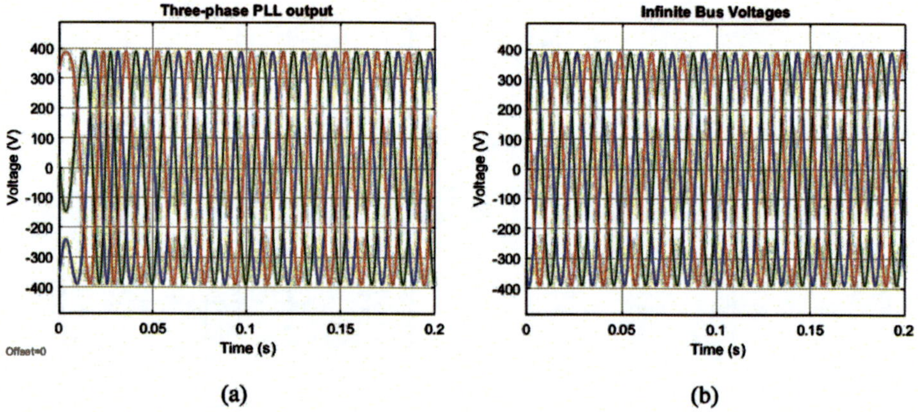

Fig. 6. Comparison of output voltages with the infinite bus

Next, the implementation of the controller in the current mode was included in the MATLAB/Simulink computational environment, as shown in Fig. 7, to control the power flow from a current signal.

The current mode controller response shown in Fig. 8 shows the behavior of the reference currents I_q and I_d and the current dynamics when there is a change in their respective references. To check the performance of the controller, a reference variation step was inserted at times 0.5 s and 0.7 s for the current I_d, where the current values varied from 0.7 p.u. to 0.5 p.u. and 0.5 p.u. to 0.9 p.u., then the variation of I_q with variation from 0 to 0.25 p.u. in 0.9 s. It can be observed that for each current variation, Fig. 7, the power flow transferred from the converter to the electric network also varies, as shown in Fig. 9.

In Fig. 8, it is possible to observe the compensation of all the reference variations of the direct and quadrature axis currents; it is important mentioning that the controller has met the demanded performance requirements.

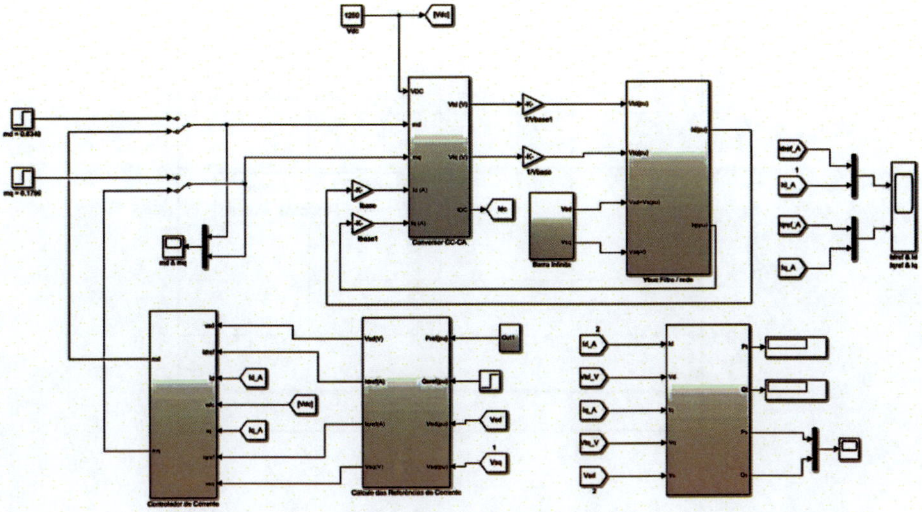

Fig. 7. Implementation of the current controller in the MATLAB/Simulink computing environment

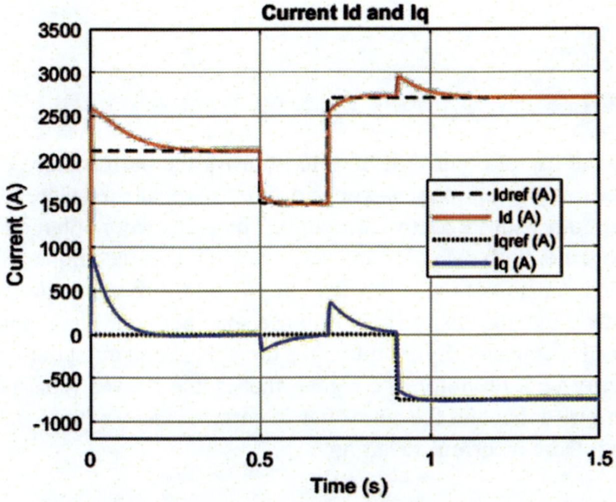

Fig. 8. Straight-axis and quadrature axis current

However, the power flow change occurs as the current references change, though, the power controller has not yet shown the desired performance, requiring adjustments for its proper operation, since the output error in the power is too high.

One of the conditions to connect the photovoltaic system to the grid is electrical is to establish the synchronism of the signals of the photovoltaic system with the grid. Without this verified condition, the use of the panels is limited to the off-grid system, restricting its potential of use. The current mode control has the advantage of protecting

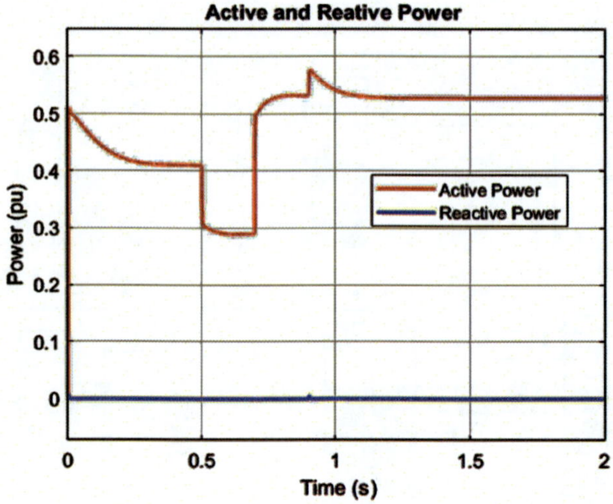

Fig. 9. Active and Reactive Powers

the system against overcurrent, as well as presenting better dynamic performance and accuracy.

5 Conclusion

It was proposed the use of control devices to improve the performance of voltage converters connected to a conventional power grid. Controls evaluated were PLL and control in current mode, both with PI control structure. The gains were obtained using the root locus method, together with the poles and zeros cancellation technique. With this, it was observed that the inclusion of the controllers improved the dynamic performance of the converter since they met the desired performance specification. Thus, it is observed that the PLL fulfills that expected by maintaining the synchronism of the output signal of the voltage inverter with the reference signal, that is, the conventional power network, it must be emphasized that and the controller in current mode, adjusts the output power flow to the output from a current variation.

Highlights
This work presents two control techniques applied in the photovoltaic system converter connected to the grid, increasing the use of systems with renewable sources.

References

1. Tolmasquim, M.T.: Energia Renovável Hidráulica, Biomassa, Eólica, Solar, Oceânica In: EPE: Rio de Janeiro (2016)
2. Dincer, I.: Renewable energy and sustainable development: a crucial review. In: Renewable and Sustainable Energy Reviews, pp. 157–175. Elsevier (2000). https://doi.org/10.1016/s1364-0321(99)00011-8

3. Barreto, E.J.F., Pinho, J.T., Tiago, G.L., Rendeiro, G., Nogueira, M., Gonzalez, W.A.: Tecnologias de Energia Renováveis Soluções Energéticas para a Amazônia. In: Ministério de Minas e Energia. Brasília (2008)
4. Villalva, M.G., Gazoli, J.R.: Conceitos básicos. In: Energia Solar Fotovoltaica - Conceitos e Aplicações - Sistemas Isolados e Conectados à Rede. 1st edn. ERICA, São Paulo (2012)
5. Hart, D.W.: Power Eletronics, 1 edn., 494 p. McGraw-Hill, New York (2011)
6. Barbi, I., Martins, D.C.: Introdução ao estudo dos conversores CC-CA, 2nd edn. Florianópolis: Ed dos Autores, 490 p. (2008)
7. Souza, W.G., Rocha, M.A., Serni, P.J.A., Alves, A.F., Andreoli, A.L., Silva, P.S.: Estudo de inversor aplicado a um Sistema Fotovoltaico para Compensação de Reativos. In: The 12th Latin-American Congress on Eletricity Generation and Transmission – CLAGTEE (2017)
8. Junior, H.G.: Estudos De Controladores Aplicados À Inversores Para UPS Com Operação Ilhada. Mestrado em engenharia elétrica, Universidade Estadual Paulista (2013)
9. Yazdani, A., Iravani, R.: Voltage-Sourced Converters in Power Systems. Wiley, IEEE Press (2010)
10. Hsieh, G.C., Hung, J.C., Phase-locked loop techniques – a survey. IEEE Trans. Ind. Electron. **43**(6), December 1996
11. D'azzo, Houpis: Análise e Projeto de Sistemas de Controle Lineares, ed. Guanabara Dois (1978)
12. Ferreira, R.J., Araújo, R.E., Peças Lopes, J.A.: A comparative analysis and implementation of various PLL techniques applied to single-phase grids. In: 3rd International Youth Conference on Energetics, July 2011

Investigation of the Performance of the Control Methodologies to Mitigate the Undesirable Oscillation in a Smart Grid System

Isaías V. de Bessa and Renan Landau Paiva de Medeiros(✉)

Universidade Federal do Amazonas, Manaus, Brazil
isaias.97.ib@gmail.com, renanlandau@ufam.edu.br

Abstract. Nowadays, the use of renewable energy has made it possible creating new ways to generate electricity like photovoltaic generation, wind generation, hydro generation, and others. It is common to connect this generation system to a power storage device like batteries promoting the voltage regulation of the system using DC-DC power converters. By connection these converters to other voltage converters, it has some undesirable phenomena such as the reduction of the system stability margin, the smart grid controllers dynamic response degradation, the voltage stability problems and undesirable oscillations due to the connection of multiple DC-DC power converters. This work proposes the analysis of some smart grid prototypes and through control techniques like pole placement, root locus and linear-quadratic regulator (LQR) to mitigate oscillations due to the connection of one or more converters on the same DC bus when this bus is responsible for supplying a constant power load (CPL). Finally, power and voltage reference variation tests are performed to evaluate the performance of the controllers designed for the smart grid.

Keywords: Renewable energy · DC-DC converters · Smart grid · Classic pole placement · Classic root locus technique · LQR technique

1 The Problem's Background

The increase in the means of generating electric energy by renewable sources resulted in the use of power converters for the system stabilization. Among these forms of generation, it is sometimes common to use DC-DC converters to stabilize adequate voltage levels in the DC bus power supply [1]. Modeling of converters, especially buck converters, is explored in the literature [2, 3].

In [5, 6], the author discusses the buck converter modeling and control through the modern control strategy, the linear-quadratic regulator (LQR). In these works, the author performs dynamic analysis of the converters for a voltage variation. In the paper [1], the author shows the behavior of a CPL in a photovoltaic system powered by a buck converter by building active damping. Already in [6–8], the authors comment on the dynamic behavior of CPL.

L. Pereira et al. (Eds.): IDEAS 2018, SIST 198, pp. 406–415, 2021.
https://doi.org/10.1007/978-3-030-55374-6_40

This work aims to present the simulation of a smart grid, through the construction of the feeder feeding the CPL.

This paper is structured as follows: Sect. 2 comments on the construction of a smart grid and the problem of CPL. Section 3 presents the main features of the LQR controller. Section 4 defines strategies for the proposed smart grid design, showing system modeling, and controller design. Section 5 shows the results of tests performed on the smart grid.

2 Stability of DC Smart Grid

The DC system is divided into two parts: the first provides constant voltage to the main bus, the feeder. The second represents the load demand elements, among them the CPL, as shown in Fig. 1a.

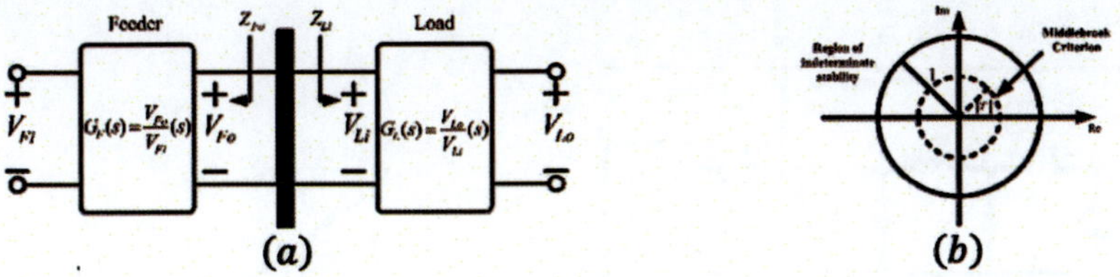

Fig. 1. (a) Subsystems of a DC system and (b) Stability by the Middlebrook criterion. Source: Elaborated by authors.

Let $G_F(s)$ and $G_L(s)$ be the transfer functions of the analyzed subsystems. The global transfer function of the cascaded system is given by (1)

$$G(s) = G_F G_L(s) \frac{1}{1+T} \tag{1}$$

where T is called the minor loop gain (MLG), given by the output impedance ratio, Z_{Fo} of the feeder system and load impedance Z_{Li} as shown in (2) [9].

$$T = \frac{Z_{Fo}}{Z_{Li}} \tag{2}$$

Performing the system stability analysis defined in (1) by the Nyquist diagram (see Fig. 1b) [11] where T defines the overall system stability considering that the functions $G_F(s)$ and $G_L(s)$ are stable. The Middlebrook criterion, see [9], guarantees the overall system stability for the relationship shown in (3).

$$|T| < 1 \tag{3}$$

As can be seen in [11], looking at the system gain margin (GM) shown in (1), determines the relationship between GM and MLG in (4), see details in [9].

$$GM = \frac{1}{|T|} > 1 \tag{4}$$

Thus, if T is within the region delimited in Fig. 1b, the system is considered stable. Given the introduction of the subject can the effects of CPL on the system.

2.1 The Problem of Constant Power Load (CPL)

There are two types of basic loads in electrical systems. The first is the resistive load, which operates with a constant voltage supply. The second operates with a constant power load (CPL) [1, 6, 9, 10] that can be represented by the DC motor operating at the constant speed [6] or by a solar panel capable of providing energy for a residence.

Connecting the CPL to the DC bus causes instability observed in the form of oscillations between the feeder connection and the CPL. The feeder-CPL set features to the smart grid to be observed and ensure that the effects of the oscillation do not lead to system instability is fundamental to the proper functioning of the smart grid.

This paper proposes that a buck converter represents the feeder, where the input voltage of the system represents the input voltage from a power converter. This converter will provide constant voltage to the DC bus of the system where the CPL is connected, as shown in Fig. 2a.

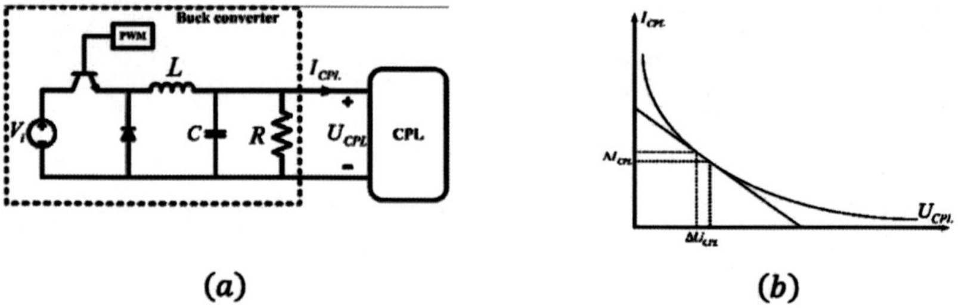

Fig. 2. (a) Buck converter powering the CPL and (b) Graphic behavior of the CPL. Source: Elaborated by authors.

Because of the characteristic of the CPL, for the maintenance of a constant power due to a voltage variation in the CPL, the current varies to correct the voltage change. Thus, for a small voltage variation due to a current variation, as shown in Fig. 2b, we have the relationship shown in (5), for more details see [8]

$$\frac{\partial U_{CPL}}{\partial I_{CPL}} = -\frac{P_{CPL}}{I_{CPL}^2} \tag{5}$$

where the term $-\frac{P_{CPL}}{I_{CPL}^2}$ represents a negative resistance that causes the problem of smart grid instability in the occurrence of a CPL oscillation.

2.2 Design of DC Microgrids

Given the buck converter is shown in Fig. 2a, operating with duty cycle D, resulting in a voltage in the capacitor V_{Co} and current in the inductor I_{Lo}. Modeling the converter through the use of the average state model [3] and linearizing around the point $X = (D, I_{Lo}, V_{Co})$ by the expansion of Taylor's series in the form of the Jacobian matrix [11], we have the state-space model shown in (6)

$$\begin{bmatrix} \Delta \dot{I}_L \\ \Delta \dot{V}_C \end{bmatrix} = \begin{bmatrix} 0 & -\frac{1}{L} \\ \frac{1}{C} & -\frac{1}{RC} \end{bmatrix} \begin{bmatrix} \Delta I_L \\ \Delta V_C \end{bmatrix} + \begin{bmatrix} \frac{V_i}{L} \\ 0 \end{bmatrix} \Delta D \tag{6}$$

where ΔD is the duty cycle variation in time. Thus the complete signal is given by (7), see details in [3].

$$D(t) = D_o + \Delta D \tag{7}$$

For the smart grid feeder operation, the voltage output is linearized around point X. For CPL, also constructed from the buck converter, the linearized power output is used around point X, as shown in (7)–(8).

$$\Delta V_o = \Delta V_C \tag{8}$$

$$\Delta P_o = \frac{2V_C}{R}\Delta V_C \tag{9}$$

With the context of the problem and the buck converter model, the design of the controllers for feeder regulation and the smooth operation of the CPL allows the smart grid design.

3 The Linear-Quadratic Regulator

The linear-quadratic regulator (LQR) is an optimal control methodology capable of allowing the designer to choose between minimizing energy or minimizing time by developing the quadratic performance index shown in (10).

$$I = \int_t \left(x^T Qx + u^T Zu\right)dt \tag{10}$$

For this, consider an observable S system described by (11) and represented for the block diagram of the system is showing in Fig. 3a.

$$S : \begin{cases} \dot{x} = Ax + Bu \\ y = Cx \end{cases} \tag{11}$$

The LQR methodology uses the state feedback structure is, as shown in Fig. 3b. The solution of the optimal control problem is to determine the gain value K for the state feedback, using the solution of the Riccati equation, see [11], shown in (12) resulting from the development of (10).

$$PA + A^T P - PBZ^{-1}B^T P = -Q \tag{12}$$

where Q is symmetric semi-definite positive matrix and Z is a scalar defined by the designer. The solution of (12) results in the definition of the vector P. The values of the gains of the vector K are given by (13).

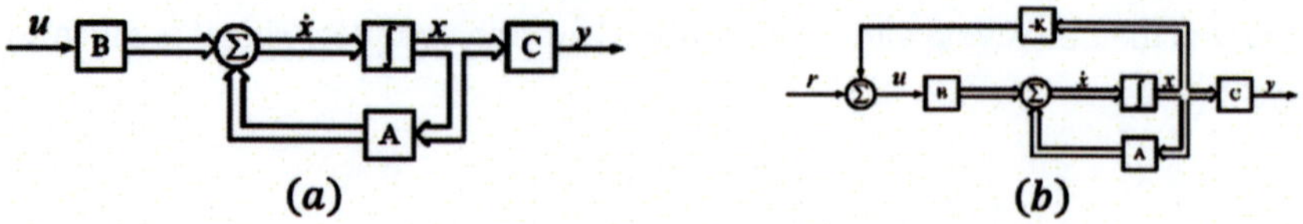

Fig. 3. (a) Block diagram of the system and (b) System with state feedback. Source: Elaborated by authors.

$$K = B^T P \tag{13}$$

The weighting of Q and Z are related to the project's need to prioritize better performance or less control effort.

4 Methodology for Smart Grid Design

4.1 Converters Design

By basing the buck converter design on its static equations [2], the R, L and C parameters can be determined through voltage ΔV_o and current ΔI_L variation specifications for an input voltage V_i and a switching frequency f.

For the feeder and CPL design, a 1 mH inductor is used. The feeder operates at a switching frequency of 1 kHz, while the CPL operates at 2 kHz. Equation (14) shows the relationship between V_o and V_i.

$$V_o = DV_i \tag{14}$$

With the base power of the CPL being calculated for $D = 1$, we obtain (15) and the power in pu shown in (16).

$$P_b = \frac{V_i^2}{R} \tag{15}$$

$$P_o = D^2 \tag{16}$$

Thus, the main relations are shown in (17)–(18) for the determination of buck converter, see details in [2]. The calculated parameters are shown in Table 1.

$$\Delta I_o = \frac{V_i D(1-D)}{fL} \tag{17}$$

$$R \leq \frac{fL}{1-D} \tag{18}$$

$$C \geq \frac{1-D}{8f^2 L(\Delta V_o / V_o)} \tag{19}$$

Table 1. Buck converters parameters

Feeder				CPL			
Parameter	Symbol	Value	Unit	Parameter	Symbol	Value	Unit
Input voltage	V_{i1}	15.0	V	Input voltage	V_{i2}	12.0	V
Duty cycle	D_1	0.8	–	Duty cycle	D_2	0.6	–
Switching frequency	f_1	1.0	KHz	Switching frequency	f_2	5.0	KHz
Load resistance	R_1	2.0	Ω	Load resistance	R_2	2.0	Ω
Inductor	L_1	1.0	mH	Inductor	L_2	1.0	mH
Capacitor	C_1	2.0	mF	Capacitor	C_2	2.0	mF
Output voltage	V_{o1}	12.0	V	Output power	P_{o2}	0.36	V

Source: Elaborated by authors.

4.2 Feeders Controllers

The feeder controller design is developed in this paper through two classical methodologies: pole placement and root locus [11]. For the realization of this design, the PID control structure with voltage feedback is used, as shown in Fig. 4a and Fig. 4b. In addition to the classical methodologies, a modern methodology will be used, LQR presented in Sect. 3. For this, the state feedback structure in Fig. 3b. To ensure zero error, an integrator at the system input, closing the system external loop, as shown in Fig. 4b.

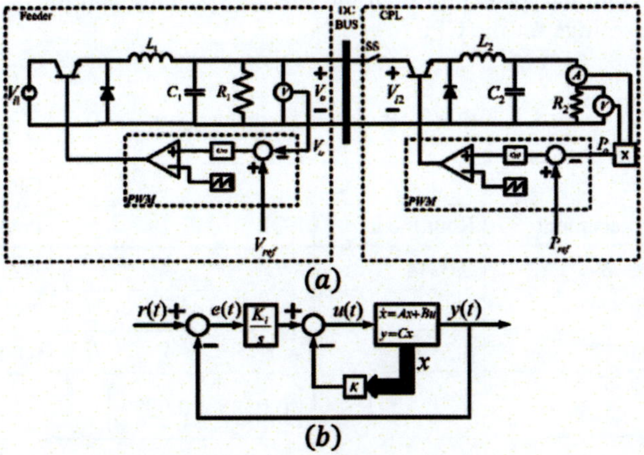

Fig. 4. (a) Smart grid designed with the feeder on the left and CPL on the right. (b) Structure for the LQR controller design used. Source: Elaborated by authors.

The feeder controller design is developed to meet the settling time specifications t_{ss}, overshoot ovs and steady-state error at step e_{ss}, as it exhibits the following desired specifications.

$$\begin{cases} t_{ss} \leq 2ms \\ ovs \leq 5\% \\ e_{ss} = 0 \end{cases} \tag{20}$$

The pole placement design is performed by comparing the desired polynomial and a polynomial characteristic of the closed-loop system, as shown in (21) [12].

$$(s + p_o)\left(s^2 + 2\zeta\omega_n s + \omega_n\right) \tag{21}$$

Solving the polynomial equality in (21), the gains of the controllers are obtained as shown in (22).

$$\begin{cases} K_P = \frac{LC(\omega_n^2 + 2\zeta p_o) - 1}{V_i} \\ K_i = \frac{LC p_o \omega_n^2}{V_i} \\ K_d = \frac{LC}{V_i}\left(2\zeta\omega_n + p_o - \frac{1}{RC}\right) \end{cases} \tag{22}$$

The root locus design has graphically defined the region of interest of the system in closed-loop, as can be seen in [12].

Finally, the Riccati equation solution determines the design of the controller by LQR, as shown in the previous section. For this work, the solution of the Riccati equation was made by computational means through the MATLAB command lqr (). Table 2 shows the calculated earnings for the CPL.

Table 2. Projected controller gain

Gain	Feeder			CPL
	Pole placement	Root locus	LQR	Pole placement
K_p	−0.0042	0.0244	–	0.2729
K_i	13.9265	15.7153	10	258.2576
K_d	$1.2 \cdot 10^{-4}$	$6.3 \cdot 10^{-5}$	–	0.0023
K	–	–	$\begin{bmatrix} 0.0385 & -0.0099 \end{bmatrix}$	–

4.3 CPL Controller Design

The CPL controller design is performed using the pole placement methodology. To perform the power feedback one has the indirect measurement as shown in Fig. 4a. The procedures used are analogous to those shown in the previous section for feeder design. Table 2 shows the calculated gains for the CPL.

4.4 Description of Test Performed

- **Feeder voltage reference variation:** For the disconnected CPL and the feeder at the operating point, the voltage reference is increased from +5 V to +6 V at time $t = 5$ s. Returning to the original reference at time t = 6 s. At time $t = 7$ s the voltage reference decreases from +6 V to +5 V and at time $t = 8$ s it returns to the original reference. This test aims to verify that the feeder controllers meet design specifications.
- **CPL power reference variation:** This test aims to verify that the CPL controller has met the requested specifications. In this test the CPL is connected to an ideal DC bus, represented by a DC source. With the CPL at the closed-loop operation point, the power reference change from 0.36 pu to 0.56 pu is performed at time $t = 5$ s and at time $t = 6$ s returns to the original reference. Then, at time $t = 7$ s, the reference is changed from 0.36 pu to 0.16 pu by returning to the original reference at time $t = 8$ s.
- **Power Reference Variation for the interconnected feeder-CPL system:** In this test, the stability of the smart grid is verified with the feeder supplying the DC bus with voltage. At the input of the feeder, the voltage source will represent the output of a power generation system from the grid or any other alternative means. The CPL connects to the feeder at the point of operation. With CPL at the operating point and in closed-loop, the power reference range is as follows: from 0.36 pu to 0.46 pu at instate $t = 5$ s, from 0.46 pu to 0.56 pu at time $t = 5.5$ s; the system returns to the original reference at $t = 6$ s. From 0.36 pu to 0.26 pu at time $t = 6.5$ s and from 0.26 pu to 0.16 pu at time $t = 7$ s.

5 Designed Microgrid Performance Analysis

This section shows the analysis of the designed controllers. The feeder controller designed by pole placement (PP), root locus (RL), LQR are analyzed. Also, the CPL controller designed by PP is analyzed. Figure 5a shows the feeder voltage in the first test, while Fig. 5b shows the behavior of the CPL for power variations. Figure 5c and Fig. 5d show the smart grid response to variations in CPL.

5.1 Feeder Voltage Reference Variation

In this test the controllers designed for the feeder are validated by a voltage reference variation without connection to the CPL. Figure 5a shows the DC bus output voltage, where the LQR controller has intense oscillation.

The purpose of this test is to verify that the controls designed for the feeder are capable of regulating the DC bus supply voltage. For a well-designed smart grid, regulation is critical to system reliability. In performance index charts [13], the RL control has better performance.

5.2 CPL Power Reference Variation

In this test the CPL is connected to an ideal DC bus, where power supplies remain stable for CPL power variations.

Using the PP methodology to simulate the dynamic behavior of the CPL, shown in Fig. 6, controller action is fast, greater than 2 ms as projected and with low overshoot as requested at the beginning of the project.

With the CPL project evaluated, one can now connect it to the feeder-fed bus and observe the effects of this connection.

5.3 Power Reference Variation for the Interconnected Feeder-CPL System

In this test connecting the CPL to the feeder and varying the CPL, reference causes oscillations between the feeder and the CPL. Controllers designed for the feeder were able to suppress these oscillations as shown in Fig. 5c.

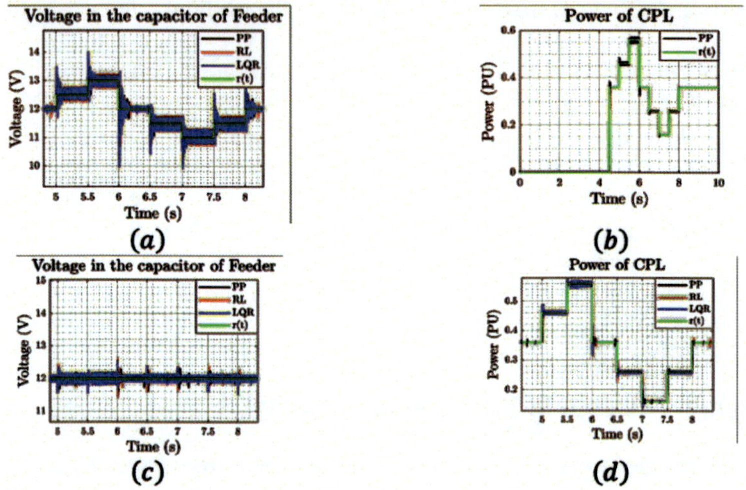

Fig. 5. (a) Feeder response for voltage reference change disconnected from CPL (b) CPL response for power reference variation connected to an ideal source (c) With the complete smart grid, the DC link voltage is displayed for (d) Power reference change in CPL. Source: Elaborated by authors.

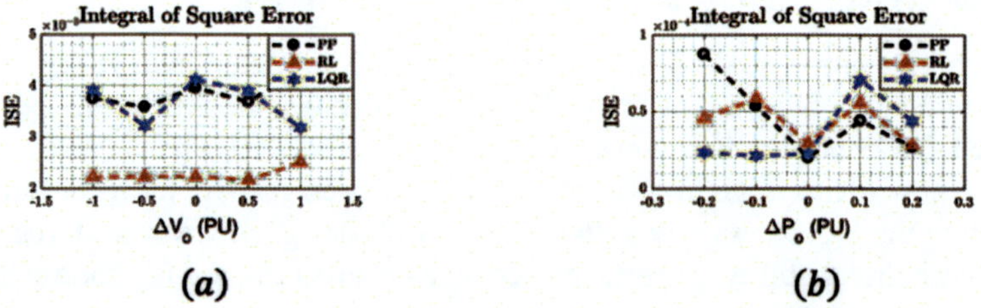

Fig. 6. (a) Feeder Controller Performance for Voltage Reference Range. (b) Smart grid feeder Controller Performance for CPL Power Variation. Source: Elaborated by authors.

With the performance calculation of each controller during the oscillations shown in the graph of Fig. 6b. In the first test (without CPL), the RL controller loses performance for the PP and LQR controllers. Also, for negative load variations, the LQR controller maintains performance while PP and RL controllers lose performance. This test is used to validate the proposed smart grid simulation by validating the design of controllers for feeder voltage regulation.

6 Conclusion

The objective of this work was to present control methodologies to ensure the stability of smart grids. Thus both the classical methodology and the modern LQR methodology were able to maintain system stability in the event of oscillations.

Through a brief review of the stability of DC systems, it was proposed to build a smart grid composed of constant power supply from an energy converter element, and through DC-DC converters it was possible to regulate this system.

Finally, it was possible to analyze the control methodologies proposed in this work observing controller performance loss by RL with CPL connection.

References

1. Jusoh, A., Baamode, H., Mekhilef, S.: Active damping network in DC distributed power system driven by photovoltaic system. Sol. Energy **87**, 254–267 (2013)
2. Hart, D.W.: Power Electronics. McGraw-Hill, New York (2011)
3. Erickson, R.W., Maksimovic, D.: Fundamentals of Power Electronics. Kluwer Academic Publishers, New York (2004)
4. Maccari, L., Montagner, V.F., Ferreira A.A.: A linear-quadratic control applied to buck converters with H-infinity constraints. In: 2013 Brazilian Power Electronics Conference, pp. 339–344 (2013)
5. Maccari Jr., L.A., Valle, R.L., Ferreira, A.A., Barbosa, P.G., Montagner, V.F.: A LQR design with rejection of disturbance and robustness to load variations applied to a buck converter. Revista Eletrônica de Potência **1**, 7–15 (2016)
6. Emadi, A., Khaligh, A., Rivetta, C.H., Williamson, G.A.: Constant power loads and negative impedance instability in automative systems: definition, modeling, stability and control of power electronics and motor drives. IEEE Trans. Veh. Technol. **55**(4), 1112–1125 (2006)
7. Singh, S., Gautam, A.R., Fulwani, D.: Constant power load and their effects in DC distributed power systems: a review. Renew. Sustain. Energy Rev. **72**, 408–421 (2017)
8. Elsayed, A.T., Mohamed, A.A., Mohammed, O.A.: DC microgrids and distribution systems: an overview. Electric Power Syst. Res. **119**, 407–417 (2015)
9. Riccobono, A., Santi, E.: Comprehensive review of stability criteria for DC distribution systems. IEEE Trans. Ind. Appl. **50**, 3525–3535 (2014)
10. Marcillo, K.E.L., Guingla, D.A.P., Barra Jr., W., Medeiros, R.L.P., Nogueira, F.G.: Interval robust controller to minimize oscillations effects caused by constant power load in a DC multi-converter buck-buck system. IEEE Access **7**, 26324–26342 (2019)
11. D'Azzo, J.J., Houpius, C.H.: Linear Control System Analysis and Design, 3rd edn. McGraw-Hill, Illinois (1981)
12. Astrom, K.J., Wittenmark, B.: Computer-Controlled Systems, 3rd edn. Dove Publications, Rio de Janeiro (1997)
13. Duarte-Mermoud, M.A., Pietro, R.A.: Performance index for quality response of dynamical systems. ISA Trans. **43**, 133–151 (2004)

Low-Cost Photovoltaic Maximum Power Point Tracking Project for Autonomous Electric Vehicle Prototype

Matias Alles Hubert[1](✉), Antonio Carlos Valdiero[2], Roberta Goergen[1], Edmilton Stein[1], Rosângela Rommel Regner[1], and Ben-Hur Maciel[1]

[1] Regional University of the Northwest of the State of Rio Grande do Sul, Panambi, RS NJ 98280-000, Brazil
matias.a.hubert@gmail.com
[2] Horizontina Faculty, Horizontina, RS NJ 98920-000, Brazil

Abstract. The storage of electric energy is still one of the major bottlenecks found in the evolution of the electric vehicle industry, but with the evolution of material technology, batteries have been increasing their ratio capacity/mass, directly impacting the growth of this technology. As this is still a problem, alternatives are sought to increase the autonomy of electric vehicles, the main one being the implementation of embedded photovoltaic generation systems, which convert solar energy into usable electric energy in the vehicle. These embedded photovoltaic systems need a power flow control system because, due to the variation of irradiation that the panel receives, there are variations on voltage and current sent to the vehicle, which is a problem to his electric system. In order to maintain the correct operation and achieve a greater power efficiency of the photovoltaic system, it is necessary to study and design a Maximum Power Point Tracking system. Within this power control model, will be used in the prototype, one of the most simple methods is this the Perturb and Observe, which is a simple comparison of samples and can be easily implemented in low-cost Arduino hardware. Thus, the energy efficiency of systems is crucial because the lack of a project and adequate control can generate instability, and low power efficiency in general, impairing the operation and autonomy. Therefore, this paper presents the design of a photovoltaic control system to be implemented in an autonomous vehicle prototype applied to engineer education and, in the future, precision agriculture.

Keywords: Photovoltaic system · MPPT · Autonomous vehicle

1 Introduction

One of the main problems caused by combustion vehicles is the long-term pollution generated. It is a problem that has been discussing for a long time, and finally, some solutions are emerging. Electric vehicles and sources of electricity generation by photovoltaic panels are rapidly growing technologies, and together these two technologies have been emerging as a solution to the problem of pollution and the energy crisis. Due

L. Pereira et al. (Eds.): IDEAS 2018, SIST 198, pp. 416–424, 2021.
https://doi.org/10.1007/978-3-030-55374-6_41

to the current capacity of storage of electric energy in batteries, the autonomy of electric vehicles is reduced, being this one of the main problems of the technology, therefore, in some cases it is necessary to implement technologies that help in the increase of autonomy, thus, the implementation of a photovoltaic system to the vehicle guarantees an increase in the autonomy and reduction of costs of battery recharge [1].

According to [1], photovoltaic energy sources have some disadvantages, even providing outstanding energy efficiency. Among the main developments, the low power transfer efficiency and the low power density available, combined with high cost on a large scale, are not the best choice for a generation when large volumes are required.

Knowing this disadvantage, several types of research have looked for ways to improve the transfer of power in photovoltaic systems. One form that has gained much space is based on the charging of batteries in a pulsed way, improving the efficiency of the system. This model works by tracking the maximum power point (MPPT) of the solar photovoltaic array. As photovoltaic generation does not remain constant over time because it varies according to the intensity of the solar rays, temperature, among other factors, the power curve (V $\times$ I) generated changes. Thus, through a pulsed CC-CC static converter, one can control the power flow, leaving it as close to the maximum as possible [2].

Lead-acid batteries are generally used to store energy. Even though it has a reduced shelf life compared to batteries of different chemical compositions, it is an attractive device due to the low-cost [3]. Therefore, this work looks for the design of an MPPT system for use in works and researches that make use of batteries and can be installed on the loading system, more specifically, in a prototype of an autonomous electric vehicle, shown in Fig. 1.

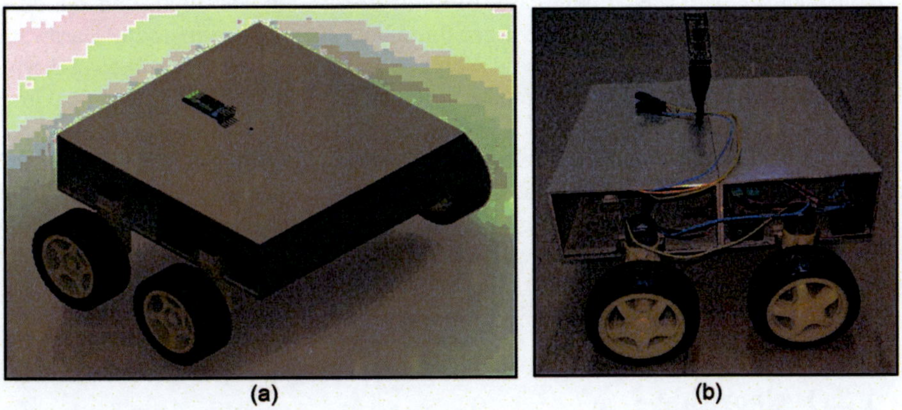

Fig. 1. Prototype of electric vehicle: (a) CAD drawing; (b) Construction of the prototype (1st version). Source: Elaborated by authors.

The prototype is said to be low-cost since the total estimated price for development is around US$ 85.00 (direct conversion from R$ to US$), considering that the prototype in the current phase is used as an engineer learning tool, integrating the control, drive and computation disciplines are mainly considered low-cost due to their teaching potential when compared to other tools and machines used in control and drive laboratories.

2 Needs Analysis

The Challenge-Oriented Mechatronics Research Project of the NIMASS research group seeks the development and construction of modular, innovative, and creative solutions for intelligent machines. The prototype of an autonomous electric vehicle of the low-cost applied to agriculture is one of the works in development, introducing to this one, concepts of mathematical modeling, control, acquisition of signals, power management, and electromechanical drives.

From the mathematical modeling of the electric subsystem and the mechanical subsystem of the prototype, it is possible to implement more efficient displacement and speed control techniques. However, in order for the drive signal to match the desired one, it is necessary that the power supply source of the prototype is able to maintain the voltage and current levels requested by the system, thus, it is sought to maintain the load of the set of batteries as best as possible, through a charge control system with MPPT by P&O.

In the NIMASS research group, several projects use low-cost hardware to control mechatronics systems and, as sustainability becomes more and more necessary, ways are sought to improve energy efficiency, in this case through integration to the system already developed, a photovoltaic plate. We are looking for the development of a low-cost electric vehicle that can be used in the learning of control and automation disciplines, as well as its implementation in agriculture. Thus, it is prioritized the design of an algorithm that is easily understood and easily redesigned, so that it can be used in other projects without the need for a great re-study of the same.

From these needs, we seek the development of a control algorithm that is compatible with the control board used, in this case, Arduino UNO. Compatibility is understood as the relation of the complexity of the algorithm about its application and of hardware. Therefore, it was opted for the study and design of an MPPT algorithm of the P&O type, since even being one of the simplest methods, it attends to the theoretical-technical level and the energy needs of the prototype [1] (Fig. 2).

The flow chart of the figure above represents the general power flow generation and control system for both the battery pack and the prototype itself. The photovoltaic panel is equipped with the function of charging the vehicle batteries or supplying power to the modules directly through the DC conversion circuits Buck and load controller. With the DC Buck circuit, the maximum utilization of the panel power and a load control circuit is sought in order to limit the load current supported by the batteries.

3 P&O Algorithm

According to [4], the MPPT algorithm of the P&O type is widely used, due to the ease of implementation of its logic, requiring no in-depth study or high-cost hardware. The logic of the P&O algorithm is based on the periodic perturbation of the control signal of the DC conversion circuit, so that, if this disturbance results in a power value above the previously measured value, the disturbance signal follows in the same direction. If the power variation is negative, the disturbance direction is inverted in order to reach the maximum powerpoint. The P&O algorithm has the advantage of the ease of implementation since it

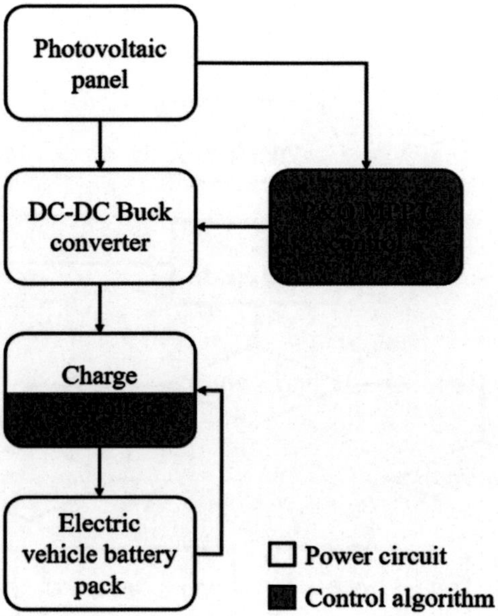

Fig. 2. General flow chart of the photovoltaic panel system, MPPT, and charge of the battery pack. Source: Elaborated by authors.

does not depend on specific characteristics of the panel it is controlling, it depends only on the instantaneous power measurements obtained by the current and voltage sensors. However, this algorithm has some disadvantages, the main one of which is the slow adaptation to a sudden change in climatic variables, which can be corrected through the implementation of an adaptive algorithm by a proportional-integrative-derivative controller (PID) search in future works [4].

The operating logic diagram of the MPPT algorithm based on the P&O method is presented, as shown in Fig. 3. It starts by performing the output current and voltage measurement of the PV panel, values that can be obtained through the law of Ohm instantaneous power, $P(k) = V(k)I(k)$. If the power variation is positive, increase the cyclic ratio of the DC converter in the same direction, if the variation is negative, it will vary in the opposite direction.

4 DC-DC Buck Converter

A DC-DC converter has the purpose of controlling the power flow of a circuit through passive elements such as resistors, capacitors and inductors, and semiconductor elements operating as switches. The ratio of the switching period of the switch element characterizes the cyclic ratio of the converter circuit; thus, by varying the cyclic ratio, the output voltage [6].

The Buck converter performs the voltage lowering, characterized by voltage input and current output, precisely what is obtained through the photovoltaic panel (voltage) and what is objectively in the output, charging current of the battery pack. The Buck

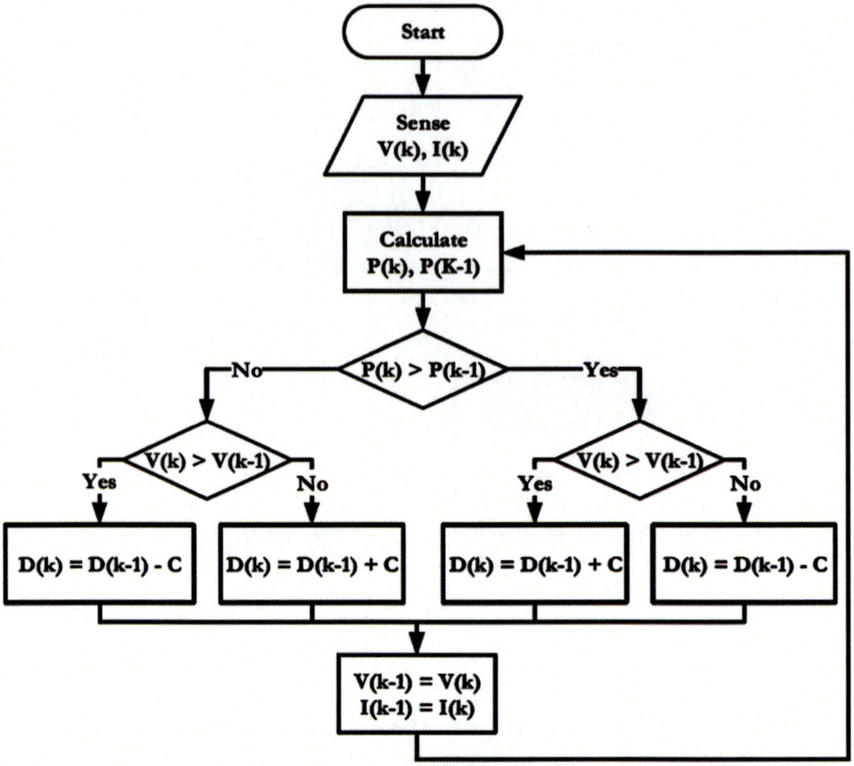

Fig. 3. P&0 flowchart. Source: [5].

converter can operate in three distinct modes: continuous driving, discontinuous driving, and critical driving. In this work, continuous mode operation is sought, which is characterized by the output current I_{out} does not reach the null value; only a current ripple appears as a function of the switching time. The equation that relates the input and output voltage of the buck converter to the continuous drive mode is given below [6]:

$$D = \frac{V_{out}}{V_{in}} \tag{1}$$

Initially, the design and simulation of a circuit for the application of the MPPT P&O method is sought. From the values to be determined for charge voltage of the battery pack, solar panel power, among others, the DC-DC conversion circuit (Buck) can be designed for operation in continuous conduction mode. In order to dimension the components of the circuits, the following specifications:

$$W_p = 10\,\mathrm{W}$$

$$V_{in} = 15 - 22\,\mathrm{V}$$

$$V_{out} = 10.5 - 13.5\,\mathrm{V}$$

$$f_s = 50\,\mathrm{kHz}$$

With the above specifications, you can carry out the calculations for the design of the components of the Buck circuit for the maximum and minimum values of voltage and current of input and output, according to. By [7], one can determine the inductance and capacitance value of the Buck circuit:

$$L = \frac{V_{in}(1-D)D}{\Delta I_L f_s} \tag{2}$$

$$L = 1\,mH$$

$$C = \frac{V_{in}(1-D)D}{8L\Delta V_{out} f_s^2} \tag{3}$$

$$C = 5\,\mu F$$

Then, the topology of the Buck circuit of this work is presented, a variation to the circuits found in most of the bibliographies is observed because, in this, the switch element will switch the earth of the circuit; thus, the voltage V_{GS} of the MOSFET is reduced, simplifying its activation.

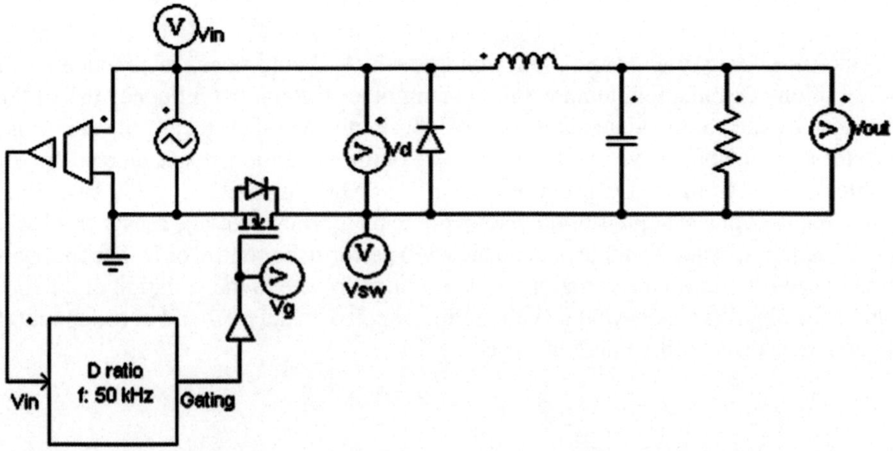

Fig. 4. DC-DC Buck converter circuit. Source: Elaborated by authors.

5 Simulation

In this section, we intend to carry out simulations in the PSIM software to verify the specified capacitance and inductance values, to determine if they satisfy the voltage input and output objectives. Thus, the circuit of Fig. 4 is simulated, using a voltage sensor placed at the input of the Buck circuit that performs the voltage reading of a voltage source with triangular variation, representing the voltage variation of the photovoltaic panel. The sensed voltage signal is read by a block of language C, which carries out the change in the cyclic ratio of the Buck converter in function of the value of voltage read in relation to the stipulated value for the output, which in this case is 10.5 V. Next an image of the Buck circuit simulation with input voltage values varying triangularly from 15–22 V and control of the output voltage from the variation of the cyclic ratio (D) (Fig. 5).

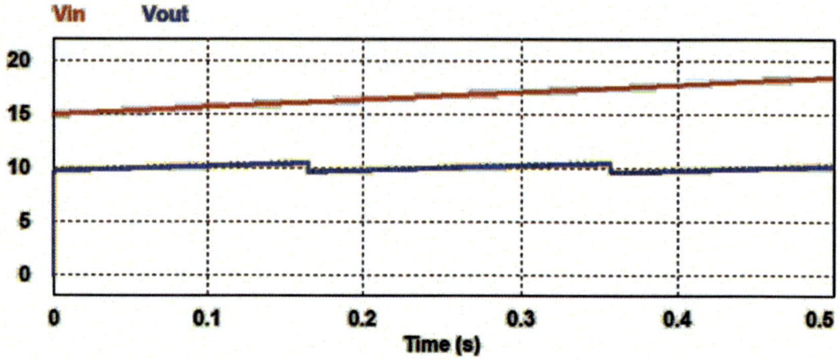

Fig. 5. Buck converter input voltage ranging from 15 V to 22 V (red) and stipulated output voltage to 10.5 V. Source: Elaborated by authors.

Due to the variation of the input voltage, which should occur in practice due to the variation of irradiation, temperature, among other factors that influence the voltage generated by the photovoltaic panel, the MPPT circuit should correct of the conversion from the variation of the cyclic ratio. In the case of the simulation, this is relayed through the block of programming C programmed in the PSIM, in practice, it will be realized by reading the voltage signals of the photovoltaic plate from a voltage sensor MAX471-GY471 acquired by the Arduino plate, which will perform the control of D. The following is the image that shows the variation of the cyclic ratio, due to the variation of the input voltage. In Fig. 6 (a) we have the cyclic ratio at approximately 0.7 and at (b) about 0.5, showing the correct functioning of the code.

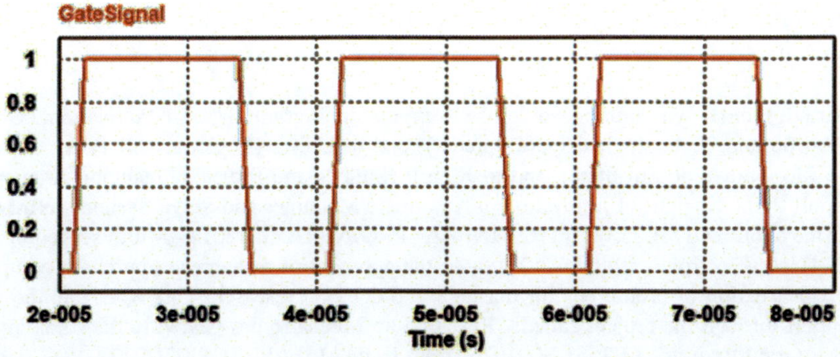

(a)

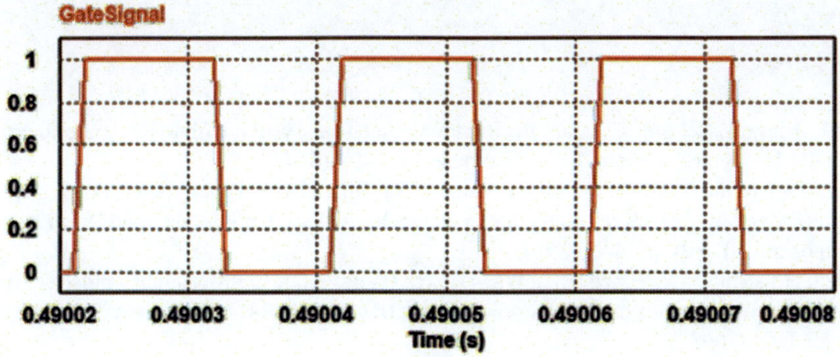

(b)

Fig. 6. Signal sent to the MOSFET gate of the circuit simulation of Fig. 3. Source: Elaborated by authors.

6 Conclusion

The initial study of the requirements for the autonomous electric vehicle prototype demonstrates the need to build an MPPT system for the best use of the photovoltaic panel power in the charging of the vehicle battery pack.

The understanding of how MPPT methods work shows that the implementation of the P&O method is the simplest of all and that it can also be easily exploited in several projects with few changes. This work seeks to determine the type of MPPT method, determination of parameters of the buck DC-DC converter, and simulation of the values found to verify the correct operation of the system. All these objectives have already been reached, following the work, we seek to implement the P&O method in an algorithm on the Arduino board, construction of the Buck converter from the determined values, it is still necessary to determine the maximum current values of input and output because the current prototype consists of a set of batteries that total 6 V low amperage, however, we know that the most common voltage of batteries to be found in vehicles is 12 V and high amperage. Thus, possibly, this project should follow in the direction of the construction of a system for charging of batteries of 12 V, since this is the most used in several projects to be developed in the research group.

Acknowledgments. The authors would like to express their gratitude to CNPq (National Council for Scientific and Technological Development), and to Unijuí (Regional University of Northwestern Rio Grande do Sul State), and to FINEP (Studies and Projects Financing Agency) for the support to the Nucleus of Innovation in Automatic Machines and Servo Systems (NIMASS), through the public call MCTI/FINEP/CT-INFRA - PROINFRA - 02/2014 – Multiusers equipment, Ref.: 0141/16 (Electronic protocol: 124), with the approval of resources to purchase equipment for the construction of prototypes for master's and doctoral research. This work had the financial support through the project called: "Research on Mechatronics geared to the Challenges of Society" (Grant term no. 17/2551-0001014-0) no EDITAL FAPERGS 02/2017 - PqG (Gaúcho Researcher Program).

References

1. Xu, J., Kang, L., Zhong, C., Cao, B.: Photovoltaic power system with MPPT functionality for a small-size electric vehicle. Int. J. Photoenergy, 1–9 (2014)
2. Kose, E., Mühürcü, G., Muhurcu, A., Sevim, B.: SFLA based PI parameter optimization for optimal controlling of a Buck converter's voltage. In: Artificial Intelligence and Data Processing Symposium (IDAP), pp. 1–5 (2017)
3. Hsieh, H.I., Shih, S.F., Hsieh, J.H., Wang, C.H.: Photovoltaic high-frequency pulse charger for lead-acid battery under maximum power point tracking. Int. J. Photoenergy **2013**, 8 (2013). Article ID 687693
4. Harrag, A., Messalti, S.: Variable step size modified P&O MPPT algorithm using GA-based hybrid offline/online PID controller. Renew. Sustain. Energy Rev. **49**, 1247–1260 (2015)
5. Kofinas, P., Doltsinis, S., Dounis, A.I., Vouros, G.A.: A reinforcement learning approach for an MPPT control method of photovoltaic sources. Renew. Energy **108**, 461–473 (2017)
6. Ivo, B., Denizar, C.M.: Conversores CC-CC Básicos Não Isolados. Florianópolis, Santa Catarina – Brasil (2000)
7. Ashfaq, A.: Power Electronics for Technology. Prentice-Hall, New Jersey (1999)

Smart Cities with Sustainable Management of Natural Resources, Green Mobility, and Entrepreneurship

Collaboration Networks for Social Innovation in the Context of Social Incubators: Constitutive Elements

Márcia Aparecida Prim(✉) and Gertrudes Aparecida Dandolini

Federal University of Santa Catarina, Florianópolis, Brazil
marcia.prim@ufsc.br

Abstract. Social innovation is a concept that arises in a complex environment. Problems related to climate change, the global epidemic of chronic diseases and social inequality are challenges that the current political structure has not treated in an egalitarian approach. Experiences in social innovation point to collaborative work. Several types of players relate to finding solutions to the problems inherent in the less favored population. In this sense, the objective of this article is to identify the constituent elements of the collaboration network, which emerge in the context of social innovation. This is a qualitative and descriptive research, of the case study type, applied in a social incubator. The authors used documentary analysis and semi-structured interviews to collect the data. Based on the thematic analysis of the primary data, it was concluded that the constituent elements of the collaboration network in the context of social incubators are: partnerships (networks of players and types of partnerships); collaboration (mutual aid, commitment and trust); self-management (shared leadership, joint decision-making and shared processes); resources (financial, material and human); learning (training, lecture and sharing of experiences). It is also worth noting that economic, social and environmental sustainability, and empowerment (recognition and construction of identity) are recognized as a result of the collaboration network.

Keywords: Social innovation · Collaboration networks · Social incubator

1 Introduction

Social Innovation (SI) has attracted the attention of several organizations as a new vision of addressing social problems[1] [1, 2].

This model emerges with the purpose of seeking innovative alternatives to the most complex social challenges, such as social inequality (in education, health, hunger and poverty), climate change, environmental pollution, and global epidemics of chronic diseases, which the current social structure and policies have not prioritized [3–5].

[1] *Social* has the prospect of being something concerning a community, a human society, the relationship between individuals [7]. A social problem is a phenomenon, a situation, or a condition that, from the perspective of certain groups within a society, does not work as it should [7].

L. Pereira et al. (Eds.): IDEAS 2018, SIST 198, pp. 427–436, 2021.
https://doi.org/10.1007/978-3-030-55374-6_42

To Morgan, Richardson, Marques [6] it is a complex and multidisciplinary concept. Among the various SI definitions raised in the literature by [7], this diversity is evident since academic publications correlate it with various sectors of the world economy. However, it is latent and at the center of the discussions the concern with meeting the basic social needs and the involvement of several players in the search for viable alternatives

Current publications such as [8] and [6] still point to the need for more SI studies, given the significant growth of the theme. These authors state that innovation, in its original definition, has been a key construct in management and economics studies; however, the addition of the adjective "social" brings a differentiated character. It is a similarly complex concept that combines all the quirks of innovation processes with the confusing nature of social issues and outcomes.

To Borges et al. [9], SI is created from the combination of existing knowledge in the community itself, involves different players, through a collaborative process, in the search for a sustainable and beneficial social change to a collective. It should be emphasized, therefore, that SI development focuses on meeting a social demand and fostering local development.

The balance between local development in a sustainable way and economic growth becomes a point of discussion in universities, in social innovation centers, in government, in private companies and in the third sector [3, 6].

Thus, social innovation transcends the boundaries between public, private, and non-profit sectors. It fosters the exchange of ideas and values, the change of roles and new relationships, and it is acknowledged as an innovative structure of governance that aims to meet the needs of marginalized or excluded groups [10]. Every day several SI initiatives emerge to promote alternatives to the construction of more egalitarian scenarios, with fairer income distribution and better living conditions for the beneficiary population [2].

Examples of such initiatives are social incubators [11]. Social incubators are institutions that work collaboratively, through different players, in order to meet the demands generated by society, government, or academia. They are considered an essential element of social development [12]. These diverse players involved are "coordinated to achieve common goals through their respective efforts" [13], and strengthened when working in a network [14].

A network can be understood as the set of players connected in favor of a common goal [15]. Collaboration networks are characterized as critical elements for the SI process, and their use provides an effective aggregate value [1, 16]. In this sense, [17] thinks that understanding these elements is relevant to the SI study since collaboration is a determining factor in the creation of new knowledge and new possibilities for the community [7].

After realizing the relevance and the need for more research on this topic, the authors of this study sought to answer the following research question: what elements constitute the collaboration network for SI, in the context of a social incubator? In order to achieve this goal, an integrative review of the literature was carried out in the Scopus and Web of Science databases and complemented with a case study, through document analysis and semi-structured interviews.

This study is structured in five sections: the first and present section is the introduction of this work. The second section presents the theoretical reference. Next, we describe the methodological procedures used in the data collection. The fourth section presents the search results. The fifth and final section presents the final considerations.

2 Theoretical Framework

The theoretical framework presents the main concepts related to SI, collaboration networks and social incubators, which will serve as a foundation for the case study.

2.1 Social Innovation

By putting together concepts already recognized in the literature as innovation and social, it is understandable that the definitions become a little diverse. The theme has expanded in the last decades [18, 19] and arouses interests of several centers of studies by the world [1, 2]. Its concept still does not find in the academy a consensus, given the diverse areas of knowledge that approach it [3, 6], such as: infrastructure and urbanism [16], agricultural production and solidarity economy [20], health and education sector [5], social movements [21], among others.

The European Commission states that social innovations relate to the development of new projects, services or models that better address social issues. It also points out that they offer innovative responses to these needs, with productive social relationships and new forms of collaboration. SI can provide sustainable solutions to today's major challenges, ranging from problems of social inequality (hunger, poverty, health, and education) to chronic diseases and other global epidemics [1, 9, 19].

To Murray, Caulier-Grice, Mulgan [22], SI is the creation of new ideas (product, service or a model) that seek to satisfy the social needs and develop new relationships or social collaborations, simultaneously. In other words, they are innovations that, at the same time, are good for the community and the society and increase their capacity to act. These actions, when carried out collaboratively, involve various players of civil society, who organize themselves for a common goal. SI players are individuals, organizations, social movements, and the government [23, 24].

From the perspective of collaborative processes, [9] states that social innovation is the creation or the combination of new knowledge through an intentional, systematic, planned, and coordinated process. To these authors, SI derives from the collaboration and the sharing of knowledge among several agents. It occurs when several players decide to work together and collaboratively, resulting in a tangible improvement for all involved.

These players connect in interest groups [3], and when they interact employing partnerships, they form a network of collaboration [16, 22]. The established partnerships contribute to social innovation through the collaboration network, in order to achieve greater efficiency in results.

2.2 Collaborative Networks

Collaboration is a key element for the development of SI and can be expanded through networks [7, 11]. To these authors, the sharing of information and the exchange of experiences among the players form a continuous basis of learning and a collective construction of knowledge. Mance [25] observes that each networked player, with his specific knowledge, allows integration to the others and allows the emergence of complex processes and experiences with previously non-existent quality.

To Borbinha [13], "collaboration networks are multi-stakeholder structures, which are coordinated to achieve common goals, by combining their respective efforts." When these exchanges occur, relationships, learning, and possibilities of partnerships emerge [17].

This exchange is a real possibility of creating alternatives and solutions, which individually would not be possible [16]. To Mance [25], collaborative networks offer great potential to generate social impact far beyond what an individual could achieve independently. However, for this network to be efficient, some essential elements are necessary.

Through a systematic review of the literature, Prim [7] has identified some elements that constitute the collaboration network and are essential for the development of SI:

Network of players: the diversity of players connected with common goals;

Collaboration: process of work done jointly and in group;

Commitment: to believe in the importance of the established relationship or partnership and not to mitigate efforts to comply with the agreement;

Trust: to believe in the work of the other and to perceive common values and goals in the group;

Partnerships: new relationships with common goals;

Leadership: indispensable to the process of collective creation since it is recognized as a north to the players involved;

Empowerment: new power relations and the fact of making the individual empowered to do their activities with autonomy;

Communities of Social Innovation: new social arrangements as a means of fostering social inventions, community governance, and collective power;

Sharing and transfer of knowledge: learning process and creation of new knowledge.

When interactively connected, these elements enable greater synergy between the processes and make the network more collaborative. To Sanzo et al. [17], collaborative participation in the network makes possible the construction of greater cohesion among the collaborators.

2.3 Social Incubator

Historically, the concept of incubator has been applied for economic purposes. It is now also recognized as an opportunity to generate solutions to solve social problems [26].

Social incubators are institutions that work collaboratively to meet the demands generated by society, government, or academia. They are considered an essential element of social development [12] since their work focuses on the development of social projects [7]. They carry out their activities in a multidisciplinary way (with psychology, administration, sociology, law, accounting, among others) to meet the diversity of social demands. In this context, they seek to bring together the various players, in order to have a collective construction or a diversified view of their demands.

The authors [27, p. 262] affirm that "the articulation between popular knowledge and technical-scientific interpretations is essential for social development". To Pérezgrovas and Cervantes [26], the role of a social incubator is to support projects incubated in the various activities performed by them. In this sense, promoting social inclusion, seeking improvement in the quality of life of marginalized communities, and enabling the production and socialization of knowledge are routine actions performed by social incubators.

3 Methodology

Based on [28], a research project can be classified as to its nature, purposes, and means of searching the data. The approach of this study is qualitative in nature since it seeks a better understanding of the facts investigated. As for the objectives, it is descriptive, since it aims to describe how the phenomenon investigated occurs and seeks to "provide an overview of the approximate type" [28, p. 43]. As for the means, this is a case study, carried out at the Technological Incubator of Popular Cooperatives belonging to the Regional University of Blumenau (ITCP/FURB) in Santa Catarina, Brazil.

The data were collected using semi-structured interviews and bibliographic techniques. For that, the literature review and the objective of this research were taken into account. Table 1 lists the keywords as well as the number of documents found.

Table 1. Scientific production

Base	Search strategy	Production selected
Search 01 Scopus	TITLE-ABS-KEY "Social Innovation" AND (Collaboration OR Alliance OR Partnership OR Cooperation OR Network*)	116
Search 01 Web of science	TOPICS "Social Innovation" AND (Collaboration OR Alliance OR Partnership OR Cooperation OR Network*)	80
Search 02 Scopus	TITLE-ABS-KEY "Social Innovation" AND Incubat*	04
Search 02 Web of science	TOPICS "Social Innovation" AND Incubat*	03
Total:		**203/26**

In search criterion 01, one hundred and sixteen documents were found in the Scopus database and eighty-three in the Web of Science, totaling one hundred and ninety-six documents with the themes "*Social Innovation*" and "*Collaboration*", "*Alliance*", "*Partnership*", "*Cooperation" and "Network*".

In search criterion 02, four documents were located in the Scopus database and three in the Web of Science, totaling seven documents with the themes "*Social Innovation*" and "*Incubat**", totaling two hundred and three documents for analysis.

A reading of the summary of the abstracts and keywords of all the documents was carried out, and twenty-six documents were considered relevant to this research. These papers served to base the theoretical part of the research and as the basis for the case study.

Parallel to the literature review, the authors contacted the incubator, and the possibility of carrying out the case study was verified. Once the research was approved, visits and interviews were made in the incubator and in its three selected projects.

An interview script was prepared, consisting of twenty-four questions. The purpose of the interviews was to identify, in the speeches, some of the elements that constitute the SI collaboration networks. According to Yin [29], a semi-structured interview allows, from the pre-elaborated script, an opening for points not thought. For that, the literature review and the objective of this research were taken into account.

A total of four interviews were carried out, with three responsible for the incubated social projects and the Incubator coordinator. The interviewees remain anonymous.

For the analysis of the data, the thematic analysis method of [30] was used. It is a clear and precise method for identifying, analyzing, and reporting patterns (themes) within the primary data in order to organize and describe them in rich detail.

4 Results

4.1 The Case Study

ITCP/FURB was chosen because it is a referenced institution in the development of projects with excluded and vulnerable people. This incubator is a reference in works aimed at society and it was even honored in the State Legislative Assembly of Santa Catarina, in 2014, for its work of social inclusion developed with women.

At the time of this research, ITCP/FURB was working with ten incubated projects in several sectors of the local economy. After analyzing the documents available and interviewing the incubator coordinator, three projects were selected for the case study.

ITCP/FURB is a "university extension program created to implement alternatives to work and income, since 1999, in the perspective of the solidarity economy in Blumenau and its region" [31]. It consists of a coordination body and a team of teachers, students, and administrative technicians. In order to meet their demands, it carries out its activities guided by the principle of collectivity collaboration.

The analysis units, in addition to the ITCP, were:

Project COOPERRECIBLU: this is a cooperative of solid waste pickers, comprised of people excluded from society and the formal labor market. It currently has sixty members;

Project ENLOUCRESCER: this is the Association of Family, Friends and Users of the Mental Health Service of the Municipality of Blumenau. It is a nonprofit civil society organization. It was created in 1996 to ensure the rights of people with mental disorders, who have family members, friends, and users of Psychosocial Care Centers (CAPs). It is composed of twenty-five members;

Project VERBO TECER: this is the Blumenau Association of Weaving. It was created in 2004 and has nineteen groups of associated artisans nowadays.

4.2 Elements of the Collaboration Network

After applying the analysis method in the primary data, it was found that all the elements that constitute the collaborative network described in the literature were present in the researched projects, being: a network of players, commitment, partnerships, leadership, empowerment, social innovation community, sharing and knowledge transfer.

However, new elements emerged from the interview. Due to their influence on the social projects studied, they were considered relevant. From the analysis of the data, it was observed that even with different languages, in many contexts, they had the same meaning. Table 2 presents these elements with their correlated themes and characteristics.

Table 2. Elements of the network and their correlated themes found in the interviews

Elements	Correlated themes	Characteristics
Partners	1) Networks of players 2) Types of partnerships	1) Individuals, government, organizations, social movements, university, incubator, community 2) Intersectoral and inter-organizational; community-involving actions
Resources	1) Financial 2) Material 3) Humane	1) Monetary 2) Equipment, infrastructure, movable property, raw materials 3) Psychological support, technical advice, training, volunteering
Collaboration	1) Self-help 2) Trust and commitment 3) Interpersonal relationship	1) Mutual help 2) Integrity, honesty, bonding 3) Involvement with the cause
Self-management	1) Types of leadership 2) Participatory processes	1) Shared leadership 2) Group decisions
Learning	1) Training and lectures 2) Sharing of experiences	1) Training, courses, workshops 2) Meetings, study visits, articulation meetings with partners
Sustainability	1) Economic 2) Social 3) Environmental	1) Means of obtaining resources 2) Valorization of the individual 3) Respect for the environment
Empowerment	1) Recognition 2) Construction of the identity	1) Perception of the importance 2) Personal achievement

Partners: those who develop some kind of relationship with the projects and allow the creation of a network.

Resources: the partnerships with players from different sectors contribute to the collaboration network in a wide variety of ways. In addition to access to financial resources, other types of resources can be observed, such as informal bonds (voluntary), advisory services, therapies, workshops, knowledge and information, relationship network, infrastructure, technologies, medicines.

Collaboration: recognized as synonymous with joint work. The phrase "performed by several hands" expresses this collaborative feeling.

Self-management: is the utmost form of leadership. This form of management is present in all projects and indicates the fact that project management is carried out by the cooperative/associates themselves. Leadership is exercised collectively and shared, backed by what the company's bylaws determine.

Learning: is something that makes possible the improvement of daily activities, through new knowledge. It is an alternative to individual, group and community growth, as well as social reintegration. Learning can be observed in the execution of handwork, cultural activities, exchange of experience, and knowledge sharing.

Sustainability: the interviewees understand sustainability as the way they seek to survive, focusing on the three pillars: economic, environmental, and social.

Empowerment: Empowerment was an element that emerged from the interviews. It is observed that the construction of identity and individual recognition are relevant aspects of the empowerment of individuals. Personal fulfillment is linked to the fact that people like what they do. Individual recognition is based on the principle of perception that his/her actions are recognized and valued. The projects experience travel possibilities to present the works, and also receive the financial resources resulting from the sales of their crafts. The recognition also occurs because of the work done by ITCP/FURB.

All these interlinked elements enable creating a collaborative network that fosters actions of social innovations, to provide a change in the communities involved.

5 Final Considerations

When working on the perspective of a social incubator, it was concluded that collaboration networks are essential in the creation and development of SI. These networks provide new social, economic, financial, cultural, and political arrangements. Besides, they enable opportunities for the community to grow, through of participatory work and the sharing of knowledge. It is concluded that the network is composed of the following elements: partners, resources, collaboration, self-management, learning, sustainability, and empowerment. It is emphasized that these elements, when worked systemically, provide new social arrangements, and foster initiatives of a social nature.

It should be noted that ITCP/FURB management represents an innovative way of working because it focuses on collaborative and inclusive processes, where knowledge is built on a two-way path between the academic and the popular environments.

Finally, it is pointed out that the collaboration network formed through incubated social projects, the incubator, and its partners, create strong links and lasting bonds, and create various opportunities for improvement for the beneficiary community.

Acknowledgments. This work was carried out with the support of the Brazilian National Coordination for the Improvement of Higher Education Personnel (CAPES).

Ethics declaration
Conflict of Interest
The authors declare that they have no conflict of interest.
Compliance with standards involving humans as subjects
All procedures performed in studies involving human participants were in accordance with the ethical standards of the institutional and/or national research committee and with the 1964 Helsinki Declaration and its later amendments or comparable ethical standards.
Informed consent was obtained from all individual participants involved in the study.

References

1. Phillips, W., Lee, H., Ghobadian, A., James, P.: Social innovation and social entrepreneurship: a systematic review. Group Organ. Manag. **40**, 428–461 (2015). https://doi.org/10.1177/1059601114560063
2. Păunescu, C.: Current trends in social innovation research: social capital, corporate social responsibility, impact measurement. Manag. Mark. **9**(2), 105–118 (2014)
3. Bignetti, L.P.: As inovações sociais: uma incursão por ideias, tendências e focos de pesquisa. Ciências Sociais Unisinos **47**(1), 3–14 (2011)
4. Hean, S., et al.: Using social innovation as a theoretical framework to guide future thinking on facilitating collaboration between mental health and criminal justice services. Int. J. Forensic Mental Health **14**(4), 280–289 (2015). https://doi.org/10.1080/14999013.2015.1115445
5. Salim-Saji, B., Ellingstad, P.: Social innovation model for business performance and innovation. Int. J. Prod. Perform. Manag. **65** (2016). https://doi.org/10.1108/ijppm-10-2015-0147
6. Morgan, K., Richardson, R., Marques, P.: Social innovation in question: the theoretical and practical implications of a contested concept. Environ. Plann. C Politics Space **36**(3), 496–512 (2018). https://doi.org/10.1177/2399654417717986
7. Prim, M.A.: Elementos constitutivos das redes de colaboração para inovação social no contexto de incubadoras sociais. Dissertação de mestrado do Departamento de Engenharia e Gestão do Conhecimento (2017)
8. Haxeltine, A., et al.: Transformative social innovation: a sustainability transitions perspective on social innovation. In: Social Frontiers: The next Edge of Social Innovation Research (2013)
9. Borges, M.A., et al.: Inovação social: uma gênese a partir da visão sistêmica e teoria da ação comunicativa de Habermas. In: Fourth International Conference on Integration of Design, Engineering and Management for innovation - IDEMI 2015, Florianópolis (2015)
10. MacCallum, D., Moulaert, F., Hillier, J., Haddock, S.: Social Innovation and Territorial Development. Ashgate Publishing Limited, Aldershot (2009)
11. Nicolopoulou, K., et al.: An incubation perspective on social innovation: the London Hub – a social incubator. R&D Manag. **47**(3), 368–384 (2015)
12. Guimarães, G.: Incubadoras tecnológicas de cooperativas populares: contribuição para um modelo alternativo de geração de trabalho e renda. In: Singer, P., de Souza, A.R. (Org.) A economia solidária no Brasil: a autogestão como resposta ao desemprego, pp. 111–122. Contexto, São Paulo (2000)
13. Borbinha, J.: Redes de colaboração: alguns elementos para análise e reflexão. Caderno de Biblioteconomia, Arquivística e Documentação. Lisboa (1), 73–88 (2004)
14. Klein, J.L., Fontan, J.M., Harrisson, D., Lévesque, B.: The Quebec system of social innovation: a focused analysis on the local development field. Finisterra **94**, 9–28 (2012)
15. Castells, M.C.: A Sociedade em Rede: Do Conhecimento à Acção Política (2005)

16. Swilling, M.: Africa's game changers and the catalysts of social and system innovation. Ecol. Soc. **21**, 1–37 (2016). https://doi.org/10.5751/ES-08226-210137
17. Sanzo, M.J., Álvarez, L.I., Rey, M., García, N.: Business–nonprofit partnerships: a new form of collaboration in a corporate responsibility and social innovation context. Serv. Bus. **9**(4), 611–636 (2015). https://doi.org/10.1007/s11628-014-0242-1
18. Cunha, J., Benneworth, P.: Universities' contributions to social innovation: towards a theoretical framework. University of Twente, School of Management and Governance - IGS, pp. 1–31 (2013). https://doi.org/10.1108/ejim-10-2013-0099
19. Cajaiba-Santana, G.: Social innovation: moving the field forward. A conceptual framework. Technol. Forecast. Soc. Change **82**, 42–51 (2014). https://doi.org/10.1016/j.techfore.2013.05.008
20. Kolk, A., Lenfant, F.: Cross-sector collaboration, institutional gaps, and fragility: the role of social innovation partnerships in a conflict-affected region. J. Public Policy Mark. **34**(2), 287–303 (2015). https://doi.org/10.1509/jppm.14.157
21. Vos, J., Wagenaar, H.: The munchhausen movement: improving the coordination of social services through the creation of a social movement. Am. Rev. Public Adm. **44**(4), 409–439 (2014). https://doi.org/10.1177/0275074012468224
22. Murray, R., Caulier-Grice, J., Mulgan, G.: The open book of social innovation. The Young Foundation, London (2010)
23. André, I., Abreu, A.: Dimensões e espaços da inovação social. Finisterra **XLI**(81), 121–141 (2006). https://doi.org/10.18055/Finis1465
24. Mulgan, G.: The process of social innovation. Innov. Technol. Gov. Glob. **1**(2), 145–162 (2006). https://doi.org/10.1162/itgg.2006.1.2.145
25. Mance, E.A.: Redes de Colaboração solidária: aspectos econômico-filosóficos: complexidade e libertação. Petrópolis. Vozes, RJ (2002)
26. Pérezgrovas, V., Cervantes, E.: Evaluación de los benefícios actuales y el potencial para el combate a la pobreza de la participación en redes de comercio justo de café en la Unión Majomut. Colorado State University. Fair Trade Research Group - FTRG. San Cristóbal de las Casas, Chiapas (2002)
27. Souza, A.L., Barbosa, M.J.S., Reis, A.: Incubadoras universitárias: inovação social e desenvolvimento. In: Bocayuva, P.C.C., Varanda, A.P.M. (Org.). Tecnologia social, economia solidária e políticas públicas, 1st edn. FASE, IPPUR-UFRJ, Rio de Janeiro (2009)
28. Gil, A.C.: Como elaborar projetos de pesquisa, 5th edn. Atlas, São Paulo (1999)
29. Yin, R.K.: Estudo de Caso: planejamento e métodos, 3rd edn. Bookman, Porto Alegre (2005)
30. Braun, V., Clarke, V.: Using thematic analysis in psychology. Qual. Res. Psychol. **3**, 77–101 (2006). https://doi.org/10.1191/1478088706qp063oa
31. Marchi, R.C., Prim, L.F., Andrade, E.T.: Economia Solidária na ITCP/FURB: Reflexões e experiências em busca da inclusão social. Blumenau. SC, Meta (2013)

Comparison of Alternative Wastewater Treatment Plants Using Life Cycle Assessment (LCA)

Anna Carolina Dall'Aneze Ferreira(✉), Sueli Oliveira, and Roseli Frederigi Benassi

Universidade Federal do ABC, Santo André, SP 09210-580, Brazil
anna.ferreira@aluno.ufabc.edu.br

Abstract. This work aims to assess potential negative impacts of two non-conventional wastewater treatment plants (WwTP) through Life Cycle Assessment (LCA) methodology. Since such systems are designed to minimize environmental impacts of sanitary or industrial sewage disposal, they must be chosen and engineered taking into account overall impacts of its own construction, operation and maintenance. In this context, a powerful decision making tool to be used is LCA, since it carries a detailed study of potential emissions generated by a process or product. Collected data regarding two pilot WwTP led to construction of life cycle inventory (LCI) for a constructed wetland system and a system consisting of septic tank, submerged aerated filter and secondary decanter. Design of both systems aimed at secondary treatment of sewage. Chosen tools were *ecoinvent* 3.4 database, open source software openLCA 1.7 and ReCiPe 2019 Midpoint (H) method for data interpretation. Considering all 18 impact categories available, the wetland system generated higher potential emissions to the environment and in more categories when compared to the septic tank-filter-decanter system. Main categories of impact were terrestrial ecotoxicity, climate change, scarcity of fossil resources, human (non-carcinogenic) toxicity and water consumption for both systems, although not of the same relevance. However, the septic tank-filter-decanter system had significantly greater impact on terrestrial ecotoxicity when compared to all other categories of impact, being 3 times higher than the second most significant category (climate change) for this system. When compared to the wetlands, impact was 3.5 time higher. Results showed that materials used in the construction of both systems caused greater environmental impacts, especially glass fiber used for septic tank, filter and decanter, responsible for 79% of the impact in terrestrial ecotoxicity. Another significantly contributing flow was electricity consumption, responsible for impacts on water consumption, which was more significant in the septic tank-filter-decanter system due to pumping and aeration equipment.

Keywords: Environmental impact · Constructed wetlands · Engineered ecosystem

1 Introduction

In principle, wastewater treatment plants (WwTP) are designed to minimize the environmental impacts of sanitary or industrial sewage disposal and should, therefore, be defined

L. Pereira et al. (Eds.): IDEAS 2018, SIST 198, pp. 437–446, 2021.
https://doi.org/10.1007/978-3-030-55374-6_43

and engineered to take into account not only economic and infrastructural resources, but also impacts of its own construction, operation, maintenance and deactivation. Different technologies have distinct performances and can – potentially - cause various environmental impacts. That said, use of tools and methods that support decision-making in a given scenario is essential.

For most conventional activated sludge WwTPs, impacts arise mainly from high energy consumption and sludge generation. On the other hand, technologies that use natural soil depuration capacity, such as waste stabilization ponds and filtering ditches, require the use of large land areas [1, 2].

Life Cycle Assessment (LCA) can be used for decision-making and evaluation of the overall environmental performance of a WwTP. This tool - standardized by ISO 14040: 2009 [3] - allows evaluation of potential environmental impacts of a process or product, taking into account all stages of its life cycle: from the raw material to final disposal, that is, from cradle to grave.

Through LCA, it is possible to evaluate potential environmental impacts caused by different sewage treatment technologies. A series of indicators, calculated through material and energy flows, relate to impact categories. Those measure quantitatively the effect of emissions. Examples are global warming, eutrophication and scarcity of natural resources [4].

Several WwTP LCA studies - whether dealing with traditional activated sludge or non-conventional technologies – determined environmental performance of such systems. Comparative studies usually choose activated sludge technology as reference, as studied by Yıldırım and Topkaya [5], Kalbar, Karmakar and Asolekar [6], Lopsik [7] and Garfí, Flores and Ferrer [8].

These studies show that alternatives with higher energy consumption - as of activated sludge process - have poorer environmental performance considering categories such as acidification and toxicity due to atmospheric emissions for power generation. In addition, disposal of the sludge is a great impact generator. In the case of non-conventional technologies, most significant impacts refer to land use, due to large area requirements to carry out treatment.

Finally, some authors, such as Dixon, Simon and Burkitt [9], who studied the use of artificial reeds and aerobic biological filters, compared non-conventional technologies. The study considered energy use, CO_2 and solid emissions, and land use. In general, both systems presented similar results regarding energy consumption. It was also possible to conclude that highest CO_2 emissions were due to transportation of construction materials and periodic site visits for maintenance.

In this context, contribution of this work is to compare the environmental performance of two non-conventional WwTPs.

2 Materials and Methods

Two different treatment systems were chosen to build the Life Cycle Inventory (LCI) and perform life cycle assessment of WwTPs. It was decisive that each system had sufficient data regarding process variables, were similarly constructed, and sewage affluent to the systems had comparable characteristics. A dotted line in Figs. 1 and 2 highlight process steps considered in Life Cycle Assessment.

2.1 Horizontal Subsurface Flow Constructed Wetland

The first plant was a wetland system studied by Sanchez [10], who built, operated and monitored a horizontal subsurface flow constructed wetland in the facilities of the WwTP of Arujá, State of São Paulo.

System received sewage previously treated by screens and grit chamber. Then, four high-density polyethylene (HDPE) tanks received wastewater, being one non-vegetated control tank and three tanks vegetated with *Typha* sp. and *Eleocharis* sp., as seen on Fig. 1. Each tank had an average flow of 4.4 L/h. Although the study compared four differently vegetated wetlands, results did not show significant statistical differences in treatment efficiencies. Thus, this work considered simultaneous operation of all tanks, making up a total average flow of 17,6 L/h.

Monitored parameters were: dissolved oxygen, conductivity, pH, total and dissolved solids, alkalinity, $BOD_{5,20}$, COD, total suspended solids and their fixed and volatile fractions, apparent color and turbidity. One can find details on design, data and statistical treatment in Sanchez's original work [10].

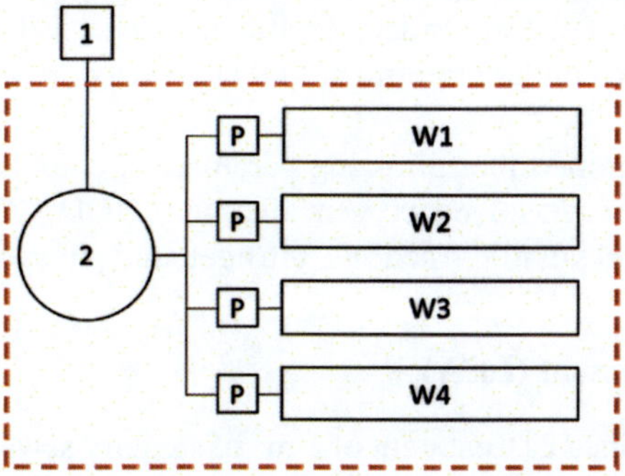

Fig. 1. Horizontal subsurface flow wetland treatment plant, as designed by Sanchez [10], being (1) pre-treatment facilities, (2) feed tank, (P) peristaltic pumps and (W1, W2, W3, W4) the four constructed wetland tanks. Elaborated by authors.

2.2 Engineered Ecosystem for Wastewater Treatment

Salomão [11] built, operated and monitored the second non-conventional WwTP. He evaluated the performance of a complex system, consisting of a flow controller, grease trap, septic tank, submerged aerated filter, biodecanter and three tanks vegetated by different species (Fig. 2). System was located in Ilha Grande, Angra dos Reis municipality. Average flow was 52 L/h.

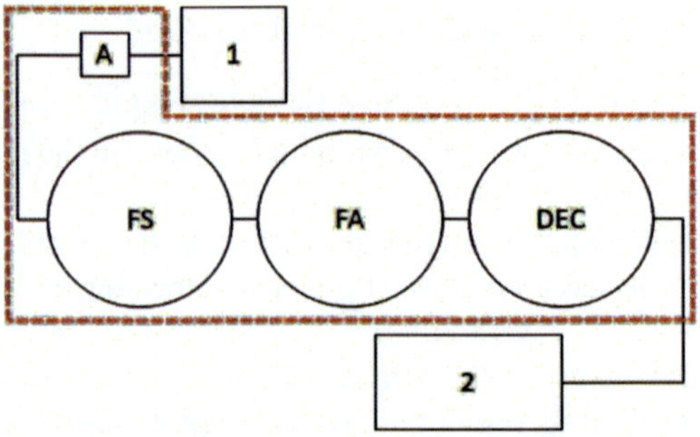

Fig. 2. Engineered Ecosystem as designed by Salomão [11], being (1) pre-treatment operations, (A) pump, (FS) septic tank, (FA) aerated filter, (DEC) decanter, (2) tertiary treatment. Elaborated by authors

The complete system performed preliminary (flow control box and grease trap), primary (septic tank), secondary (submerged aerated filter and secondary decanter) and tertiary (three wetlands and one algae tank) treatments. This work considered only data obtained until the end of the secondary treatment, since wetland system studied by Sanchez [10] aimed at secondary treatment. All tanks used in the system were made of glass fiber.

Salomão [11] determined the following parameters: color, turbidity, pH, electrical conductivity, temperature, dissolved oxygen, alkalinity, BOD, COD, suspended and total solids and their fixed and volatile fractions, nitrogen and phosphorus.

2.3 Life Cycle Assessment (LCA)

Functional unit was defined as treatment of 1 m^3 of sanitary sewage with initial $BOD_{5,20}$ in the range of 200 to 400 mg L^{-1}. Wastewater must be treated to the limit established by CONAMA Resolution 430/2011 of maximum $BOD_{5,20}$ equal to 120 mg L^{-1} [12]. Time horizon was 1 year. Although systems could operate for a much longer time, authors monitored systems for approximately 7 months. Due to lack of information, this work considered a shorter time horizon and disconsidered system maintenance.

Information obtained in original works, bibliographic research and specifications of similar products based the estimation of values for most construction materials and electricity consumed (Tables 1 and 2). When information on size and volume of materials used were not available, estimates considered material properties and most conservative scenarios, i.e. use of greater volume of material. Chosen process parameters were BOD, COD, total nitrogen and total phosphorus, being all data taken from the works previously mentioned (Table 3). Calculation of reference flows for matter and energy during construction and operation considered average system flows.

Table 1. Total mass, land area and electricity used in construction phase of wetland WwTP and respective reference flows calculated for one year of operation. Elaborated by authors.

Wetlands – sewage treated in 1 year = 154,18 m^3		
Material	Total	Reference flow (unit/m^3)
HDPE tanks	55,3 kg	0,359 kg/m^3
Feed tank	12,5 kg	0,081 kg/m^3
#1 gravel	7140 kg	46,1 kg/m^3
Area	10 m^2	0,06 m^2/m^3
Peristaltic pump	1584 kWh	10,27 kWh/m^3

Table 2. Total mass, land area and electricity used in construction of engineered ecosystem WwTP and respective reference flows calculated for one-year operation. Elaborated by authors.

Engineered ecosystem – sewage treated in 1 year = 455,52 m^3		
Material	Total	Reference flow (unit/m^3)
Glass fibre tanks	90 kg	0,197 kg/m^3
Plastic biorings	72,8 kg	0,160 kg/m^3
Area	25,9 m^2	0,06 m^2/m^3
½cv pump and air compressor	17664 kWh	38,78 kWh/m^3

Tools used to support lifecycle assessment were openLCA software, *ecoinvent* database (version 3.4) and ReCiPe 2016 Midpoint (H) method.

At first, all eighteen impact categories were considered, in order to determine most relevant ones for each treatment plant. Five major impact categories were evaluated in more detail, as well as most significant material and energy flows for each impact category. System's operation phase considered pruning of macrophytes used in wetland tanks and their final destination. Estimated aerial biomass production of *Typha* sp. was 7059 kg ha^{-1} of dry mass [13], making up a total of 1,7 kg per month. Another assumption was disposal of all biomass.

Table 3. Reference flows for operational parameters considered in the scope of this work. Elaborated by authors.

	Affluent (kg/m^3)					Effluent (kg/m^3)				
	$DBO_{5,20}$	DQO	N_{NH4}	Nitrato	P_{Total}	$DBO_{5,20}$	DQO	N_{NH4}	Nitrato	P_{Total}
Wetlands	0,206	0,446	0,068	0,001	0,008	0,017	0,051	0,028	0,000	0,004
Eng. Ecos.	0,226	0,512	0,021	0,002	0,023	0,030	0,044	0,02	0,003	0,021

3 Results and Discussion

Table 4 shows top ten results for wetland and engineered ecosystem WwTPs, and freshwater and marine eutrophication categories. For most impact categories, engineered ecosystem system proposed by Salomão [11] generated less potentially negative environmental impacts. However, it should be noted that terrestrial ecotoxicity category for this system was the most significant of all evaluated. Result was approximately three times higher than the calculated impact for climate change. Following is fossil resources scarcity, human toxicity and water consumption.

Table 4. Overall results for top ten impact categories available in the ReCiPe 2016 Midpoint (H) method. Results marked in red are higher than the ones in green. Elaborated by authors.

Impact category	Unit	Result	
		Eng. Ecosystem	*wetlands*
Terrestrial ecotoxicity	kg 1,4-DCB	4,09	1,18
Global warming	kg CO_2 eq	1,45	1,74
Fossil resource scarcity	kg oil eq	$3{,}95.10^{-1}$	$8{,}17.10^{-1}$
Human non-carcinogenic toxici	kg 1,4-DCB	$3{,}95.10^{-1}$	$4{,}43.10^{-1}$
Water consumption	m^3	$2{,}14.10^{-1}$	$1{,}33.10^{-1}$
Human carcinogenic toxicity	kg 1,4-DCB	$3{,}16.10^{-2}$	$5{,}80.10^{-2}$
Ionizing radiation	kBq Co-60 eq	$2{,}60.10^{-2}$	$4{,}31.10^{-2}$
Marine ecotoxicity	kg 1,4-DCB	$1{,}71.10^{-2}$	$2{,}43.10^{-2}$
Freshwater ecotoxicity	kg 1,4-DCB	$1{,}19.10^{-2}$	$1{,}63.10^{-2}$
Land use	m^2a crop eq	$5{,}97.10^{-2}$	$2{,}16.10^{-2}$
Freshwater eutrophication	kg P eq	$1{,}18.10^{-4}$	$-3{,}98.10^{-2}$
Marine eutrophication	kg N eq	$1{,}84.10^{-5}$	$1{,}62.10^{-5}$

As for the constructed wetland system by Sanchez [10], most significant impact was climate change, followed by terrestrial ecotoxicity, fossil resources scarcity and human toxicity. Figure 3 shows the top five potential impacts generated by both processes.

Wetland system showed negative result for freshwater eutrophication, indicating removal of phosphorus by the system and small contribution of other processes to this flow. On the other hand, treatment proposed by Salomão [11] presents a positive result, explained by inadequate system design. As noted by the author, this issue resulted in low phosphorous removal.

At first, wetland system appears as a treatment option with poorer environmental performance, but it does not present major discrepancies between impact categories. This can present environmental advantages in its implementation when compared to the engineered ecosystem, which has a major negative impact on terrestrial ecotoxicity. Further studies should consider a more detailed study of flows and processes responsible for such emissions.

Although eutrophication and ecotoxicity of fresh and marine waters impact categories yielded low results, this work presents detailed analysis of both. It is important to determine if there was any significant contribution of secondary processes to these impacts.

Figures 4 and 5 show in detail process contributions for each impact category. For wetlands system, production of HDPE (used in manufacture of tanks), had a significant contribution to impacts on climate change and fossil resources scarcity. It corresponds to 52% and 84% of these impacts, respectively.

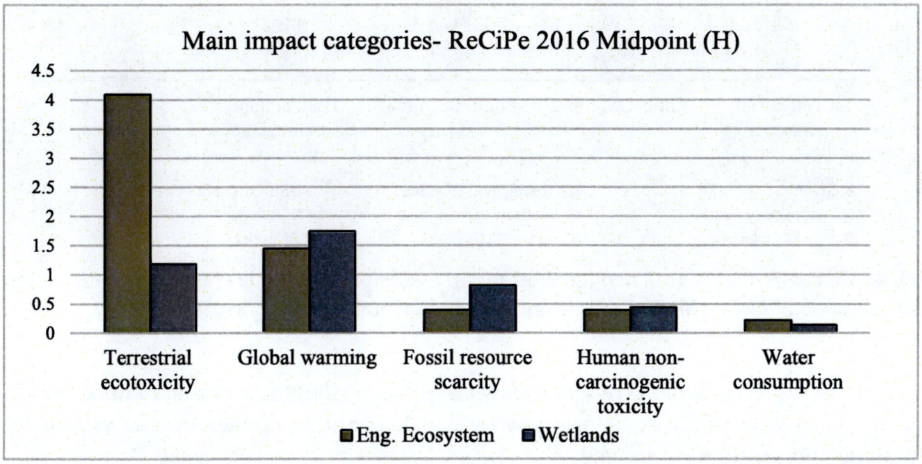

Fig. 3. Comparison between the five main potential impacts generated by the two processes. Elaborated by authors

Impacts generated by use of gravel as a support medium come mainly from production and treatment of the mining residues. These processes contribute significantly to human toxicity generated by non-carcinogenic substances.

As for the Engineered Ecosystem, main impact category relates to fiberglass production. It's production contributes to 79% of terrestrial ecotoxicity impacts. This result was 3.5 times higher than the second most relevant impact category. For the climate change category, main contributors were hydroelectric power generation and production of polypropylene used as support medium for submerged aerated filter. This result is justified due to the use of equipment for pumping and system aeration.

A significant contribution of construction materials to environmental impacts is noted. Although this is valuable information, it is important to note that this work considered only 1 year of system operation. A more in-depth study, considering the entire life of systems, would probably result in less significant impacts when compared to operation and maintenance.

Flows with greatest contribution to each impact category were also evaluated. Tables 5 and 6 show most significant flows. Main impacts on terrestrial ecotoxicity were associated to copper and antimony, both toxic and potentially carcinogenic. In addition, the wetlands are responsible for higher emissions of fossil CO_2, crude oil and zinc ions.

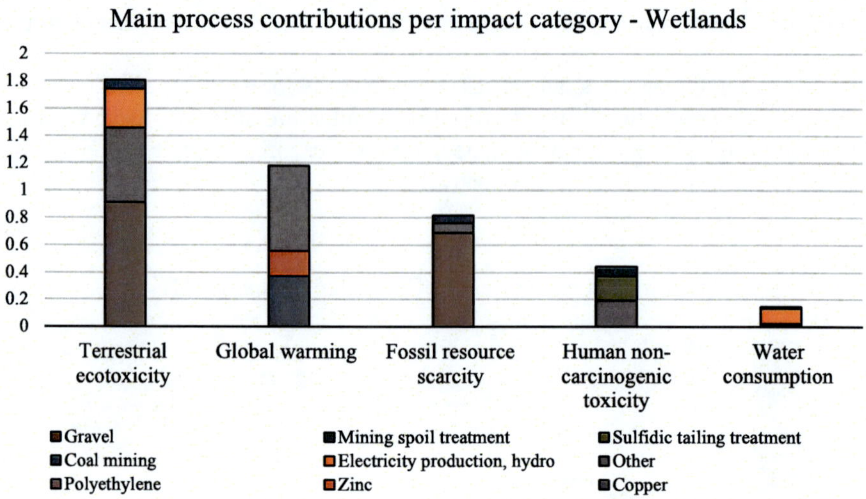

Fig. 4. Processes that contributed the most to top five impact categories related to construction and operation of the engineered ecosystem system. Elaborated by authors.

The only flow in which wetlands showed less contribution was water consumption by the generation of hydroelectric power, which confirms greater impact caused by pumping and aeration equipment.

Finally, flows and processes with highest contributions to impact categories related to ecotoxicity and eutrophication of fresh and marine water were evaluated. No secondary process had significant contribution to these impact categories. This analysis demonstrates that no process associated with the construction or operation of systems generate significant impact on ecotoxicity and eutrophication. Given that WwTPs aim to reduce such impacts, it would be contradictory that processes involved caused negative environmental impacts in these categories.

Table 5. Flows that contributed the most to top five impact categories related to construction and operation of the wetlands system. Elaborated by authors.

Impact category	Unit	Wetlands	
		Flow	Result
Terrestrial ecotoxicity	kg 1,4-DCB	Copper	0,60
Global warming	kg CO_2 eq	CO_2, fossil	1,17
Fossil resource scarcity	kg oil eq	Oil, crude	0,43
Human non-carcinogenic toxicity	kg 1,4-DCB	Zinc, ion	0,39
Water consumption	m^3	Turbine use	30,44

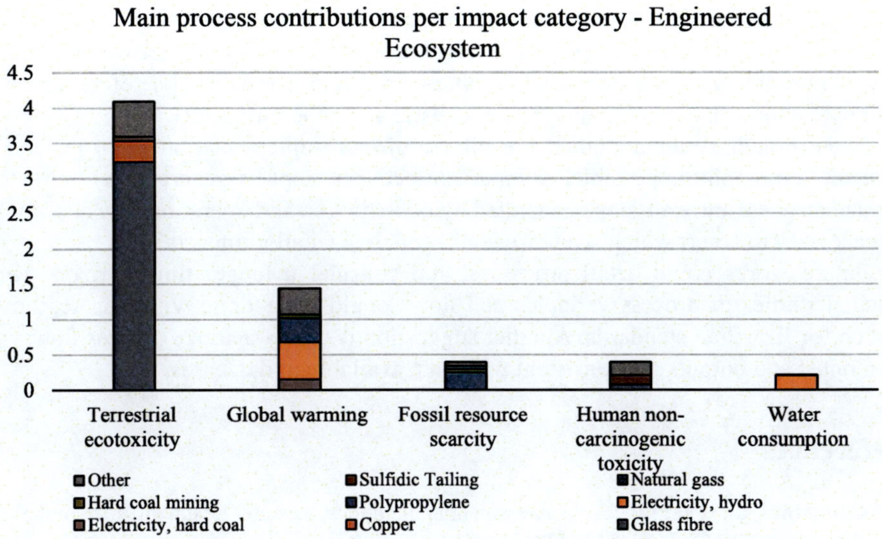

Fig. 5. Processes that contributed the most to top five impact categories related to construction and operation of the wetland system. Elaborated by authors

Table 6. Flows that contributed the most to top five impact categories related to construction and operation of the engineered ecosystem system. Elaborated by authors.

Impact category	Unit	Eng. Ecosystem	
		Flow	Result
Terrestrial ecotoxicity	kg 1,4-DCB	Antimony	3,04
Global warming	kg CO_2 eq	CO_2, fossil	0,76
Fossil resource scarcity	kg oil eq	Oil, crude	0,19
Human non-carcinogenic toxicity	kg 1,4-DCB	Zinc, ion	0,28
Water consumption	m^3	Turbine use	57,43

4 Conclusions

Both processes presented higher impacts in the same five categories, showing that systems do not differ much when it comes to environmental impact. It is important noticing the much higher impact in ecotoxicity caused by emission of Antimony. Any decision-making process between technologies must definitely consider this flow.

LCA was an adequate method to determine potential environmental impacts of both WwTPs, and chosen tools were suitable. A reservation is made: many database does not present output flows in terms of BOD, COD, N or P, leading to possibly underestimated calculations. Two possible approaches would be interesting: characterize processes in terms of BOD, COD and other parameters of environmental interest, or perform a more detailed effluent characterization.

Another drawback was database regionalization. Many processes are Europe standards and generalized to the rest of the world. Many processes consider energy matrix or waste treatment technologies that do not match Brazilian reality. Therefore, process regionalization is key to obtaining more realistic and coherent results.

Even though systems studied are pilot projects, implementation for a residence or small community is possible. Comparison between impacts generated by untreated sewage disposal and emissions generated by construction and operation WwTPs would be necessary to assert which scenario would generate smaller impacts.

Future works could detail processes, and consider a longer time horizon. Very detailed studies of process variables and flows could collaborate with data regionalization for Brazilian standards. Another suggestion is to characterize process flows, so that inputs and outputs are consistent with data available in databases.

References

1. Corominas, L., et al.: Life cycle assessment applied to wastewater treatment: state of the art. Water Res. **47**(15), 5480–5492 (2013)
2. de Sampaio Lopes, T.A., et al.: Revisão crítica da literatura sobre aplicação da Avaliação de Ciclo de Vida ao tratamento de esgotos. Revista DAE **65**(208), 47–55 (2017)
3. ABNT. Associação Brasileira de Normas Técnicas. ABNT NBR ISO 14040: Gestão Ambiental – Avaliação de Ciclo de Vida – Principios e Estrutura. Rio de Janeiro (2009)
4. Hauschild, M., Rosenbaum, R.K., Olsen, S.: Life Cycle Assessment - Theory and Practice, 1st edn. Springer, Heidelberg (2018)
5. Yildrim, M., Topkaya, B.: Assessing environmental impacts of wastewater treatment alternatives for small-scale communities. CLEAN Soil Air Water **40**(2), 171–178 (2012)
6. Kalbar, P.P., Karmakar, S., Asolekar, S.R.: Assessment of wastewater treatment technologies: life cycle approach. Water Environ. J. **27**(2), 261–268 (2013)
7. Lopsik, K.: Life cycle assessment of small-scale constructed wetland and extended aeration activated sludge wastewater treatment system. Int. J. Environ. Sci. Technol. **10**(6), 1295–1308 (2013)
8. Garfí, M., Flores, L., Ferrer, I.: Life Cycle Assessment of wastewater treatment systems for small communities: activated sludge, constructed wetlands and high rate algal ponds. J. Clean. Prod. **161**, 211–219 (2017)
9. Dixon, A., Simon, M., Burkitt, T.: Assessing the environmental impact of two options for small-scale wastewater treatment: comparing a reedbed and an aerated biological filter using a life cycle approach. Ecol. Eng. **20**(4), 297–308 (2003)
10. Sanchez, A.A.: Desempenho de sistema piloto de alagados construídos de fluxo subsuperficial horizontal no tratamento secundário de efluente sanitário. Universidade Federal do ABC, Santo André (2017)
11. Salomão, A.L.S.: Ecossistema engenheirado no tratamento descentralizado de águas residuárias de pequenos geradores: a engenharia ecológica na Ilha Grande, RJ. Rio de Janeiro (2009)
12. CONAMA - Conselho Nacional do Meio Ambiente. Resolução nº 430, de 13 de maio de 2011 Dispõe sobre as condições e padrões de lançamento de efluentes, complementa e altera a Resolução no 357, de 17 de março de 2005
13. Brasil, M.S., et al.: Plantio e desempenho fenológico da taboa (Typha sp.) utilizada no tratamento de esgoto doméstico em sistema alagado construído. Revista de Engenharia Sanitária e Ambiental **12**(3), 266–272 (2007)

The Diagnosis of Solid Residues Produced in Free Trade Shows: A Case Study

Glaucia Oliveira, Deivison Molinari, Marcelo Albuquerque de Oliveira, Gabriela Veroneze(✉), and Dércio Luiz Reis

Federal University of Amazonas, Manaus, AM 69050, Brazil
gabidmv@gmail.com

Abstract. The high production of solid waste in urban fairs despite being a fact known and visible to anyone, is little studied in gravimetric terms. In this sense, this article aims to identify the gravimetric characterization of the Modern Banana Fair, located in the central area of Manaus, in the Amazon region, to characterize the types of waste generated and the environmental impacts. In methodological terms, the following steps were carried out: Bibliographical research, Data collection, Composition analysis based on the Brazilian Regulatory Standards: NBR-10007 - waste sampling: procedure, and NBR-13463 referring to solid waste collection: classification. The results allow to affirm that the predominant gravimetric fraction was that of organic matter (86%), coherent for a large fair, as studied, followed by paper/cardboard (6%) and plastic (4%). These data are in line with the activities of generating such waste, called: retail and wholesale marketing of products (fruits, legumes and vegetables). The major concern becomes the destination of these wastes, since they currently go to common public collection, and are destined to the sanitary landfill of the city. In addition, according to the results of this research, the residues generated have the potential to generate energy (bio gas), which can be obtained through stimulated decomposition of the material. The possibility of generating energy from the waste is still neglected and not exploited by the region.

Keywords: Solid waste · Gravimetric composition · Solid waste management

1 Introduction

Currently, among many of the problems faced by Brazilian urban centers, waste management is considered one of the most critical. Since its effects are assorted in everyday situations, such as: floods, overload of public basic sanitation, diseases to the most socially vulnerable population, and even degradation of the urban landscape.

The waste management process consists in the effective adoption of a robust system, composed of a set of actions, covering all stages of the waste disposal cycle, such as: collection, transportation, transshipment, treatment (physical or chemical), the destination and the correct final disposal thereof. Since each Brazilian discards 1 kilo of waste per day, having a total annual generation of Urban Solid Waste of approximately 78.6 million tons [1]. Of this total, approximately 31.9% is composed of recyclable materials, such as: paper/cardboard, plastic, metals (copper, aluminum), and glass. It also has

L. Pereira et al. (Eds.): IDEAS 2018, SIST 198, pp. 447–456, 2021.
https://doi.org/10.1007/978-3-030-55374-6_44

16.7% of other materials, and the largest and most worrying part is attributed to organic matter, about 51.4% of the total [2, 3].

In the current scenario, only 3% of nationally produced wastes are recycled (CEMPRE, 2013). Thus, it is estimated that 74 tons of waste generated have inadequate final disposal (landfills, rivers, landfills, etc.) [1]. The result of population unconsciousness, lack of structure, lack of basic sanitation, and, finally, lack of effective management, cause the country an annual loss of approximately R $ 8 billion (IPEA, 2012).

According to Brazilian standard NBR 10004 - Solid Waste – classification [4], Solid Waste can be found in the solid and semi-solid states, and are everything that results from human activities, such as: industrial, domestic, health services, agricultural, sweeping, and yet, commercial. Popularly, the incorrect differentiation of solid waste and trash is commonplace. However, there are differences between the two, since:

1. Solid waste has economic value and recycling potential, which generates employment and income. That is, there is reuse.
2. Garbage has no economic value whatsoever and should be discarded. So they are regarded as 'bad', as causing major environmental problems, since their.

However, a preponderant factor for the problems that happen today, is the incorrect and irregular disposition of Solid Waste like rubbish. Such a practice carries great environmental damages, such as contamination of soil, air, groundwater and damage to collective health.

Concern about this theme led the Federal Government to create a National Policy on Solid Waste (Law No. 12,305/2010), established in the country nine years ago. Regarding the law, the proposal is the guarantee of maximum reuse and, if possible, the recycling of waste, applicable to waste processes that have technical and economic viability.

It is salutary that the absence, or the bad management of waste, causes, like other sectors, the acceleration of emissions of greenhouse gases (GHGs). Gross Brazilian GHG emissions increased from 1.86 billion tons of carbon dioxide to 1.92 GtCO2, an increase of 3.5%. The overall picture of gross GHG emissions in Brazil is reported periodically, including: Land Use Change (46%), Energy (24%), Agriculture (22%), Industrial Processes (5%) and Waste 3%) [2].

As portrayed, the Waste sector accounts for the smallest share of emissions in Brazil, with 64.5 million tons of GHG. Although this is the smallest share, this represents an alarming growth of 500% since the year 1970 of 48% between the years 2000 and 2014 [2]. The emission estimates for this sector are directly associated with public policies at the municipal, state and federal levels. In this context, the possibility of reuse of GHG as an energy matrix could be studied as an alternative to attenuation or mitigation of the problem.

According to the law, the responsibility for the management of Urban Solid Waste is shared between the Public Power, the Business Sector and the whole community. That is, each generator is responsible for the generated waste. Generators are expected to minimize waste generation and promote recycling in order to ensure an environmentally sound disposal.

The Government's main tool is Law No. 12,305/2010, mentioned above, which contains specific guidelines and recommendations for waste management. Once implemented, they can be used to reduce GHG emissions, with the waste being used in a variety of ways, among which the induction of composting of the largest portion (organic matter), and the generation of energy through the use of gases from bio-digestion of organic compounds and gases generated in landfills, or treatment plants for the treatment of biogas [3].

The city of Manaus has a total population of 2,145,444 million people distributed in an area of 11,401,092 km^2. Of this total, approximately 99.5% of the population resides in the urban sector occupying only 4% of the municipal territorial extension, which, therefore, has a high demographic concentration of 158.06 inhabitants/km^2 [4]. This population makes the city the 7th most populous municipality in Brazil [2]. In this context, the historic center is one of the places where the management problem is more latent due to the intense commercial activity and the existing free fairs. It is known that the production of garbage is inevitable, because what has been discarded and that, after the use of certain processes or not, can be useful and exploited by man [5]. Waste pollution, in the specific case of free trade fairs, can be easily observed. Generally, they are characterized by the permanent production of solid residues, especially the organic ones, originating from the activity of wholesale and retail sale of fruits, cereals, vegetables, herbs, spices, vegetables, meats and so on. The logistics of work (receiving and shipping) of the commercialized materials, favors the accumulation of waste.

2 Methodology

2.1 Identification of the Study Site

Fairs are generally human traditions, and represent a sociocultural and economic phenomenon, coming from agglomerations of tents, people and benches, where the most diverse types of products are marketed. The first fair that has been registered, dates from 500 BC [6].

It is noted that in spite of the human and urban development, the fairs remain and are reborn in our environment, as cultural heritage and object usual in the present day.

The term "fair" derives from the Latin "fair" and means, holy day, holiday or day of rest, since the merchants, preoccupied with selling the surplus of production, gathered close to the Churches on Sundays. Marketing their products, since they were the places that had the largest flow of people [7]. It can be observed, therefore, that the fairs were nothing more than the immediate and low-cost supply (leftovers) of an existing and constant demand. Of course, behind the sales and demands, there is a historical-cultural load typical of each place.

In the case of the Banana Modern Fair, it is no different. According to the Supply Policy of the city of Manaus, through Law 123/2004 [8], article 22, it is stated that: "Art. 22-The policy of supply in the city of Manaus has as fundamental objectives: I - the stimulus to direct commercial practice between producer and consumer". Therefore, the public political reinforcement clarifies the intentions of incentive to the practices of Fairs and Markets and the commercial and cultural practice of the municipality.

The Modern Banana Fair is located in the center of the city of Manaus near the Port of Manaus, with access to the Pedro Botelho, Miranda Leão, Barés and Manaus Moderna avenues, in which there is also another covered fair of the same name, according to Fig. 1, below:

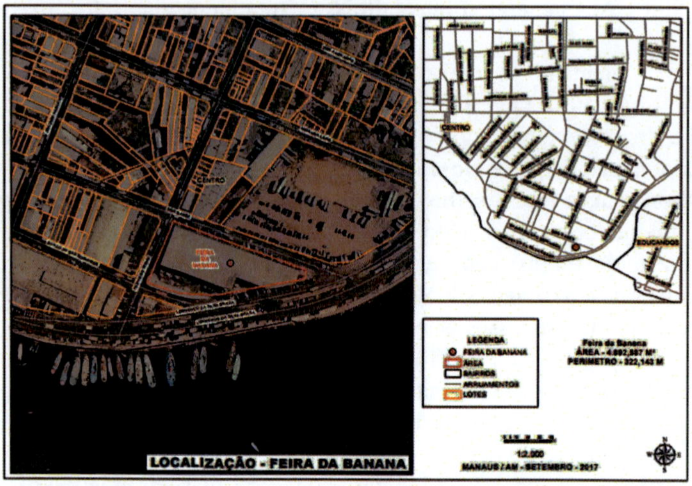

Fig. 1. Location Modern Fair of Banana, Manaus – AM. Source: Authors, 2017.

The fair was created in an attempt to expand the Adolpho Lisboa Municipal Market in the 1990s, which proved to be insufficient for the commercialization of agricultural production in the interior of the state of Amazonas and other Brazilian states.

2.1.1 Internal Structure

The Banana Manaus Modern Fair is organized by a large area and two annexes. In all of them containing "boxes" where there is the assorted trade of fruits, vegetables, vegetables, vegetables in general, among others. According to Article IV of Law 123/2004 [9], "Box" is defined as: "IV - box: physical division of markets for the implementation of a single commercial nucleus".

It is understood that the Fair of Manaus Moderna of Banana is a fair of the covered type, with wide corridors, which facilitates the movement of people, devoid of physical divisions (walls), as can be seen in the field, and shown in Fig. 2, below:

The Annex 1, popularly known as the Modern Watermelon Fair contains boxes that are dedicated exclusively to the sales of specific products, such as: watermelon, pineapple and regional products.

The field data and surveyed allow us to state that the fair has 138 boxes (Table 1). Being 128 boxes of the main area, that commercialize goods of diverse types, between eggs, cheese and diversity of greenery and others. The 10 boxes of Annex 1 are dedicated exclusively to the sale of banana, watermelon and pineapple. In general, the biggest follow-up of the fair is the banana, fruit, commerce and culture that gave rise to the name of the fair.

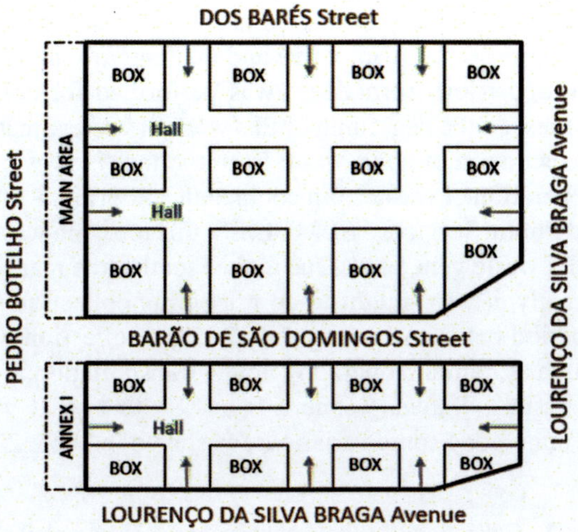

Fig. 2. Infrastructure of the Banana Modern Fair. Source: Authors, 2017.

Table 1. Mapping of found boxes.

Area	Quantity of boxes
Main area	128 boxes
Annex I	10 boxes

Source: Elaborated by authors, 2018.

These boxes are, in addition to point of trade, receiving locations (loading and unloading), transportation, shipping and wholesale and retailing of merchandise. That is, waste is generated daily in these places, coming from the most diverse activities. As seen, the Banana Modern Fair does not have the disposal of standardized dumps throughout its structure, so that improvised containers are used and a service of constant circulation of waste collection is adopted, making impossible the ideal destination of the materials, as for example selective collection.

2.2 Characterization of Solid Residues

The characterization of the solid waste produced at the fair was carried out in four stages, according to the methodology described in [10] and specifically adapted:

1. **Bibliographical research** - it is constituted by a survey and deepening of the bibliography, especially the researches of the competent public agencies, considering that works on the subject solid residues in the cut, are scarce;
2. **Data Collection** - refers to the collection of primary data from the permission holders, and especially data from competent agencies. In practical terms, forms were

prepared with a script of interviews, and also for the organization of the material under study, for purposes of quantitative and qualitative analysis;
3. **Analysis of Composition** - procedure was performed for collection and sample analysis of residues. At the beginning of this work, it was estimated to collect 0.05% of the weekly generation of solid waste from the fair. However, due to the lack of information (generation of waste from competent authorities - Municipal Secretariat of Cleaning and Public Services - SEMULSP), this research was limited to a sample of 0.5 ton of solid waste generated. Due to the fact that the material commercialized did not vary greatly, it was sought to do a monthly collection work, so that every month, in the period between January-July/2018, a collection was performed. Due to limitations such as: rainfall, operating hours, transportation, logistics and security at the collection site - Banana Modern Fair - So that what was planned for the collections has undergone some variations, as shown in Table 2 below:

Table 2. Collection of solid residues generated vs activity time.

Date	Quantity collected
04/01/2018	50 kg
24/02/2018	80 kg
31/03/2018	60 kg
08/04/2018	40 kg
30/04/2018	90 kg
31/05/2018	50 kg
02/06/2018	80 kg
08/07/2018	50 kg
06 Meses	500 kg

Source: Elaborated by authors, 2018.

The samples were collected as shown in Table 2, with the use of containers (containers) of hard plastic, fiber bags and a cart for displacement, according to Fig. 3a), a large part of the rubbings mapped in this work had their waste collected and analyzed.

After the collection of the residues, they were transported by car to an accessible location for immediate analysis (still on the same day of work, due to the expressive amount of organic matter). At the site of segregation, they were unpacked, according to Figs. 3b) and 2a), and subsequently segregated, manually, Fig. 3c).

The segregation of the residues was performed, and the residues were arranged separately, for later weighing, according to their classification in: Organic matter Fig. 4b), Paper/Cardboard Fig. 4c), Plastic Fig. 4d), Wood Fig. 4e), Metals Fig. 4f), Textile residue Fig. 4g), and Others.

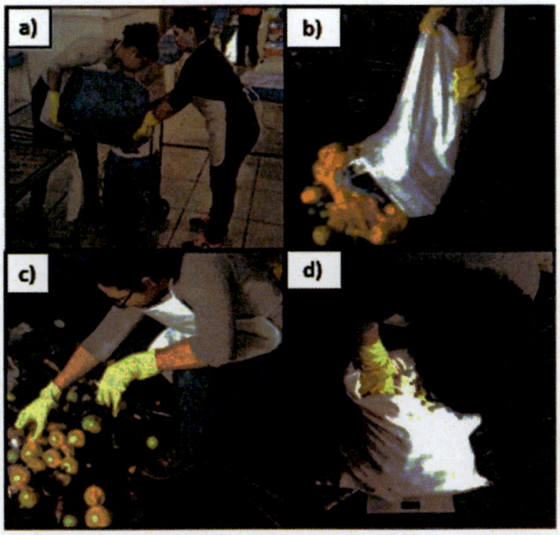

Fig. 3. a) collection of solid waste generated at Banana Modern Fair b) unpacking of solid waste c) separation/segregation of solid waste d) weighing of solid waste. Source: Authors, 2018. Source: Authors, 2018.

After segregation of the residues, these were isolated and weighed according to Fig. 3d). The weighing of the waste was done by using two scales, with different capacities: Scale 1, capacity 150 kg, resolution 0.5 g. And, the scale 2, capacity 15 kg, resolution 0.1 g.

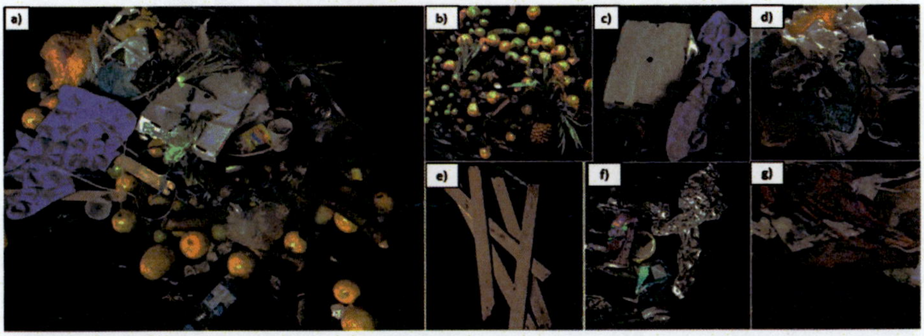

Fig. 4. a) waste collected at the Banana Modern Fair after unpacking b) segregated organic matter c) segregated paper/cardboard d) segregated plastic residue e) segregated wood f) segregated metal residue g) segregated textile residue. Source: Authors, 2018.

After weighing, we recorded the values found for the waste generated, using a form. After that, the analysis of the generated residues was followed, which was limited to the qualitative merit, since there was not enough time for the possibility of extending to a deep gravimetric study, that makes use of greenhouse, room for isolated unpacking, incinerator, and etc. The collection procedures were performed based on the Brazilian

Regulatory Standards: NBR-10007 - waste sampling: procedure [11], and NBR-13463: solid waste collection: classification [12]. The objective of this step is to know the composition of the waste, classifying them, and;

4. **Integrated Analysis -** consists of the data tabulation of the forms, a form was used. In addition, consultations were held with the Municipal Secretariat of Production and Supply - SEMPAB, and the Institute of Agricultural and Forestry Development Sustainable of the State of Amazonas - IDAM. The application of the forms and interviews (permission holders), equivalent to 5% of the boxes identified in the Banana Modern Fair.

3 Results and Discussions

The analysis and classification of the residues was performed according to [13], which refers us that the gravimetric composition is a way of knowing the composition of the solid waste generated in a given locality, since it is in itself the relation between quantity of each component of a sample in relation to the total amount sampled. Thus, the results found after analysis of 0.5 tol of residues were arranged according to Fig. 5, below:

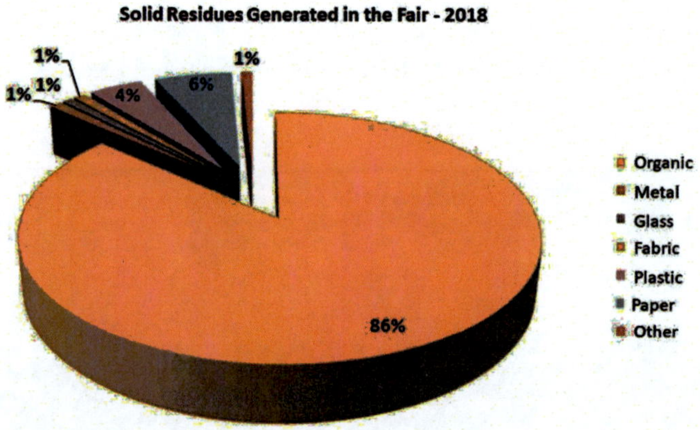

Fig. 5. Graphical representation of the incidence of Solid Waste in Banana Modern Fair. Source: Authors, 2018.

The high generation of organic waste (86%), followed by paper (6%), and a significant proportion of plastic (4%) were identified after analysis of the gravimetric composition of several samples, according to methodological description. affirmed by permission holders in the parallel development of this research.

The alarming data of this study is the high rate of organic matter. These data differ from the works of [14] and [15] hat indicate a number of divergences from those that were identified, as shown in Fig. 6. The work of Andrade (1997) carried out a general gravimetric analysis in the whole city of Manaus, on solid waste produced. On the other

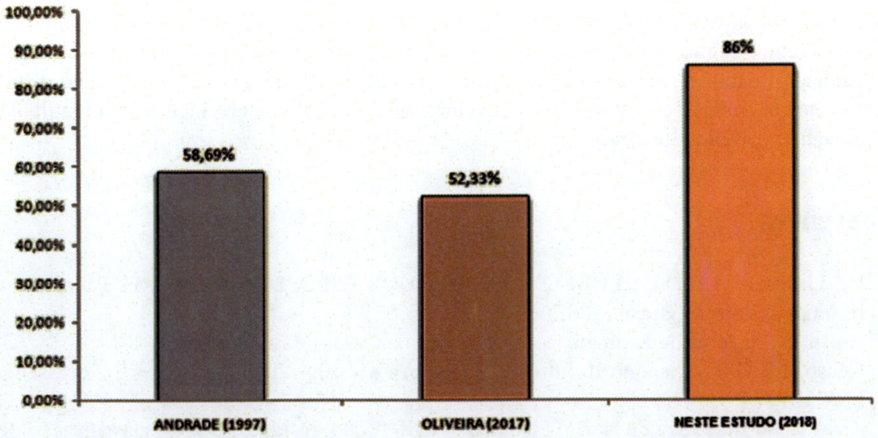

Fig. 6. Comparison between scientific studies. Source: Authors, 2018.

hand, the work of Oliveira (2017) brought to the optics solid residues domiciliates in the cut of the sources of the Igarapé of Mestre Chico.

The difference of approximately 30% (proportional) can be justified because of the change of location studied. For, according to [10], the composition of the residues can vary from one locality to another in function of social, economic, cultural, geographic and climatic characteristics. Thus, it is confirmed that in the gravimetric composition of fairs, or places where there are commercialization of products, such as: fruits, vegetables, vegetables, etc., tends to present a higher index of organic matter. Such an identification is of great importance, since this expressive amount of organic waste (435.05 kg in a 500 kg sample) has an energy potential (biogas), if properly treated and manipulated.

4 Final Considerations

It is important to note that the Banana Modern Fair is only one fair, among the 134 approved in the Municipal Secretariat of Production and Supply in 2018, and there are still other informal and/or seasonal.

It is impossible not to point out the possibility of energetic use of some gases such as Methane (CH4) and Carbon Dioxide (CO2) that are generated with the chemical decomposition of waste like those studied. The potential for energy generation from these materials (biogas) is therefore under-exploited and overlooked by public resource managers in the region.

As a sequence of studies, it is necessary to investigate the biogas generation capacity of the residues generated at Banana's Modern Manaus Fair, as well as the exploration of the dimensioning of energy generation from these resources.

Ethics Declaration

Conflict of Interest. The authors declare that they have no conflict of interest.

Compliance with Standards Involving Humans as Subjects. All procedures performed in studies involving human participants were in accordance with the ethical standards of the institutional and/or national research committee and with the 1964 Helsinki Declaration and its later amendments or comparable ethical standards. Informed consent was obtained from all individual participants involved in the study.

References

1. De Limpeza, A.B.D.E.: PÚBLICA E RESÍDUOS ESPECIAIS-ABELPRE. Panorama de resíduos sólidos no Brasil (2007)
2. Brasil, S.: Sistema de Estimativas de Emissões de Gases de Efeito Estuda (2017)
3. Censo, I.B.G.E.: Disponível em (2010). http://www.censo2010.ibge.gov.br/. Accessed 15 May 2019
4. Associação Brasileira de Normas técnicas. NBR 10004: resíduos sólidos: classificação. Rio de Janeiro (2004)
5. França, K.M., Teles, M.A.P.: Pacto pela saúde e gestão da atenção básica em Manaus. Anais do Encontro Internacional e Nacional de Política Social, vol. 1, no. 1 (2018)
6. Topan, Cláudia de Oliveira. Lixo: curiosidades e conceitos. Manaus: EDUA (2002)
7. Ferreira, T.B.: Comunicação e marketing: um estudo das interações comunicacionais entre feirantes e fregueses na feira livre de Paripiranga-BA (2017)
8. Toda Matéria: Disponível em (2015). https://www.todamateria.com.br/historia-e-origem-das-feiras/. Accessed 27 Jan 2018
9. Brasil. Lei no 123, de 25 de novembro de 2004. Dispõe sobre a organização e o funcionamento dos mercados e feiras no município de Manaus, e dá outras providências
10. Matos e Silva Júnior, I., Prost, C.: Olhar Geográfico de Resíduos Sólidos Urbanos Um estudo comparativo das representações socioespaciais das feiras livres dos bairros de George Américo e Cidade Nova em Feira de Santana – BA. Revista eletrônica de Gestão e Tecnologias Ambientais (GESTA), Bahia, no. 2, vol. 1, pp. 286–304 (2013)
11. Associação Brasileira de Normas Técnicas. NBR 10007: amostragem de resíduos: procedimento. Rio de Janeiro (1986)
12. Associação Brasileira de Normas Técnicas. NBR 13463: coleta de resíduos sólidos: classificação. Rio de Janeiro (1995)
13. Monteiro, J.H.P., Zular, V.: Manual de Gerenciamento Integrado de resíduos sólidos. Rio de Janeiro: IBAM, 2001. Disponível em: Acesso em: 15 de mar. de 2018
14. de Andrade, J.B.L., Schalch, V.: Determinação da composição gravimétrica, peso específico e teor de umidade dos resíduos sólidos produzidos na cidade de Manaus. Revista Limpeza Pública, São Paulo, no. 44, pp. 27–31 (1997)
15. Oliveira, G.: Gerenciamento de Resíduos Sólidos nas Residências das Nascentes da Bacia Hidrográfica do Mestre Chico, Manaus – AM. Projeto de Iniciação Científica. UFAM (2017)

Bicycle as a Mode of Transport in Brazil: Joint Action of Society, Sustainability and Innovation

Thiago Caliari[1](✉), Ernesto Cordeiro Marujo[1], and Thais Marzola Zara[2]

[1] Aeronautics Institute of Technology, São José dos Campos, Brazil
caliari@ita.br
[2] Rosenberg Associados, São Paulo, Brazil

Abstract. The use of bicycles as a mode of transport is widely disseminated around the world, being its choice dependent on several factors as culture, socioeconomic, environmental aspects, public policies, among others. In this paper, we highlight features of Brazilian bicycle market (demand and supply), showing (i) that demand responds mainly to increasing on infrastructure and traffic safety and (ii) improvements on scale and technological content in productive firms are important to warrant national supply. In a nutshell, this economic approach is used to affirm that a joint action through policy makers, specialists and society tends to be the main policy to improve bicycle as modal transport.

Keywords: Bicycle market · Innovation · Sustainability · Brazil

1 Introduction

The use of bicycles as a mode of transport is widely disseminated around the world, being its choice dependent on several factors as culture, socioeconomic, environmental aspects, public policies, among others. In sum, economic growth generates new demands, social pressures of space, and government policies need to rethink the occupation of public space in order to create sustainability. In this sense, the European Union's positioning in public policies to encourage the use of the bicycle is always emblematic, allocating resources, creating specific programs for popularization [1–3] and, at the same time, establishing policies to reduce the use of road transport.

Considering this, the following work presents aspects regarding the use of bicycle as a modal transport in Brazil, highlighting its distinct market characteristics as a developing country. Our text will focus on demand and supply features, considering this economic approach is useful to affirm that a joint action through policy makers, specialists and society tends to be the main policy to improve bicycle as a modal transport.

2 Demand, Supply and Economic Well-Being

Demand and supply are crucial in microeconomic analysis, in order to understand market and welfare results. A demand curve represents the amount of demand (quantity) for

L. Pereira et al. (Eds.): IDEAS 2018, SIST 198, pp. 457–466, 2021.
https://doi.org/10.1007/978-3-030-55374-6_45

a good given its price. According law of demand, *ceteris paribus* (all other features remain constant), the higher the price, the lower the quantity demanded. On the other way, the supply curve presents the amount of product offered in a market at given prices. In supply side, when all other features remain constant, the higher the price, the higher the quantity offered. In a market, pressures from both sides leads to an equilibrium in price and quantity [2].

When defined demand and supply, it is possible to identify the demand and supply surplus. These concepts are important to study economics welfare. In short, Consumer Surplus (CS) is a measure of how well consumers are, given by the willingness to pay minus the price actually paid. Analogously, Producer Surplus (PS) is a measure of how well producers are, given by the price received minus the cost of production. The Fig. 1 below highlights those measures [2].

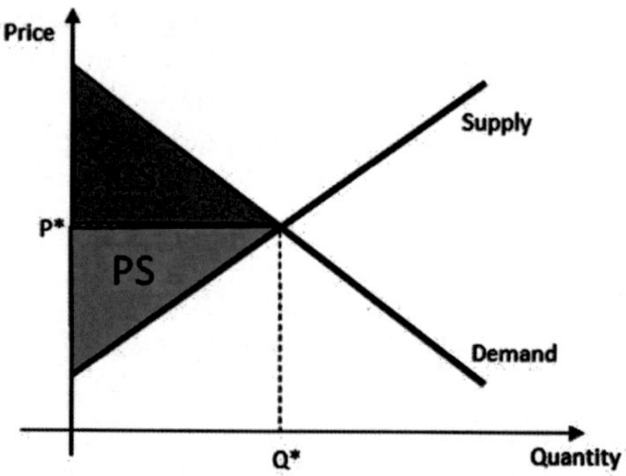

Fig. 1. Demand, Consumer Surplus, Supply and Producer Surplus. Source: Own elaboration.

In this outline, the sum of consumer surplus plus producer surplus defines market welfare, which means the satisfaction of both demanders and producers given market price and quantity [2]. Studies considering welfare economics uses those approach to calculate gains and losses for market participants from changes in market conditions and economic policies. For that, it is important to define features (elasticities) of demand and supply in order to understand how public policies may affect preferences.

3 Bicycle Market in Brazil

3.1 Demand

The use of bicycle as a modal transport is directed by a bunch of drivers, being the main ones the agility [3], health [1, 3–5, 6], low acquisition and maintenance cost [7–11], energy efficiency [3], flexibility [12], infrastructure costs [12], equity [13] and safety [3, 14–16]. Despite those drivers, which may give preference to the use of bicycles when considering the pressing global environmental issues, the observation of the demand for bicycles in Brazil allows us to affirm the following points:

a) **Low participation of bicycles in the number of trips by type of modal:** according to a study by the National Association of Public Transport [14], bicycles are the mode of transportation with the lowest number of journeys per kilometer.
b) **Price-sale relationship:** the evolution of the bicycle market in the period 2007–2017 shows a considerable reduction in domestic sales in the Brazilian market (approximately 37%). This drop was also followed by a decrease in the relative price of bicycles compared to the Consumer Price Index (IPCA). This information shows that relative reductions in bicycle prices were not sufficient to generate a positive impact on demand, strengthening the thesis that price is not the main determinant of demand.
c) **Low Impact on Budget:** according to data obtained from the Family Budget Survey (POF 2008–2009), the average value of the bicycle corresponded to only 1.2% of annual GDP per capita in 2009 and to 3.2% of the annual income of a person who, in 2009, received a monthly minimum wage. As it is a durable good-average of 7 years, this would imply an annual cost of about 0.45% of their income. To corroborate this finding, an exercise shows the average price of bicycles on sale at the site Zoom (R$ 1,070) compared to the average GDP per capita of the Brazilian population in 2016 (approximately R$ 30,000), and considering the durability of the good also of 7 years, show an impact of approximately 0.51% on income.
d) **Income-demand relationship:** again according to POF 2008–2009, people who usually move on foot or bicycle to work normally are people with lower income. As income rises, therefore, the use of bicycles and walking decreases in favor of public transport, firstly, and individual motorized transport (motorcycles, in the first instance, and cars, in the second instance), after. There would be, therefore, evidence of 'social ascension' through the migration of the mode of transport.
e) **Motivations and Obstacles:** data from the research 'Perfil do Ciclista 2018' [15] provides information on the main issues considered in order to demand bicycle as a modal transport (Figs. 2 and 3).

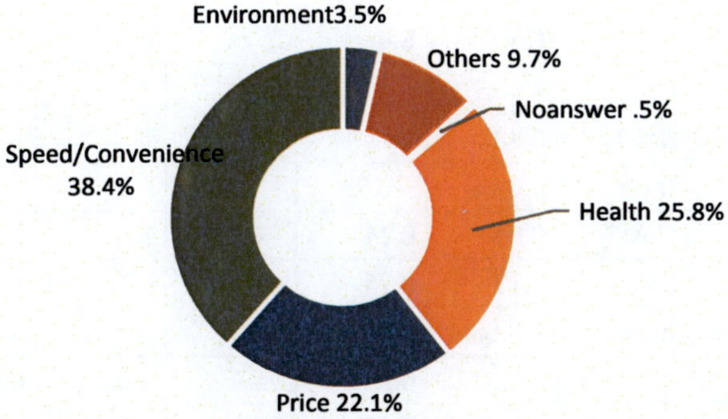

Fig. 2. Motivations to use bicycle as a daily modal transport

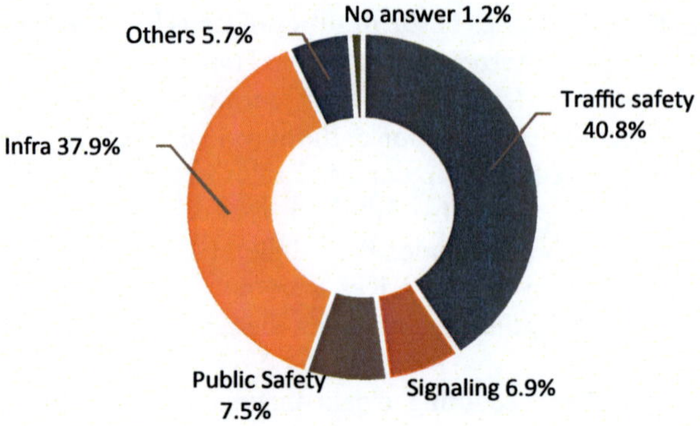

Fig. 3. Obstacles to use bicycle as a daily modal transport Source: [15].

'Price' is a motivation for only 22% of the applicants, being 'speed/convenience' and 'health' the most important issues when considering the motivations. Also, the main obstacles are infrastructure and traffic safety. A very large range of studies, using a variety of analyzes and methodologies, corroborates this position, inserting the convenience/use variable as one of the important pillars for bicycle demand [7–10, 16–21].

f) **Demand differences by size of city:** the smaller the city, the greater the participation of bicycles in the modal distribution of urban trips [14]. Using data from 'Perfil do Ciclista' [15] and considering some city profiles (size, income, economic development), we are able to analyze how demand behavior varies depending on these regional socio-economic variables (Table 1).

Table 1. Correlation among motivations, obstacles and socio-economic data.

		Potential demand			Current demand		
		Cost	Safety	Infra	Cost	Safety	Infra
Potential demand	Cost	1					
	Safety	0.18	1				
	Infra	−0.16	−0.73	1			
Current demand	Cost	0.94	0.24	−0.18	1		
	Safety	0.17	0.75	−0.60	0.19	1	
	Infra	−0.08	−0.21	0.46	−0.16	−0.50	1
Population		−0.31	0.23	0.04	−0.25	0.09	0.23
GDP		−0.30	0.17	0.01	−0.26	0.07	0.21
GDP *per capita*		−0.33	0.17	0.08	−0.33	0.11	0.14

Source: Own elaboration with data extracted from 'Perfil do Ciclista' [15].

We can highlight:

- Negative correlation cost-population, cost-GDP and cost-GDP per capita;
- Importance of infrastructure;
- Traffic safety is more important as bigger and wealthier the city, mainly for potential demand.

From all those findings presented above, we can summarize main features of bicycle demand as a modal transport in Brazil. First, there is a **low price-elasticity of demand**, since relative reductions in bicycle prices were not enough to generate a positive impact on demand. Besides, the average price of a bicycle corresponds to a relatively small portion of income (topic c); thus, if the purpose was to acquire a bicycle for use as an urban transport medium, price would hardly be the major determinant. In addition, there is a wide range of prices and products offered with easy accessibility to the consumer budget;

Secondly, **bicycle seems to qualify as an inferior good in developing countries**, when considered as a modal transport. That is, as income grows, its consumption falls, signaling the importance of policies that encourage people, even in the face of a higher level of purchasing power, to use the modal. On this matter, **deficiency in infrastructure and traffic safety seems to be crucial in the decision of the consumer to use the modal**. Different characteristics of regional demand intensify those findings.

These observations allow us to infer, in the Brazilian case, a demand curve with low elasticity for urban bicycles (little effect of the price on the purchase of bicycles) and the need to establish strategies that modify the perception and preference of the consumers, in order to consider bicycle as an effective mode of transport in urban centers. This includes the observation of the differentiated characteristics of bicycle perception and preferences in developed countries in Europe, where well-established public policies to increase infrastructure and safety, coupled with the insertion of greater technological content in products, was important for the success of the modal.

3.2 Supply

According with data from *Ministério do Trabalho e Emprego* (RAIS-MTE) [22], there is a continuous increase in the number of bicycle companies operating in Brazil during the period 2006–2017, approximately 5.5% growth per year. This movement was also followed by the increase in the number of employees up to the year 2013, but which has been reversed since 2014 and intensified in the last two years, a clear reflection of economic crisis experienced in Brazil. In 2017, a new reversal occurs in the number of employees, with a 7.8% growth rate.

In spite of these movements, during the analyzed period the average size of a bicycle company was never higher than 34 employees, reaching the value of 24.7 employees in 2017. Also, at least as from 2013, the evolution of the sector in the number of companies was focused on a decrease in the average size of the manufacturing unit, considering the decrease in the number of employees in the sector. A closer look at the size of the facility shows that there is a very large concentration of small businesses, as can be seen in Fig. 4 below for the year 2017.

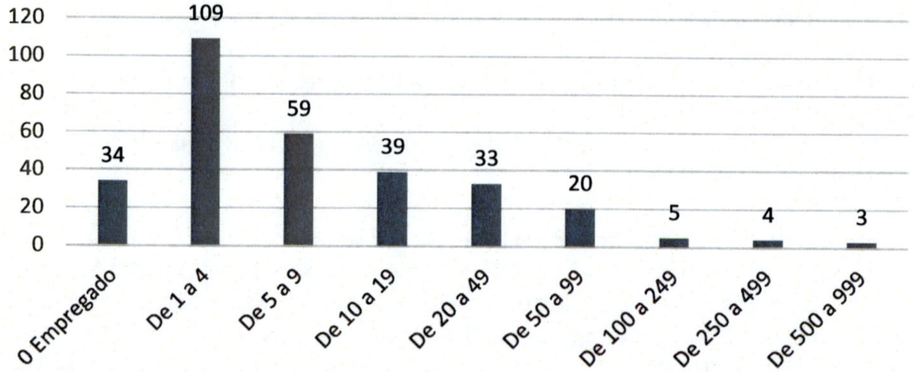

Fig. 4. Number of Companies by size (number of employees), 2017, Brazil. Source: Own elaboration with data from [22].

The major problem of a small-scale production structure is the importance of scale and scope for increasing efficiency and productivity, which are essential factors in the manufacturing industry. To prove this, the Fig. 5 below presents the construction of a measure of productivity (gross value of production per worker) separated by average size of companies and State of the Federation.

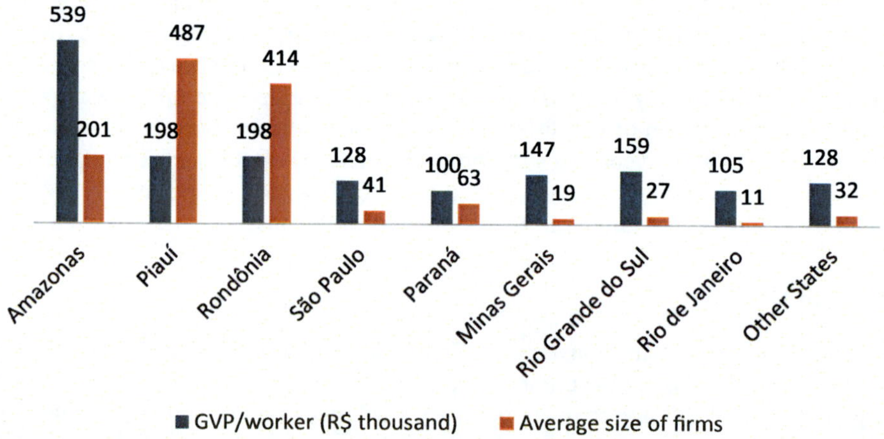

Fig. 5. Productivity and Average Size of firms in Brazil. Source: Own elaboration with data from [22–24].

The measurement points out important information regarding the relationship between size of companies and labor productivity. While production in Amazonas, Piauí and Rondônia is based on large-scale production plants, supply in São Paulo and other states is based on small-scale enterprises. This difference is reflected in the productive efficiency; while the average productivity obtained by the extrapolation of PIA and RAIS to States in the Southeast and South Region averaged approximately R$ 128k, for the Amazonas' companies, the measured productivity was around R$ 539k, and for Piauí and Rondônia, R$ 198k, indicating the existence of economies of scale. To reaffirm, scale is important for competition purposes, since it allows prices reduction to final consumer (through the sharing of fixed costs, specially investment in new technologies).

An exercise to test this assumption is made in the next Table, when final prices (discounted all taxes) of products from Amazonas' firms, other Brazilian firms and imported products are presented by distinct segment of products.

Table 2. Price comparison (R$)

Description		Amazonas		Small Brazilian companies		Medium/high Brazilian companies		Imported bicycles	
Up to 26 ring		N	Aver. Price (no tax)	N	Aver. Price (no tax)	N	Aver. Price (no tax)	N	Aver. Price (no tax)
Urban	Steel	4	612.9	8	741.4	8	673.3	8	517.0
	Carbon Steel	15	464.8	55	542.3	55	492.5	55	378.1
	Aluminum	29	981.4	51	1355.5	51	1231.1	51	945.2
	Carbon			2	13470.9	2	12234.4	2	9393.3
MTB *	Steel	2	787.7						
	Carbon Steel	5	599.6	24	659.4	24	598.9	24	459.8
	Aluminum	49	1330.9	52	1308.8	52	1188.7	52	912.7
	Carbon	3	8221.4						

Source: Own elaboration with data from www.zoom.com.br.
* MTB: Mountain Bike.

Table 2 shows that there is a wide range of bicycles offered by the Amazonas firms, and the average sale price is lower than that practiced by other domestic firms in all analyzes (except one) when deducted the taxes paid for firms. This competitiveness is lost, however, when compared to imported bicycles, since these have lower average values in all analyzes. In this regard, one observation that can be deduced is the empirical existence of lower prices practiced by companies with scale and scope on production at national level, but the need of more improvements in production to achieve productivity of foreign firms (being Chinese firms the main competitors).

However, economies of scale are not the only explanation for the greater productivity of companies in Amazonas. It should also be highlighted the value added in the productive differentiation of products with greater technological content, which is important to support product developments and spillovers for products with lesser technological content. Observing supply structure by market segments, we could find those firms (Amazonas' firms) are mainly producing in segments with high technological content, different from its Brazilian competitors.

3.3 Economics Welfare

The observation of the bicycle market from an economic view allows us to raise considerations that signal the need for a joint multidisciplinary action to promote the modal of bicycles in Brazil. For this, we remember that bicycle demand has a low price-elasticity, which assures us that there are small changes in quantity demanded in reaction to price changes.

Still, in the supply side, results show an expressive number of firms, low entry barriers and an import pressure; all these features allow us to conclude that supply curve presents high elasticity.

Given that, demand-supply analysis considering consumer and producer surplus helps to provide understanding on a public policy focused in structural aspects of the infrastructure and traffic security besides the improvement of industry efficiency (Fig. 6).

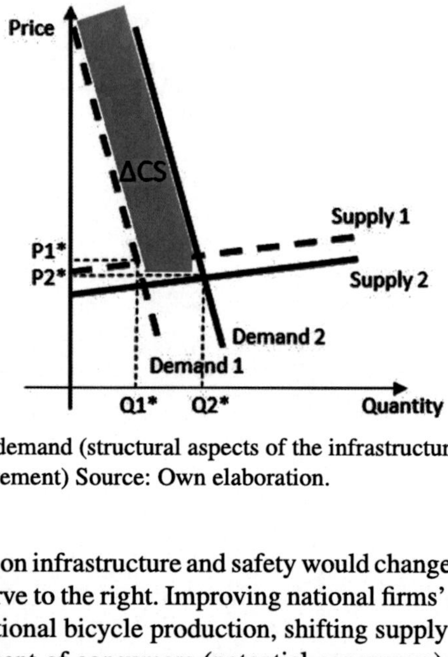

Fig. 6. Public Policy on demand (structural aspects of the infrastructure and traffic security) and supply (efficiency improvement) Source: Own elaboration.

Movement focused on infrastructure and safety would change consumer preferences, shifting the demand curve to the right. Improving national firms' capabilities on production would increase national bicycle production, shifting supply curve also to the right. The result is an increment of consumers (potential consumers) in the market and gain on welfare for current consumers, allowing an expressive increase on consumer surplus and a possible positive impact on prices (decrease on prices).

4 Conclusions

This study aimed to establish an analysis on Brazilian bicycle market, highlights features of demand and supply. Through economic welfare analysis, the results ob-served demonstrates the relevance of a joint perspective of policy makers to reaffirm the importance of bicycles through strategies for improvements in infrastructure and safety, involving the whole society, and a careful observation of national capabilities to offer those products.

We point out that, in addition to policies that focus on demand, careful observation of the productive capacities of national companies cannot (and should not) be ruled out, otherwise the sectoral trade balance (which is already quite negative, US$ 25 million in 2017) tends to suffer major negative impacts, mainly from competitive pressure of Chinese products. Currently, Chinese Firms' competition is very damaging to industries in other countries, due to the scale and technological capabilities of those companies. For these reasons the European Union sets an anti-dumping tariff on Chinese products of 48.5% [25, 26], and Brazil has introduced an import tariff of 35% since 2011. In this sense, the establishment of the import tariff has allowed the creation of a group of companies operating in a productive cluster in Amazonas, which has managed to increase production by increasing scale and technological content in products.

Thus, as the main conclusion and recommendation of this article, there is a need to consider the bicycle market as a whole, understanding that the increase in demand must be coupled to national productive capabilities promotion, with the growth of companies that have shown to be competitive by both increasing scale and technological content. There are many ways to pursue this last target, that should be object of further research, such as scale gains by exports promotion, reduction/simplification in tributary load, improvement in logistic/infrastructure, among others.

This path, however, cannot be established by unconnected policies. It is imperative to take joint actions, aligning the relevance of social sustainability and the potential to expand local production at lower costs.

References

1. Davis, A., Cavill, N., Rutter, H., Crombie, H.: Making the case: improving health through transport. NHS Agency (2005)
2. Varian, H.: Intermediate Microeconomics: A Modern Approach, 9th edn. 798 p. W.W. Norton & Company (2016)
3. ECMT: Implementing Sustainable Urban Travel Policies: Moving Ahead (2004)
4. Oja, P., Titze, S., Bauman, A., de Geus, B., Krenn, P., Reger-Nash, B.: Health benefits of cycling: a systematic review. Scand. J. Med. Sci. Sports **21**(4), 496–509 (2011)
5. Kienteka, M., Fermino, R., Reis, R.: Fatores individuais e ambientais associados com o uso de bicicleta por adultos: uma revisão sistemática. Revista Brasileira de Atividade Física & Saúde **19**(1), 12–24 (2014)
6. Wen, L.M., Rissel, C.: Inverse associations between cycling to work, public transport, and overweight and obesity: findings from a population based study in Australia. Prev. Med. **46**, 29–32 (2008)
7. Andrade, V., et al.: Mobilidade por Bicicleta no Brasil. Rio de Janeiro: [s.n.] (2016)
8. Kienteka, M., Fermino, R.C., Reis, R.S.: Fatores individuais e ambientais associados com o uso de bicicleta por adultos: uma revisão sistemática. Revista Brasileira de Atividades Físicas e Saúde **19**(1), 12–24 (2014)
9. da Silveira, M.O., Maia, M.L.A.: Variáveis que influenciam no uso da bicicleta e as crenças da teoria do comportamento planejado. Transportes **23**(1), 24–36 (2015)
10. Ritta, L.A.S.: Motivos de Uso e Não-uso de Bicicletas em Porto Alegre: um Estudo Descritivo com Estudantes da UFRGS. 2012. 113 f. Master Dissertation, UFRGS (2012)
11. Coelho, M.: Custo por quilômetro da bicicleta como transporte diário. 2011, Rio de Janeiro: [s.n.] (2011)

12. Andrade, V.: Economia da bicicleta no Brasil. Rio de Janeiro: LABMOB-UFRJ, 155 p. (2018)
13. IBGE. Pesquisa Nacional por Amostra de Domicílios. Brasília: [s.n.] (2008)
14. ANTP: Sistema de Informações da Mobilidade Urbana da Associação Nacional de Transportes Público - Simob/ANTP. [S.l: s.n.] (2018). http://files.antp.org.br/simob/simob-2016-v6.pdf. Accessed Aug 2018
15. Transporte Ativo: Labmob-Ufrj. Perfil do Ciclista 2018. [S.l: s.n.] (2018)
16. Vieira, D.J.F., Carvalho, K.R.F.: Infraestrutura cicloviária e seu impacto no uso de bicicleta em Goiânia, 77 f. UFG (2016)
17. de Sousa, P.B., Kawamoto, E.: Análise de fatores que influem no uso da bicicleta para fins de planejamento cicloviário. Transportes **23**(4), 79 (2015)
18. de Paiva, M.: Fatores que influenciam no uso de bicicleta de forma integrada com o metrô, 206 f. UnB (2013)
19. Pezzuto, C.C., Sanches, S.P.: Identificação dos Fatores que influenciam no uso de bicicleta. Florianópolis: [s.n.] (2004)
20. Providelo, J.K., Sanches, P.: Percepções de indivíduos acerca do uso da bicicleta como modo de transporte. Transportes **18**(2), 53–61 (2010)
21. da Silveira, M.O.: Mobilidade Sustentável: a Bicicleta como um meio de transporte integrado, 168 f. COPPE/UFRJ (2010)
22. MINISTÉRIO DO TRABALHO E EMPREGO. Relação Anual das Informações Sociais. http://bi.mte.gov.br/bgcaged/login.php. Accessed July 2018
23. INSTITUTO BRASILEIRO DE GEOGRAFIA E ESTATÍSTICA: Pesquisa Industrial Anual (PIA). https://www.ibge.gov.br/estatisticas-novoportal/economicas/industria/9042-pesquisa-industrial-anual.html?=&t=o-que-e. Accessed July 2018
24. Abraciclo: Anuário da Industria Brasileira de Duas Rodas, São Paulo (2018)
25. Reuters: EU looks into extending dumping duties on Chinese bicycles. https://www.cnbc.com/2018/06/04/reuters-america-eu-looks-into-extending-dumping-duties-on-chinese-bicycles.html. Accessed October 2018
26. Toll, M.: EU joins the e-bike trade war with new tariffs to prevent Chinese electric bicycle dumping - Electrek. https://electrek.co/2018/07/20/europe-e-bike-china-dumping-tariffs/. Accessed October 2018

Management and Entrepreneurship Orientation in the Fisheries of the Micro-region of Rio Preto da Eva – AM

Patrícia dos Santos(✉), Marcelo Albuquerque de Oliveira(✉), Joaquim Maciel da Costa Craveiro(✉), Dercio Luiz Reis(✉), and Ely Senna de Almeida(✉)

Graduate Program in Engineering Production, Federal University of Amazonas, UFAM, Av. Gen. Rodrigo Octávio Jordão Ramos, 3000, Coroado I, University Campus, Senador Arthur Virgílio Filho, Setor Norte. CEP, Manaus, AM 69077-000, Brazil
suppat@hotmail.com, {marceloliveira,elysenaa}@ufam.edu.br, jgabael26@gmail.com

Abstract. Fish farming is an aquaculture activity that is characterized by the creation of freshwater fish in rural properties in a planned and controlled manner. The regions of the State of Amazonas have plenty of water resources, a favorable condition for the development of fish farming and available and needed labor but has been surpassed by the production of the other states of Brazil. The micro-region of Rio Preto da Eva has a profile of small family producers, with the favorable nature conditions for fish farming with emphasis on the production of tambaqui fish *(colossoma macropomum)* with export potential, needing to mitigate blocking factors for the expansion of the scale production. The objective of this work was to evaluate the factors that prevented the expansion of the scale of production and, based on the bibliography researched on the subject, were proposed techniques of management and entrepreneurship. The methodology used in this bibliographic research collected public information provided by official organizations and previous scientific publications. The results of the research showed that the same difficulties have been reported in scientific works developed from the year of 2010 to the year of 2018. We concluded that the central problem has been searched untiringly in the diversified researches on the activity has always become catalogs of problems to be solved. In this way, we suggested that the development entrepreneurial orientation in the fish farming activity and the grouping of the small producers of the region into cooperatives to mitigate production and distribution costs and make feasible for the increase of the scale of production.

Keywords: Fish farming · Management · Entrepreneurship · Rio Preto da Eva Micro-region · AM

1 Introduction

The abundance of water resources and different species of fish makes the State of Amazonas a high potential in large-scale fishing. The Micro-region of Rio Preto da Eva is part

L. Pereira et al. (Eds.): IDEAS 2018, SIST 198, pp. 467–479, 2021.
https://doi.org/10.1007/978-3-030-55374-6_46

of the Amazon Center region and two municipalities, Rio Preto da Eva and Presidente Figueiredo compose the region [1]. The two municipalities are the largest producers of tambaqui fish and Pirarucu fish in the State of Amazonas and promising for large-scale production in fish farming.

Aquaculture is the most widely used culture medium in the State of Amazonas and is the cultivation usually occurs in a confined and controlled space of aquatic organisms such as fish, crustaceans, mollusks, algae, reptiles and any other aquatic life of productive economic interest. The productive activity has different modalities as fish farming; shrimp farming; frog breeding; cultivation of mollusks (oysters and mussels); algae cultivation and other species with less commercial appeal, such as a creation of turtles and tracajás and the creation of alligators [2].

Fish farming is part of aquaculture and cell is the activity and fish farming in fresh and marine waters and involves production planning and forecasting of direct and indirect costs. Fish farming has a high social and economic relevance on job creation and maintaining living in rural areas, accounting for sustainable solutions in food production and in the conservation of the biological diversity of natural environments [3].

Brazilian fish farming is represented in 43% by tilapia (Oreochromis niloticus), 23% by tambaqui fish (Colossoma macropomum) and 15% by tambacu fish (Piaractus mesopotamicus) and tambatiga (colossoma macropomum). Only tilapia contributes 35% to national aquaculture production [4]. According to a survey by the IBGE's Paraná Municipal Livestock Piling - 2017 the state of Paraná was the largest producer in 2017 followed by the states of São Paulo, Rondônia and Mato Grosso. The production of native fish the state of Rondônia leads to the production of 77, 000 tons and the State of Amazonas is in third place with 28,000 tons. According to Brazilian Association of Fish Farming [5], in 2017 Brazil was the fourth position in largest producer of Tilapia in the world, accounting for over 51% of national production.

The Tilapia production was responsible for the positive index of Brazilian fish farming it pulling production to 700 000 tons in 2017 [6]. The native fish of the Amazon are highly appreciated and demanded even by international markets. To achieve the growth of demand the state of Amazonas needs to acquire fish from the state of Rondônia even though it has the excellent conditions to work with a larger production scale and to meet all demand.

The question to which this work pursued the answer was: What prevents the Micro-region of Rio Preto da Eva to develop the activity and increase their fish production scale in captivity and as a consequence provides the state of Amazonas sustainability in the fish market and boosts its economy and competitiveness in fish farming?

The search problem is the possible that the blocking factors for the development of fish farming in the Micro-region of Rio Preto da Eva is the lack of knowledge of management tools and entrepreneurial orientation to the producers (Fig. 1).

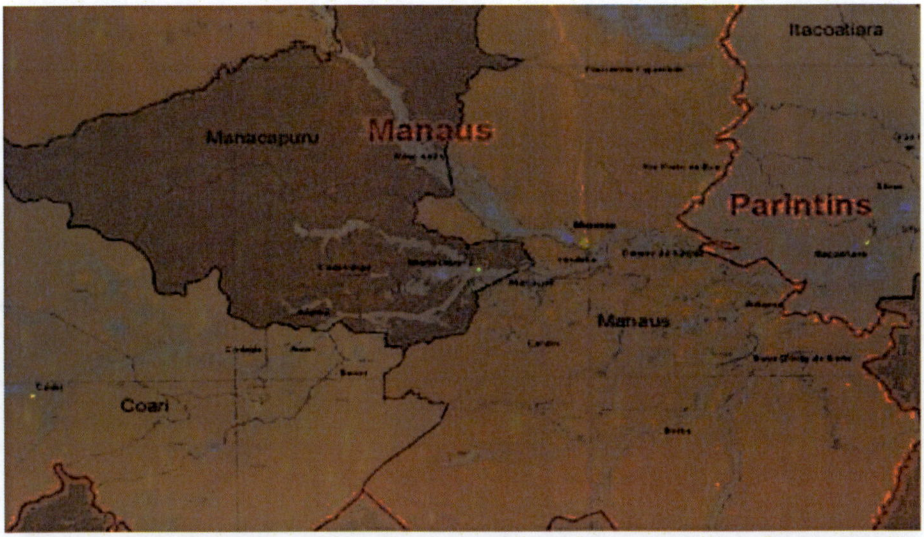

Fig. 1. Clipping of the map of the Micro-regions of the state of Amazonas, Brazilian Institute of Geography and Statistics https://portaldemapas.ibge.gov.br/portal.php#mapa876, extracted in 11/13/2018.

2 Materials and Methods

This work has a bibliographic characteristic and it has a descriptive nature, primary research and qualitative approach. First of all, we selected the relevant bibliography on the subject of fish farming referring to the period from 2010 to 2018, based on queries to periodicals available on online platforms, such as Capes, Google Academic and directory of the Federal University of Amazonas.

Technical and statistical information was selected through publication's consultation of official research and development agencies such as Brazilian Agricultural Research Corporation (Embrapa), Brazilian Institute of Geography and Statistics (IBGE), the Manaus Free Trade Zone Superintendence (SUFRAMA), Brazilian Support Service for Micro and Small Enterprises (SEBRAE), The organization Food and Agriculture Organization (FAO), Department of Rural Production (SEPROR), the Executive Secretariat for Fisheries and Agriculture (SEPA), the Institute for Sustainable Agricultural and Forestry Development of the State of Amazonas (IDAM) and Brazilian Institute for the Environment and Renewable Natural Resources (IBAMA).

The information gathered was selected according to their relevance to the work described briefly through literature review and finally, it was listed in a spreadsheet the significant factors that are blocking the expansion of fish farming in Micro-region of Rio Preto da Eva, detected by previous works developed and published by authors of the region.

3 Results and Discussion

3.1 Fish Farming Market

Fisheries and aquaculture are being outlined as crucial activities for the future's nutrition, employability and sustainability. Aquaculture is growing more quickly than other food sectors, although it has grown in the decades of 1980 in 11.30% and 1990 in 10% (1). Official statistics indicate that in 2016 59.6 million people worked in the capture and aquaculture, of this total 85% was from the Asian continent (1). Many challenges still need to be overcome for the growth of fish production in captivity, but it can already be verified the advancement, mainly of the activity of Pisciculture. The graph below shows how the capture activities and the production of captive fish are advancing (Fig. 2).

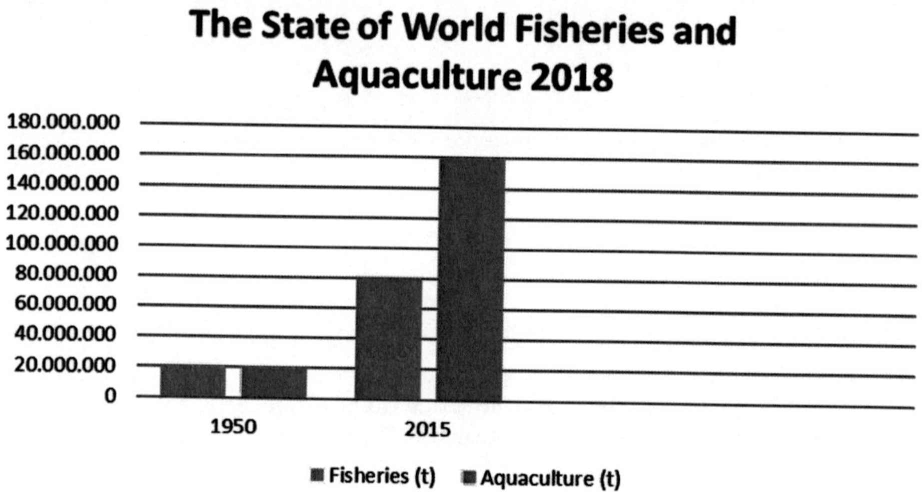

Fig. 2. The State of World Fisheries and Aquaculture 2018 (Adaptation FAO-2017)

China has been leading fish production since 2002 and has also been the largest exporter of fish and fish products. Headed by China, the leading exporters in 2016 were Norway, Vietnam and Thailand. In 2016 the overall production of fish in captivity was 80 million of fish for food, of this total 51.4 t were produced in inland aquaculture, especially in nurseries (1).

Fish from fish farming has an ecological quality of water and is an activity in constant monitoring. Fish farming possesses precisely the same necessary nutrients – protein, carbon moisturized, fats, vitamins and minerals – that fish caught in the river or sea. “What can vary is the number of nutrients and that depends on food.” Fish bred in captivity eat special rations, while those living in natural environments have feed based on algae, plankton and other small species.

In Brazil the fish farming market is always presented as promising due to the water and environmental conditions of the country and with the growing demand for a healthier diet and with less consumption of red meat. Even with this positive trend, Brazil does not figure among the ten largest fish producers in the world. In 2018 Brazil 722,560 tons of cultivation fish against 691,700 tons in 2017 the growth compared to the previous year was 4.5% (2).

"The United States was the most important destination of Brazilian products, importing 8,208,791 tons, representing 26% of total volume and 44% of the fish exports by Brazil. This data shows that exports to the American market are products with high commercial value and added value. Other countries that imported from Brazil significant quantities were Thailand, Spain, South Korea and China (1)".

"The production of growing fish in Brazil reached 722,560 tons in 2018, growth of 4,5% compared to the year 2017 because tilápia fish production promote it, which increase11% in 2018. Among the native fish the tambaqui fish led by the state of Rondônia with the production of 72,800 followed by Mato Grosso with 52,000 tons but the cultivation in Brazilian north shrank 7% in 2018. The state of Amazonas lost positions in Brazilian ranking falling from eighth to sixteenth in 2018 (2).

The fish farming activity had gross revenue of R$ 4.7 billion in 2017 originating from fish farming. According to information from the organization FAO, last year 697,000 tons of growing fish were produced, a result 8% higher than the previous year 640,500 tons (2). The most cultivated species in Brazil today is tilapia fish, which represents 51.7% of the national fish farming with increased about 357,600 tons produced last year. The number makes Brazil reach fourth position in largest producer of tilapia fish in the world. Tambaqui fish is in the next position followed by carp and trout fishes.

Although the has a wild catalog of species Brazil has more imports than exports fish. In 2017 it exported about $150 million in fish which includes fishing and aquaculture and the projection for 2019, according to the Department of Foreign Trade of the Ministry of Industry (MDIC) the country expects to reach 10% growth in export according to Brazil Portal (2).

3.2 Potential of the Fishing and Aquaculture Industry

Asia produces reach 91% of the world's cultivated fish and is responsible for 50% of the fish produced on the planet. China's management initiatives in the fisheries and aquaculture sector should have a remarkable impact globally, with changes in prices, production and consumption. For Brazil the projection for 2030 in the activity of fish production in captivity is 1.097 billion tons, growth around 89% in comparison to the year of 2016.

As previously described in this paper, Brazil has high potential for the production of captive fish. The table below demonstrates the environmental advantages that Brazil gathers for the development of fish farming:

3.3 Challenges for the Development of Fish Farming in the Micro-region of Rio Preto da Eva

The Micro-region of Rio Preto da Eva has favorable conditions potential for the expansion of the scale of production in fish farming but runs into difficulties listed in other scientific works and that last in time as if there was a big bubble that prevents its growth.

Table 1. Brazil's environmental advantages for the development of fish farming

01	11,000 rivers, streams and streams cataloged at the National Water Agency (ANA)
02	219 Hydroelectric reservoirs located in 22 states of the Federation
03	8.5 million km^2 (territory)
04	The largest country in size in South America and the 5th largest in the world
05	The Climate and ecosystem diversity
06	The Climatic regions of equatorial and subtropical characteristics; a large semiarid area in the northeast
07	The Coastal Zone of 3.5 km^2
08	The country's coast has 8,500 km
09	The estimate of 13.7% of all water on the planet
10	More than 2 million hectares of favorable areas for the development of aquaculture
11	An estimated 30 to 35,000 km^2 the flooded area
12	About 400,000 km of irrigated canals in the Northeast region

Elaborated by Patrícia dos Santos - adapted from the text of PEIXE BR-2019).

Previous research has pointed out that fish farming in the region is characterized mainly by the administration by low-income families and with small-scale production [5].

This work collected in previous scientific works published between the years of 2010 to 2018 the relevant information about difficulties for the development of fish farming in the State of Amazonas. We suggest for the future research the application of entrepreneurial orientation using the training and implementation variables of a cooperative business system (Table 2).

3.4 Management Indicators or Monitoring for the Growth of Production

This paper proposes that the adoption of the actions of entrepreneurship and management with the implementation of Balanced Scorecard adapted the activity of fish farming. Balanced Scorecard is a management tool developed by Robert S. Kaplan and David P. Norton to solve measurement problems (17) and is based on critical success factors in four dimensions: 1) performance Financial 2) Customer satisfaction; 3) internal operational processes; 4) Innovation and learning. This tool provides the basis for a complete analysis of the business and can drive any size or type of organization.

3.5 Percentage of Amazon Fishery Consumption Compared to the Total Consumption of Fish or Food

Brazil consumes 14 kg of fish per person, a quantity considered lower than the world average, which is 20 kg (1). No updated data on Amazon fishery consumption were found compared to the total of fish or other foods in Brazil but local research data indicate that the local market consumes all state production.

Table 2. Significant scientific papers published in the period 2010 to 2018 on the fish farming in the Amazon State

Article	Author	Conclusion	Means of publication
Modeling of Production from Fisheries and Artisanal Commercial Fishing in the Municipality of Manaus, Amazonas	[3]	Need to organize the sector	Brazilian Journal of Fisheries Engineering
Characterization and Diagnosis of the Socioeconomic Profile of Fish farming in the State of Amazonas	[4]	"Deficiency in the production chain; insufficient technical expertise and technical assistance; bureaucracy in the release of financial resources from banking institutions and support to producers; technological deficiency for the production and optimization of processes; lack of public policies for incentives and subsidies to fish farming; lack of a collective organization (associations and cooperatives); little investment in research; high prices of the rations and lack of cheaper alternatives for the producer"	Thesis, UFAM
Risk of Profitability in Tambaqui fish Fisheries in the States of Amazonas, Rondônia and Roraima	[5]	"Lack of technical direction; lack of adequate planning; High Cost of Production	Article, PECEGE Magazine
Family Fisheries, Technical Assistance and Management Practices Harvest and Post-harvest: Case Study of the Metropolitan Region of Manaus - Amazonas, Brazil	[6]	"Difficulty in safely determining recommended handling practices; the need for improved approaches by technical assistance agents to provide information to the rural producer on appropriate and proper managed management practices	Article, Observatory of The Latin American Economy

(*continued*)

Table 2. (*continued*)

Article	Author	Conclusion	Means of publication
Fish Production and Fishing Effort in the Municipality of Manaus (Amazonas-Brazil): Analysis and Proposal for Improvements to Enable Industrialization	[7]	"Need for rapid commercialization of fish because of the highly perishable nature of the product; lack of storage infrastructure and difficulties in disposing of production; need to operationalize the existing Fishing Terminal; lack of incentive to fish in the region for the constant fish production	Dissertation UFAM
Management model for the Creation of Amazonian species in a Semi-Intensive System: A Tambaqui fish Case Study (*Colossoma macropomum* CUVIER, 1818)	[8]	"Difficulty in managing the production process and economic decision-making practice short-cycle spending resulting in a reduction in operating margins and affecting its profitability	Doctoral thesis Postgraduate Program in Fisheries Science in the Tropics - PPG-CIPET
Fish Farming Characteristics in Presidente Figueiredo, Amazonas	[9]	"Distance from the headquarters of urban centers; little professionalism; low technological level; difficulty in logistics and infrastructure on production systems	Article, Igapó Magazine
Table 1 Livestock Tambaqui fish, The Manauara Family Fish	[10]	"Providing quality feed in adequate quantities; processing industry for export of tambaqui fish in frozen or chilled products dispatch to various markets, including international markets	Article, Field Day Portal
Metropolitan Region of Manaus: Characteristics and Dilemmas of the Development of a Metropolitan Region in the Western Amazon	[11]	Lack of connections to help structure the green belt that supplies the capital	Thesis, UFAM

(*continued*)

Table 2. (*continued*)

Article	Author	Conclusion	Means of publication
Strategic Plan of Embrapa Western Amazon for Aquaculture	[12]	"Low availability and quality of larvae and fry of matrinxã and Pirarucu fish; low availability of specific rations to meet the nutritional needs of species; increased occurrence of parasitic and bacterial diseases; increasing demand for productive systems that will foster production growth; lack of recommended tank-network culture system; increased demand for tambaqui fish fingerlings with better zootechnical performance	Publication EMBRAPA, Strategic Plan for the 2013 Empraba
Characterization of fish farming activity in the mesoregions of the State of Amazonas, Brazilian Amazonia	[13]	"Need for production and transfer of technical and scientific information to the producing communities	Article, Revista Colombiana cienc. Anim. 2012
Fish and Fishing Solimões-Amazonas: An Integrated Assessment	[14]	"Direct governmental intervention for the responsible and productive use of the fishing resources; understanding of the importance of statistical records for the scientific community, activities developed sporadically and temporarily discontinued	Publication IBAMA/ Pro-Várzea
Links and Connections: The Competitiveness Challenge in the Amazonian Fish Sector	[15]	"Poor technology transfer; lack of infrastructure logistics; professional qualification	Article, XXXI National Meeting of Production Engineering

(*continued*)

3.6 Adverse Impacts on the Fish Farming Environment

The Developing fish farming benefits fishing cause decreases impact on fisheries stocks and contributes to the general income and employment in many rural communities.

Table 2. (*continued*)

Article	Author	Conclusion	Means of publication
The Fish Market of the Metropolitan Region from Manaus	[16]	"Difficulty in obtaining the necessary information due to the absence of statistics in the region; fisheries sector still poorly organized; high cost of fuel and ice; structural disorganization of the sector; discontinuity between capture and sale of the product	Publication INFOPESCA

Elaborated by Patrícia dos Santos.

Besides water consumption, there are other factors influence that cause adverse impacts on the environment of fish farming as issues related to the Fingerlings, with the rations and with the conception of the productive system, which can influence the failure of the activity.

"The production system adopted may generate high or lesser environmental interferences, according to its conception. These systems can, in a simplistic way, be gather in: productivity function – extensive, semi-intensive-intensive or intense; Number of species involved – monoculture or polyculture; and sharing – consortia with other species other than those exclusively aquatic (18)".

To mitigate these adverse effects is a necessary investment in technical guidelines and business management.

3.7 Discussion About Merits, Risks and Success Probabilities

According to (19) there is little effort in investing in improvements in the production process, as well as dedication in the administration and financial control of the projects in the regions of the state of Amazonas. The author concluded that the difficulties encountered may be linked to the treatment of the activity as of secondary importance within the rural establishment and that it does not attribute, concerning to the other activities, a condition of financial dependence for its relevance among fish growers (19).

Lima concluded that the fish farming as has been developed in the region has little probability of economic success in Amazonas and that the high cost of the ration and infrastructure below the necessary for the excellent progress of the business Hinder this Development.

The Amazonian Pisciculture can advance and achieve the scale of production of higher proportions if management and development actions are established, generating occupation and real income source to attract supporters and boost, by the profit function it Offers, Some growth (4).

3.8 A Strategic Plan

The development plan for fish farming in Rio Preto da Eva is composed of three moments:

1) Application of knowledge of entrepreneurship and the technique of Balanced Scorecard with adaptation to the activity of fish farming;
2) Elaboration of indicators for monitoring the activity and generation of managerial information;
3) Study of the public power of the impact of cooperative formation for cost mitigation of the activity and implementation of cooperative-based beneficiation and distribution centers;

The action of the public power responsible for the policy of encouraging fish farming to assist fish farmers in the formation of the cooperative and fundraising with the business agents interested in the development of fish farming.

3.9 China - The International Success Story

The Asian continent is the largest producer in aquaculture activity. This continent contributes 89% of world production in agriculture. Although China has decreased its production in aquaculture, it is the world's largest producer in the fish farms of 1991 (1).

"Transformation within the sector includes higher attention to environmental responsibility and sustainability; quality improvement and product diversity; improved economic efficiency and benefits to fish farmers; and strengthened business integration along the value chain and economies of scale" (1).

China can to feed its large population and still export its production in fish farming. To reach this level, the country develops strategic plans and public policies with particular attention to sustainability and the environment. The "Thirteenth Five-Year Plan for Fisheries Development" has already been introduced, which, with other public policies, intends to introduce robust changes in the country's Aquaculture. Of the 89% of captive fish allocated to Asia, 61.5% were produced by China in 2016 (1).

4 Conclusion

This research concluded that the limiting factors in the Rio Preto da Eva fisheries, reported in papers presented since the year of 2010, are still reported in the year of 2018 and further research are indispensable to obtain robustness in the data collection and the application of tests using the variables of entrepreneurship orientation and management and implementation of a test program of a cooperative business system.

This work did not identify a strategic plan, an action management and entrepreneurship orientation or monitoring indicators for growth of production scale fish farming in the Rio Preto da Eva micro-region, which could be made feasible through public policies. It has been observed through previous studies that actions still occur in an isolated way, focusing only on technical orientations and without continuity in the cities that aggregate the Micro-region which is made up of the municipalities of Rio Preto da Eva and Presidente Figueiredo which own the largest productive chain in the fish farming of the State of Amazonas.

For future research, we suggest an implementation of 1) application of entrepreneurship orientation and management with emphasis on tools of the total quality adapted to fish farmer's conditions; 2) implementation of a cooperative business system program to test the development of fish farming; 3) monitoring actions through performance indicators in order to create conditions for expanding the scale of production and sustainability of fish farming in the Rio Preto da Eva Micro-region; 4) Planned actions of the public power responsible for the policy of encouraging fish farming to formation of cooperative system and fundraising with the business agents interested in the development of fish farming.

Acknowledgments. To the Foundation for Research Support of the State of Amazonas - FAPEAM for the master's scholarship: N. 02/2018 - POSGRAD 2018 - UFAM.

References

1. FAO: The State of Fisheries and Aquaculture, Rome (2018). www.fao.org/3/i9540en/I9540EN.pdf. Accessed 23 June 2019
2. Peixe BR (2019). https://www.peixebr.com.br/Anuario2018/AnuarioPeixeBR2018.pdf. Accessed 08 May 2019
3. Nascimento, S.C.B., Souza, L.A., Falcão, W.O.: Amazonas modeling of production from fisheries and artisanal. Braz. J. Fishing Engineering **11**(2), 44–56 (2019)
4. Lima, C.A.: Characterization and diagnosis of the socioeconomic profile of pisciculture in the state of Amazonas. Dissertation, Federal University of Amazonas (2018). www.ufam.edu.br. Accessed 08 Mar 2019
5. Feitoza, D.L.S., Sonoda, D.Y., de Souza, L.A.: Risk of profitability in fish farmers in the states of Amazonas, Rondônia and Roraima. Rev IPecege. **4**(4), 40–53 (2019)
6. Estevão-Rodrigues, T.T.: Family fish farming, technical assistance and harvesting and postharvest management practices: case study of the metropolitan Region of Manaus – Amazonas, Brasil. Observatorio de la Latin American economy, 1–15 (2017)
7. Viana, S.: Fisheries production and fishing effort in the municipality of Manaus. Dissertation, Federal University of Amazonas (2017). https://tede.ufam.edu.br/handle/tede/5929
8. da Craveiro J.: Management model for the creation of Amazonian species in semi-intensive system: a case study tambaqui fish. Dissertation, Federal University of Amazonas (2016)
9. Barbosa, H.T.B.B., Lima, J.P.: J. Educ. Sci. Technol. IFAM. Amazonas Acta **31**(4), 15–30 (2017). http://periodicos.ifap.edu.br/index.php/JBFS/article/view/96%0Ahttp://www.scielo.br/scielo.php?script=sci_arttext&pid=S1413-41522015000400589. Accessed 23 June 2019
10. Inoue, L.A.A.: Tambaqui fish, the fish of the Manauara family-Portal Day field. Amazonas, pp. 7–9 (2014)
11. Cavalcante, K.V., Franchi, T., Lopes, R.H.: Metropolitan region of Manaus: characteristics and dilemmas of the development of a metropolitan region in the western Amazon, vol. YEAR XV – N, online magazine Urban Research, pp. 1–19 (2015)
12. Izel, A., de Boijink, C.L., Chagas, E.C., O'Sullivan, F.L.D.A., Dairiki, J.K., Inoue, L.A.K.A., et al.: Strategic plan of Embrapa western Amazon for aquaculture, p. 98 (2013)
13. De Oliveira, A.M., Paula, M.D.N., De Almeida-Val, V.M.F., Val, A.L.: Characterization of the activity of fish farming in the mesoregions of the state of Amazonas, Brazilian Amazon. Rev Colombia Animal Science **4**(1), 154–162 (2012)

14. Batista, V.S., Isaac, V.J., Fabré, N.M., Almeida, O.T., Gonzalez, J.C.A., Ruffino, M.L., et al.: The state of fishing in the Amazon. Fish and fisheries in Solimões-Amazonas an Integral evaluation, pp. 13–29 (2012)
15. Moraes, S.C.S., Neto, P.L.D.O.C.: Links and connections: the challenge of competitiveness in the fish farming sector of Amazonas. Enegep (2011)
16. Gandra, A.L.: The fish market in the metropolitan region of Manaus (2010). http://www.infopesca.org/sites/default/files/complemento/publilibreacceso/282/Manaus-completo.pdf17. Accessed 02 April 2019
17. Kaplan, R.S., Norton, D.P.: Strategy-oriented Organization. Rio de Janeiro (2001)
18. Oliveira, R.: The Panorama in aquaculture in Brazil: the practice with a focus on the sustainability. Revinter-Intertox Magazine, pp. 72–88 (2009)

The Interrelationship Between Sustainable Development and Social Innovation: A Bibliometric Study

Ricardo Pereira(✉), Daniela de Oliveira Massad, Fábio Lorensi do Canto, and Gertrudes Aparecida Dandolini

Federal University of Santa Catarina, Florianópolis, SC, Brazil
rikardop@gmail.com

Abstract. Although studies on sustainable development are long-standing, we realized that most research on the subject favors the focus on environmental preservation. It is important to emphasize that sustainable development must consider three perspectives: environmental, social, and economic. The same goes for the social innovation construct, whose impact is also analyzed from different conceptions. Studies that seek to identify research that interrelates social innovation and sustainable development constructs are of high relevance. The present work, having this premise, carried out a bibliometric study. As a result of the work, it was found, from the perspective of the interrelation between sustainable development and social innovation, that this type of research is recent, beginning to expand in 2012, and that the main focus is the environmental issue. Thus, there is a lack of research that correlates these constructs with the triad bias: environmental, social, and economic.

Keywords: Sustainable development · Social innovation · Bibliometrics

1 Introduction

The number of studies on sustainable development has grown in recent times, but the idea of harmonizing society, economy, and the environment is not new [1]. However, there is still no precise definition of how to achieve sustainable development. Some understand that one should not be strict in its definition since imprecision allows the conciliation of people from different sectors [2–4].

We note that most research on sustainable development favors a focus on environmental preservation, although it is considered in three dimensions: environmental, social, and economic.

Social innovation seeks through new ideas to meet social needs so far neglected. Social innovation actions can help promote the sustainable development of a place or community.

However, the lack of fundamentals of sustainable development and social innovation hinders their potential for action and research, making it difficult to answer some questions, such as: what exactly is the sustainable development? To what extent can social

L. Pereira et al. (Eds.): IDEAS 2018, SIST 198, pp. 480–489, 2021.
https://doi.org/10.1007/978-3-030-55374-6_47

innovation contribute to this? How to delineate the perimeter of innovations that contribute to this process and, for this reason, can be called social innovations? Moreover, how can this kind of social innovation be stimulated? [5].

In this sense, in order to understand the dimension of the researches that study the interrelation between the constructs social innovation and sustainable development, the present study had the objective of performing a bibliometric analysis of the articles published in the Scopus database, since it does not we found bibliometric studies that addressed both terms. It was decided to perform a bibliometric study to verify through the quantitative analysis the dynamics of the scientific productions on the subject.

2 Sustainable Development

"Sustainable development is a development that meets the needs of the present without compromising the ability of future generations to meet their own needs" [6] (p. 54). This definition, published in 1987 in Our Common Future, a report developed by the World Commission on Environment and Development, also known as The Brundtland Report, referring to the name of the Norwegian Prime Minister who chaired the committee, Gro Harlem Brundtland, is widely referenced when it comes to sustainable development [4]. Over time the concept has developed, and the perspectives of economic, social, and environmental development have been more contemplated.

In 1990, the first Human Development Report of the United Nations Development Program (UNDP) was published and brought people as a starting point for development. One of its central conclusions addresses the issue of sustainable development, stressing that people and their "future choices need to be protected" and that "any form of financial debt, personal neglect debt or environmental degradation debt - is like borrowing from the next generations " [7] (p. 7).

Developed nations need to be the first to change their environmental policies, providing resources to ensure environmental protection, as they account for the bulk of environmental degradation [7]. Developing countries, on the other hand, need to decide between imitating industrialized countries and developing themselves by polluting and squandering or incorporating efficient technologies so as not to replicate the same mistakes of developed countries [8].

The word sustainable is related to several sectors and applications, such as agriculture, fishing, forestry, tourism, urban and rural communities, among others, without criticism of sustainable development and is widely accepted [4]. The world needs to be proactive and responsible to consider sustainable development with total seriousness [1]. However, there is a contradiction between his words since development gives the idea of a change from one state to another even better and sustainable, it refers to the concept of something lasting [4].

Although there are several definitions, the concept of sustainability is still imprecise in its application among the various sectors and disciplines involved because there is no exactness in how the problems should be solved [3, 4].

It is important to note that, despite all the recent discussions on sustainable development, several cultures throughout history have recognized the need for harmony between the environment, society, and the economy [1]. The researches and reflections on the theme are that they have grown as the numbers show. By searching the Scopus database for "sustainable development" in the title of the document, we obtain about twenty thousand articles on the subject.

Considering that development will occur only when the social dimension is understood and integrated into the economic aspect, social innovation arises as a way of developing solutions to social problems [9].

3 Social Innovation

The literature on social innovation (SI) shows a lack of consensus on definition issues [10–12]. In its purest form, social innovation represents new ideas that meet unmet social needs [13], or it refers to processes and results that develop a new approach to dealing with a problem or need social [14].

Thus, there is a diversity of scales, contexts, and diffusion processes between what is social innovation, which implies a variety of definitions, conceptualizations, and disciplinary fields [11]. It results in ambiguity, conceptual mismatch, and operational difficulties in promoting social innovation.

Social innovations are new social practices, created by collective actions, intentional and goal-oriented, aiming to stimulate the macro-quality of life through the reconfiguration of how to achieve social objectives and minimize these ambiguities [5].

Some authors also differentiate between sustainable innovation, sustainability innovation, eco-innovation, and social innovation. "Sustainable innovation is a process in which sustainability considerations (environmental, social, financial) integrate company's systems, from idea generation to research and development (R & D) and marketing" [15] (p. 9). The innovation directed towards sustainability "is the creation of new market spaces, products, and services or processes driven by social, environmental, or sustainability issues" [16] (p. 3). Eco-innovation, in turn, would be the production, assimilation or exploitation of a product, production process, service or management or business, a method that is new to the organization and which results, throughout its life cycle, in a reduction of environmental impact, pollution and other negative impacts of resource use, compared to relevant alternatives [17]. On the other hand, "social innovations are new ideas that meet social needs, create relationships, and form new collaborations. These innovations can be products, services, or models that meet the most unmet needs" [18].

Finally, social innovation can help meet the needs of society not reached by the government [19] and solve sustainable development problems. Therefore, it is necessary, before deepening the research of the interrelationship between the two constructs, to perform a bibliometric analysis to understand the context of these researches.

4 Methodological Procedures

It is a research with a bibliometric approach, characterized as an analysis of the quantitative aspects related to the publication of scientific articles on social innovation and sustainable development. The database used was Scopus, considered as one of the main referential bases of articles of international scientific journals. We chose this base for its comprehensiveness and level of structure data that make it possible to export.

The authors used the search strategy ("social innovation" AND "sustainable development" OR sustainability) with occurrence in the title, abstract, and keywords. The selection took place by articles, event works, books, and book chapters and reviews.

We export statistical data from the database itself. The authors used the techniques of citation analysis and analysis of social networks for data collection and analysis. The graphs, elaborated with the VosWiewer software, represent the networks formed through the collaboration between authors of different countries, cocitation of authors and periodicals and relation of keywords.

According to the technique of analysis of social networks (ARS), the points or nodes represent the elements of analysis (authors, countries, periodicals, keywords), while the nodes or links represent the relationship between these elements (citation, co-occurrence). The higher the point size, the higher the density of this element in the network. The more center the position of the point on the network, the higher the number of links (connections) that it establishes with other points (centrality) Clusters are called clusters, which represent points with more significant connection to each other (proximity).

5 Bibliometric Analysis

There were 379 scientific articles, and no exclusion criteria were adopted, and there was no temporal delimitation. The variables are analyzed below.

5.1 Number of Articles Per Year

Figure 1 shows the distribution of publications with the research topic. We can see that between 2001 and 2011, there was no significant variation in the number of articles published, a fact that may indicate the embryonic stage of the studies, especially related to social innovation.

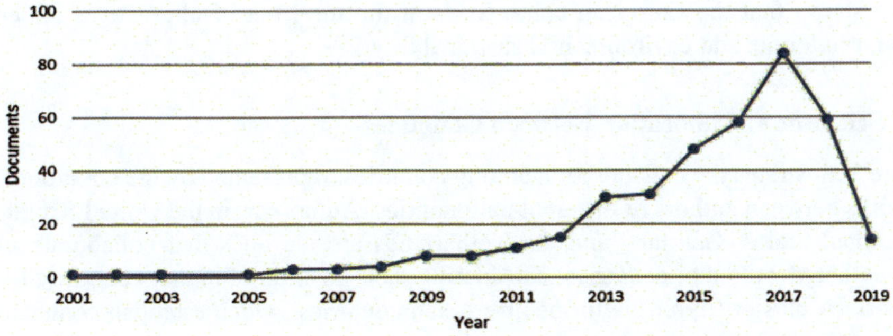

Fig. 1. The number of articles published per year. Source: Elaborated by authors in Feb 2019

As of 2012, however, there is a significant increase in the number of published works, which shows an abrupt increase of 15 articles this year, to 84 articles in 2017, when it reaches the apex in publications.

From the result of this variable, we observe that social innovation and sustainable development are rising themes in science, with significant growth in the number of scientific articles published in the last four years in journals indexed in the Scopus database.

5.2 Most Productive Countries

The graph of Fig. 2 shows the ten countries with the highest scientific production. The list consists of six European countries, one Asian, one North American, one from Oceania and Brazil, representing the South American continent with 13 documents. However, Canada, France, and India do not appear on the chart, but they have the same number of publications in Brazil. The top five countries concentrate more than half of all jobs.

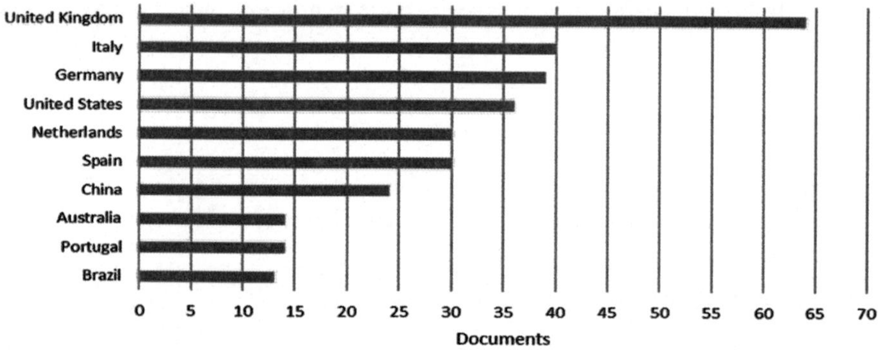

Fig. 2. Documents by country or territory. Source: Elaborated by authors in Feb 2019

The United Kingdom is the most productive country with 64 publications, almost twice the Germany that produced 39 articles. Of the top five, four are from Europe, which shows that the European continent is at the forefront of alternatives aimed at social, economic, and environmental sustainability.

5.3 Scientific Collaboration Between Countries

Figure 3 illustrates the collaboration among countries, represented by the co-authorship of works between authors of different nationalities. Authors from the United Kingdom, the United States, Germany and the Netherlands are the ones that collaborate most with authors from other countries, considering their position of higher centrality in the network. A cluster (in red) is formed by Asian countries, which establish connections between themselves and with the United States.

The European countries form two other clusters (green and purple) that establish connections between them. The most isolated countries in the network are Brazil, Argentina, and Russia.

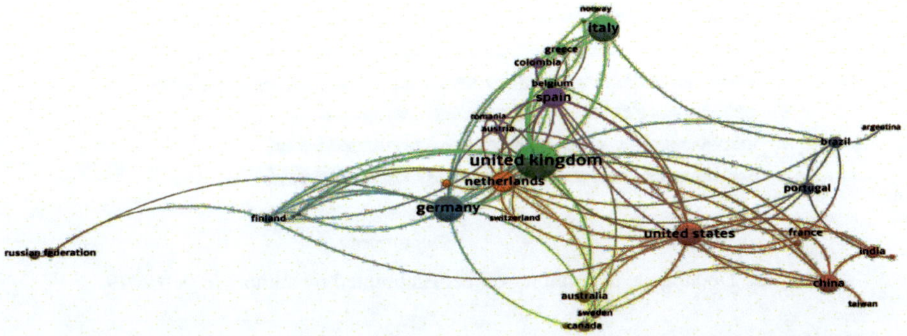

Fig. 3. Collaboration between countries. Source: Elaborated by authors in Feb 2019

5.4 More Productive Journals

The graph of Fig. 4 presents a comparison of the five journals with the highest number of articles published year by year. It also shows the total of publications of each journal. The journal Sustainability (Switzerland) stands out with 23 publications.

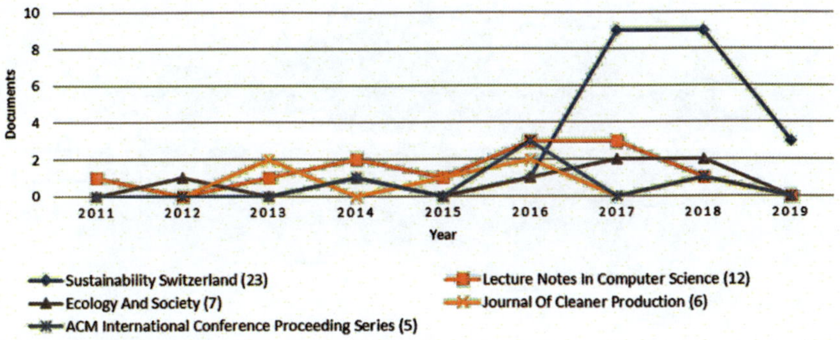

Fig. 4. Documents per year per period. Source: Elaborated by authors in Feb 2019

This variable is relevant for researchers wishing to submit articles for publication since it indicates periodicals tending to accept works related to the subject under analysis.

5.5 More Productive Authors

The graph of Fig. 5 shows the most productive authors, maintaining a high level of scientific production on the concept. Authors Alex Haxeltine and Gill Seyfang are from the University of East Anglia, while Marta Peris-Ortiz belongs to the Polytechnic University of Valencia, the two most productive institutions based on the data surveyed.

The recent growth of articles addressing social innovation and its relation to sustainable development concentrated between 2013 and 2017 may be an explanation for the low number of publications per author.

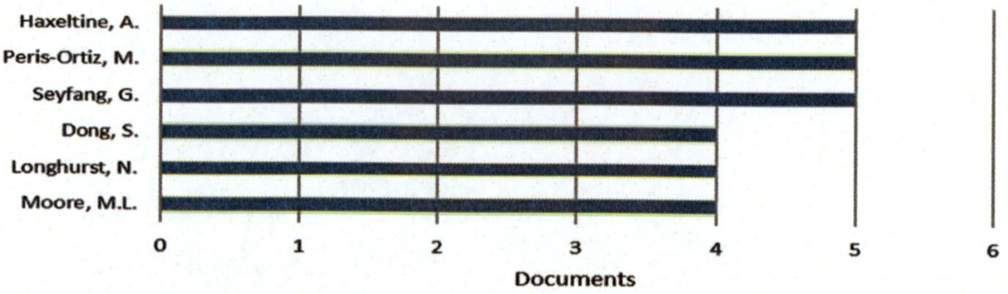

Fig. 5. Documents by author. Source: Elaborated by authors in Feb 2019

5.6 Occurrence and Relationship of Keywords

In Fig. 6, we present the graph that presents the network of keywords with the highest occurrence in the set of articles. Four clusters form the network, represented by the colors blue, red and green and yellow.

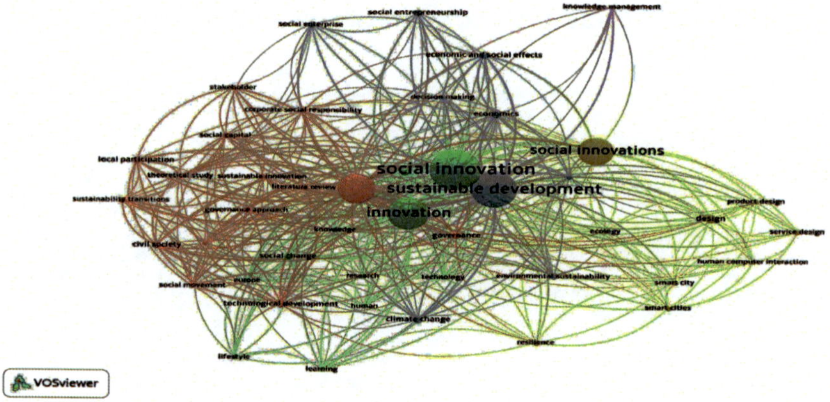

Fig. 6. The occurrence of keywords. Source: Elaborated by authors in Feb 2019

We observed that the blue cluster, which represents sustainable development, presents particular interaction with keywords related to the environmental area, such as climate change, environmental sustainability. However, it addresses other themes such as social entrepreneurship, social enterprises, social and economic effects. This finding indicates the tripartition of sustainable development, which must be addressed by the social, economic, and environmental perspectives.

The cluster of red color brings words underlying others. The green cluster presents a significant interaction between the words innovation and social innovation and greater interaction with the blue cluster that represents sustainable development. The yellow cluster represents concrete examples of social innovations, such as intelligent cities, design, human-computer interaction, among others.

5.7 Most Popular Articles

The 379 articles received a total of 3087 citations, with an average of 8.14 citations per article. The index *h* is 26; that is, 26 articles received 26 or more citations.

Table 1. Most popular articles.

Title (Authors, year)	Cited by
Business models for sustainable innovation: State-of-the-art and steps towards a research agenda (Boons, Lüdeke-Freund, 2013)	448
Growing grassroots innovations: Exploring the role of community-based initiatives in governing sustainable energy transitions (Seyfang, Haxeltine, 2012)	306
Designing long-term policy: Rethinking transition management (Voß, Smith, Grin, 2009)	184
Studying the complexity of change: Toward an analytical framework for understanding deliberate social-ecological transformations (Moore et al., 2014)	80
Desperately seeking niches: Grassroots innovations and niche development in the community currency field (Seyfang, Longhurst, 2013)	75
Towards sustainable households: A methodology for developing sustainable technological and social innovations (Green, Vergragt, 2002)	73
LIVING LAB: User-driven innovation for sustainability (Liedtke et al., 2012)	59
Resilience of and through urban ecosystem services (McPhearson et al., 2015)	56
Leadership of organizational change toward an integrated model of leadership for corporate responsibility and sustainable development: A process model of corporate responsibility beyond management innovation (D'amato, Roome, 2009)	51
Social innovation: Blurring boundaries to reconfigure markets (Nicholls, Murdock, 2011)	49
Strategic niche management of social innovations: The case of social entrepreneurship (Witkamp, Raven, Royakkers, 2011)	49
Commercial orientation in grassroots social innovation: Insights from the sharing economy (Martin, Upham, Budd, 2015)	43
What influences the diffusion of grassroots innovations for sustainability? Investigating community currency niches (Seyfang, Longhurst, 2016)	40
'Don't call me resilient again!': the New Urban Agenda as immunology or what happens when communities refuse to be vaccinated with 'smart cities' and indicators (Kaika, 2017)	38
City regions and place development (Healey, 2009)	36

Source: survey data (2019).

Table 1 shows the fifteen most cited articles. It is interesting to note that the author Seyfang appears in three most cited publications, being also one of the authors who produced more on the subject, as can be observed in Fig. 6. Still, among the authors who produced more on the subject, Haxeltine has only one article among the most cited and Peris-Ortiz has none.

From the reading of the abstracts, we observe that the works that explore social innovation and sustainable development have a bias for the environmental segment. So much, so that of the 20 most cited articles, 13 are of journals related to the environmental area. The most cited article is from the Journal of Cleaner Production.

6 Conclusion

The Future We Want, the final document of the United Nations Conference on Sustainability Development 2012 [20], included some references to innovation. In its preamble, it calls for the continuation and strengthening of international cooperation in the field of innovation in order to achieve sustainable development. It also recognizes the critical role of promoting innovation, mainly in developing countries, and calls on governments to create enabling structures that promote environmentally sound innovation, including in support of the green economy.

This excerpt above is only one of the signs that innovation, especially social, and sustainable development interconnect, in a certain way. In particular, if we consider that social innovations are a set of actions aimed at filling gaps left by governments in the pursuit of public policies, aiming at social justice, environmental preservation, and economic development.

This paper aimed to understand the dimension of the researches that study the interrelation between the constructs social innovation and sustainable development. The bibliometric data analyzed bring us some relevant information, especially the signs that, despite the fact that the constructs sustainable development and social innovation are already well developed in the literature, the interconnection of the two themes is still incipient and when done is given in the wrong way, perceiving a study bias tending towards an environmental perspective.

The relationship between innovation and sustainability occurs in two ways: innovation as a driving force for sustainability (role of innovation in sustainable development) or sustainability as an engine of innovation (sustainability as a new paradigm and a guiding concept for innovation) [21].

For future studies it is suggested to increase the number of bases studied, seeking to analyze the state of the art of the studied subject and to carry out an integrative revision of the same.

References

1. Song, C.: Fuel processing for low-temperature and high-temperature fuel cells: Challenges, and opportunities for sustainable development in the 21st century. Catal. Today **77**(1–2), 17–49 (2002)
2. Lele, S.M.: Sustainable development: a critical review. World Dev. **19**(6), 607–621 (2002)
3. Broman, G.I., Robèrt, K.H.: A framework for strategic sustainable development. J. Clean. Prod. **140**, 17–31 (2017)
4. Wall, G.: Beyond sustainable development. Tourism Recreation Res. **43**(3), 390–399 (2018)

5. Periac, F., David, A., Roberson, Q.: Clarifying the interplay between social innovation and sustainable development: a conceptual framework rooted in paradox management. Eur. Manag. Rev. **15**(1), 19–35 (2018)
6. World Commission on Environment and Development: Our Common Future. Oxford University Press, Oxford (1987)
7. United Nations Development Programme.: Human Development Report 1990. Oxford University Press, New York (1990)
8. United Nations Development Programme: Human Development Report. Oxford University Press, New York (1998)
9. Correia, S.N., De Oliveira, V.M., Feitosa, M.J., Goméz, C.R.P.: Inovação Social para o Desenvolvimento Sustentável: um caminho possível. Administração Pública e Gestão Social **10**(3), 199–212 (2018)
10. Pol, E., Ville, S.: Social innovation: buzz word or enduring term? J. Socio Econ. **38**(6), 878–885 (2009)
11. Rüede, D., Lurtz, K.: Mapping the various meanings of social innovation: towards a differentiated understanding of an emerging concept (2012)
12. Cajaiba-Santana, G.: Social innovation: moving the field forward. A conceptual framework. Technol. Forecasting Soc. Change **82**(1), 42–51 (2014)
13. Mulgan, G., Tucker, S., Ali, R., Sanders, B.: Social innovation: what it is, why it matters and how it can be accelerated (2007)
14. Nicholls, A., Dees, J.: Social Innovation, International Encyclopedia of the Social & Behavioral Sciences, 2nd edn., pp. 355–361 (2015)
15. Charter, M.., Clark, T.: Sustainable Innovation - Key conclusions from Sustainable Innovation Conferences 2003–2006 organised by The Centre for Sustainable Design. The Centre for Sustainable Design (2007)
16. Little, A.D.: How Leading Companies are Using Sustainability-Driven Innovation to Win Tomorrow's Customers. Arthur D. Little (2005)
17. Kemp, R., Pearson, P.: Final report MEI project about measuring eco-innovation". Maastricht (2007). https://www.oecd.org/env/consumption-innovation/43960830.pdf
18. European Commission. https://ec.europa.eu/growth/industry/innovation/policy/social_en. Accessed 24 Feb 2019
19. Beckmann, M.: The impact of social entrepreneurship on societies. In: Volkmann, C.K., Tokarski, K.O., Ernst, K. (eds.) Social Entrepreneurship and Social Business: An Introduction and Discussion with Case Studies, pp. 235–254. Springer, Gabler (2012)
20. UN 2012: The Future We Want. Outcome document of the United Nations Conference on Sustainable Development. Rio de Janeiro, Brazil; pp. 20–22 (2012). https://sustainabledevelopment.un.org/content/documents/733FutureWeWant.pdf
21. El Bilali, H.: Relation between innovation and sustainability in the agro-food system. Italian J. Food Sci. **30**(2), 200–225 (2018)

Author Index

A
Amaral, Christian, 13
Amoedo, Andres, 268
Azinheira, José R., 13

B
Baldo, Crhistian Raffaelo, 34
Benassi, Roseli Frederigi, 437
Bueno, Samuel S., 13, 363

C
Caliari, Thiago, 377, 457
Carvalho, José Reginaldo H., 13, 76, 363
Cavalcante, Karina, 44
Cesar, Patrícia Gonzaga, 280

D
da Costa Craveiro, Joaquim Maciel, 23, 44, 467
da Costa Dias, José Eduardo, 52, 179
da Costa, Eminy Laís Silva, 386
da Fonseca, Roger Pamponet, 95
da Silva Bento, Mateus, 149
da Silva Falcão, Franciane, 168, 322
da Silva Misucochi, Luciana Kurack, 386
da Silva, Lina Maria Moreira Garai, 280
Dandolini, Gertrudes Aparecida, 3, 115, 251, 427, 480
de Almeida, Ely Senna, 467
de Andrade, Alexandre Acácio, 52, 126, 136, 179, 280
de Bessa, Isaías V., 406
de Carvalho Santos, Bruno Raphael, 86
de Carvalho, Marenice Melo, 396
de Castro Filho, Fernando Gonçalves, 52, 179
de Medeiros, Ana Carolina Correa, 95, 386
de Medeiros, Renan Landau Paiva, 396, 406
de Melo Barreto, Sicy Rusalka Goes, 23
de Oliveira, Ana Nubia dos Santos, 23
de Oliveira, Marcelo Albuquerque, 23, 44, 447, 467
de Oliveira Massad, Daniela, 3, 115, 480
de Paiva, Ely C., 13
de Souza Costa, Luiz H., 377
de Souza Pacheco, Almir, 149
de Souza Silva, Deise Carolina, 314
de Souza, Gabriela Costa Arouck, 168
de Souza, João Arthur, 3, 115
de Souza, Jocarly Patrocínio, 160
de Souza, Marianna Gioppo, 160
Dietrich, Álvaro Batista, 136
do Canto, Fábio Lorensi, 480
do Nascimento Fernandes, Tito Lívio, 322
dos Santos Figueiredo, Yohani Dominik, 251
dos Santos, Patrícia, 467

F
Facó, Júlio Francisco Blumetti, 52, 126, 136, 179, 298
Falavigna, Gregory, 115
Falcão, Franciane, 189, 219
Fernandes, Aline, 219
Fernandes, Rodrigo Bastos, 197, 228
Ferreira, Anna Carolina Dall'Aneze, 437
Figueiredo, Marcos, 261
Fischer, Bruno, 65
Fonseca, Gustavo A., 13
Fragal, Ana Carolina, 34

L. Pereira et al. (Eds.): IDEAS 2018, SIST 198, pp. 491–492, 2021.
https://doi.org/10.1007/978-3-030-55374-6

G
Gaspar, Ricardo, 126
Goergen, Roberta, 160, 239, 356, 416
Gomes, Matheus, 197, 228

H
Hubert, Matias Alles, 160, 239, 356, 416

J
Jorge, Ricardo Reolon, 136

K
Kamensky, Andrea, 261
Koyama, Mauro F., 13
Krus, Petter, 208

L
Lapolli, Édis, 115

M
Maciel, Ben-Hur, 356, 416
Mafalda, Rovilson, 261
Mantovani, Ivan, 356
Marques, Karuliny, 197, 228
Marujo, Ernesto Cordeiro, 457
Matsumae, Akane, 335
Mirisola, Luiz G. B., 13
Modesto, Josivaldo, 363
Molinari, Deivison, 447
Mori, Shintarou, 335
Moutinho, Alexandra, 13

N
Nagai, Yukari, 335
Nogueira, Lucas A. C. O., 13
Nunes-Silva, Carlos Gustavo, 76

O
Oberdörfer, Maurício, 160
Ogliari, André, 197, 228
Okada, Karla, 76
Oliveira, Glaucia, 447
Oliveira, Matheus, 44
Oliveira, Sueli, 437

P
Pacheco, Bruno Perdigão, 149
Pereira, Luciana, 298
Pereira, Ricardo, 115, 480
Prim, Márcia Aparecida, 251, 427

Q
Quirino, Magnólia Grangeiro, 149

R
Ramos, Josué Jr. G., 13
Rannov, Carla, 356
Ranzolin Junior, Jatyr, 314
Rasia, Luiz Antônio, 160, 239, 356
Regner, Rosângela Rommel, 416
Rego, Ronnie Rodrigo, 290
Reis, Dércio Luiz, 23, 447, 467
Ribeiro, Anderson Orzari, 34
Rocha, Augusto César Barreto, 95
Rueda, Miguel Á. C., 13
Ruschival, Claudete Barbosa, 86, 105, 386

S
Santos, Alef, 105
Santos, Eulanda M., 76
Santos, Úlima, 105
Saraiva, Luana Bittencourt, 86
Schaeffer, Paola Rücker, 65
Scorsatto, Filipe, 65
Silva, Giovanna, 126
Silva, Paula, 189
Souza, Marianna, 239
Stein, Edmilton, 239, 416

T
Teixeira, Clarissa Stefani, 314
Tombini Wittmann, Tatian, 3

U
Uriona-Maldonado, Mauricio, 377

V
Valdiero, Antonio Carlos, 160, 239, 356, 416
Valdiero, Carlos, 239
Vasconcelos, Leandro, 268
Vasconcelos, Max, 76
Vaz, Caroline Rodrigues, 377
Veroneze, Gabriela, 44, 447
Villares, Luiz, 346

W
Wojslaw, Mauricio, 298

Z
Zara, Thais Marzola, 457